Springer Texts in Statistics

Advisors:
George Casella   Stephen Fienberg   Ingram Olkin

Springer

New York
Berlin
Heidelberg
Barcelona
Hong Kong
London
Milan
Paris
Singapore
Tokyo

Springer Texts in Statistics

(continued after index)

E.L. Lehmann

Elements of
Large-Sample Theory

With 10 Figures

Springer

E.L. Lehmann
Department of Statistics
367 Evans Hall #3860
University of California
Berkeley, CA 94720-3860
USA

Library of Congress Cataloging-in-Publication Data
Lehmann, E.L. (Erich Leo), 1917–
    Elements of large-sample theory / Erich Lehmann.
        p.   cm. — (Springer texts in statistics)
    Includes bibliographical references and index.

    1. Sampling (Statistics).   2. Asymptotic distribution
(Probability theory)   3. Law of large numbers.
    I. Title.   II. Series.
    QA276.6.L45   1998
    519.5′2—dc21   98-34429

ISBN 978-1-4419-3136-8   e-ISBN 978-0-387-22729-0

Printed in the United States of America.   (BPR/EB)

9 8 7 6 5 4 3 (Corrected third printing, 2004)

Springer-Verlag is a part of *Springer Science+Business Media*

springeronline.com

Preface

The subject of this book, first order large-sample theory, constitutes a coherent body of concepts and results that are central to both theoretical and applied statistics. This theory underlies much of the work on such different topics as maximum likelihood estimation, likelihood ratio tests, the bootstrap, density estimation, contingency table analysis, and survey sampling methodology, to mention only a few. The importance of this theory has led to a number of books on the subject during the last 20 years, among them Ibragimov and Has'minskii (1979), Serfling (1980), Pfanzagl and Wefllmeyer (1982), Le Cam (1986), Rüschendorf (1988), Barndorff-Nielsen and Cox (1989, 1994), Le Cam and Yang (1990), Sen and Singer (1993), and Ferguson (1996).

These books all reflect the unfortunate fact that a mathematically complete presentation of the material requires more background in probability than can be expected from many students and workers in statistics. The present, more elementary, volume avoids this difficulty by taking advantage of an important distinction. While the proofs of many of the theorems require a substantial amount of mathematics, this is not the case with the understanding of the concepts and results nor of their statistical applications.

Correspondingly, in the present introduction to large-sample theory, the more difficult results are stated without proof, although with clear statements of the conditions of their validity. In addition, the mode of probabilistic convergence used throughout is convergence in probability rather than strong (or almost sure) convergence. With these restrictions it is possible to present the material with the requirement of only two years of calculus and, for the later chapters, some linear algebra. It is the purpose of the book, by these means, to make large-sample theory accessible to a wider audience.

It should be mentioned that this approach is not new. It can be found in single chapters of more specialized books, for example, Chapter 14 of Bishop, Fienberg, and Holland (1975) and Chapter 12 of Agresti (1990). However, it is my belief that students require a fuller, more extensive treatment to become comfortable with this body of ideas.

Since calculus courses often emphasize manipulation without insisting on a firm foundation, Chapter 1 provides a rigorous treatment of limits and order concepts which underlie all large-sample theory. Chapter 2 covers the basic probabilistic tools: convergence in probability and in law, the central limit theorem, and the delta method. The next two chapters illustrate the application of these tools to hypothesis testing, confidence intervals, and point estimation, including efficiency comparisons and robustness considerations. The material of these four chapters is extended to the multivariate case in Chapter 5.

Chapter 6 is concerned with the extension of the earlier ideas to statistical functionals and, among other applications, provides introductions to U-statistics, density estimation, and the bootstrap. Chapter 7 deals with the construction of asymptotically efficient procedures, in particular, maximum likelihood estimators, likelihood ratio tests, and some of their variants. Finally, an appendix briefly introduces the reader to a number of more advanced topics.

An important feature of large-sample theory is that it is nonparametric. Its limit theorems provide distribution-free approximations for statistical quantities such as significance levels, critical values, power, confidence coefficients, and so on. However, the accuracy of these approximations is not distribution-free but, instead, depends both on the sample size and on the underlying distribution. To obtain an idea of the accuracy, it is necessary to supplement the theoretical results with numerical work, much of it based on simulation. This interplay between theory and computation is a crucial aspect of large-sample theory and is illustrated throughout the book.

The approximation methods described here rest on a small number of basic ideas that have wide applicability. For specific situations, more detailed work on better approximations is often available. Such results are not included here; instead, references are provided to the relevant literature.

This book had its origin in a course on large-sample theory that I gave in alternate years from 1980 to my retirement in 1988. It was attended by graduate students from a variety of fields: Agricultural Economics, Biostatistics, Economics, Education, Engineering, Political Science, Psychology, Sociology, and Statistics. I am grateful to the students in these classes, and particularly to the Teaching Assistants who were in charge of the associated laboratories, for many corrections and other helpful suggestions. As the class notes developed into the manuscript of a book, parts were read at various stages by Persi Diaconis, Thomas DiCiccio, Jiming Jiang, Fritz Scholz, and Mark van der Laan, and their comments resulted in many improvements. In addition, Katherine Ensor used the manuscript in a course on large-sample theory at Rice University and had her students send me their comments.

In 1995 when I accompanied my wife to Educational Testing Service (ETS) in Princeton, Vice President Henry Braun and Division Head Charles Davis proposed that I give a course of lectures at ETS on the forthcoming book. As a result, for the next 2 years I gave a lecture every second week to an audience of statisticians from ETS and the surrounding area, and in the process completely revised the manuscript. I should like to express my thanks to ETS for its generous support throughout this period, and also for the help and many acts of kindness I received from the support staff in the persons of Martha Thompson and Tonia Williams. Thanks are also due to the many members of ETS who through their regular attendance made it possible and worthwhile to keep the course going for such a long time. Special appreciation for their lively participation and many valuable comments is due to Charles Lewis, Spencer Swinton, and my office neighbor Howard Wainer.

I should like to thank Chris Bush who typed the first versions of the manuscript, Liz Brophy who learned LaTeX specifically for this project and typed the class notes for the ETS lectures, and to Faye Yeager who saw the manuscript through its final version.

Another person whose support was crucial is my Springer-Verlag Editor and friend John Kimmel, who never gave up on the project, helped it along in various ways, and whose patience knows no bounds.

Many thanks are due to David Hunter, Ramani Pilla and their students for their very careful reading of the book and for finding a large number of errors.

My final acknowledgment is to my wife Juliet Shaffer who first convinced me of the need for such a book. She read the early drafts of the manuscript, sat in on the course twice, and once taught it herself. Throughout, she gave me invaluable advice and suggested many improvements. In particular, she also constructed several of the more complicated figures and tables. Her enthusiasm sustained me throughout the many years of this project, and to her this book is gratefully dedicated.

Erich L. Lehmann
Berkeley, California

Contents

1

Mathematical Background

Preview

The principal aim of large-sample theory is to provide simple approximations for quantities that are difficult to calculate exactly. The approach throughout the book is to embed the actual situation in a sequence of situations, the limit of which serves as the desired approximation.

The present chapter reviews some of the basic ideas from calculus required for this purpose such as limit, convergence of a series, and continuity. Section 1 defines the limit of a sequence of numbers and develops some of the properties of such limits. In Section 2, the embedding idea is introduced and is illustrated with two approximations of binomial probabilities. Section 3 provides a brief introduction to infinite series, particularly power series. Section 4 is concerned with different rates at which sequences can tend to infinity (or zero); it introduces the o, $\asymp$, and O notation and the three most important growth rates: exponential, polynomial, and logarithmic. Section 5 extends the limit concept to continuous variables, defines continuity of a function, and discusses the fact that monotone functions can have only simple discontinuities. This result is applied in Section 6 to cumulative distribution functions; the section also considers alternative representations of probability distributions and lists the densities of probability functions of some of the more common distributions.

1.1   The concept of limit

Large-sample (or asymptotic*) theory deals with approximations to probability distributions and functions of distributions such as moments and quantiles. These approximations tend to be much simpler than the exact formulas and, as a result, provide a basis for insight and understanding that often would be difficult to obtain otherwise. In addition, they make possible simple calculations of critical values, power of tests, variances of estimators, required sample sizes, relative efficiencies of different methods, and so forth which, although approximate, are often accurate enough for the needs of statistical practice.

Underlying most large-sample approximations are limit theorems in which the sample sizes tend to infinity. In preparation, we begin with a discussion of limits. Consider a sequence of numbers a_n such as

$$(1.1.1) \qquad a_n = 1 - \frac{1}{n}(n = 1, 2, \ldots): \ 0, \frac{1}{2}, \frac{2}{3}, \frac{3}{4}, \frac{4}{5}, \frac{5}{6}, \cdots,$$

and

$$(1.1.2) \qquad a_n = 1 - \frac{1}{n^2}(n = 1, 2, \ldots): \ 0, \frac{3}{4}, \frac{8}{9}, \frac{15}{16}, \frac{24}{25}, \frac{35}{36}, \cdots,$$

or, more generally, the sequences

$$(1.1.3) \qquad a_n = a - \frac{1}{n} \text{ and } a_n = a - \frac{1}{n^2}$$

for some arbitrary fixed number a.

Two facts seem intuitively clear: (i) the members of both sequences in (1.1.3) are getting arbitrarily close to a as n gets large; (ii) this "convergence" toward a proceeds faster for the second series than for the first. The present chapter will make these two concepts precise and give some simple applications. But first, consider some additional examples.

The sequence obtained by alternating members of the two sequences (1.1.3) is given by

$$(1.1.4) \qquad a_n = \begin{cases} a - \frac{1}{n} & \text{if } n \text{ is odd,} \\ \\ a - \frac{1}{n^2} & \text{if } n \text{ is even:} \end{cases}$$

$$a - 1, a - \frac{1}{4}, a - \frac{1}{3}, a - \frac{1}{16}, a - \frac{1}{5}, a - \frac{1}{36}, \cdots.$$

---

*The term "asymptotic" is not restricted to large-sample situations but is used quite generally in connection with any limit process. See, for example, Definition 1.1.3. For some general discussion of asymptotics, see, for example, DeBruijn (1958).

For this sequence also, the numbers get arbitrarily close to a as n gets large. However, they do so without each member being closer to a than the preceding one. For a sequence $a_n, n = 1, 2, \ldots$, to tend to a limit a as $n \to \infty$, it is not necessary for each a_n to be closer to a than its predecessor a_{n-1}, but only for a_n to get arbitrarily close to a as n gets arbitrarily large.

Let us now formalize the statement that the members of a sequence $a_n, n = 1, 2, \ldots$, get arbitrarily close to a as n gets large. This means that for any interval about a, no matter how small, the members of the sequence will eventually, i.e., from some point on, lie in the interval. If such an interval is denoted by $(a - \epsilon, a + \epsilon)$ the statement says that from some point on, i.e., for all n exceeding some n_0, the numbers a_n will satisfy $a - \epsilon < a_n < a + \epsilon$ or equivalently

$$(1.1.5) \qquad |a_n - a| < \epsilon \text{ for all } n > n_0.$$

The value of n_0 will of course depend on ϵ, so that we will sometimes write it as $n_0(\epsilon)$; the smaller ϵ is, the larger is the required value of $n_0(\epsilon)$.

Definition 1.1.1 The sequence $a_n, n = 1, 2, \ldots$, is said to tend (or converge) to a limit a; in symbols:

$$(1.1.6) \qquad a_n \to a \text{ as } n \to \infty \quad \text{or} \quad \lim_{n \to \infty} a_n = a$$

if, given any $\epsilon > 0$, no matter how small, there exists $n_0 = n_0(\epsilon)$ such that (1.1.5) holds.

For a formal proof of a limit statement (1.1.6) for a particular sequence a_n, it is only necessary to produce a value $n_0 = n_0(\epsilon)$ for which (1.1.5) holds. As an example consider the sequence (1.1.1). Here $a = 1$ and $a_n - a = -1/n$. For any given ϵ, (1.1.5) will therefore hold as soon as $\frac{1}{n} < \epsilon$ or $n > \frac{1}{\epsilon}$. For $\epsilon = 1/10, n_0 = 10$ will do; for $\epsilon = 1/100, n_0 = 100$; and, in general, for any ϵ, we can take for n_0 the smallest integer, which is $\geq \frac{1}{\epsilon}$.

In examples (1.1.1)–(1.1.4), the numbers a_n approach their limit from one side (in fact, in all these examples, $a_n < a$ for all n). This need not be the case, as is shown by the sequence

$$(1.1.7) \qquad a_n = \left\{ \begin{array}{ll} 1 - \frac{1}{n} & \text{if } n \text{ is odd} \\[2mm] 1 + \frac{1}{n} & \text{if } n \text{ is even} \end{array} \right\} = 1 + (-1)^n \frac{1}{n}.$$

It may be helpful to give an example of a sequence which does not tend to a limit. Consider the sequence

$$0, 1, 0, 1, 0, 1, \ldots$$

given by $a_n = 0$ or 1 as n is odd or even. Since for arbitrarily large n, a_n takes on the values 0 and 1, it cannot get arbitrarily close to any a for all sufficiently large n.

The following is an important example which we state without proof.

Example 1.1.1 The exponential limit. For any finite number c,

(1.1.8) $$\left(1 + \frac{c}{n}\right)^n \to e^c \text{ as } n \to \infty.$$

To give an idea of the speed of the convergence of $a_n = \left(1 + \frac{1}{n}\right)^n$ to its limit e, here are the values of a_n for a number of values of n, and the limiting value $e(n = \infty)$ to the nearest $1/100$.

TABLE 1.1.1. $\left(1 + \frac{1}{n}\right)^n$ to the nearest $1/100$

n	1	3	5	10	30	50	100	500	∞
a_n	2.00	2.37	2.49	2.59	2.67	2.69	2.70	2.72	2.72

To the closest $1/1000$, one has $a_{500} = 2.716$ and $e = 2.718$. $\square$

The idea of limit underlies all of large-sample theory. Its usefulness stems from the fact that complicated sequences $\{a_n\}$ often have fairly simple limits which can then be used to approximate the actual a_n at hand. Table 1.1.1 provides an illustration (although here the sequence is fairly simple). It suggests that the limit value $a = 2.72$ shown in Table 1.1.1 provides a good approximation for $n \geq 30$ and gives a reasonable ballpark figure even for n as small as 5.

Contemplation of the table may raise a concern. There is no guarantee that the progress of the sequence toward its limit is as steady as the tabulated values suggest. The limit statement guarantees only that *eventually* the members of the sequence will be arbitrarily close to the limit value, not that each member will be closer than its predecessor. This is illustrated by the sequence (1.1.4). As another example, let

(1.1.9) $a_n = \begin{cases} 1/\sqrt{n} & \text{if } n \text{ is the square of an integer } (n = 1, 4, 9, \dots) \\ 1/n & \text{otherwise.} \end{cases}$

Then $a_n \to 0$ (Problem 1.7) but does so in a somewhat irregular fashion. For example, for $n = 90, 91, \dots, 99$, we see a_n getting steadily closer to the limit value 0 only to again be substantially further away at $n = 100$. In sequences encountered in practice, such irregular behavior is rare. (For a statistical example in which it does occur, see Hodges (1957)). A table such as Table 1.1.1 provides a fairly reliable indication of smooth convergence to the limit.

Limits satisfy simple relationships such as: if $a_n \to a, b_n \to b$, then

(1.1.10) $$a_n + b_n \to a + b \quad \text{and} \quad a_n - b_n \to a - b,$$

(1.1.11) $$a_n \cdot b_n \to a \cdot b$$

and

(1.1.12) $a_n/b_n \to a/b$ provided $b \neq 0$.

These results will not be proved here. Proofs and more detailed treatment of the material in this section and Section 1.3 are given, for example, in the classical texts (recently reissued) by Hardy (1992) and Courant (1988). For a slightly more abstract treatment, see Rudin (1976).

Using (1.1.12), it follows from (1.1.8), for example, that

(1.1.13) $$\left(\frac{1 + \frac{a}{n}}{1 + \frac{b}{n}}\right)^n \to e^{a-b} \text{ as } n \to \infty.$$

An important special case not covered by Definition 1.1.1 arises when a sequence tends to ∞. We say that $a_n \to \infty$ if eventually (i.e., from some point on) the a's get larger than any given constant M. Proceeding as in Definition 1.1.1, this leads to

Definition 1.1.2 The sequence a_n tends to ∞; in symbols,

(1.1.14) $a_n \to \infty$ or $\lim_{n\to\infty} a_n = \infty$

if, given any M, no matter how large, there exists $n_0 = n_0(M)$ such that

(1.1.15) $a_n > M$ for all $n > n_0$.

Some sequences tending to infinity are

(1.1.16) $a_n = n^\alpha$ for any $\alpha > 0$

(this covers sequences such as $\sqrt[3]{n} = n^{1/3}, \sqrt{n} = n^{1/2}, \ldots$ and $n^2, n^3, \ldots$);

(1.1.17) $a_n = e^{\alpha n}$ for any $\alpha > 0$;

(1.1.18) $a_n = \log n,\ a_n = \sqrt{\log n},\ a_n = \log\log n$.

To see, for example, that $\log n \to \infty$, we check (1.1.15) to find that $\log n > M$ provided $n > e^M$ (here we use the fact that $e^{\log n} = n$), so that we can take for n_0 the smallest integer that is $\geq e^M$.

Relations (1.1.10)–(1.1.12) remain valid even if a and/or b are $\pm\infty$ with the exceptions that $\infty - \infty, \infty \cdot 0$, and ∞/∞ are undefined.

The case $a_n \to -\infty$ is completely analogous (Problem 1.4) and requires the corresponding restrictions on (1.1.10)–(1.1.12).

Since throughout the book we shall be dealing with sequences, we shall in the remainder of the present section and in Section 4 consider relations between two sequences a_n and $b_n, n = 1, 2, \ldots$, which are rough analogs of the relations $a = b$ and $a < b$ between numbers.

Definition 1.1.3 Two sequences $\{a_n\}$ and $\{b_n\}$ are said to be (asymptotically) equivalent as $n \to \infty$; in symbols:

(1.1.19) $$a_n \sim b_n$$

if

(1.1.20) $$a_n/b_n \to 1.$$

This generalizes the concept of equality of two numbers a and b, to which it reduces for the sequences $a, a, a, \ldots$ and $b, b, b, \ldots$.

If b_n tends to a finite limit $b \neq 0$, (1.1.20) simply states that a_n tends to the same limit. However, if the limit b is 0 or $\pm\infty$, the statement $a_n \sim b_n$ contains important additional information. Consider, for example, the sequences $a_n = 1/n^2$ and $b_n = 1/n$, both of which tend to zero. Since their ratio a_n/b_n tends to zero, the two sequences are not equivalent. Here are two more examples:

(1.1.21) $$a_n = n + n^2, \ b_n = n$$

and

(1.1.22) $$a_n = n + n^2, \ b_n = n^2,$$

in both of which a_n and b_n tend to ∞. In the first, $a_n/b_n \to \infty$ so that a_n and b_n are not equivalent; in the second, $a_n/b_n \to 1$ so that they are equivalent.

A useful application of the idea of equivalence is illustrated by the sequences

(1.1.23) $$a_n = \frac{1}{n} + \frac{3}{n^2} + \frac{1}{n^3}, \ b_n = \frac{1}{n}.$$

Both a_n and b_n tend to zero. Since their ratio satisfies

$$\frac{a_n}{b_n} = 1 + \frac{3}{n} + \frac{1}{n^2} \to 1,$$

the two sequences are equivalent. The replacement of a complicated sequence such as a_n by a simpler asymptotically equivalent sequence b_n plays a central role in large-sample theory.

Replacing a true a_n by an approximating b_n of course results in an error. Consider, for example, the two equivalent sequences (1.1.22). When $n = 100$,

$$a_n = 10,100, \ b_n = 10,000,$$

and the error (or absolute error) is $|a_n - b_n| = 100$. On the other hand, the

(1.1.24)   relative error $= \left| \dfrac{a_n - b_n}{a_n} \right|$

is $\dfrac{100}{10,100}$ which is less than .01. The small relative error corresponds to the fact that, despite the large absolute error of 100, b_n gives a pretty good idea of the size of a_n.

As the following result shows, asymptotic equivalence is closely related to relative error.

Lemma 1.1.1 *The sequences $\{a_n\}$ and $\{b_n\}$ are asymptotically equivalent if and only if the relative error tends to zero.*

Proof. The relative error

$$\left| \frac{a_n - b_n}{a_n} \right| = \left| 1 - \frac{b_n}{a_n} \right| \to 0$$

if and only if $b_n/a_n \to 1$.   ∎

The following is a classical example of asymptotic equivalence which forms the basis of the application given in the next section.

Example 1.1.2 Stirling's formula. Consider the sequence

(1.1.25) $a_n = n! = 1 \cdot 2 \cdots n.$

Clearly, $a_n \to \infty$ as $n \to \infty$, but it is difficult from the defining formula to see how fast this sequence grows. We shall therefore try to replace it by a simpler equivalent sequence b_n. Since n^n is clearly too large, one might try, for example, $(n/2)^n$. This turns out to be still too large, but taking logarithms leads (not obviously) to the suggestion $b_n = (n/e)^n$. Now only a relatively minor further adjustment is required, and the final result (which we shall not prove) is Stirling's formula

(1.1.26) $n! \sim \sqrt{2\pi n}\,(n/e)^n.$

The following table adapted from Feller (Vol. 1) (1957), where there is also a proof of (1.1.26), shows the great accuracy of the approximation (1.1.26) even for small n.

It follows from Lemma 1.1.1 that the relative error tends to zero, and this is supported by the last line of the table. On the other hand, the absolute error tends to infinity and is already about 30,000 for $n = 10$.   □

The following example provides another result, which will be used later.

TABLE 1.1.2. Stirling's approximation to $n!$

n	1	2	5	10	100
$n!$	1	2	120	3.6288×10^6	9.3326×10^{157}
(1.26)	.922	1.919	118.019	3.5987×10^6	9.3249×10^{157}
Error	.078	.081	1.981	$.0301 \times 10^6$	$.0077 \times 10^{157}$
Relative Error	.08	.04	.02	.008	.0008

Example 1.1.3 Sums of powers of integers. Let

$$(1.1.27) \qquad S_n^{(k)} = 1^k + 2^k + \cdots + n^k \ (k \text{ a positive integer})$$

so that, in particular,

$$S_n^{(0)} = n, \ S_n^{(1)} = \frac{n(n+1)}{2}, \quad \text{and} \quad S_n^{(2)} = \frac{n(n+1)(2n+1)}{6}.$$

These formulas suggest that perhaps

$$(1.1.28) \qquad S_n^{(k)} \sim \frac{n^{k+1}}{k+1} \text{ for all } k = 1, 2, \ldots,$$

and this is in fact the case. (For a proof, see Problem 1.14).   □

Summary

1. A sequence of numbers a_n tends to the *limit* a if for all sufficiently large n the a's get arbitrarily close [i.e., within any preassigned distance ε] to a. If $a_n \to a$, then a can be used as an approximation for a_n when n is large.

2. Two sequences $\{a_n\}$ and $\{b_n\}$ are *asymptotically equivalent* if their ratio tends to 1. The members of a complicated sequence can often be approximated by those of a simpler sequence which is asymptotically equivalent. In such an approximation, the relative error tends to 0 as $n \to \infty$.

3. Stirling's formula provides a simple approximation for $n!$. The relative error in this approximation tends to 0 as $n \to \infty$ while the absolute error tends to ∞.

1.2 Embedding sequences

The principal aim of the present section is to introduce a concept which is central to large-sample theory: obtaining an approximation to a given situation by embedding it in a suitable sequence of situations. We shall illustrate this process by obtaining two different approximations for binomial probabilities corresponding to two different embeddings.

The probability of obtaining x successes in n binomial trials with success probability p is

(1.2.1) $$P_n(x) = \binom{n}{x} p^x q^{n-x} \quad \text{where } q = 1 - p.$$

Suppose that n is even and that we are interested in the probability $P_n\left(\frac{n}{2}\right)$ of getting an even split between successes and failures. It seems reasonable to expect that this probability will tend to 0 as $n \to \infty$ and that it will be larger when $p = 1/2$ than when it is $\neq 1/2$.

To get a more precise idea of this behavior, let us apply Stirling's formula (1.1.26) to the three factorials in

$$P_n\left(\frac{n}{2}\right) = \frac{n!}{\left(\frac{n}{2}\right)! \left(\frac{n}{2}\right)!} (pq)^{n/2}.$$

After some simplification, this leads to (Problem 2.1)

(1.2.2) $$P_n\left(\frac{n}{2}\right) \sim \sqrt{\frac{2}{\pi}} \cdot \frac{1}{\sqrt{n}} \left(\sqrt{4pq}\right)^n.$$

We must now distinguish two cases.

Case 1. $p = 1/2$. Here we are asking for an even split between heads and tails in n tosses with a fair coin. The third factor in (1.2.2) is then 1, and we get the simple approximation

(1.2.3) $$P_n\left(\frac{n}{2}\right) \sim \sqrt{\frac{2}{\pi}} \cdot \frac{1}{\sqrt{n}} \quad \text{when } p = 1/2.$$

This result confirms the conjecture that the probability tends to 0 as $n \to \infty$. The exact values of $P_n\left(\frac{n}{2}\right)$ and the approximation (1.2.3) are shown in Table 1.2.1 for varying n.

TABLE 1.2.1. $P_n\left(\frac{n}{2}\right)$ for $p = 1/2$

n	4	20	100	500	1,000	10,000
Exact	.375	.176	.0796	.0357	.0252	.00798
(1.2.3)	.399	.178	.0798	.0357	.0252	.00798

A surprising feature of the table is how slowly the probability decreases. Even for $n = 10,000$, the probability of an exactly even 5,000–5,000 split

is not much below .01. Qualitatively, this could have been predicted from
(1.2.3) because of the very slow increase of $\sqrt{n}$ as a function of n. The table
indicates that the approximation is highly accurate for $n > 20$.

Case 2. $p \neq 1/2$. Since $\sqrt{4pq} < 1$ for all $p \neq 1/2$ (Problem 2.2), the approximate probabilities (1.2.3) for $p = 1/2$ are multiplied by the nth power of a
number between 0 and 1 when $p \neq 1/2$. They are therefore greatly reduced
and tend to 0 at a much faster rate. The exact values of $P_n \left(\frac{n}{2} \right)$ and the
approximation (1.2.2) are shown in Table 1.2.2 for the case $p = 1/3$. Again,

TABLE 1.2.2. $P_n \left(\frac{n}{2} \right)$ for $p = 1/3$

n	4	20	100	1,000	10,000
Exact	.296	.0543	.000220	6.692×10^{-28}	1.378×10^{-258}
(2.2)	.315	.0549	.000221	6.694×10^{-28}	1.378×10^{-258}

the approximation is seen to be highly accurate for $n > 20$.

A comparison of the two tables shows the radical difference in the speed
with which $P_n \left(\frac{n}{2} \right)$ tends to 0 in the two cases.

So far we have restricted attention to the probability of an even split,
that is, the case in which $\frac{x}{n} = \frac{1}{2}$. Let us now consider the more general
case that x/n has any given fixed value $\alpha \, (0 < \alpha < 1)$, which, of course,
requires that αn is an integer. Then

$$P_n (x) = \binom{n}{\alpha n} \left(p^{\alpha} q^{1-\alpha} \right)^n$$

and application of Stirling's formula shows in generalization of (1.2.2) that
(Problem 2.3)

(1.2.4) $$P_n (\alpha n) \sim \frac{1}{\sqrt{2\pi \alpha (1 - \alpha)}} \cdot \frac{1}{\sqrt{n}} \gamma^n$$

with

(1.2.5) $$\gamma = \left(\frac{p}{\alpha} \right)^{\alpha} \left(\frac{q}{1 - \alpha} \right)^{1-\alpha}.$$

As before, there are two cases.

Case 1. $p = \alpha$. This is the case in which p is equal to the frequency of
success, the probability of which is being evaluated. Here $\gamma = 1$ and (1.2.4)
reduces to

(1.2.6) $$P_n (\alpha n) \sim \frac{1}{\sqrt{2\pi \alpha (1 - \alpha)}} \cdot \frac{1}{\sqrt{n}} \quad \text{when } p = \alpha.$$

TABLE 1.2.3. $P_n\left(\dfrac{n}{100}\right)$ for $p = .01$

$x = \dfrac{n}{100}$	1	2	5	10	100
n	100	200	500	1,000	10,000
Exact	.370	.272	.176	.126	.0401
(1.2.6) for $\alpha = .01$	.401	.284	.179	.127	.0401
(1.2.8)	.368	.271	.176	.125	.0399
λ	1	2	5	10	100

The exact values of $P_n(\alpha n)$ and the approximation (1.2.6) are shown in the first two rows of Table 1.2.3 for the case $\alpha = .01$. The results are similar to those of Table 1.2.1, except that the approximation (1.2.6) does not become satisfactory until $n = 500$. We shall return to this difficulty below.

Case 2. $p \neq \alpha$. Here formula (1.2.4) applies, and it can be shown that $0 < \gamma < 1$ when $p \neq \alpha$ (Problem 2.4), so that the approximate probabilities will be much smaller than when $p = \alpha$. The exact values of $P_n(\alpha n)$ and the approximation (1.2.4) are shown in the first two rows of Table 1.2.4 for the case $\alpha = .01$, $p = .02$. Again, the results are similar to those for

TABLE 1.2.4. $P_n\left(\dfrac{n}{100}\right)$ for $p = .02$

$x = \dfrac{n}{100}$	1	2	5	10	100
n	100	200	500	1,000	10,000
Exact	.271	.146	.0371	.00556	1.136×10^{-15}
(1.2.4)	.294	.152	.0377	.00560	1.137×10^{-15}
(1.2.8)	.271	.146	.0378	.00582	1.880×10^{-15}
λ	2	4	10	20	200

$\alpha = 1/2$ (Table 1.2.2) except that it requires much larger values of n before the approximation becomes satisfactory.

The diminished accuracy for $n = 100$ and 200 in Tables 1.2.3 and 1.2.4 is explained by the small values of x in these cases, since the approximations were based on the assumption that not only n but then also $x = \alpha n$ is large. The table entries in question correspond to situations in which n is large and x is small but, nevertheless, the probability $P_n(x)$ of observing x is appreciable. To obtain a better approximation for such cases, note that $P_n(x)$ is a function of the three variables x, p, and n. If p is fixed, it is nearly certain that only large values of x will occur when n is large. To obtain situations in which small values of x have appreciable probability for large n, it is necessary to allow p to become small as n gets large.

For large n, we expect x/n to be close to p; so small values of x, say between 1 and 5, are most likely when np is in this range. This suggests letting p be proportional to $1/n$, say

(1.2.7)
$$p = \frac{\lambda}{n}.$$

Then

$$P_n\left(x\right) = \frac{n!}{x!\left(n-x\right)!}\left(\frac{\lambda}{n}\right)^x\left(1-\frac{\lambda}{n}\right)^{n-x}.$$

Now

$$\frac{n!}{n^x\left(n-x\right)!} = \frac{n\left(n-1\right)\cdots\left(n-x+1\right)}{n^x} \to 1 \text{ as } n \to \infty$$

and by (1.1.8)

$$\left(1-\frac{\lambda}{n}\right)^{n-x} \sim \left(1-\frac{\lambda}{n}\right)^n \to e^{-\lambda},$$

so that

(1.2.8)
$$P_n\left(x\right) \to \frac{\lambda^x}{x!}e^{-\lambda}.$$

The terms on the right side of (1.2.8) are the probabilities $P\left(X = x\right)$ of the Poisson distribution with parameter $\lambda = E\left(X\right)$, and (1.2.8) is the Poisson approximation to the binomial distribution.

To see how well the approximation works for small x and large n, consider once more Table 1.2.4. The third row shows that for small values of x, the Poisson approximation is considerably more accurate than the earlier approximation (1.2.4) which was based on Stirling's formula. On the other hand, as x increases, the Poisson approximation becomes less accurate.

Formulas (1.2.4) and (1.2.8) provide two different approximations for binomial probabilities with large n. Both are obtained by considering sequences $\left(n, p_n, x_n\right)$ for increasing n, and determining the limit behavior of $P_n\left(x\right) = f\left(n, x, p\right)$ for these sequences. A given situation can be viewed as a member of either sequence. To be specific, suppose we want an approximation for the case that $n = 100$, $x = 1$, and $p = 1/100$. We can embed this situation in the two sequences as follows.

Sequence 1: p is fixed, $x = \dfrac{n}{100}$, $n = 100, 200, 300, \ldots$

$n = 100$, $x = 1$, $p = .01$

$n = 200$, $x = 2$, $p = .01$

$n = 300, \ x = 3, \ p = .01$

This is the sequence leading to (1.2.4) with $\alpha = .01$.

Sequence 2: x is fixed, $p = 1/n$, $n = 100, 200, 300, \ldots$

$n = 100, \ x = 1, \ p = .01$

$n = 200, \ x = 1, \ p = .005$

$n = 300, \ x = 1, \ p = .00333\ldots$

This is the sequence leading to (1.2.8) with $\lambda = 1$.

Table 1.2.3 shows that in this particular case, the approximation derived from the second sequence is better.

Note: The two stated sequences are of course not the only ones in which the given situation can be embedded. For another example, see Problem 2.14.

The present section illustrates a central aspect of large-sample theory: the interplay and mutual support of theoretical and numerical work. Formulas such as (1.2.4) and (1.2.8) by themselves are of only limited usefulness without some information about their accuracy such as that provided by Tables 1.2.1–1.2.4. On the other hand, the numbers shown in these tables provide only snapshots. It is only when the two kinds of information are combined that one obtains a fairly comprehensive and reliable picture.

Summary

1. For fixed p and large n, the binomial probabilities $P_n(x)$ can be well approximated (with the help of Stirling's formula) for values of x which are not too close to 0 or n by (1.2.4) and (1.2.5).

2. For small x, large n, and appreciable $P_n(x)$, a better approximation is obtained by letting $p \to 0$ in such a way that np tends to a finite limit.

3. These two examples illustrate the general method of obtaining approximations by embedding the actual situation in a sequence of situations, the limit of which furnishes the approximation. Among the many possible embedding sequences, the most useful are those that are simple and give good approximations even for relatively small n.

1.3 Infinite series

For a sequence of numbers $a_1, a_2, \ldots$ consider the partial sums

$$s_n = a_1 + a_2 + \cdots + a_n, \quad n = 1, 2, \ldots.$$

If the sequence of the sums $s_n, n = 1, 2, \ldots$, tends to a limit l, we write

(1.3.1)
$$\sum_{i=1}^{\infty} a_i = a_1 + a_2 + \cdots = l.$$

We call the left side an *infinite series* and l its sum.

Example 1.3.1 Three modes of behavior.

(i) Consider the sum

$$\frac{1}{2} + \frac{1}{4} + \frac{1}{8} + \cdots = \sum_{i=1}^{\infty} \frac{1}{2^i}.$$

Here the partial sums are

$$s_1 = \frac{1}{2}, \ s_2 = \frac{1}{2} + \frac{1}{4} = \frac{3}{4}, \ s_3 = \frac{1}{2} + \frac{1}{4} + \frac{1}{8} = \frac{7}{8}$$

and quite generally

$$s_n = \frac{1}{2} + \frac{1}{4} + \cdots + \frac{1}{2^n}.$$

This sum is equal to (Problem 3.9)

$$\frac{1}{2^n} \left[1 + 2 + \cdots + 2^{n-1} \right] = \frac{2^n - 1}{2^n} = 1 - \frac{1}{2^n}.$$

Thus $s_n \to 1$ and we can write

(1.3.2)
$$\sum_{i=1}^{\infty} \frac{1}{2^i} = 1.$$

(ii) If $a_i = 1/i$, the resulting series $\sum 1/i$ is the *harmonic series*. It is easily seen that

$$s_n = 1 + \frac{1}{2} + \frac{1}{3} + \frac{1}{4} + \frac{1}{5} + \frac{1}{6} + \frac{1}{7} + \frac{1}{8} + \cdots + \frac{1}{n}$$

tends to infinity by noting that

$$\frac{1}{3} + \frac{1}{4} > \frac{1}{4} + \frac{1}{4} = \frac{1}{2},$$

that

$$\frac{1}{5} + \frac{1}{6} + \frac{1}{7} + \frac{1}{8} > \frac{4}{8} = \frac{1}{2},$$

that the sum of the next 8 terms exceeds $\dfrac{8}{16} = \dfrac{1}{2}$, and so on.

A series for which s_n tends to a finite limit l is said to converge to l. A series for which $s_n \to \infty$ is said to *diverge*. A series of *positive* terms $a_1 + a_2 + \cdots$ will always tend to a limit which may be finite or infinite. It converges to a finite limit if the a's tend to 0 sufficiently fast, as in (1.3.2). It will diverge if the a's do not tend to 0 or if they tend to 0 sufficiently slowly, as in the harmonic series.

(iii) If the a's are not all positive, the series may not tend to any limit, finite or infinite. For example, if the a's are $+1, -1, +1, -1, \ldots$, then

$$s_n = \begin{cases} 1 & \text{if } n \text{ is } odd \\ 0 & \text{if } n \text{ is } even \end{cases}$$

And $\lim s_n$ does not exist.   □

Example 1.3.2 Sums of powers of integers. The harmonic series of Example 1.3.1 (ii) is the special case $\alpha = 1$ of the sum of negative powers

(1.3.3) $$\sum_{i=1}^{\infty} \frac{1}{i^{\alpha}}.$$

Since this series diverges for $\alpha = 1$, it must also diverge when $\alpha < 1$ since then the terms tend to zero even more slowly. In fact, it can be shown that the partial sums of (1.3.3) satisfy

(1.3.4) $$s_n = \sum_{i=1}^{n} \frac{1}{i^{\alpha}} \sim \frac{n^{1-\alpha}}{1-\alpha} \text{ if } \alpha < 1$$

and

(1.3.5) $$s_n = \sum_{i=1}^{n} \frac{1}{i} \sim \log n \text{ if } \alpha = 1,$$

both of which tend to ∞ as $n \to \infty$. On the other hand, when $\alpha > 1$, the terms tend to 0 fast enough for (1.3.3) to converge to a finite limit.

Formula (1.3.4) holds not only for negative but also for positive powers of the integers, so that (in generalization of (1.1.28))

(1.3.6) $$s_n = \sum_{i=1}^{n} i^{\alpha} \sim \frac{n^{1+\alpha}}{1+\alpha} \text{ for all } \alpha > 0.$$

(For proofs of these results, see, for example, Knopp (1990).)   □

Example 1.3.3 The geometric series. The series (1.3.2) is the special case of the *geometric series* $\sum x^i$ when $x = 1/2$. If we start the series at $i = 0$, we have

$$s_n = 1 + x + \cdots + x^{n-1} = \frac{1 - x^n}{1 - x}.$$

For $|x| < 1$, this tends to the limit

(1.3.7)
$$\sum_{i=0}^{\infty} x^i = \frac{1}{1 - x}.$$

For $x = 1$, the series diverges since $s_n = n \to \infty$. For $x = -1$, we have the case (iii) of Example 1.3.1 and s_n therefore does not tend to a limit. Similarly, the series diverges for all $x > 1$, while for $x < -1$, the series does not tend to a limit. □

The geometric series is a special case of a *power series*

(1.3.8)
$$\sum_{i=0}^{\infty} c_i x^i.$$

In generalization of the result for the geometric series, it can be shown that there always exists a value r $(0 \le r \le \infty)$, called the *radius of convergence* of the series, such that (1.3.8) converges for all $|x| < r$ and for no $|x| > r$; it may or may not converge for $x = \pm r$. The case $r = \infty$ corresponds to the situation that (1.3.8) converges for all x. An example is provided by the exponential series in Example 1.3.4 below. For an example with $r = 0$, see Problem 3.4.

The following are a number of important power series. For each, the sum and the radius of convergence will be stated without proof. These series will be considered further in Section 2.5.

Example 1.3.4 The exponential series. In Example 1.1.1, an expression for any power of e was given in (1.1.8). An alternative expression is provided by the infinite series

(1.3.9)
$$1 + \frac{x}{1!} + \frac{x^2}{2!} + \cdots = e^x,$$

which converges for all x. For $x = \lambda > 0$, (1.3.9) shows that the Poisson distribution (1.2.8) with probabilities

$$P(X = i) = \frac{\lambda^i}{i!} e^{-\lambda}, \; i = 0, 1, 2, \ldots$$

really is a distribution, i.e., that the probabilities for $i = 0, 1, 2, \ldots$ add up to 1. □

Example 1.3.5 Binomial series. For any real number $\alpha < 1$, the series

(1.3.10) $$1 + \frac{\alpha}{1!}x + \frac{\alpha(\alpha-1)}{2!}x^2 + \cdots = (1+x)^\alpha$$

has radius of convergence 1.

Some special cases are worth noting.

(i) If α is a positive integer, say $\alpha = n$, the coefficient of x^i is zero for all $i > n$ and (1.3.10) reduces to the binomial theorem

$$(1+x)^n = \sum_{i=0}^{n}\binom{n}{i}x^i.$$

(ii) If α is a negative integer, say $\alpha = -m$, (1.3.10) with $x = -q$ reduces to

(1.3.11) $$(1-q)^{-m} = 1 + \binom{m}{1}q + \binom{m+1}{2}q^2 + \cdots .$$

This shows that the probabilities

(1.3.12)
$$P(X = i) = \binom{m+i-1}{m-1}p^m q^i, \ i = 0,1,2,\ldots \quad (0 < p < 1, q = 1 - p)$$

of the negative binomial distribution (listed in Table 1.6.2) add up to 1.

(iii) For $m = 1$, (1.3.11) reduces to the geometric series (1.3.7) (with $x = q$) and the distribution (1.3.12) reduces to the geometric distribution with probabilities

(1.3.13) $$P(X = i) = p \cdot q^i, \ i = 0, 1, 2, \ldots .$$

$\square$

Example 1.3.6 The logarithmic series. The series

(1.3.14) $$x + \frac{x^2}{2} + \frac{x^3}{3} + \cdots = -\log(1-x)$$

has radius of convergence 1. When $x = 1$, it reduces to the harmonic series, which in Example 1.3.1(ii) was seen to diverge. For $0 < \theta < 1$, (1.3.14) shows that the probabilities

(1.3.15) $$P(X = i) = \frac{1}{-\log(1-\theta)}\frac{\theta^i}{i}, \ i = 1, 2, \ldots$$

of the *logarithmic distribution* add up to 1. (For a discussion of this distribution, see Johnson, Kotz, and Kemp (1992), Chapter 7. $\square$

Summary

1. An infinite series is said to converge to s if the sequence of its partial sums tends to s.

2. For any power series $\sum c_i x^i$, there exists a number $0 \leq r \leq \infty$ such that the series converges for all $|x| < r$ and for no $|x| > r$.

3. Important examples of power series are the geometric, the exponential, the binomial, and the logarithmic series.

1.4   Order relations and rates of convergence

In Section 1.1, two sequences a_n, b_n were defined to be asymptotically equivalent if $a_n/b_n \to 1$. This relation states that for large n, the numbers of the two sequences are roughly equal. In the present section, we consider three other relationships between sequences, denoted by o, $\asymp$, and O, which correspond to a_n being of smaller, equal, or less or equal, order than b_n, respectively. In a related development, we shall study the behavior of sequences exhibiting respectively slow, moderate, and rapid growth.

We begin by defining the relationship $a_n = o(b_n)$, which states that for large n, a_n is an order of magnitude smaller than b_n.

Definition 1.4.1 We say that $a_n = o(b_n)$ as $n \to \infty$ if

$$(1.4.1) \qquad\qquad a_n/b_n \to 0.$$

When a_n and b_n both tend to infinity, this states that a_n tends to infinity more slowly than b_n; when both tend to 0, it states that a_n tends to zero faster than b_n. For example,

$$(1.4.2) \qquad\qquad \frac{1}{n^2} = o\left(\frac{1}{n}\right) \quad \text{as } n \to \infty$$

since $\dfrac{1}{n^2} \Big/ \dfrac{1}{n} = \dfrac{n}{n^2} \to 0$. As is shown in Table 1.4.1, it makes a big difference whether a sequence tends to 0 at rate $1/n^2$ or $1/n$.

TABLE 1.4.1. Speed of convergence of $1/n$ and $1/n^2$

n	1	2	5	10	20	100	1,000
$1/n$	1	.5	.2	.1	.05	.01	.001
$1/n^2$	1	.25	.04	.01	.0025	.0001	.000001

The following provides a statistical illustration of this difference.

Example 1.4.1 Bayes estimator for binomial p. Suppose that X has the binomial distribution $b(p, n)$ corresponding to n trials with success probability p. The standard estimator X/n is unbiased (i.e., satisfies $E(X/n) = p$) and has variance pq/n. An interesting class of competitors is the set of estimators

(1.4.3)
$$\delta(X) = \frac{a + X}{a + b + n}.$$

($\delta(X)$ is the Bayes estimator which minimizes the expected squared error when p has a beta prior with parameters a and b.)

Since
$$E[\delta(X)] = \frac{a + np}{a + b + n},$$

the bias of δ is
$$E[\delta(X)] - p = \frac{a + np}{a + b + n} - p,$$

which simplifies to

(1.4.4)
$$\text{bias of } \delta = \frac{aq - bp}{a + b + n} \text{ where } q = 1 - p.$$

Similarly, the variance of δ is

(1.4.5)
$$\text{Var}[\delta(X)] = \frac{npq}{(a + b + n)^2}.$$

The accuracy of an estimator $\delta(X)$ of a parameter $g(\theta)$ is most commonly measured by the expected squared error $E[\delta(X) - g(\theta)]^2$, which can be decomposed into the two terms (Problem 4.1)

(1.4.6) $E[\delta(X) - g(\theta)]^2 = (\text{bias of } \delta)^2 + \text{variance of } \delta.$

When $\delta(X)$ is given by (1.4.3), the two components of the *bias-variance decomposition* (1.4.6) make very unequal contributions to the squared error since

$$\frac{(\text{bias of } \delta)^2}{\text{Var}[\delta(X)]} \sim \frac{(aq - bp)^2}{npq} \to 0 \text{ as } n \to \infty,$$

and hence

$$(\text{bias of } \delta)^2 = o[\text{Var}\,\delta(X)].$$

Both terms tend to zero, but the square of the bias tends much faster: at the rate of $1/n^2$ compared to the rate $1/n$ for the variance. The bias therefore contributes relatively little to the expected squared error. As a numerical illustration, Table 1.4.2 shows the squared bias, the variance, and the contribution of the squared bias to the total, for the case $a = b = 1/2, p = 1/3$.

TABLE 1.4.2. Squared bias and variance of a biased estimator

n	10	50	100
(1) Variance	.018	.0043	.0022
(2) $(\text{bias})^2$	.00023	.000011	.0000027
(2) / [(1) + (2)]	.012	.0025	.0012

$\square$

The results of this example are typical for standard estimation problems: The variance and bias will each be of order $1/n$, and so therefore will be the expected squared error; the relative contribution of the squared bias to the latter will tend to 0 as $n \to \infty$. An important exception is the case of nonparametric density estimation, which will be considered in Section 6.4.

The following example illustrates an important use of the o notation.

Example 1.4.2 Order of a remainder. Suppose that

$$(1.4.7) \qquad a_n = \frac{1}{n} - \frac{2}{n^2} + \frac{4}{n^3}.$$

An obvious first approximation to a_n for large n is

$$(1.4.8) \qquad b_n = \frac{1}{n}.$$

Clearly, $a_n \sim b_n$. Let us denote the remainder in this approximation by R_n so that

$$(1.4.9) \qquad a_n = \frac{1}{n} + R_n,$$

where

$$(1.4.10) \qquad R_n = \frac{-2}{n^2} + \frac{4}{n^3} = o\,(1/n).$$

The information provided by (1.4.9) and the right side of (1.4.10) is frequently written more compactly as

$$(1.4.11) \qquad a_n = \frac{1}{n} + o\left(\frac{1}{n}\right),$$

where $o\,(1/n)$ denotes any quantity tending to 0 faster than $1/n$. We can refine the approximation (1.4.8) by taking into account also the $1/n^2$ term, and approximate (1.4.7) by

$$(1.4.12) \qquad b'_n = \frac{1}{n} - \frac{2}{n^2};$$

we can then write

$$(1.4.13) \qquad a_n = \frac{1}{n} - \frac{2}{n^2} + o\left(\frac{1}{n^2}\right).$$

Since the remainder is of order $1/n^2$ in the approximation (1.4.8) and of order $1/n^3$ in (1.4.12), we expect the second approximation to be more accurate than the first. This is illustrated by Table 1.4.3, which generally supports this expectation but shows that for very small n, the first approximation is, in fact, closer in this case.   □

TABLE 1.4.3. The accuracy of two approximations

n	1	2	3	5	10	50
$\frac{1}{n} - \frac{2}{n^2} + \frac{4}{n^3}$	3.0	.5	.259	.152	.84	.0192
$1/n$	1.0	.5	.333	.200	1.00	.0200
$\frac{1}{n} - \frac{2}{n^2}$	-1.0	0.0	.111	.120	.80	.0192

Note: The approximation (1.4.8) which takes account only of the principal term $(1/n)$ of (1.4.7) is a *first order approximation* while (1.4.12), which also includes the next $(1/n^2)$ term, is of *second order*.

While higher order approximations tend to be more accurate, they also suffer from some disadvantages: They are more complicated and, for the probability calculations to be considered later, they typically require more knowledge of the mathematical model. In addition, the higher order theory is more difficult mathematically. For these reasons, in this book we shall restrict attention mainly to first order approximations and shall make only a few brief excursions into higher order territory.

The following are some simple properties of the o relation (Problem 4.3).

Lemma 1.4.1

(i) *A sequence of a's satisfies*

$$(1.4.14) \qquad a_n = o(1) \ \ \textit{if and only if } a_n \to 0.$$

(ii) *If $a_n = o(b_n)$ and $b_n = o(c_n)$, then $a_n = o(c_n)$.*

(iii) *For any constant $c \neq 0$,*

$$(1.4.15) \qquad a_n = o(b_n) \ \textit{implies } ca_n = o(b_n).$$

(iv) For any sequence of numbers c_n different from 0,

$$a_n = o(b_n) \text{ implies } c_n a_n = o(c_n b_n).$$

(v) If $d_n = o(b_n)$ and $e_n = o(c_n)$, then $d_n e_n = o(b_n c_n)$.

Example 1.4.3 The order relationship of powers of n. If $a_n = n^\alpha, b_n = n^\beta$, then $a_n = o(b_n)$, provided $\alpha < \beta$. To see this, note that the ratio

$$\frac{a_n}{b_n} = \frac{n^\alpha}{n^\beta} = \frac{1}{n^{\beta-\alpha}}$$

tends to zero when $\alpha < \beta$. Thus $\sqrt[3]{n}, \sqrt{n}, n, n^2, n^3, \ldots$ constitute sequences which tend to infinity as $n \to \infty$ with increasing rapidity.   □

Example 1.4.4 Exponential growth. For any $k > 0$ and $a > 1$, we have

(1.4.16) $$n^k = o(a^n).$$

To prove (1.4.16), write $a = 1 + \epsilon \, (\epsilon > 0)$, and consider the case $k = 2$. Instead of showing that $n^2/a^n \to 0$, we shall prove the equivalent statement that

$$a^n/n^2 \to \infty.$$

For this purpose, we use the expansion

$$a^n = (1 + \epsilon)^n = 1 + n\epsilon + \binom{n}{2}\epsilon^2 + \binom{n}{3}\epsilon^3 + \cdots > \binom{n}{3}\epsilon^3,$$

which shows that

$$\frac{a^n}{n^2} > \frac{n(n-1)(n-2)}{n^2} \cdot \frac{\epsilon^3}{6} = \left(n - 3 + \frac{2}{n}\right) \cdot \frac{\epsilon^3}{6}.$$

Since the right side tends to ∞ as $n \to \infty$, this completes the proof for the case $k = 2$. The argument for any other integer k is quite analogous. If k is not an integer, let k' be any integer greater than k. Then

$$n^k = o\left(n^{k'}\right) \text{ and } n^{k'} = o(a^n),$$

and the result follows from part (ii) of Lemma 1.4.1.

More generally, it is seen that (Problem 4.7)

(1.4.17) $$c_0 + c_1 n + \cdots + c_k n^k = o(a^n)$$

for any integer k, any $a > 1$, and any constants $c_0, c_1, \ldots, c_k$. Sequences that are asymptotically equivalent either to (1.4.17) or to $a^n \, (a > 1)$ are said to grow at polynomial or exponential rate, respectively. Exponential growth is enormously much faster than polynomial growth, as is illustrated in the first two rows of Table 1.4.4. This explains the surprising growth due to compound interest (Problem 4.9).

Note: Because of this very great difference, it has been an important issue in computer science to determine whether certain tasks of size n require exponential time (i.e., a number of steps that increases exponentially) or can be solved in polynomial time. □

Let us next consider some sequences that increase more slowly than n^k for any $k > 0$.

Example 1.4.5 Logarithmic growth. Let $a_n = \log n, b_n = n^k$. Then $a_n = o(b_n)$ for every $k > 0$. To show that $\dfrac{\log n}{n^k} \to 0$, it is enough by Definition 1.1.1 to show that for any $\epsilon > 0$, we have $-\epsilon < \dfrac{\log n}{n^k} < \epsilon$ for all sufficiently large n. Since $\log n/n^k$ is positive, it is therefore enough to show that for all sufficiently large n

$$n < e^{\epsilon n^k},$$

which follows from Problem 4.7(ii). □

The three scales a^n, n^k, and $\log n$ are the most commonly used scales for the growth of a sequence tending to infinity, representing fast, moderate, and slow growth, respectively. (Correspondingly their reciprocals a^{-n}, n^{-k}, and $1/\log n$ provide standard scales for convergence to 0.) One can get an idea of the differences in these rates by looking at the effect of doubling n. Then, for example,

$a_n = a^n$ becomes $a_{2n} = a^{2n} = a^n a_n$,

$a_n = n^2$ becomes $a_{2n} = 4n^2 = 4a_n$,

$a_n = \log n$ becomes $a_{2n} = \log(2n) = a_n + \log 2$.

Thus doubling n multiplies a_n in the first case by a factor which rapidly tends to ∞; in the second case, by the fixed factor 4; finally in the third case, doubling n only adds the fixed amount $\log 2$ to a_n. Table 1.4.4 illustrates these three rates numerically. While the first sequence seems to explode, the last one barely creeps along; yet eventually it too will become arbitrarily large (Problem 4.10(i)).

If a_n/b_n tends to 0 or ∞, a_n is respectively of smaller or larger order than b_n. In the intermediate case that a_n/b_n tends to a finite non-zero limit, the two sequences are said to be of the same order or to tend to 0 or ∞ at the same rate. (This terminology was already used informally at the end of

TABLE 1.4.4. Three rates of invergence

n	1	2	5	10	15	20
e^n	2.7	7.4	148.4	22,026	3,269,017	4,851,165,195
n^3	1	8	125	1000	3375	8000
$\log n$	0	.7	1.6	2.3	2.7	3.0

Example 1.4.1 when it was stated that the bias and variance are typically "of order $1/n$.")

The concept of two sequences being of the same order can be defined somewhat more generally without requiring a_n/b_n to tend to a limit.

Definition 1.4.2 Two sequences $\{a_n\}$ and $\{b_n\}$ are said to be of the same order, in symbols,

$$(1.4.18) \qquad a_n \asymp b_n,$$

if $|a_n/b_n|$ is bounded away from both 0 and ∞, i.e., if there exist constants $0 < m < M < \infty$ and an integer n_0 such that

$$(1.4.19) \qquad m < \left|\frac{a_n}{b_n}\right| < M \text{ for all } n > n_0.$$

As an example, let

$$a_n = 2n + 3, \ b_n = n.$$

Then

$$\frac{a_n}{b_n} < 3 \text{ holds when } 2n + 3 < 3n,$$

i.e., for all $n > 3$, and

$$\frac{a_n}{b_n} > 2 \text{ holds for all } n.$$

Thus (1.4.19) is satisfied with $m = 2$, $M = 3$, and $n_0 = 3$.

It is easy to see that if a_n/b_n tends to any finite non-zero limit, then $a_n \asymp b_n$ (Problem 4.11).

Lemma 1.4.2 *If* $a_n \asymp b_n$, *then* $ca_n \asymp b_n$ *for any* $c \neq 0$.

The symbol $\asymp$ denotes equality of the orders of magnitude of a_n and b_n and constitutes a much cruder comparison than the asymptotic equivalence $a_n \sim b_n$. This is illustrated by the statement of Lemma 1.4.2 that the relationship $a_n \asymp b_n$ is not invalidated when a_n is multiplied by any fixed constant $c \neq 0$. Definition 1.4.2 of equality of order is motivated by the extreme difference in the speed of convergence of sequences of different order shown in Table 1.4.4. In the light of such differences, multiplication by a fixed constant does little to change the basic picture.

In addition to o and $\asymp$, it is useful to have a notation also for a_n being of order smaller than or equal to that of b_n. This relationship is denoted by $a_n = O(b_n)$. It follows from Definitions 1.4.1 and 1.4.2 that $a_n = O(b_n)$ if $|a_n/b_n|$ is bounded, i.e., if there exist M and n_0 such that

$$(1.4.20) \qquad \left| \frac{a_n}{b_n} \right| < M \text{ for all } n > n_0.$$

Note in particular that

$$(1.4.21) \qquad a_n = O(1) \text{ if and only if the sequence } \{a_n\} \text{ is bounded.}$$

Example 1.4.6 Information provided by $\sim$, $\asymp$, O, and o. To illustrate the distinction among $\sim$, $\asymp$, O, and o and the information they provide about a remainder, suppose that

$$a_n = \frac{1}{n} + \frac{b}{n\sqrt{n}} + \frac{c}{n^2} + \frac{d}{n^2\sqrt{n}}$$

is approximated by $1/n$, so that

$$a_n = \frac{1}{n} + R_n$$

with

$$R_n = \frac{b}{n\sqrt{n}} + \frac{c}{n^2} + \frac{d}{n^2\sqrt{n}}.$$

Then the following implications hold (Problem 4.20):

(i) $R_n \sim \dfrac{1}{n^2}$ if and only if $b = 0, c = 1$;

(ii) $R_n \asymp \dfrac{1}{n^2}$ if and only if $b = 0, c \neq 0$;

(iii) $R_n = O\left(\dfrac{1}{n^2}\right)$ if and only if $b = 0$;

(iv) $R_n = o\left(\dfrac{1}{n^2}\right)$ if and only if $b = c = 0$.

$\square$

Summary

1. For two sequences $\{a_n\}$, $\{b_n\}$, we consider the possibilities that

(a) a_n is of smaller order than b_n; in symbols, $a_n = o\,(b_n)$,

(b) a_n is of the same order as b_n; in symbols, $a_n \asymp b_n$,

(c) a_n is of order smaller than or equal to b_n; in symbols, $a_n = O\,(b_n)$,

and we discuss some properties of these order relations.

2. Three of the most commonly encountered growth rates are the logarithmic ($\log n$), polynomial (n^k), and exponential $(a^n, a > 1)$, and they satisfy

$$\log n = o\,(n^k) \text{ for all } k > 0,$$
$$n^k = o\,(a^n) \text{ for all } k > 0, a > 1.$$

3. When approximating a complicated a_n by a simpler b_n with remainder R_n, the three orders described in 1 can be used to characterize our knowledge about the order of the remainder.

1.5   Continuity

So far we have been concerned with the behavior of sequences $a_n, n = 1, 2, \ldots$. If the notation is changed from a_n to $a\,(n)$, it becomes clear that such a sequence is a function defined over the positive integers. In the present section, some of the earlier ideas will be extended to functions of a continuous variable x. Such a function is a rule or formula which assigns to each value of x a real number $f\,(x)$, for example, $x^2 + 1$, e^x, or $\sin x$. The function will be denoted by f, the value that f assigns to x by $f\,(x)$.

To define the concept of limit, we begin with the limit of $f\,(x)$ as $x \to \infty$, which is completely analogous to the Definition 1.1.1 of the limit of a sequence.

Definition 1.5.1 As $x \to \infty, f\,(x)$ tends (or converges) to a; in symbols,

(1.5.1) $$f\,(x) \to a \text{ as } x \to \infty \text{ or } \lim_{x \to \infty} f\,(x) = a,$$

if, given any $\epsilon > 0$, there exists $M = M\,(\epsilon)$ such that

$$|f\,(x) - a| < \epsilon \text{ for all } x > M.$$

A new situation arises when x tends to a finite value x_0, because it can then tend to x_0 from the right (i.e., through values $> x_0$) or from the left (i.e., through values $< x_0$). We shall denote these two possibilities by $x \to x_0+$ and $x \to x_0-$, respectively.

We shall say that $f\,(x) \to a$ as $x \to x_0+$ if $f\,(x)$ gets arbitrarily close to a as x gets sufficiently close to x_0 from the right. This is formalized in

Definition 1.5.2 As $x \to x_0+$, $f(x)$ tends to a; in symbols,

(1.5.2) $\qquad f(x) \to a$ as $x \to x_0 +$ or $\lim\limits_{x \to x_0+} f(x) = a$,

if, given any $\epsilon > 0$, there exists $\delta > 0$ such that

(1.5.3) $\qquad |f(x) - a| < \epsilon$ for all $x_0 < x < x_0 + \delta$.

The limit of $f(x)$ as $x \to x_0-$ is defined analogously.

Example 1.5.1 Let

(1.5.4) $\qquad f(x) = \begin{cases} x^3 + 1 & \text{if } x > 0 \\ x^3 - 1 & \text{if } x < 0. \end{cases}$

Then

(1.5.5) $\qquad \begin{array}{ll} f(x) \to 1 & \text{as } x \text{ tends to } 0 \text{ from the right,} \\ f(x) \to -1 & \text{as } x \text{ tends to } 0 \text{ from the left.} \end{array}$

To see the first of these statements, note that for $x > 0$,

$$|f(x) - 1| = x^3 < \epsilon \text{ when } 0 < x < \sqrt[3]{\epsilon},$$

so that (1.5.3) holds with $\delta = \sqrt[3]{\epsilon}$.   □

The one-sided limits $f(x_0+)$ and $f(x_0-)$ need not exist as is shown by the classical example

(1.5.6) $\qquad f(x) = \sin\dfrac{1}{x}$

Here there exist positive values of x arbitrarily close to 0 for which $f(x)$ takes on all values between -1 and $+1$ (Problem 5.3) so that $f(x)$ does not tend to a limit as $x \to 0+$, and the same is true as $x \to 0-$.

Two-sided limits are obtained by combining the two one-sided concepts.

Definition 1.5.3 We say that

$$f(x) \to a \text{ as } x \to x_0$$

if

(1.5.7) $\qquad \lim\limits_{x \to x_0+} f(x) = \lim\limits_{x \to x_0-} f(x) = a,$

i.e., if given any $\epsilon > 0$, there exists $\delta > 0$ such that

(1.5.8) $\qquad |f(x) - a| < \epsilon$ for all $x \neq x_0$ satisfying $|x - x_0| < \delta$.

In Example 1.5.1, the limit as $x \to 0$ does not exist since the two one-sided limits are not equal. On the other hand, if $f(x) = x^3$, it is seen that $f(x) \to 0$ as $x \to 0+$ and as $x \to 0-$, and that, therefore, $\lim_{x \to 0} f(x)$ exists and is equal to 0.

It is interesting to note that the definitions of $\lim_{x \to x_0+} f(x)$, $\lim_{x \to x_0-} f(x)$, and hence $\lim_{x \to x_0} f(x)$ say nothing about the value $f(x_0)$ that f takes on at the point x_0 itself but involves only values of f for x close to x_0.

The relations $\sim$, $\asymp$, O, and o extend in the obvious way. For example,

$$f(x) = o[g(x)] \text{ as } x \to x_0 \quad (x_0 \text{ finite or infinite})$$

if $f(x)/g(x) \to 0$ as $x \to x_0$.

The limit laws (1.1.10)–(1.1.12) also have their obvious analogs; for example, if $f(x) \to a$ and $g(x) \to b$ as $x \to x_0$, then $f(x) + g(x) \to a + b$. Similarly, the results of Examples 1.4.3 to 1.4.5 continue to hold when n is replaced by x; for example,

(1.5.9)
$$\frac{\log x}{x^a} \to 0 \text{ for any } \alpha > 0 \text{ as } x \to \infty.$$

Definition 1.5.4 A function f is said to be continuous at x_0 if $f(x_0+)$ and $f(x_0-)$ exist and if

(1.5.10)
$$f(x_0+) = f(x_0-) = f(x_0).$$

Any polynomial $P(x)$ is continuous for all x; so is any rational function, i.e., the ratio $P(x)/Q(x)$ of two polynomials, at any point x for which $Q(x) \neq 0$. The exponential function e^x is continuous for all x, and $\log x$ for all $x > 0$ (the latter is not defined for $x \leq 0$).

Continuity can be violated in a number of ways:

(a) Even if $f(x_0+)$ and $f(x_0-)$ exist and are equal, $f(x_0+) = f(x_0-) = a$, say, it may happen that $f(x_0) \neq a$.

As an example, let

$$f(x) = \begin{cases} x^2 & \text{if } x \neq 2 \\ 0 & \text{if } x = 2. \end{cases}$$

(b) A more interesting possibility is that $f(x_0+) \neq f(x_0-)$, as was the case in Example 1.5.1.

Definition 1.5.5 The function f is said to be continuous on the right at x_0 if $f(x_0+)$ exists and

(1.5.11)
$$f(x_0) = f(x_0+).$$

It is continuous on the left at x_0 if $f(x_0-)$ exists and

(1.5.12)
$$f(x_0) = f(x_0-).$$

If $f(x_0+)$ and $f(x_0-)$ exist, whether f is continuous on the right, on the left, or neither depends on the value $f(x_0)$. The function f given by (1.5.4) is continuous on the right at 0 if $f(0) = 1$ and continuous on the left if $f(0) = -1$. If $f(0)$ has a value different from ± 1, it is continuous neither on the right nor left at 0.

A function that is continuous on both the left and right at x_0 is continuous at x_0.

Discontinuities of both types (a) and (b) are called *simple discontinuities* or *discontinuities of the first kind*.

A classical example of the third possibility in which $f(x_0+)$ and/or $f(x_0-)$ do not exist is provided by the function (1.5.6).

As this example shows, discontinuities that are not simple can occur. They can, however, be ruled out for the important class of monotone functions, i.e., functions that are either non-decreasing or non-increasing. This follows from the following theorem which for the sake of simplicity is stated for the non-decreasing case.

Theorem 1.5.1

(i) If a_n is a non-decreasing sequence (i.e., satisfies $a_n \leq a_{n+1}$ for all n), then either

   (a) a_n is bounded above and tends to a finite limit as $n \to \infty$

   or

   (b) $a_n \to \infty$ as $n \to \infty$.

(ii) Let f be a function defined for all $a < x < b \, (-\infty \leq a < b \leq \infty)$ which is non-decreasing (i.e., satisfies $f(x) \leq f(y)$ when $x < y$). If $a \leq x_0 \leq b$, then as x tends to x_0 from the left (or the right), $f(x)$ tends either to a finite limit or to $\infty \, (-\infty)$.

The proof (see Problem 5.11) depends on the following basic property of real numbers, discussed, for example, in Rudin (1976) and Hardy (1908, 1992): Let S be a set of real numbers which is bounded above, i.e., there exists a constant M such that $x \leq M$ for all x in S. Then among all the upper bounds M, there exists a smallest one, which will be denoted by $\sup S$. Analogously, if S is bounded below, there exists a largest lower bound, denoted by $\inf S$.

The least upper or greatest lower bound of a set S is a member of S, or a limit of points in S, or both (Problem 5.12).

Example 1.5.2 Let S be the set of numbers $1 - \dfrac{1}{n}, n = 1, 2, 3, \ldots$. Then $\inf S = 0$ and $\sup S = 1$. The first of these is a member of S; the second is not. $\qquad \square$

Summary

1. The definition of a limit of a sequence of numbers is extended to one- and two-sided limits of a continuous variable.

2. A function f is *continuous* at a point x_0 if its limit as $x \to x_0$ exists and is equal to $f(x_0)$.

3. A function f has a simple discontinuity at x_0 if the limits $f(x_0-)$ and $f(x_0+)$ as x tends to x_0 from the left or right exist but are either not equal or are equal but $\neq f(x_0)$.

4. If a function f is monotone, the limits $f(x_0-)$ and $f(x_0+)$ exist for all x_0, and f can therefore have only simple discontinuities.

1.6   Distributions

The principal properties of random variables are reflected in their distributions and the principal aspects of the limiting behavior of a sequence of random variables in the limiting behavior of their distributions. Probability distributions can be represented in a number of different ways, and we shall in the present section consider several such representations, particularly cumulative distribution functions, probability densities, and quantile functions.

Definition 1.6.1 The cumulative distribution function (cdf) of a random variable X is defined as

(1.6.1)
$$F(x) = P(X \leq x).$$

Cumulative distribution functions have the following properties.

Theorem 1.6.1

(i) $0 \leq F(x) \leq 1$ *for all* x;

(ii) F *is non-decreasing;*

(iii) $\lim_{x \to -\infty} F(x) = 0$ *and* $\lim_{x \to \infty} F(x) = 1$.

(iv) *All discontinuities of* F *are simple, that is,* $F(x-)$ *and* $F(x+)$ *exist for all* x.

(v) F *is continuous on the right, that is,* $F(x) = F(x+)$ *for all* x.

(vi) *The jump of* $F(x)$ *at a discontinuity point* x *is equal to*

(1.6.2)
$$F(x) - F(x-) = P(X = x).$$

Proof. Parts (i) and (ii) are obvious; part (iv) follows from (ii) and Theorem 5.1. For proofs of the remaining parts, see, for example, Hoel, Port, and Stone (1971) or Parzen (1960, 1992). ∎

It follows from (vi) that F is continuous at x if and only if

$$P(X = x) = 0.$$

Such a point x is called a *continuity point* of F.

In this book, we shall be concerned primarily with two types of distributions:

(i) Continous distributions F which, in addition, are differentiable, that is, possess a probability density f satisfying

(1.6.3)
$$F(x) = \int_{-\infty}^{x} f(t)\, dt.$$

(ii) Discrete distributions, that is, distributions of a random variable X for which there exists a countable set A of values a_1, a_2, ... such that

(1.6.4)
$$\sum_{i=1}^{\infty} P(X = a_i) = 1.$$

TABLE 1.6.1. Some common probability densities

Density	Name	Notation
$\dfrac{1}{\sqrt{2\pi}b} e^{-(x-a)^2/b^2}$	Normal	$N\left(a,\ b^2\right)$
$\dfrac{1}{b} \dfrac{e^{(x-a)/b}}{\left[1 + e^{(x-a)/b}\right]^2}$	Logistic	$L\left(a,\ b\right)$
$\dfrac{1}{b} e^{-(x-a)/b}$ if $x > a$	Exponential	$E\left(a,\ b\right)$
$\dfrac{1}{2b} e^{-|x-a|/b}$	Double Exponential	$DE\left(a,\ b\right)$
$\dfrac{1}{b}$ if $a - \dfrac{b}{2} < x < a + \dfrac{b}{2}$	Uniform	$U\left(a - \dfrac{b}{2},\ a + \dfrac{b}{2}\right)$
$\dfrac{b}{\pi} \dfrac{1}{b^2 + (x-a)^2}$	Cauchy	$C\left(a,\ b\right)$

Examples of the continuous case are the normal, exponential, χ^2, uniform, and Cauchy distributions. Discrete distributions of particular importance are *lattice distributions*, i.e., distributions for which the set A is the set $\{a+k\Delta,\ k=0,\ \pm 1,\ \pm 2,\ \ldots\}$ for some a and some $\Delta > 0$. Examples of lattice distributions are the binomial and Poisson distributions. Distributions that are continous but not differentiable or that are partly continuous and partly discrete are of lesser interest, although they are included in some results concerning general classes of distributions.

The following result states some distributional facts that will often be used.

Theorem 1.6.2 *Let X_i $(i=1,\ldots,n)$ be independently distributed according to F_i. Then in the notation of Tables 1.6.1 and 1.6.2:*

(i) *If F_i is the Poisson distribution $P(\lambda_i)$, $Y = \sum X_i$ is distributed according to $P(\sum \lambda_i)$.*

(ii) *If F_i is the normal distribution $N(\xi_i,\ \sigma_i^2)$, $Y = \sum X_i$ is distributed according to $N(\sum \xi_i,\ \sum \sigma_i^2)$.*

(iii) *If F_i is the Cauchy distribution $C(a_i,\ b_i)$, $Y = \sum X_i$ is distributed according to $C(\sum a_i,\ \sum b_i)$.*

So far we have represented a probability distribution by its cdf (1.6.1); alternatively, we might have used

$$(1.6.5) \qquad\qquad F^*(x) = P(X < x).$$

This alternative satisfies (i)–(iv) and (vi) of Theorem 1.6.1, but condition (v), continuity on the right, is replaced by continuity on the left.

In the continuous case (1.6.3), a distribution is often represented by its density f; in the discrete case, by its probability function

$$(1.6.6) \qquad\qquad P(X = a_i),\ i = 1,\ 2,\ \ldots.$$

Tables 1.6.1 and 1.6.2 list some of the more common densities and probability functions.

Still another representation is by means of the inverse F^{-1} of F, the so-called *quantile* function.

(a) If F is continuous and strictly increasing, F^{-1} is defined by

$$(1.6.7) \qquad\qquad F^{-1}(y) = x \text{ when } y = F(x).$$

(b) If F has a discontinuity at x_0, suppose that $F(x_0-) < y < F(x_0) = F(x_0+)$. In this case, although there exists no x for which $y = F(x)$, $F^{-1}(y)$ is defined to be equal to x_0.

(c) Condition (1.6.7) becomes ambiguous when F is not strictly increasing. Suppose that

$$(1.6.8) \qquad\qquad F(x) \begin{cases} < y & \text{for } x < a \\ = y & \text{for } a \le x \le b \\ > y & \text{for } x > b. \end{cases}$$

TABLE 1.6.2. Some common discrete distributions

Probability function $P(X=x)$	Values	Name	Notation
$\binom{n}{x}p^x(1-p)^{n-x}$	$x = 0, 1, \ldots, n$	Binomial	$b(p, n)$
$\frac{1}{x!}\lambda^x e^{-\lambda}$	$x = 0, 1, \ldots$	Poisson	$P(\lambda)$
$\binom{m+x-1}{m-1}p^m q^x$	$x = 0, 1, \ldots$	Negative Binomial	$Nb(p, m)$
$\dfrac{\binom{D}{x}\binom{N-D}{n-x}}{\binom{N}{n}}$	$\max(0, n-(N-D))$ $\le x \le \min(n,\ D)$	Hypergeometric	$H(N, D, n)$

Then any value $a \le x \le b$ could be chosen for $x = F^{-1}(y)$. The convention in this case is to define

$$(1.6.9) \qquad F^{-1}(y) = a,$$

i.e., as the smallest value x for which $F(x) = y$. This causes F^{-1} to be continuous on the left.

The three cases can be combined into the single definition

$$(1.6.10) \qquad F^{-1}(y) = \min\{x\colon F(x) \ge y\}.$$

With these conventions, the graph of F^{-1} is obtained by rotating the graph of $y = F(x)$ around the line $y = x$ as the axis, so that the x-axis points up and the y-axis to the right. (A simple way of accomplishing this is to graph $y = F(x)$ on semi-transparent paper and then to reverse the sheet to obtain the graph of $x = F^{-1}(y)$.) The resulting picture, illustrated in Figure 1.6.1, shows that F^{-1} is defined on $(0,1)$, and that discontinuities of F become converted into flat stretches of F^{-1} and flat stretches of F into discontinuities of F^{-1}.

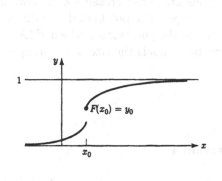

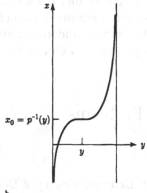

FIGURE 1.6.1(a): $y = F(x)$ FIGURE 1.6.1(b): $x = F^{-1}(y)$

The term *quantile function* reflects the close connection of F^{-1} with the quantiles of F. In case (a), the p-quantile of F is defined as the point x such that the probability $F(x)$ to the left of x is equal to p, that is, as $x = F^{-1}(p)$. In case (b), when $p = y$, the p-quantile is the unique point x for which $F(x-) < p < F(x)$.

Note: The quantiles corresponding to $p = 1/2$, $p = 1/4$, and $p = 3/4$ are called the *median*, *first quartile*, and *third quartile*, respectively.

In case (c), there is no uniquely agreed upon definition of quantile. If (1.6.8) holds, some authors call any x in the interval $[a, b]$ a y-quantile, some define the y-quantile uniquely as a, while at least in the case of the median, still others define it to be $\frac{1}{2}(a + b)$, the midpoint of the interval $[a, b]$.

Cumulative distribution functions, probability densities, and quantile functions constitute three possible ways of representing the distribution of a random variable. Still other representations (which we shall not consider here) are by means of moment generating functions or characteristic functions.

Summary

1. The distributions of a random variable can be represented in various ways: for example, by its cumulative distribution function, by its probability density function, and by its quantile function.

2. Of particular interest are (i) the continuous case in which the cdf is differentiable, its derivative being the probability density; and (ii) the discrete case. Two tables exhibit some of the most common densities in case (i) and probability functions in case (ii).

3. Cumulative distribution functions are non-decreasing and hence any of their discontinuities must be simple. The points of discontinuity of the cdf of a random variable X are the points x for which $P(X = x)$ is positive and this probability then equals the size of the jump at x.

1.7 Problems

Section 1

1.1 (i) Determine $n_0(\epsilon)$ of Definition 1.1.1 when a_n is given by (1.1.2).

(ii) Compare the smallest possible values of $n_0(\epsilon)$ for the cases that a_n is given by (1.1.1) or (1.1.2) when $\epsilon = 1/1,000,000$.

1.2 (i) Show that the sequence a_n defined by (1.1.7) is also given by the single formula

$$a_n = 1 + (-1)^n \cdot \frac{1}{n}.$$

(ii) If a_n is the nth member of the sequence (1.1.9), find a formula for a_n analogous to that of (i).

1.3 Determine $n_0 (M)$ of Definition 1.1.2 when

(a) $a_n = \sqrt{n}$,

(b) $a_n = n^2$,

and compare their values for the case $M = 10^4$.

1.4 Define the limit $a_n \to -\infty$ in analogy with Definition 1.1.2.

1.5 If $f(a) = c_0 + c_1 a + \cdots + c_k a^k$ for all a, use (1.1.10) and (1.1.11) to show that $f(a_n) \to f(a_0)$ when $a_n \to a_0$.

1.6 If $a_n \to \infty$, then

(i) $\sqrt{a_n} \to \infty$;

(ii) $\log a_n \to \infty$;

(iii) $1/a_n \to 0$.

1.7 (i) Show that the sequence (1.1.9) tends to 0 as $n \to \infty$.

(ii) Make a table of a_n given by (1.1.9) for $n = 75, 76, \ldots, 100$.

1.8 Let $a_n \to \infty$ and let $b_n = a_{n+1}$. Determine whether $a_n \sim b_n$ for the following cases:

(i) $a_n = n^k$, k is a positive integer;

(ii) $a_n = e^{\alpha n}$, $\alpha > 0$;

(iii) $a_n = \log n$.

[**Hint (iii)**: Apply Lemma 1.1.1 with $b_n = a_{n+1}$.]

1.9 Replace each of the following quantities by a simpler one that is asymptotically equivalent as $n \to \infty$:

(i) $n + \frac{1}{2}n^2 + \frac{1}{3}n^3$;

(ii) $n + 2 + \frac{3}{n}$;

(iii) $\frac{1}{n} + \frac{2}{n^2} + \frac{3}{n^3}$.

1.10 (i) If $a_n \sim b_n$, $b_n \sim c_n$, then $a_n \sim c_n$;

(ii) if $a_n \sim b_n$, then $\dfrac{1}{a_n} \sim \dfrac{1}{b_n}$.

(iii) If $a_n \sim b_n$, $c_n \sim d_n$, then $a_n c_n \sim b_n d_n$.

(iv) If $a_n \sim b_n$, then $a_n^\alpha \sim b_n^\alpha$ for any $\alpha > 0$.

1.11 If $a_n \sim b_n$, $c_n \sim d_n$, then

(i) $a_n + c_n \sim b_n + d_n$, provided a_n, b_n, c_n, and d_n are all > 0;

(ii) The result of (i) need not hold without the restriction to positive values.

1.12 If $a_n \to \infty$, determine whether $a_n \sim b_n$ implies that $a_n^2 + a_n \sim b_n^2 + b_n$.
[**Hint:** $a_n^2 + a_n = a_n (a_n + 1)$.]

1.13 Calculate the absolute and relative error in (1.1.26) for $n = 7$.

1.14 (i) Determine the sum $\displaystyle\sum_{i=1}^{n} i^3$ and check that (1.1.28) holds for $k = 3$.

(ii) Prove (1.1.28) for general k.

[**Hint:** (i) Add both sides of the equation $(i + 1)^4 - i^4 = 4i^3 + 6i^2 + 4i + 1$ from $i = 1$ to $i = n + 1$. (ii) Generalize the method of (i) and use induction.]

1.15 If k remains fixed and $N \to \infty$, use Stirling's formula to show that

(1.7.1) $$\binom{N}{k} \sim \frac{1}{k!} N^k$$

and make a table giving an idea of the accuracy of this approximation.

1.16 (i) For fixed α $(0 < \alpha < 1)$, if $k = \alpha N$ is an integer and $N \to \infty$, show that

$$\binom{N}{k} \sim \frac{1}{\sqrt{2\pi\alpha(1-\alpha)N}} \cdot \frac{1}{\alpha^k (1-\alpha)^{N-k}}.$$

(ii) Compare this approximation with that of the preceding problem for some representative values of k and N.

1.17 If $0 < \gamma < 1$, then $a_n = \gamma^n \to 0$.

[**Hint:** Since $\gamma > 0$, it is enough to show that for any $\epsilon > 0$ there exists n_0 such that $\gamma^n < \epsilon$ for all $n > n_0$. Take logarithms on both sides of this last inequality.]

1.18 (i) For any positive γ,

(1.7.2) $\qquad\qquad a_n = \sqrt[n]{\gamma} \to 1 \text{ as } n \to \infty.$

(ii) For $\gamma = 1/8,\ 1/4,\ 1/2,\ 2,\ 4,\ 8$, make a table showing the convergence of a_n to 1.

[**Hint**: Suppose $\gamma > 1$. Then it is enough to show that for any ϵ, there exists n_0 such that $\sqrt[n]{\gamma} < 1 + \epsilon$; take logarithms. The case $\gamma < 1$ can be treated analogously.]

1.19 Make a table showing the absolute and relative error resulting from approximating $S_n^{(2)}$ (defined by (1.1.27)) by $n^3/3$.

Section 2

2.1 Verify formula (1.2.2).

2.2 Show that $\sqrt{4pq} < 1$ for all $p \neq \dfrac{1}{2}$ $(0 \le p \le 1)$.

2.3 Verify formula (1.2.4).

2.4 Show that γ defined by (1.2.5) is < 1 for $p \neq \alpha$.

2.5 For fixed x as $n \to \infty$, the probability (1.2.1) satisfies

$$P_n(x) \sim n^x p^x q^{n-x}/x! = \left(\frac{np}{q}\right)^x q^n/x!.$$

2.6 Let X be distributed according to the hypergeometric distribution shown in Table 1.6.2. If n is held fixed and D and N tend to infinity in such a way that $\dfrac{D}{N} = p$ remains fixed, use (1.7.1) to show that

(1.7.3) $\qquad\qquad P(X = x) \to \dbinom{n}{x} p^x q^{n-x}.$

2.7 Make tables similar to Table 1.2.2 for $p = 1/4$ and $p = 1/10$.

2.8 Let X be binomial $b(p, n)$ with $p = 1/2$ and n even. Generalize formula (1.2.3) to determine whether for fixed d

(1.7.4) $\qquad\qquad P\left(X = \dfrac{n}{2} + d\right) \sim \dfrac{c}{\sqrt{n}}$

continues to hold for some c.

2.9 For $x = 0$ and $x = 1$, make a table showing the accuracy of the Poisson approximation (1.2.8) for $\lambda = .1,\ .5,\ 1$ and $n = 5,\ 10,\ 20,\ 50,\ 100$.

2.10 Let X have the Poisson distribution $P(\lambda)$ shown in Table 1.6.2.

(i) Show that for $x = \alpha\lambda$ (α fixed, $\lambda \to \infty$),

(1.7.5)
$$P(X = x) \sim \frac{1}{\sqrt{2\pi\alpha\lambda}} \left(\frac{e^\alpha}{e\alpha^\alpha}\right)^\lambda.$$

Note that this reduces to

(1.7.6)
$$P(X = x) \sim \frac{1}{\sqrt{2\pi\lambda}}$$

when $\alpha = 1$.

(ii) Show that $\gamma = e^\alpha/e\alpha^\alpha$ satisfies $0 < \gamma \le 1$ for all $\alpha > 0$, and that $\gamma = 1$ if and only if $\alpha = 1$.

(iii) Make a table showing the accuracy of the approximations (1.7.5) and (1.7.6) for selected values of α and λ.

2.11 Let X have the negative binomial distribution $Nb(p, m)$ shown in Table 1.6.2. For $p = p_m$ satisfying

$$m(1 - p_m) \to \lambda > 0 \text{ as } m \to \infty,$$

show that the negative binomial distribution tends to the Poisson distribution $P(\lambda)$.

2.12 Obtain approximations analogous to those of Problem 2.10 for the probability $P\left(\dfrac{X}{m} = \alpha\dfrac{q}{p}\right)$ when $\alpha = 1$ and $\alpha \ne 1$.

2.13 Provide an approximation analogous to (1.2.8) for the case that p is close to 1 rather than to 0.

[**Hint:** If p is close to 1, $q = 1 - p$ is close to 0.]

2.14 The sequences $p_n = 1/100$, $x_n = n/100$, $n = 100, 200, 300, \ldots$ and $x_n = 1$, $p_n = 1/n$, $n = 100, 200, 300, \ldots$ provide two embeddings for the triple $n = 100$, $x = 1$, $p = 1/100$.

(i) Show that the following provides another such embedding.

$$p_n = \frac{1}{\frac{n}{2} + 50}, \quad x_n = \frac{n^2}{10,000}, \quad n = 100, 200, 300, \ldots.$$

(ii) Give examples of two additional such embeddings.

Section 3

3.1 (i) If $0 \leq a_i \leq b_i$ for all i and if $\sum b_i$ converges, then so does $\sum a_i$ and $\sum a_i \leq \sum b_i$.

(ii) If $\sum a_i$ and $\sum b_i$ converge, then so do $\sum (a_i + b_i)$ and $\sum (a_i - b_i)$ and $\sum (a_i \pm b_i) = \sum a_i \pm \sum b_i$.

3.2 If $\sum |a_i|$ converges, so does $\sum a_i$

[**Hint:** Let

$$b_i = \left\{ \begin{array}{ll} a_i & \text{if } a_i \geq 0 \\ 0 & \text{otherwise} \end{array} \right. \quad \text{and} \quad c_i = \left\{ \begin{array}{ll} -a_i & \text{if } a_i \leq 0 \\ 0 & \text{otherwise.} \end{array} \right.$$

Then $\sum b_i$ and $\sum c_i$ converge by part (i) of Problem 3.1 and hence $\sum a_i = \sum (b_i - c_i)$ converges by part (ii).]

3.3 If $\sum a_i$ converges, then $a_i \to 0$ as $i \to \infty$.

[**Hint:** Consider first the case that all $a_i \geq 0$. Then if a_i does not tend to 0, there exists $\epsilon > 0$ such that infinitely many of the a's are $> \epsilon$, and hence $\sum a_i = \infty$. The case of arbitrary a_i can be reduced to that of non-negative a_i by the device used in the preceding problem.]

3.4 If $a_i = i^i$, the sum $\sum a_i x^i$ converges for no $x \neq 0$, i.e., the radius of convergence of the series is 0.

[**Hint:** $a_i x^i = (ix)^i$ and the result follows from the preceding problem.]

The next set of problems utilize the following result.

Theorem 1.7.1 *If*

$$(1.7.7) \qquad f(x) = \sum_{i=0}^{\infty} a_i x^i \quad \text{converges for all} \quad |x| < d,$$

then for all $|x| < d$, the derivative of f is given by

$$(1.7.8) \qquad f'(x) = \sum_{i=1}^{\infty} i a_i x^{i-1}$$

and the right side of (1.7.8) converges for all $|x| < d$.

3.5 Use Theorem 1.7.1 to show that if (1.7.7) holds, then the second derivative of f is

$$(1.7.9) \qquad f''(x) = \sum_{i=2}^{\infty} i(i-1) a_i x^{i-2} \quad \text{for all} \quad |x| < d.$$

3.6 (i) If X has the Poisson distribution $P(\lambda)$, determine $E(X)$ and $E[X(X-1)]$.

(ii) Use (i) to find $\text{Var}(X)$.

(iii) Determine the 3rd central moment $\mu_3 = E(X-\lambda)^3$.

[**Hint for (i)**: Use the fact that $E(X) = \sum_{i=1}^{\infty} \dfrac{i\lambda^i}{i!}$, the exponential series (1.3.9), and (1.7.8).]

3.7 (i) Obtain $E(X)$, $\text{Var}(X)$, and $E[X - E(X)]^3$ when X has the negative binomial distribution (1.3.12) instead of the Poisson distribution.

(ii) Specialize the results of (i) to the geometric distribution (1.3.13).

3.8 Solve Problem 3.7(i) when X has the logarithmic distribution (1.3.15).

3.9 Show that

$$1 + x + \cdots + x^{n-1} = \frac{x^n - 1}{x - 1}.$$

[**Hint**: Multiply both sides by $x - 1$.]

Section 4

4.1 Prove the bias-variance decomposition (1.4.6).

[**Hint**: On the left side add and subtract $E[\delta(X)]$ inside the square bracket.]

4.2 In Table 1.4.4, add the rows n^{10} and n^{20}.

4.3 Prove parts (i)–(v) of Lemma 1.4.1.

4.4 (i) If $a_n = o(c_n)$ and $b_n = o(c_n)$, then $a_n + b_n = o(c_n)$.

(ii) If a_n, b_n, c_n, d_n are all > 0, then $a_n = o(c_n)$ and $b_n = o(d_n)$ implies $a_n + b_n = o(c_n + d_n)$.

(iii) The conclusion of (ii) is no longer valid if the a's, b's, c's, and d's are not all positive.

(iv) If $a_n = o(b_n)$, then $a_n + b_n \sim b_n$.

4.5 Consider the three statements:

(1) $b_n \sim a_n$,

(2) $b_n = a_n + o(a_n)$,

(3) $b_n = a_n + o(1)$.

For each pair (i, j), determine whether (i) implies (j), (j) implies (i), or whether neither implication holds.

4.6 Replace each of the following quantities by a simpler one which consists of only one of the terms and is asymptotically equivalent:

(i) $\log n + \dfrac{1}{2}n$;

(ii) $\log n + \log(\log n)$;

(iii) $n^2 + e^n$.

4.7 (i) Prove (1.4.17).

(ii) Prove that $n^k = o\left(e^{an^b}\right)$ for any $a, b > 0$.

4.8 Show that

$$\left[1 + \frac{c}{n} + o\left(\frac{1}{n}\right)\right]^n \to e^c \text{ as } n \to \infty.$$

[**Hint**: Given any $\epsilon > 0$, there exists n_0 such that $1 + \dfrac{c - \epsilon}{n} < 1 + \dfrac{c}{n} + o\left(\dfrac{1}{n}\right) < 1 + \dfrac{c + \epsilon}{n}$ for all $n > n_0$.]

4.9 A sum of S dollars is deposited in a checking account which pays $p\%$ interest per annum. If the interest is compounded monthly, determine the value of the account after n months.

4.10 Show that the following sequences tend to infinity as $n \to \infty$:

(i) $\log n$;

(ii) $\log \log n$.

4.11 If $a_n/b_n \to l \ (0 < l < \infty)$, then $a_n \asymp b_n$.

4.12 State and prove the analogs of the five parts of Lemma 1.4.1 when $a_n = o(b_n)$ is replaced by $a_n \asymp b_n$.

4.13 For each of the following functions f, determine whether $f(cn) \asymp [f(dn)]$ if c and d are any positive constants:

(i) $f(n) = 2n$,

(ii) $f(n) = n^2$,

(iii) $f(n) = e^n$,

(iv) $f(n) = \log n$.

4.14 If $b_n \to 0$, and a_n and b_n are both of the same order as c_n, what can you say about the relation of a_n and b_n?

4.15 Examine the validity of the four parts of Problem 4.4 with o replaced by $\asymp$.

4.16 Show that $a_n \asymp b_n$ if and only if both $a_n = O(b_n)$ and $b_n = O(a_n)$.

4.17 State and prove the analogs of the five parts of Lemma 1.4.1 with o replaced by O.

4.18

$$\text{Let}\quad \begin{array}{l|l} a_{1n} \text{ and } A_{1n} \text{ be } o(1/n) & a_{4n} \text{ and } A_{4n} \text{ be } O(1/n) \\ a_{2n} \text{ and } A_{2n} \text{ be } o(1) & a_{5n} \text{ and } A_{5n} \text{ be } O(1) \\ a_{3n} \text{ and } A_{3n} \text{ be } o(n) & a_{6n} \text{ and } A_{6n} \text{ be } O(n) \end{array}$$

For each of the sums $a_{in} + A_{jn}$, determine the best possible order statement.

[**Hint**: For example, $a_{1n} + A_{1n} = o(1/n)$.]

4.19 In the preceding problem, assume that all the a_{in} and A_{jn} are positive and determine the best possible order statement for each product $a_{in}A_{jn}$.

4.20 In Example 1.4.6, verify the statements (i)–(iv).

4.21 For the two statements

$$a_n = o(n^\alpha), \quad a_n = O(n^\beta),$$

determine whether either implies the other, and hence which is more informative when

(i) $0 < \alpha < \beta$,

(ii) $\beta < \alpha < 0$.

4.22 Determine the order relation between

$$a_n = \sqrt{\log n} \text{ and } b_n = \log\left(\sqrt{n}\right)$$

Section 5

5.1 Extend Definition 1.5.1, and show that $\sqrt{x} \to \infty$ as $x \to \infty$.

5.2 Use Definition 1.5.2 to prove the second statement of (1.5.5).

5.3 There exist values $x > 0$ arbitrarily close to 0 for which $\sin(1/x)$

(i) equals $+1$,

(ii) equals -1.

[**Hint**: $\sin y = 1$ for all $y = \dfrac{\pi}{2} + k \cdot 2\pi, \quad k = 1, 2, \dots$]

5.4 Determine the behavior of $\cos(1/x)$ as $x \to 0+$.

5.5 Let $f(x) = [x]$ denote the greatest integer $\leq x$ (sometimes called the integral part of x). Determine at which values of x, f is continuous, continuous from the right, and continuous from the left.

5.6 A function f is said to be differentiable at x if

$$(1.7.10) \qquad \lim_{\Delta \to 0} \frac{f(x+\Delta) - f(x)}{\Delta}$$

exists, and this limit is then called the derivative of f at x and denoted by $f'(x)$. If f is differentiable at x, then

(i) it is continuous at x,

(ii) an alternative expression for $f'(x)$ is

$$(1.7.11) \qquad f'(x) = \lim_{\Delta \to 0} \frac{f(x+\Delta) - f(x-\Delta)}{2\Delta}.$$

5.7 The following example shows that the existence of the limit (1.7.11) does not necessarily imply that of (1.7.10). Let $f(x) = \min(x, 1-x)$. Determine for each $0 < x < 1$ whether

(i) f is continuous,

(ii) the limit (1.7.10) exists,

(iii) the limit (1.7.11) exists.

5.8 Define

(i) $f(x) \sim g(x)$ as $x \to x_0$,

(ii) $f(x) = O[g(x)]$ as $x \to x_0$.

5.9 Show that the results of Examples 1.4.3–1.4.5 remain valid if n is replaced by a continuous variable x.

5.10 (i) If X has the binomial distribution (1.2.1), graph the function $F(x) = P(X \leq x)$ for the case $n = 2$, $p = 1/2$. At its discontinuities, is F continuous on the right, the left, or neither?

(ii) Solve part (i) if $F(x) = P(X < x)$.

5.11 If f is non-decreasing on $a < x < x_0$, then as x tends to x_0 from the left, $f(x)$ tends either to a finite limit or to ∞.

[**Hint**: (i) If f is bounded above for $x < x_0$ and if m is the least upper bound of $f(x)$ for $x < x_0$, then $f(x) \to m$ as $x \to x_0$ from the left.

(ii) If $f(x)$ is not bounded above for $x < x_0$, then $f(x) \to \infty$ as $x \to x_0-$.]

5.12 Give an example of each of the following three possibilities for a set S with least upper bound m:

(i) m is a member of S but not a limit point of S;

(ii) m is a limit point of S but not a member of S;

(iii) m is both a member and a limit point of S.

5.13 Determine the inf and the sup of each of the following sets and determine whether it is in the set.

(i) $S = \left\{1 + \dfrac{1}{n}, \ n = 1, 2, \dots \right\}$;

(ii) $S = \left\{1 + (-1)^n \dfrac{1}{n}, \ n = 1, 2, \dots \right\}$;

(iii) $S = \left\{\dfrac{1}{\sqrt[n]{2} - 1}, \ n = 1, 2, \dots \right\}$.

5.14 If $\{a_n\}$ and $\{b_n\}$ are two sequences of numbers with $a_n \le b_n$ for all n and $a_n \to a$, $b_n \to b$, then $a \le b$.

5.15 Let

$$f(x) = \begin{cases} 1 & \text{if } x = 1/n, \ n = 1, 2, \dots \\ 0 & \text{for all other values of } x. \end{cases}$$

Determine for which values of x the function f is continuous.

Section 6

6.1 Let $F^*(x) = \dfrac{1}{2}[P(X \le x) + P(X < x)]$. At a point of discontinuity of F, determine whether F^* is continuous on the left, on the right, or neither.

6.2 For any cdf F, the quantile function F^{-1} defined by (1.6.10) is

(i) nondecreasing;

(ii) continuous on the left.

6.3 For any cdf F, we have

(i) $F^{-1}[F(x)] \le x$ for all $-\infty < x < \infty$;

(ii) $F[F^{-1}(y)] \ge y$ for all $0 < y < 1$.

6.4 Use the preceding problem to show that for any F,

(1.7.12) $F(x) \ge y$ if and only if $x \ge F^{-1}(y)$.

6.5 If F is continuous at x and $0 < y = F(x) < 1$, then

(i) $F\left[F^{-1}(y)\right] = y$ but

(ii) $F^{-1}\left[F(x)\right]$ may be $< x$.

[**Hint (ii)**: Let x be the right-hand end point of a flat stretch of F.]

6.6 (i) The cdf of the logistic distribution $L(0,1)$ given in Table 1.6.1 is

(1.7.13)
$$F(x) = \frac{1}{1 + e^{-x}}.$$

(ii) Give an explicit expression for F^{-1}.

[**Hint (i)**: Differentiate (1.7.13).]

6.7 (i) The cdf of the exponential distribution $E(0,1)$ given in Table 1.6.1 is

(1.7.14)
$$F(x) = 1 - e^{-x}.$$

(ii) Give an explicit expression for F^{-1}.

6.8 If X is distributed as $U(0,1)$, then $-\log X$ is distributed as $E(0,1)$.

6.9 If X has the continuous cdf F, then $F(X)$ is distributed as $U(0,1)$.

[**Hint**: Use (1.7.12).]

6.10 If the graph of F is rotated about the line $y = x$, the resulting inverse F^{-1} is continuous on the left.

[**Hint**: Consider how south-north and west-east stretches of F (corresponding to discontinuities and flat stretches, respectively) are transformed under the rotation.]

Bibliographic Notes

Among the many books providing a rigorous treatment of basic concepts such as limit, continuity, orders of magnitude, and the convergence of infinite series, two outstanding classical texts (both recently reissued) are Hardy (1908, 1952, 1992) and Courant (1927, 1937, 1982). A very accessible presentation in the more general setting of metric spaces is Rudin (1976). Knopp (1922, 1951, 1990) provides an extensive account of infinite series, and the first chapter of de Bruijn (1958) contains a general introduction to asymptotics with a detailed discussion of the o, O notation. Details on the distributions of Table 1.6.2 and others can be found in Johnson, Kotz, and Kemp (1992) and on those in Table 1.6.1 in Johnson, Kotz, and Balakrishnan (1994,1995).

2
Convergence in Probability and in Law

Preview

This chapter develops the principal probability tools for first order large-sample theory. The convergence of a sequence of real numbers is extended to that of a sequence of random variables. Convergence of X_n in probability (i.e., to a constant) and in law (i.e., the convergence of the distribution of X_n to a limit distribution) are treated in Sections 1–3. Sections 4, 7, and 8 are concerned with the central limit theorem (CLT) which gives conditions for the standardized sum of n random variables to have a normal limit distribution as $n \to \infty$: In Section 4, the terms are assumed to be i.i.d., in Section 7, independent but no longer identically distributed; Section 8 considers sums of dependent variables. The CLT is strengthened in two directions: (i) The Berry-Esseen theorem provides a bound for the error of the normal approximation; (ii) a better approximation can usually be obtained through the first terms of the Edgeworth correction (Section 4). Finally, the usefulness of the CLT is greatly increased by Slutsky's theorem given in Section 3, and by the delta method (Section 5) which extends the limit result to smooth functions of asymptotically normal variables.

2.1 Convergence in probability

In the present section, we extend the notion of the limit of a sequence of numbers to that of the convergence of a sequence of random variables to a constant. As an example let S_n be the number of successes in n binomial trials with success probability p. Then for large n, one would expect S_n/n to be close to p. However, Definition 1.1.1 will not work, for it would require that for any $\epsilon > 0$, there exists n_0 such that

$$(2.1.1) \qquad\qquad \left| \frac{S_n}{n} - p \right| < \epsilon$$

for all $n > n_0$. But if $0 < p < 1$, it is possible for $n_0 + 1$ trials to result in $n_0 + 1$ heads, in which case, $S_n/n = 1$ for $n = n_0 + 1$ and (2.1.1) would be violated. What can be said is that for any large n, it is very unlikely to get n heads in n trials (when $p < 1$), and more generally that for any sufficiently large n, it is unlikely for (2.1.1) to be violated. This suggests that the probability of the event (2.1.1) should tend to 1 as $n \to \infty$, and motivates the following definition.

Definition 2.1.1 A sequence of random variables Y_n is said to converge to a constant c *in probability*, in symbols

$$(2.1.2) \qquad\qquad Y_n \overset{P}{\to} c,$$

if for every $\epsilon > 0$,

$$(2.1.3) \qquad\qquad P\left(|Y_n - c| < \epsilon \right) \to 1 \text{ as } n \to \infty$$

or equivalently

$$(2.1.4) \qquad\qquad P\left(|Y_n - c| \geq \epsilon \right) \to 0 \text{ as } n \to \infty.$$

Thus, roughly speaking, (2.1.2) states that for large n, the probability is high that Y_n will be close to c.

Before considering any examples, we shall first obtain a sufficient condition for convergence in probability, which often is easier to check than (2.1.3) or (2.1.4). This condition is based on the following lemma.

Lemma 2.1.1 Chebyshev inequality. *For any random variable Y and any constants $a > 0$ and c,*

$$(2.1.5) \qquad\qquad E(Y - c)^2 \geq a^2 P[|Y - c| \geq a].$$

Proof. The result holds quite generally, but we shall prove it only for the case that the distribution of $Z = Y - c$ has a density $f(z)$. Then

$$\begin{aligned} E(Z^2) &= \int z^2 f(z)\,dz = \int_{|z| \geq a} z^2 f(z)\,dz + \int_{|z| < a} z^2 f(z)\,dz \\ &\geq \int_{|z| \geq a} z^2 f(z)\,dz \geq a^2 \int_{|z| \geq a} f(z)\,dz = a^2 P(|Z| \geq a). \end{aligned}$$

If the distribution of Z is discrete, the integrals are replaced by sums. A completely general proof requires a more general concept of integral. ■

Theorem 2.1.1 *A sufficient condition for $Y_n \overset{P}{\to} c$ is that*

$$(2.1.6) \qquad\qquad E(Y_n - c)^2 \to 0,$$

i.e., that Y_n tends to c in quadratic mean.

Proof. Given any $\epsilon > 0$, we have, by (2.1.5),

$$P[|Y_n - c| \geq \epsilon] \leq \frac{1}{\epsilon^2} E(Y_n - c)^2.$$

The limit relation (2.1.4) now follows from (2.1.6). ■

Example 2.1.1 Binomial. If S_n is the number of successes in n binomial trials with success probability p, let $Y_n = S_n/n$ and $c = p$. Then

$$E\left(\frac{S_n}{n} - p\right)^2 = \mathrm{Var}\left(\frac{S_n}{n}\right) = \frac{pq}{n} \to 0 \text{ as } n \to \infty$$

and it follows that $\dfrac{S_n}{n} \overset{P}{\to} p$. □

Example 2.1.1 is a special case of the following result.

Theorem 2.1.2 Weak law of large numbers. *Let $X_1, \ldots, X_n$ be i.i.d. with mean $E(X_i) = \xi$ and variance $\sigma^2 < \infty$. Then the average $\bar{X} = (X_1 + \cdots + X_n)/n$ satisfies*

$$(2.1.7) \qquad\qquad \bar{X} \overset{P}{\to} \xi.$$

Proof. The result follows from Theorem 2.1.1 and the fact that

$$E(\bar{X} - \xi)^2 = \mathrm{Var}(\bar{X}) = \sigma^2/n \to 0 \text{ as } n \to \infty.$$

■

Note: The convergence result (2.1.7) remains true even if $\sigma^2 = \infty$. For a proof of this result which is due to Khinchine, see, for example, Feller (Vol. 1) (1957), Section X.2.

Example 2.1.1 (continued). To see that Example 2.1.1 is a special case of Theorem 2.1.2, consider a sequence of binomial trials, and let

$$(2.1.8) \qquad\qquad X_i = \begin{cases} 1 & \text{if the } i^{\text{th}} \text{ trial is a success} \\ 0 & \text{otherwise.} \end{cases}$$

Then $S_n = \sum_{i=1}^{n} X_i$, $S_n/n = \bar{X}$, and $E(X_i) = p$. $\square$

The limit laws (1.1.10)–(1.1.12) of Chapter 1 have the following probabilistic analogs.

Theorem 2.1.3 *If A_n and B_n are two sequences of random variables satisfying*

$$A_n \xrightarrow{P} a \text{ and } B_n \xrightarrow{P} b$$

respectively, then

(2.1.9) $$A_n + B_n \xrightarrow{P} a + b \text{ and } A_n - B_n \xrightarrow{P} a - b,$$

(2.1.10) $$A_n \cdot B_n \xrightarrow{P} a \cdot b,$$

and

(2.1.11) $$\frac{A_n}{B_n} \xrightarrow{P} \frac{a}{b} \text{ provided } b \neq 0.$$

The proof requires the following lemma, which is also useful in other contexts.

Lemma 2.1.2 *If E_n and F_n are two sequences of events, then*

$$P(E_n) \to 1, P(F_n) \to 1 \text{ implies } P(E_n \text{ and } F_n) \to 1.$$

Proof. If $\bar{E}$ denotes the complement of E, we have

$$P(\overline{E_n \text{ and } F_n}) = P(\bar{E}_n \text{ or } \bar{F}_n) \leq P(\bar{E}_n) + P(\bar{F}_n) \to 0.$$

■

Proof of (2.1.9). From

$$|(A_n + B_n) - (a + b)| \leq |A_n - a| + |B_n - b|,$$

it follows that $P\{|(A_n + B_n) - (a + b)| < \epsilon\} \geq P\{|A_n - a| < \epsilon/2 \text{ and } |B_n - b| < \epsilon/2\}$. Since $P[|A_n - a| < \epsilon/2] \to 1$ and $P[|B_n - b| < \epsilon/2] \to 1$, Lemma 2.1.2 proves (2.1.9). (For the proof of (2.1.10) and (2.1.11), see Problem 1.1).   ■

Another useful result is

Theorem 2.1.4 *If Y_n is a sequence of random variables such that $Y_n \xrightarrow{P} c$, and if f is a function which is continuous at c, then*

$$(2.1.12) \qquad\qquad f(Y_n) \xrightarrow{P} f(c).$$

This is intuitively clear: With high probability, Y_n will be close to c, and by continuity, $f(Y_n)$ will then be close to $f(c)$. (For a formal proof, see Problem 1.3).

As an example, suppose that $X_1, \ldots, X_n$ are i.i.d. according to a Poisson distribution $P(\lambda)$. Then $\bar{X} \xrightarrow{P} \lambda$ and therefore $e^{-\bar{X}} \xrightarrow{P} e^{-\lambda}$.

An important statistical application of convergence in probability is the consistency of a sequence of estimators.

Definition 2.1.2 A sequence of estimators δ_n of a parametric function $g(\theta)$ is *consistent* if

$$(2.1.13) \qquad\qquad \delta_n \xrightarrow{P} g(\theta).$$

Example 2.1.2 Consistency of the mean. Let $X_1, \ldots, X_n$ be i.i.d. with mean ξ and variance $\sigma^2 < \infty$. Then $\bar{X}$ is a consistent estimator of ξ. (This is just a restatement of Theorem 2.1.2). $\qquad\qquad\Box$

Note: The notation and terminology used in Example 2.1.2 is rather imprecise. The estimator $\bar{X}$ depends on n and should be denoted by $\bar{X}_n$ and the convergence in probability is a statement not about a single estimator but about the sequence of estimators $\bar{X}_n$, $n = 1, 2, \ldots$. The abbreviations used in Example 2.1.2 are customary and will often be used in this book.

Theorem 2.1.5 *A sufficient condition for δ_n to be consistent for estimating $g(\theta)$ is that both the bias and the variance of δ_n tend to zero as $n \to \infty$.*

Proof. Theorem 2.1.1 and equation (1.4.6) of Chapter 1. ∎

Example 2.1.3 Consistency of sample moments. Let us now return to the case that $X_1, \ldots, X_n$ are i.i.d. with mean ξ, and suppose that $E|X_i|^k < \infty$ so that the kth central moment

$$\mu_k = E(X_i - \xi)^k$$

exists (Problem (1.9)). Then we shall show that the kth sample moment

$$M_k = \frac{1}{n} \sum_{i=1}^{n} (X_i - \bar{X})^k$$

is a consistent estimator of μ_k as $n \to \infty$.

To prove this result, suppose without loss of generality that $\xi = 0$. (If $\xi \neq 0$, replace $X_i - \xi$ by Y_i so that $\mu_k = E(Y_i^k)$ and note that $\sum(X_i - \bar{X})^k = \sum(Y_i - \bar{Y})^k$.) By the binomial theorem, we have

(2.1.14)
$$\frac{1}{n}\sum(X_i - \bar{X})^k = \frac{1}{n}\sum X_i^k - \binom{k}{1}\bar{X}\frac{1}{n}\sum X_i^{k-1}$$
$$+ \binom{k}{2}\bar{X}^2 \cdot \frac{1}{n}\sum X_i^{k-2} + \cdots + (-1)^k \bar{X}^k.$$

Now $\bar{X} \xrightarrow{P} 0$ and, more generally, $\dfrac{1}{n}\sum_{i=1}^{n} X_i^r \xrightarrow{P} E(X_i^r)$ for $r = 1, 2, \ldots, k$ by the law of large numbers. From (2.1.10), it follows that all terms in (2.1.14) tend in probability to zero, except the first term, which tends to μ_k, as was to be proved. □

The Chebychev inequality shows that if a random variable Y has a small variance, then its distribution is tightly concentrated about $\xi = E(Y)$. It is important to realize that the converse is not true: the fact that Y is tightly concentrated about its mean tells us little about the variance or other moments of Y.

The following example shows that $Y_n \xrightarrow{P} c$ does not imply either

(i) $E(Y_n) \to c$

or

(ii) $E(Y_n - c)^2 \to 0$.

Example 2.1.4 A counterexample. Let

(2.1.15)
$$Y_n = \begin{cases} 1 & \text{with probability } 1 - p_n \\ n & \text{with probability } p_n. \end{cases}$$

Then $Y_n \xrightarrow{P} 1$, provided $p_n \to 0$ (Problem 1.2(i)). On the other hand,

$$E(Y_n) = (1 - p_n) + np_n$$

which tends to $a+1$ if $p_n = a/n$ and to ∞ if, for example, $p_n = 1/\sqrt{n}$. This shows that (i) need not hold, and that (ii) need not hold is seen analogously (Problem 1.2(ii)). □

A crucial distinction.
    The example shows clearly the difference between the probability limit of Y_n and the limiting behavior of moments such as $E(Y_n)$ or $E(Y_n - c)^2$.

The limit relation $Y_n \xrightarrow{P} c$ states that for large n, Y_n is very likely to be close to c. However, it says nothing about the location of the remaining small probability mass which is not close to c and which can strongly affect the value of the mean and of other moments.

This distinction depends on the assumption that the distribution of the Y's can have positive probability mass arbitrarily far out. The situation changes radically if the Y's are uniformly bounded, and hence there exists M such that

$$P[|Y_n - c| < M] = 1 \text{ for all } n.$$

In that case

$$Y_n \xrightarrow{P} c$$

implies both (Problem 1.8)

$$E(Y_n) \to c \text{ and } E(Y_n - c)^2 \to 0.$$

Additional insight into the one-sided relationship between convergence in probability and the behavior of $E(Y - c)^2$ can be obtained by observing that a large value of $E(Y - c)^2$ can obtain in two quite different situations:

(a) When the distribution of Y is very dispersed; for example, when Y has a uniform distribution $U(-A, A)$ with a very large value of A;

(b) As the result of a small probability very far out; for example, when Y is distributed as

$$(1 - \epsilon)N(0, \sigma^2) + \epsilon N(\xi, 1)$$

with ϵ small but with a large value of ξ.

In both cases, the variance of Y will be large. However, in case (b) with small values of ϵ and σ^2, the distribution will be tightly concentrated about 0; in case (a), it will not.

Convergence in probability can be viewed as a stochastic analog of the convergence of a sequence of numbers, of which Problem 1.6 shows it to be a generalization. We shall now consider the corresponding analogs of the relations o, $\asymp$, and O for random variables.

Definition 2.1.3 A sequence of random variables A_n is of smaller order in probability than a sequence B_n (in symbols

$$A_n = o_P(B_n))$$

if

(2.1.16) $$\frac{A_n}{B_n} \xrightarrow{P} 0.$$

In particular,

(2.1.17) $A_n = o_P(1)$ if and only if $A_n \xrightarrow{P} 0.$

The properties of o stated in Lemma 1.4.1 generalize in a natural way to o_P (Problem 1.10). For example, $A_n = o_P(B_n)$ implies

(2.1.18) $C_n A_n = o_P(c_n B_n)$ and $C_n A_n = o_p(C_n B_n)$

for any numbers c_n and random variables $C_n \neq 0$.

The o_P notation permits us to write statements such as

$$A_n = B_n + R_n$$

when $R_n \xrightarrow{P} 0$ or when $n R_n \xrightarrow{P} 0$ very compactly as respectively

(2.1.19) $A_n = B_n + o_P(1)$ or $A_n = B_n + o_P(1/n).$

Here the latter statement follows from the fact that by (2.1.18),

$$n o_P(1/n) = o_P(1).$$

There are analogous extensions from O to O_p and from $\asymp$ to $\asymp_P$.

Definition 2.1.4   (i) A sequence A_n is said to be of order less than or equal to that of B_n in probability (in symbols, $A_n = O_P(B_n)$) if given $\epsilon > 0$ there exists a constant $M = M(\epsilon)$ and an integer $n_0 = n_0(\epsilon)$ such that

$$P(|A_n| \leq M |B_n|) \geq 1 - \epsilon \text{ for all } n > n_0.$$

(ii) A sequence A_n is said to be of the same order as B_n in probability (in symbols, $A_n \asymp_P B_n$) if given $\epsilon > 0$ there exist constants $0 < m < M < \infty$ and an integer n_o such that

$$P\left[m < \left|\frac{A_n}{B_n}\right| < M\right] \geq 1 - \epsilon \text{ for all } n > n_o.$$

Summary

1. The convergence of a sequence of numbers is extended to the convergence of a sequence of random variables Y_n to a constant c. This states that for large n, we can be nearly certain that Y_n will differ from c by an arbitrarily small amount, i.e., that the probability of Y_n falling within any preassigned distance ϵ of c tends to 1 as $n \to \infty$.

2. A sufficient condition for $Y_n \xrightarrow{P} c$ is that $E(Y_n - c)^2 \to 0$, which typically is much easier to check. In particular, it provides an easy proof of the weak law of large numbers, which states that the average of n i.i.d. variables with mean ξ and finite variance σ^2 tends in probability to ξ.

3. An important distinction must be made between the statements (a) $Y_n \xrightarrow{P} c$ and (b) $E(Y_n) \to c$. The first of these statements concerns only the bulk of the probability mass and is independent of the position of the small remainder. On the other hand, (b) depends crucially on the behavior of the Y's in the extreme tails.

4. A sequence of estimators δ_n of a function $g(\theta)$ is *consistent* if $\delta_n \xrightarrow{P} g(\theta)$. The sample moments are consistent estimators of the population moments.

5. The o, O, $\asymp$ notation for the comparison of two sequences of numbers is extended to the corresponding o_P, O_P, $\asymp_P$ notation for random variables.

2.2 Applications

In the present section, we shall consider whether the sample mean $\bar{X}$ is consistent for estimating the common mean ξ of the X's when these variables (a) are independent but no longer identically distributed and (b) in some situations when they are dependent. We shall also investigate the convergence in probability of the sample quantiles to their population analogs.

Example 2.2.1 Estimation of a common mean. Suppose that $X_1, \ldots,$ X_n are independent with common mean $E(X_i) = \xi$ and with variances $\mathrm{Var}(X_i) = \sigma_i^2$. (Different variances can arise, for example, when each of several observers takes a number of observations of ξ, and X_i is the average of the n_i observations of the ith observer.) Then

$$E(\bar{X} - \xi)^2 = \frac{1}{n^2} \mathrm{Var}\left(\sum X_i\right) = \frac{1}{n^2} \sum_{i=1}^{n} \sigma_i^2.$$

It follows from Theorem 2.1.1 that $\bar{X}$ will continue to be a consistent estimator of ξ provided $\sum_{i=1}^{n} \sigma_i^2/n^2 \to 0$, i.e., if

$$(2.2.1) \qquad \sum_{i=1}^{n} \sigma_i^2 = o(n^2).$$

This covers, of course, the case that $\sigma_i^2 = \sigma^2$ for all i since then $\sum \sigma_i^2 = n\sigma^2 = o(n^2)$.

When the variances σ_i^2 are not all equal, consider the cases that they are either (a) decreasing or (b) increasing.

In case (a)

$$\sum_{i=1}^{n} \sigma_i^2 \leq n\sigma_1^2 = o(n^2).$$

It follows that (2.2.1) holds and $\bar{X}$ is consistent.

In case (b), condition (2.2.1) clearly will not hold if the σ_i^2 increase too fast, e.g., if $\sigma_n^2 \geq n^2$, but will be satisfied if the increase is sufficiently slow. Suppose, for example, that

$$(2.2.2) \qquad \sigma_i^2 = \Delta i^{\alpha} \text{ for some } \alpha > 0.$$

Then (1.3.6) shows that

$$(2.2.3) \qquad \Delta \sum_{i=1}^{n} i^{\alpha} = o(n^2)$$

if and only if $\alpha + 1 < 2$, i.e., if $\alpha < 1$, and hence that $\bar{X}$ is a consistent estimator for ξ when σ_i^2 is given by (2.2.2) with $\alpha < 1$.

When $\alpha \geq 1$, the sufficient condition (2.2.3) does not hold and therefore does not tell us whether $\bar{X}$ is consistent. Consistency in this case turns out to depend not only on the variances σ_i^2 but also on other aspects of the distributions F_i of the X_i. The following result shows that there will then always exist distributions for which $\bar{X}$ is not consistent. $\qquad \square$

Theorem 2.2.1 *If $Y_n, n = 1, 2, \ldots$, is a sequence of random variables with normal distribution $N(\xi, \tau_n^2)$, then $Y_n \xrightarrow{P} \xi$ if and only if $\tau_n^2 \to 0$.*

Proof. We have

$$(2.2.4) \qquad P(|Y_n - \xi| \leq a) = P\left(\left|\frac{Y_n - \xi}{\tau_n}\right| \leq \frac{a}{\tau_n}\right).$$

Since $(Y_n - \xi)/\tau_n$ is distributed as $N(0, 1)$, the probabilities (2.2.4) tend to 1 if and only if $a/\tau_n \to \infty$, that is, if $\tau_n^2 \to 0$. $\qquad \blacksquare$

If the X_i are independent $N(\xi, \sigma_n^2)$, the distribution of $\bar{X}$ is $N(\xi, \tau_n^2)$ with $\tau_n^2 = \sum \sigma_i^2/n^2$. It then follows from the theorem that $\bar{X}$ is consistent for ξ if and only if (2.2.1) holds.

The reason for inconsistency in the normal case when the variances increase too fast is that the observations for large n (and hence large σ_n^2) provide very little information about the position of ξ.

Necessary and sufficient conditions for $\bar{X}$ to be consistent for ξ in the general case of independent non-i.i.d. X's are given, for example, in Chow and Teicher (1978) and in Petrov (1995). From these, it can be shown that when $\alpha \geq 1$, there will exist non-normal (heavy-tailed) distributions F_i for which $\bar{X}$ is consistent.

To summarize: $\bar{X}$ is a consistent estimator of ξ for all sequences of distributions F_i satisfying (2.2.1), but when (2.2.1) does not hold, $\bar{X}$ may or may not be consistent dependent on the specific nature of the F_i.

Consider next the more general situation in which $X_1, \ldots, X_n$ are independent with $E(X_i) = \xi_i$, $\mathrm{Var}(X_i) = \sigma_i^2$, and

$$(2.2.5) \qquad \delta_n = \sum w_i X_i$$

is a sequence of statistics, with the weights w_i satisfying

$$(2.2.6) \qquad \sum w_i \xi_i = \xi, \text{ independent of } n.$$

Here we shall permit the weights to depend not only on i but also on n, but we continue to denote them by w_i instead of the more correct but cumbersome $w_i^{(n)}$.

By Theorem 2.1.1, a sufficient condition for δ_n to be a consistent estimator of ξ is that

$$(2.2.7) \qquad \mathrm{Var}(\delta_n) = \sum_{i=1}^n w_i^2 \sigma_i^2 \to 0 \text{ as } n \to \infty.$$

From Theorem 2.2.1, it follows that when the X_i are distributed according to $N(\xi_i, \sigma_i)$, then (2.2.7) is not only sufficient but also necessary for the consistency of Y_n.

We shall now illustrate this condition in a number of examples.

Example 2.2.2 Best linear estimator for a common mean. Let us return to the situation of Example 2.2.1. When the variances σ_i^2 are unequal, there exists a better linear estimator of ξ than $\bar{X}$, which puts less weight on the observations with larger variance. The best (i.e., minimum variance) weighted average $\delta_n = \sum w_i X_i$ (with $\sum w_i = 1$ so that $E(\sum w_i X_i) = \xi$) assigns to X_i the weight

$$(2.2.8) \qquad w_i = \frac{1/\sigma_i^2}{\sum_{j=1}^n 1/\sigma_j^2}$$

and its variance is (Problem 2.2)

$$(2.2.9) \qquad \mathrm{Var}(\delta_n) = \frac{1}{\sum\limits_{j=1}^{n} 1/\sigma_j^2}.$$

The optimal linear estimator δ_n is therefore consistent for ξ if

$$(2.2.10) \qquad \sum_{j=1}^{n} 1/\sigma_j^2 \to \infty \text{ as } n \to \infty.$$

Since $\mathrm{Var}(\delta_n) \leq \mathrm{Var}(\bar{X})$, condition (2.2.10) is satisfied whenever (2.2.1) holds and therefore in particular δ_n is consistent for ξ when the σ_i^2 are non-increasing and also when σ_i^2 is given by (2.2.2) with $0 < \alpha < 1$. For $\alpha > 1$, $\sum\limits_{i=1}^{n} 1/\sigma_i^2$ tends to a finite limit by Example 1.3.2 of Chapter 1, and (2.2.10) therefore does not hold. Finally, when $\alpha = 1$, it follows from (1.3.5) that $\sum\limits_{i=1}^{n} 1/\sigma_i^2 \sim \log n$ and therefore tends to infinity. Thus (2.2.10) holds if and only if $\alpha \leq 1$. As before, condition (2.2.10) is also necessary when the X's are normal. With regard to consistency, the improvement of δ_n over $\bar{X}$ therefore is slight. The sufficient condition holds for $\alpha \leq 1$ instead of only for $\alpha < 1$.

However, consistency is a very weak property and a more useful comparison between $\bar{X}$ and δ_n is obtained by a direct comparison of their variances. The relative efficiency $e_{\bar{X},\delta_n}$ of $\bar{X}$ to δ_n is defined by (see Section 4.3)

$$(2.2.11) \qquad e_{\bar{X},\delta_n} = \frac{\mathrm{Var}(\delta_n)}{\mathrm{Var}(\bar{X})},$$

which is shown in Table 2.2.1 for $\alpha = 1/4$, $1/2$, $3/4$, and 1.

The table shows that the improvement of δ_n over $\bar{X}$ can be substantial, particularly for values of α close to 1. $\qquad \Box$

TABLE 2.2.1. Relative efficiency $e_{\bar{X},\delta_n}$ for $\sigma_i^2 = i^\alpha$

n	5	10	20	50	100	∞
$\alpha = 1/4$	.980	.970	.961	.952	.948	.938
$\alpha = 1/2$	.923	.886	.854	.820	.801	.750
$\alpha = 3/4$	.836	.763	.700	.633	.595	.438
$\alpha = 1$	.730	.621	.529	.436	.382	0

Example 2.2.3 Simple linear regression. Let

$$(2.2.12) \qquad X_i = \alpha + \beta v_i + E_i,$$

where α and β are unknown regression coefficients, the v's are known constants, and the E's are i.i.d. variables with expectation 0 and variance σ^2. The standard estimators of α and β (which have optimality properties when the E's are normal) are

$$(2.2.13) \qquad \hat{\beta} = \sum(v_i - \bar{v})X_i / \sum(v_j - \bar{v})^2$$

and

$$(2.2.14) \qquad \hat{\alpha} = \bar{X} - \bar{v}\hat{\beta}.$$

To apply (2.2.7) to the consistency of $\delta_n = \hat{\beta}$ with

$$(2.2.15) \qquad \xi_i = \alpha + \beta v_i, \sigma_i^2 = \sigma^2 \text{ and } w_i = (v_i - \bar{v}) / \sum(v_j - \bar{v})^2,$$

we first check that (Problem 2.4(i))

$$(2.2.16) \quad \sum w_i \xi_i = \frac{1}{\sum(v_j - \bar{v})^2} \left[\alpha \sum(v_i - \bar{v}) + \beta \sum v_i(v_i - \bar{v}) \right] = \beta,$$

independent of n. Thus, by (2.2.7), a sufficient condition for $\hat{\beta}$ to be a consistent estimator of β is that

$$\text{Var}(\hat{\beta}) = \sigma^2 \sum w_i^2 = \sigma^2 \frac{\sum(v_i - \bar{v})^2}{(\sum(v_j - \bar{v})^2)^2} = \frac{\sigma^2}{\sum(v_j - \bar{v})^2} \to 0$$

and hence that

$$(2.2.17) \qquad \sum_{j=1}^{n}(v_j - \bar{v})^2 \to \infty \text{ as } n \to \infty.$$

Similarly, it is seen (Problems 2.4(ii) and (iii)) that

$$(2.2.18) \qquad E(\hat{\alpha}) = \alpha$$

and

$$(2.2.19) \qquad \text{Var}(\hat{\alpha}) = \sigma^2 \left[\frac{1}{n} + \frac{\bar{v}^2}{\sum(v_j - \bar{v})^2} \right].$$

The sufficient condition (2.2.17) involves a sequence of coefficients v_j, but in any given situation, the sample size n and the coefficients $(v_1^{(n)}, \ldots, v_n^{(n)})$ are fixed. How then should we interpret the significance of the condition for the given situation? When in Example 2.1.2 we stated that $\bar{X} \xrightarrow{P} \xi$, the implication was that for large n, the sample mean will be close to ξ with high probability and that its value therefore provides a reliable estimate of ξ. Analogously, in the present situation, the sufficiency of (2.2.17) permits the conclusion that if $\sum(v_j - \bar{v})^2$ is large, then $\hat{\beta}$ provides a reliable estimate of β. $\qquad\square$

Let us now return to the consistency of an average $\bar{X}$ of n random variables, but consider the case that the variables are no longer independent.

Suppose that $X_1, \ldots, X_n$ have a joint distribution with common mean $E(X_i) = \xi$ and with covariances $\mathrm{Cov}(X_i, X_j) = \gamma_{ij}$. Then $E(\bar{X}) = \xi$ and

$$(2.2.20) \qquad \mathrm{Var}(\bar{X}) = \frac{1}{n^2} \sum_{i=1}^{n} \sum_{j=1}^{n} \gamma_{ij}.$$

By Theorem 2.1.1 a sufficient condition for $\bar{X}$ to be a consistent estimator of ξ is therefore that

$$(2.2.21) \qquad \sum_{i=1}^{n} \sum_{j=1}^{n} \gamma_{ij} = o(n^2).$$

A difficulty with discussing dependent variables is the great variety of possible dependence structures. We here give only a few examples.

Example 2.2.4 Estimating a lot proportion. Consider a lot Π of N items, of which D are defective and $N - D$ satisfactory. A sample of n items is drawn at random without replacement. Let

$$(2.2.22) \qquad X_i = \begin{cases} 1 & \text{if the } i^{\text{th}} \text{ item drawn is defective} \\ 0 & \text{otherwise.} \end{cases}$$

Then

$$(2.2.23) \qquad E(X_i) = p = \frac{D}{N}$$

is the probability of a single item drawn being defective. It is easy to see that (Problem 2.6)

$$(2.2.24) \qquad \mathrm{Var}\, X_i = pq, \qquad \mathrm{Cov}(X_i, X_j) = -pq/(N-1).$$

The negative covariances reflect the negative dependence of the X's, which results from the fact that when one item in the sample is defective, this decreases the chance of any other item drawn being defective.

To ask whether $\bar{X}$, the proportion of defectives in the sample, is a consistent estimator of p makes sense only if we let the population size N tend to infinity as well as the sample size n. We then have

$$\sum_{i=1}^{n} \sum_{j=1}^{n} \gamma_{ij} = npq - \frac{n(n-1)}{N-1} pq < npq = o(n^2).$$

Thus $\bar{X}$ is consistent if both N and n tend to infinity. □

Example 2.2.5 A two-sample probability. Suppose that $X_1, \dots, X_m$ and $Y_1, \dots, Y_n$ are samples from distributions F and G, respectively, and that one wishes to estimate the probability

$$(2.2.25) \qquad\qquad p = P(X < Y).$$

An unbiased estimator is

$$(2.2.26) \qquad\qquad \delta = \frac{1}{mn} \sum_{i=1}^{m} \sum_{j=1}^{n} U_{ij},$$

where

$$(2.2.27) \qquad\qquad U_{ij} = 1 \text{ or } 0 \text{ as } X_i < Y_j \text{ or } X_i \geq Y_j.$$

Since δ is an average of mn terms, (2.2.21) becomes

$$(2.2.28) \qquad\qquad \sum\sum \gamma_{ij;kl} = o(m^2 n^2),$$

where

$$\gamma_{ij;kl} = \mathrm{Cov}(U_{ij}, U_{kl}).$$

The γ's are zero if all four subscripts i, j, k, l are distinct, from which it follows that

$$\sum\sum \gamma_{ij;kl} = O(m^2 n) + O(mn^2)$$

(for an exact formula, see Section 6.1). This shows that (2.2.28) holds, and therefore δ is a consistent estimator of p, provided m and n both tend to infinity. $\qquad\qquad\qquad\qquad\qquad\qquad\qquad\qquad\qquad\square$

Example 2.2.6 A counterexample. As an example in which consistency does not hold, suppose that one of two distributions F and G is chosen with probability p and q, respectively, and then a sample $X_1, \dots, X_n$ is obtained from the selected distribution. If the means and variances of the two distributions have the known values $\xi \neq \eta$ and σ^2, τ^2, respectively, we have

$$E(\bar{X}) = p\xi + q\eta,$$

so that

$$\delta = \frac{\bar{X} - \eta}{\xi - \eta}$$

is an unbiased estimator of p. It is intuitively clear that in the present situation, no consistent estimator of p can exist. For suppose that instead of observing only a sample from F or G, we were actually told whether the selected distribution was F or G. With this additional information, we would only have a single trial with outcomes F or G having probabilities p and q and we would thus be left without a basis for consistent estimation. $\square$

Consider finally a sequence of observations $X_1, X_2, \ldots$ taken at times $t_1, t_2, \ldots$, a so-called *time series*. If there is no trend, the distribution of the X's may be invariant under time changes, i.e., have the following property.

Definition 2.2.1 The sequence $X_1, X_2, \ldots$ is said to be *stationary* if for any positive integers i and k, the joint distribution of $(X_i, X_{i+1}, \ldots, X_{i+k})$ is independent of i. The sequence is said to be weakly stationary if this time independence is assumed only for the first and second moments of $(X_i, X_{i+1}, \ldots, X_{i+k})$.

Stationarity is a natural generalization of i.i.d. when the assumption of independence is dropped.

For a stationary sequence the condition (2.2.21) for consistency of $\bar{X}$ as an estimator of $\theta = E(X_i)$ reduces to

$$(2.2.29) \qquad n\sigma^2 + 2[(n-1)\gamma_1 + (n-2)\gamma_2 + \cdots + \gamma_{n-1}] = o(n^2),$$

where

$$(2.2.30) \qquad \gamma_k = \mathrm{Cov}(X_i, X_{i+k})$$

and hence to

$$(2.2.31) \qquad \frac{1}{n} \sum_{k=0}^{n-1} \left(1 - \frac{k}{n}\right) \gamma_k \to 0.$$

Example 2.2.7 Moving averages; m-dependence. As an example of a stationary sequence, consider the moving averages

$$(2.2.32) \qquad X_1 = \frac{U_0 + \cdots + U_m}{m+1}, \qquad X_2 = \frac{U_1 + \cdots + U_{m+1}}{m+1}, \ \ldots$$

or, more generally, the sequence

$$(2.2.33) \quad X_1 = \lambda_1 U_1 + \cdots + \lambda_m U_m, \qquad X_2 = \lambda_1 U_2 + \cdots + \lambda_m U_{m+1}, \ \ldots$$

where the U's are i.i.d. with mean θ and variance $\sigma^2 < \infty$. The X's have the property that $(X_1, \ldots, X_i)$ and $(X_j, X_{j+1}, \ldots)$ are independent for any $i < j$ with $j - i > m$. Any sequence with this property is said to be *m-dependent*.

Since for any m-dependent sequence we have

(2.2.34) $$\gamma_k = 0 \text{ if } k > m$$

and hence

$$\frac{1}{n}\sum_{k=0}^{n-1}\left(1-\frac{k}{n}\right)\gamma_k = \frac{1}{n}\sum_{k=0}^{m}\left(1-\frac{k}{n}\right)\gamma_k \to 0,$$

we see that (2.2.31) holds and $\bar{X}$ is therefore consistent for estimating $E(X_i) = \theta\sum_{i=1}^{m}\lambda_i$.

For m-dependent sequences, condition (2.2.31) is satisfied because all but a finite number of the γ's are zero. This is a rather special situation. However, more generally it is often the case that

(2.2.35) $$\gamma_k \to 0 \text{ as } k \to \infty,$$

i.e., the dependence becomes negligible when the observations are very far apart in time. (For some examples, see Section 2.8.) It can be shown that (2.2.35) implies (2.2.31) (Problem 2.8), and hence that for any stationary time series satisfying (2.2.35), $\bar{X}$ is consistent for $\theta = E(X_i)$.   □

Summary

1. Sufficient conditions are obtained for $\bar{X}$ and some other linear estimators to be consistent estimators of their expectation when the X's are independent but not identically distributed.

2. The consistency problem of 1 is also considered for a number of situations in which the observations are dependent, including in particular stationary sequences.

2.3   Convergence in law

At the beginning of Chapter 1, large-sample theory was described as dealing with approximations to probability distributions and with the limit theorems underlying these approximations. Such limit theorems are based on a suitable concept of the convergence of a sequence of distributions. As we saw in Section 1.6, distributions can be characterized in many different ways, for example, by their cumulative distribution functions, their quantile functions, or their probability densities. We shall define here the convergence of a sequence of distributions in terms of their cdf's.

A natural starting point is a sequence $\{H_n, n = 1, 2, \ldots\}$ of cdf's for which the limit

(2.3.1) $$\lim_{n\to\infty} H_n(x) = H(x)$$

exists for all x, and to ask whether it follows that H itself is a cdf. For this, it must satisfy conditions (i), (ii), (iii), and (v) of Theorem 1.6.1. Conditions (i) and (ii) are easily seen to hold (Problem 3.1). However the following example shows that this is not the case for (iii), which states that

$$(2.3.2) \qquad H(-\infty) = \lim_{x \to -\infty} H(x) = 0 \text{ and } H(+\infty) = \lim_{x \to \infty} H(x) = 1.$$

Example 2.3.1 Counterexample to (iii). Let H_n be the normal distribution with mean 0 and with variance $\sigma_n^2 \to \infty$. Then if X_n has distribution H_n, we have

$$H_n(x) = P\left(\frac{X_n}{\sigma_n} \le \frac{x}{\sigma_n}\right) = \Phi\left(\frac{x}{\sigma_n}\right),$$

where Φ is the cdf of the standard normal distribution $N(0,1)$. As $\sigma_n \to \infty$, $x/\sigma_n \to 0$ and hence $\Phi(x/\sigma_n) \to \Phi(0) = 1/2$. It follows that (2.3.1) holds with $H(x) = 1/2$ for all x. Clearly, H does not satisfy (2.3.2).

What has happened is that half the probability has escaped to $-\infty$ and half to $+\infty$. $\qquad \square$

The more usual situation in which this kind of pathology does not occur is characterized by the following definition:

Definition 2.3.1 A sequence of random variables Y_n is *bounded in probability* if for any $\epsilon > 0$, there exists a constant K and a value n_0 such that

$$P(|Y_n| \le K) \ge 1 - \epsilon \text{ for all } n > n_0.$$

Note: It is seen from Definition 2.1.4 that Y_n being bounded in probability is equivalent to the statement

$$Y_n = O_P(1).$$

Example 2.3.2 Let

$$Y_n = \begin{cases} 0 & \text{with probability } 1 - p_n \\ n & \text{with probability } p_n. \end{cases}$$

Then Definition 2.3.1 shows that Y_n is bounded in probability if $p_n \to 0$ but not if $p_n \to p > 0$. In the latter case, an amount p of probability escapes to ∞ as $n \to \infty$. $\qquad \square$

A useful consequence of Definition 2.3.1 is the following lemma (Problem 3.2(ii)).

Lemma 2.3.1 *If the sequence $\{Y_n, n = 1, 2, \ldots\}$ is bounded in probability and if $\{C_n\}$ is a sequence of random variables tending to 0 in probability, then $C_n Y_n \overset{P}{\to} 0$.*

The following theorem shows that condition (2.3.2) is violated only if some probability escapes to $+\infty$ or $-\infty$.

Theorem 2.3.1 *If $\{X_n, n = 1, 2, \ldots\}$ is a sequence of random variables with cdf H_n and if (2.3.1) holds for all x, then (2.3.2) holds if and only if the sequence $\{X_n\}$ is bounded in probability.*

The result is easy to see and we do not give a formal proof.

Let us next turn to condition (v) of Theorem 1.6.1, which states that a cdf is continuous on the right. If x is a continuity point of H, then H is automatically continuous on the right at x. As the following example shows, (v) need not hold at discontinuities of H.

Example 2.3.3 Counterexample. Let X have cdf F. Let

(a)
$$X_n = X - \frac{1}{n}$$

have cdf H_n, so that

$$H_n(x) = P\left(X \le x + \frac{1}{n}\right) = F\left(x + \frac{1}{n}\right).$$

Then

$$H_n(x) \to F(x+) = F(x)$$

and the limit F of H_n satisfies (v) for all x.

(b)
$$\text{Let } X'_n = X + \frac{1}{n}$$

have cdf H'_n. Then

$$H'_n(x) = F\left(x - \frac{1}{n}\right) \to F(x-)$$

and hence the limit of H'_n violates (v) at all points of discontinuity of F. $\square$

If in case (a) of this example we say that $H_n \to F$, we would also want to say that this relation holds in case (b). This is possible only if we require (2.3.1) not for all points x but only for continuity points of H. One might be concerned that such a restriction leaves the convergence definition too weak. That this is not the case is a result of the fact (which we shall not prove) that a cdf is completely determined by its values at all continuity points. We shall therefore adopt the following definition.

Definition 2.3.2 A sequence of distributions with cdf's H_n is said to converge to a distribution function H (in symbols, $H_n \to H$) if

(2.3.3) $H_n(x) \to H(x)$ at all continuity points x of H.

If Y_n is a sequence of random variables with cdf's H_n, and Y has cdf H, we shall then also say that Y_n tends *in law* to Y, or to H; in symbols,

(2.3.4) $Y_n \overset{L}{\to} Y$ or $Y_n \overset{L}{\to} H$.

The notation $Y_n \overset{L}{\to} Y$ is somewhat misleading since it suggests that for large n, the random variable Y_n is likely to be close to Y. In fact, Definition 2.3.2 only states that for large n, the distribution of Y_n is close to that of Y but not that the random variables themselves are close. To illustrate the difference, suppose that X is uniformly distributed on $(0,1)$ and let

$$Y_n = \begin{cases} X & \text{if } n \text{ is odd} \\ 1 - X & \text{if } n \text{ is even.} \end{cases}$$

Then Y_n has the uniform distribution $U(0,1)$ for all n and hence $Y_n \overset{L}{\to} X$ and also $Y_n \overset{L}{\to} 1 - X$. In fact, Y_n converges in law to any random variable, however defined, which is distributed as $U(0,1)$. On the other hand, Y_n clearly does not get close to X when n is even, no matter how large.

A better notation than $Y_n \overset{L}{\to} Y$ would be $d[Y_n] \to d[Y]$ with $d[\ ]$ denoting the distribution of the indicated variable, or $d[Y_n] \longrightarrow H$, where H is the limit distribution. We shall use $Y_n \overset{L}{\to} H$ as a shorthand version. This, in fact, corresponds to common terminology such as "Y_n is asymptotically normal."

Note: The symbol $H_n \to H$ is meant to imply that H is a cdf and thus in particular satisfies (2.3.2).

Example 2.3.4 Normal. If Y_n is normally distributed as $N(\xi_n, \sigma_n^2)$ and if $\xi_n \to 0$, $\sigma_n \to 1$, then $Y_n \overset{L}{\to} N(0,1)$.
To see this, note that

$$H_n(x) = P\left(\frac{Y_n - \xi_n}{\sigma_n} \leq \frac{x - \xi_n}{\sigma_n}\right) = \Phi\left(\frac{x - \xi_n}{\sigma_n}\right).$$

Now $(x - \xi_n)/\sigma_n \to x$ and hence, since Φ is continuous, $H_n(x) \to \Phi(x)$ for all x.   □

Example 2.3.5 Convergence in probability. If $Y_n \overset{P}{\to} c$ and Y is a random variable with $P(Y = c) = 1$, then $Y_n \overset{L}{\to} Y$; in this sense, convergence in probability is the special case of convergence in law in which the limit distribution assigns probability 1 to a constant.

To see this result, note that the cdf of Y is

$$H(y) = \begin{cases} 0 & \text{if } y < c \\ 1 & \text{if } y \geq c. \end{cases}$$

For $y < c$, we have

$$H_n(y) = P(Y_n \leq y) \to 0 = H(y).$$

For $y > c$, analogously,

$$H_n(y) = P(Y_n \leq y) \to 1 = H(y).$$

For $y = c$, we cannot say whether $H_n(y) \to H(y) = 1$ (Problem 3.4), but we do not need to since c is a discontinuity point of H. $\square$

The following consequence of convergence in law is often useful.

Theorem 2.3.2 *If Y_n converges in law to a distribution H, then the sequence Y_n is bounded in probability.*

Proof. Given $\epsilon > 0$, we must find K and n_0 such that $P(-K \leq Y_n \leq K) > 1 - \epsilon$ for all $n > n_0$. Now (Problem 3.5) there exist continuity points K_1, K_2 of H so large that $H(K_1) > 1 - \epsilon/4$ and $H(-K_2) < \epsilon/4$ and n_0 such that for all $n > n_0$,

$$H_n(K_1) > H(K_1) - \epsilon/4 > 1 - \epsilon/2 \text{ and } H_n(-K_2) < H(-K_2) + \epsilon/4 < \epsilon/2$$

and hence that

$$P(-K_2 \leq Y_n \leq K_1) \geq H_n(K_1) - H_n(-K_2) > 1 - \epsilon.$$

The result follows by taking $K = \max(|K_1|, |K_2|)$.   ∎

A very general class of situations for convergence in law has as its starting point a sequence of random variables Y_n converging in probability to a constant c.
Then

$$P(|Y_n - c| < a) \to 1 \text{ for all } a > 0,$$

and one would expect that typically

$$(2.3.5) \qquad P(|Y_n - c| < a_n) \to \begin{cases} 0 & \text{if } a_n \to 0 \text{ sufficiently fast} \\ 1 & \text{if } a_n \to 0 \text{ sufficiently slowly.} \end{cases}$$

If $k_n = a/a_n$, (2.3.5) becomes

$(2.3.6)$

$$P_n(a) = P(k_n|Y_n - c| < a) \to \begin{cases} 0 & \text{if } k_n \to \infty \text{ sufficiently fast} \\ 1 & \text{if } k_n \to \infty \text{ sufficiently slowly.} \end{cases}$$

One might hope that there exists an intermediate rate $k_n \to \infty$ for which $P_n(a)$ tends to a limit which is strictly between 0 and 1. Such a factor k_n (called a normalizing constant) then magnifies the very small differences $|Y_n - c|$ by just the factor needed to bring them into focus.

In many situations, one can guess the order of k_n by the following argument. Suppose that

$$H_n(a) = P[k_n(Y_n - c) \le a] \to H(a)$$

and that H is a cdf with finite variance v^2. If τ_n^2 is the variance of H_n, it will typically (but not always; Problem 3.7) be the case that

$$\mathrm{Var}\,[k_n(Y_n - c)] = k_n^2 \tau_n^2 \to v^2$$

so that

(2.3.7) $$k_n \sim \frac{v}{\tau_n}$$

is the right magnification.

Note: If $P[k_n(Y_n - c) \le a]$ tends to a limit strictly between 0 and 1 for all a, this is also true for $P\,[bk_n(Y - c) \le a]$ for any $b > 0$ and hence the factor bk_n provides another suitable magnification. Thus any sequence k_n satisfying

(2.3.8) $$k_n \asymp 1/\tau_n$$

will do as well as that determined by (2.3.7).

Example 2.3.6 De Moivre's theorem. Let X_n have the binomial distribution $b(p, n)$. Then

$$Y_n = \frac{X_n}{n} \xrightarrow{P} p.$$

Since

$$\tau_n^2 = \mathrm{Var}\, Y_n = pq/n,$$

we expect $k_n \left(\dfrac{X_n}{n} - p \right)$ to tend to a limit distribution when $k_n \asymp \sqrt{n}$. It was in fact shown by De Moivre (1733) that

(2.3.9) $$\sqrt{n}\left(\frac{X_n}{n} - p \right) \xrightarrow{L} N(0, pq)$$

□

Lemma 2.3.2 *If $Y_n \overset{L}{\to} Y$, and a and b are constants with $b \neq 0$, then $bY_n + a \overset{L}{\to} bY + a$.*

Proof. Let $b > 0$ and let x be a continuity point of $bY + a$. Then $\dfrac{x - a}{b}$ is a continuity point of Y and

$$P(bY_n + a \leq x) = P\left(Y_n \leq \frac{x - a}{b}\right) \to P\left(Y \leq \frac{x - a}{b}\right) = P(bY + a \leq x);$$

the proof for $b < 0$ is completely analogous.   ∎

This lemma shows that (2.3.9) is equivalent to

(2.3.10)
$$\frac{\sqrt{n}\left(\dfrac{X_n}{n} - p\right)}{\sqrt{pq}} \overset{L}{\to} N(0, 1).$$

Example 2.3.7 Uniform. Let $X_1, \ldots, X_n$ be i.i.d. according to the uniform distribution $U(0, \theta)$. The maximum likelihood estimator (MLE) of θ is $X_{(n)}$, the largest of the X's. This is always less than θ, but tends to θ in probability since, for $0 < c < \theta$,

$$P[\theta - c < X_{(n)} < \theta] = 1 - P[X_{(n)} < \theta - c] = 1 - \left(\frac{\theta - c}{\theta}\right)^n \to 1$$

by Problem 1.1.17. The variance of $X_{(n)}$ is (Problem 3.8)

(2.3.11)
$$\tau_n^2 = \mathrm{Var}(X_{(n)}) = \frac{n\theta^2}{(n+1)^2(n+2)} \asymp \frac{1}{n^2}.$$

This suggests that an appropriate magnifying factor by which to multiply the difference $\theta - X_{(n)}$ is $k_n = n$.

Consider therefore the probability

$$P_n = P[n(\theta - X_{(n)}) \leq x] = P[X_{(n)} \geq \theta - x/n],$$

so that

$$\begin{aligned} 1 - P_n &= P\left[X_{(n)} < \theta - \frac{x}{n}\right] = P\left(X_1 < \theta - \frac{x}{n}\right)^n = \left(\frac{\theta - \dfrac{x}{n}}{\theta}\right)^n \\ &= \left(1 - \frac{x}{\theta n}\right)^n. \end{aligned}$$

It follows from Example 1.1.1 that

$$1 - P_n \to e^{-x/\theta}$$

and hence that

$$P_n \to 1 - e^{-x/\theta} = H(x),$$

where H is the cdf of a random variable Y with exponential density

(2.3.12) $$h(y) = \frac{1}{\theta} e^{-y/\theta}.$$

□

The usefulness of a convergence result $Y_n \overset{L}{\to} Y$ is often greatly enhanced by the following generalization of Lemma 2.3.2, which is commonly referred to as *Slutsky's theorem*.

Theorem 2.3.3 *If* $Y_n \overset{L}{\to} Y$, *and* A_n *and* B_n *tend in probability to constants* a *and* b, *respectively, then*

(2.3.13) $$A_n + B_n Y_n \overset{L}{\to} a + bY.$$

(For a proof see, for example, Bickel and Doksum (1977) or Cramér (1946)).

Corollary 2.3.1 *If* $Y_n \overset{L}{\to} Y$ *and* $R_n \overset{P}{\to} 0$, *then*

(2.3.14) $$Y_n + R_n \overset{L}{\to} Y.$$

Corollary 2.3.2 *If* $Y_n \overset{L}{\to} Y$ *and* $B_n \overset{P}{\to} 1$, *then*

(2.3.15) $$\frac{Y_n}{B_n} \overset{L}{\to} Y.$$

Note: In Corollaries 2.3.1 and 2.3.2, the variables A_n, B_n, and R_n are not required to be independent of Y_n.

When $Y_n \overset{P}{\to} c$, typically $\tau_n^2 = \text{Var}(Y_n) \to 0$. The heuristic argument leading to (2.3.6) suggests that one might then expect the existence of a distribution H such that

(2.3.16) $$k_n(Y_n - c) \overset{L}{\to} H$$

with

(2.3.17) $$k_n \to \infty.$$

The following is a useful converse.

Theorem 2.3.4 *If the sequence* $\{Y_n\}$ *satisfies (2.3.16) and (2.3.17), then*

(2.3.18) $$Y_n \overset{P}{\to} c.$$

The proof is left to Problem 3.17.

Following Definition 2.3.2, it was pointed out that $Y_n \overset{L}{\to} Y$ does not imply that for large n, the random variable Y_n is close to the random variable Y. Such a relationship is expressed by the following definition.

Definition 2.3.3 A sequence of random variables Y_n converges in probability to a random variable Y, in symbols,

$$(2.3.19) \qquad Y_n \overset{P}{\to} Y,$$

if $Y_n - Y \overset{P}{\to} 0$.

The following result (together with the example following Definition 2.3.2) shows that (2.3.19) is stronger than (2.3.4).

Theorem 2.3.5 *If* $Y_n \overset{P}{\to} Y$, *then also* $Y_n \overset{L}{\to} Y$.

Proof. Let $R_n = Y_n - Y$ so that $Y_n = Y + R_n$. If $Y_n \overset{P}{\to} Y$, then $R_n \overset{P}{\to} 0$ by Definition 2.3.3 and hence $Y_n \overset{L}{\to} Y$ by Slutsky's theorem.   ∎

The following characterization of convergence in law is frequently taken as its definition instead of Definition 2.3.2.

Theorem 2.3.6 *A necessary and sufficient condition for* $Y_n \overset{L}{\to} Y$ *is that*

$$(2.3.20) \qquad E f(Y_n) \to E f(Y)$$

for all bounded and continuous functions f.

For a proof see, for example, Billingsley (1986), Section 2a.

Convergence in law does not imply (2.3.20) if either the condition of boundedness or of continuity is violated.

Example 2.3.8 Let

$$Y_n = \begin{cases} Y & \text{with probability } 1 - p_n \\ n & \text{with probability } p_n \end{cases}$$

and let f be the function

$$f(y) = y \quad \text{for all } y,$$

which is unbounded. Then $Y_n \overset{L}{\to} Y$ provided $p_n \to 0$ (Problem 3.14). On the other hand, it was seen in Example 2.1.4 that (2.3.20) need not hold, nor the corresponding result for the variance.   □

Thus, convergence in law of a sequence $\{Y_n\}$ to Y does not imply the corresponding convergence of either expectation or variance. This extends the conclusion of Example 2.1.4 and the note following it from convergence in probability to convergence in law.

Example 2.3.9 The indicator function. To show that the assumption of continuity cannot be dropped in Theorem 2.3.6, suppose that $Y_n \overset{L}{\to} Y$ and that the distribution H of Y has a discontinuity at a. If f is the indicator function

$$f(y) = \begin{cases} 1 & \text{if } y \le a \\ 0 & \text{if } y > a, \end{cases}$$

then

$$E\left[f(Y_n)\right] = P(Y_n \le a) = H_n(a) \ \text{ and } \ E\left[f(Y)\right] = H(a).$$

Example 2.3.3 shows that $H_n \to H$ need not imply that $H_n(a)$ tends to $H(a)$, and hence does not imply (2.3.20). □

Summary

1. A sequence of random variables is *bounded in probability* if none of the probability mass escapes to $+\infty$ or $-\infty$ as $n \to \infty$.

2. A sequence of cdf's H_n converges to a cdf H (in symbols, $H_n \to H$) if $H_n(a) \to H(a)$ at all continuity points of H. The exception is needed to avoid various inconsistencies and other complications.

3. If random variables Y_n and Y have cdf's H_n and H satisfying $H_n \to H$, we also say that $Y_n \overset{L}{\to} Y$ and $Y_n \overset{L}{\to} H$. The first of these is somewhat misleading; it concerns only the distributions of Y_n and Y and does not imply that Y_n and Y themselves are close to each other for large n.

4. Slutsky's theorem states that if $Y_n \overset{L}{\to} Y$ and if A_n and B_n converge in probability respectively to constants a and b, then $A_n + B_n Y_n \overset{L}{\to} a + bY$; in particular, if $Y_n \overset{L}{\to} Y$ and $R_n \overset{P}{\to} 0$, then $Y_n + R_n \overset{L}{\to} Y$.

2.4   The central limit theorem

The central limit theorem (CLT) is not a single theorem but encompasses a variety of results concerned with the sum of a large number of random variables which, suitably normalized, has a normal limit distribution. The following is the simplest version of the CLT.

Theorem 2.4.1 Classical CLT. *Let* X_i, $i = 1, 2, \ldots$, *be i.i.d. with* $E(X_i) = \xi$ *and* $\mathrm{Var}(X_i) = \sigma^2 < \infty$. *Then*

(2.4.1) $$\sqrt{n}(\bar{X} - \xi)/\sigma \xrightarrow{L} N(0, 1)$$

or equivalently

$$\sqrt{n}(\bar{X} - \xi) \xrightarrow{L} N(0, \sigma^2).$$

A proof can be found in most probability texts and is sketched in Section A.5 of the Appendix.

Example 2.4.1 Binomial. It follows from the representation of a binomial variable X as a sum of i.i.d. variables X_i given by (2.1.8) that (2.3.9) is a special case of Theorem 2.4.1.   □

Example 2.4.2 Chi squared. The χ^2-distribution with n degrees of freedom is the distribution of $\sum_{i=1}^{n} Y_i^2$, where the Y_i are independent $N(0, 1)$. Since $E(Y_i^2) = 1$ and $\mathrm{Var}(Y_i^2) = 2$, it follows that

(2.4.2) $$\sqrt{n}\left(\frac{\chi_n^2}{n} - 1\right) \xrightarrow{L} N(0, 2).$$

The tendency to normality as n increases is shown in Figure 2.4.1. With increasing n, the density becomes more symmetric and closer to the normal density.   □

The gain in generality from the binomial result (2.3.9) of de Moivre (1733) to the central limit theorem (2.4.1) of Laplace (1810) is enormous. The latter theorem states that the mean of any long sequence of i.i.d. variables — no matter what their distribution F, provided it has finite variance — is approximately normally distributed. The result plays a central role in probability theory, and for this reason was named the central limit theorem by Polya (1920). A remarkable feature of the result is that it is distribution-free. This makes it possible to derive from it statistical procedures which are (asymptotically) valid without specific distributional assumptions.

The CLT permits us to approximate the probability

$$P\left[\sqrt{n}|\bar{X} - \xi|/\sigma \le a\right]$$

by the area under the standard normal curve between $-a$ and $+a$. This is a first order approximation: It uses a limit theorem to approximate the actual probability. Such approximations are somewhat crude and can be improved. (For a review of such improvements, see Johnson, Kotz and Balakrishnan (1994)). However, the resulting second (or higher order) approximations

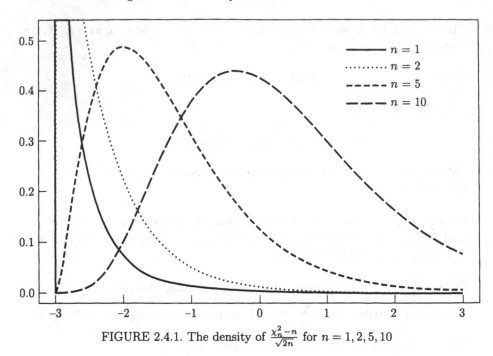

FIGURE 2.4.1. The density of $\frac{\chi_n^2 - n}{\sqrt{2n}}$ for $n = 1, 2, 5, 10$

are more complicated and require more knowledge of the underlying distribution.

While the normal limit distribution in the CLT is independent of the distribution F of the X's, this is not true of the sample size required for the resulting approximation to become adequate. The dependence of the speed of convergence on F is illustrated by Figure 2.4.2, which shows the histogram of $\sqrt{n}(X/n - p)/\sqrt{pq}$ when X has the binomial distribution $b(p, n)$ for $p = .05, .2, .5$ and $n = 10, 30, 90$. The figure shows that for each fixed p, the histogram approaches the normal shape more closely as n increases, but that the speed of this approach depends on p: it is faster the closer p gets to $1/2$. In fact, for $p = .05$, the approximation is still quite unsatisfactory, even for $n = 90$.

To get an idea of the sample size n needed for a given F, one can perform simulation studies (Problem 4.3). Then if, for example, the approximation is found to be poor for $n = 10$, somewhat better for $n = 20$, and satisfactory for $n = 50$, one may hope (although there is no guarantee*) that it will be at least as close for $n > 50$.

The applicability of the CLT can be greatly extended by combining it with Slutsky's theorem, as is shown by the following two examples.

Example 2.4.3 Student's t. When σ is unknown, the approximations suggested by Theorem 2.4.1 cannot be used, for example, for testing ξ. It

---

*For a counterexample, see Hodges (1957).

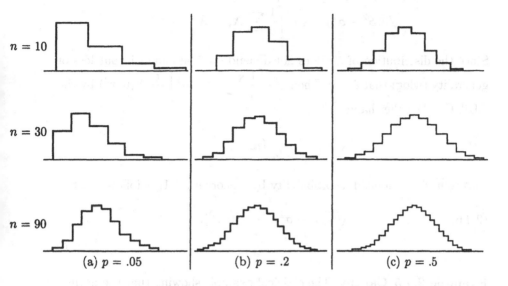

$n = 10$

$n = 30$

$n = 90$

(a) $p = .05$ (b) $p = .2$ (c) $p = .5$

FIGURE 2.4.2. Histograms of $\sqrt{n}\left(\frac{X}{n} - p\right)/\sqrt{pq}$

is then frequently useful instead to obtain an approximation for $P(\sqrt{n}|\bar{X} - \xi|/\hat{\sigma} \leq a)$, where $\hat{\sigma}$ is a consistent estimator of σ. Consider in particular the statistic

$$(2.4.3) \qquad t_n = \frac{\sqrt{n}\bar{X}}{\sqrt{\sum(X_i - \bar{X})^2/(n-1)}}.$$

When the X_i $(i = 1, \ldots, n)$ are independent $N(0, \sigma^2)$, the distribution of t_n is the t-distribution with $n-1$ degrees of freedom. Suppose now instead that the X_i are i.i.d. according to an arbitrary fixed distribution F with mean zero and finite variance σ^2. Then it follows from Theorem 2.4.1 that $\sqrt{n}\bar{X}/\sigma \xrightarrow{L} N(0,1)$. On the other hand, it follows from Example 2.1.3 that

$$(2.4.4) \qquad S^2 = \frac{1}{n}\sum(X_i - \bar{X})^2 \xrightarrow{P} \sigma^2.$$

Application of Slutsky's theorem thus shows that

$$(2.4.5) \qquad t_n \xrightarrow{L} N(0,1).$$

In particular, it follows that the t-distribution with n degrees of freedom tends to $N(0,1)$ as $n \to \infty$. $\qquad\square$

Example 2.4.4 Sample variance. Let $X_1, X_2, \ldots$ be i.i.d. with $E(X_i) = \xi$, $\mathrm{Var}(X_i) = \sigma^2$, and $\mathrm{Var}(X_i^2) = \tau^2 < \infty$, so that (2.4.4) holds. Consider now the limit behavior of

$$\sqrt{n}(S^2 - \sigma^2) = \sqrt{n}\left[\frac{1}{n}\sum X_i^2 - \bar{X}^2 - \sigma^2\right].$$

Since the distribution of S^2 does not depend on ξ, assume without loss of generality (wlog) that $\xi = 0$. Then $\sqrt{n}\left[\frac{1}{n}\sum X_i^2 - \sigma^2\right] \xrightarrow{L} N(0, \tau^2)$ by the CLT. On the other hand,

$$\sqrt{n}\bar{X}^2 = \frac{1}{\sqrt{n}}\left(n\bar{X}^2\right) \xrightarrow{P} 0$$

since $\sqrt{n}\,\bar{X}$ is bounded in probability by Theorem 2.3.1. It follows that

(2.4.6) $$\sqrt{n}(S^2 - \sigma^2) \xrightarrow{L} N(0, \tau^2).$$

$\square$

Example 2.4.5 Cauchy. The classical example showing that the normal tendency of $\bar{X}$ asserted by the CLT requires some assumptions beyond i.i.d. is the case in which the X's are i.i.d. according to the Cauchy distribution $C(0, 1)$. Here the distribution of $\bar{X}$ is the same as that of a single X, i.e., it is again equal to $C(0, 1)$, regardless of the value of n. Thus, trivially, $\bar{X}$ converges in law to $C(0, 1)$ instead of being asymptotically normal. (For an elementary proof, see, for example, Lehmann (1983, Problems 1.7 and 1.8 of Chapter 1) or Stuart and Ord (1987 p. 348)). $\square$

Even more extreme examples exist, in which the tail of the distribution is so heavy that $\bar{X}$ is more variable than a single observation. In these cases, as n increases, the sample is more and more likely to contain an observation which is so large that it essentially determines the size of $\bar{X}$. As a result, $\bar{X}$ then shares the variability of these outliers.

An example of such a distribution F is given by the distribution of $X = 1/Y^2$, where Y is $N(0, 1)$. The density of F is (Problem 4.4)

(2.4.7) $$f(x) = \frac{1}{\sqrt{2\pi x^3}} e^{-\frac{1}{2x}}, \quad x > 0.$$

In this case, it turns out for any n that $\bar{X}$ has the same distribution as nX_1 and is therefore much more variable than the single observation X_1. (For a discussion of the distribution (2.4.7), see, for example, Feller (Vol. 1) (1957, p. 231) and Feller (Vol. 2) (1966, p. 51)).

Example 2.4.6 Three distributions. The difference in behavior of the sample mean $\bar{X}_n$ from three distributions—(a) standard normal, (b) standard Cauchy, (c) the distribution (2.4.7)—is illustrated in Figure 2.4.3. The figure is based on samples drawn from these distributions and shows box-plots of the middle 50 of 100 averages of $n = 1, 10, 100$ observations for the three situations. In case (a), the distribution of the mean $\bar{X}_n$ shrinks by a factor of $\dfrac{1}{\sqrt{10}}$ as n increases from 1 to 10 and from 10 to 100; in case (b), the distribution of $\bar{X}_n$ is the same for all n; finally, in case (c), the distribution expands by a factor of 10 as n is multiplied by 10.

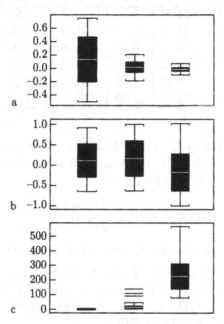

FIGURE 2.4.3. Boxplot of middle 50 of 100 averages from three distributions

The reason for restricting the boxplots to the middle 50 of 100 averages is that the full set of 100 contains observations so extreme that it seriously distorts the display. This can be seen from Table 2.4.1 which gives the values of the extremes in addition to those of the median and quartiles from the same samples. □

If the cdf of $\sqrt{n}\,(\bar{X} - \xi)/\sigma$ is denoted by

$$(2.4.8) \qquad G_n(x) = P\left[\frac{\sqrt{n}\,(\bar{X} - \xi)}{\sigma} \leq x\right],$$

the central limit theorem states that for every x (when the variance of X is finite),

$$(2.4.9) \qquad G_n(x) \to \Phi(x) \quad \text{as } n \to \infty.$$

However, it provides no indication of the speed of the convergence (2.4.9) and hence of the error to be expected when approximating $G_n(x)$ by $\Phi(x)$. The following result gives a bound for this error.

TABLE 2.4.1. Extremes, quartiles, and medians of 100 averages of means of 1, 10, and 100 observations from 3 distributions

		Min	25%	Med	75%	Max
Normal	1	-2.6	-0.51	0.13	0.77	2.5
	10	-0.77	-0.20	0.03	0.21	0.67
	100	-0.21	-0.09	0.02	0.06	0.20
Cauchy	1	-14.72	-0.72	0.15	0.93	27.07
	10	-23.67	-0.69	0.14	1.05	68.02
	100	-145.15	-0.99	-0.18	1.01	30.81
(4.7)	1	0.12	0.76	2.44	8.71	1,454.16
	10	0.00	7.05	16.01	141.22	14,421.33
	100	11.22	81.89	210.85	587.19	95,096.87

Theorem 2.4.2 (Berry–Esseen). *If $X_1, \ldots, X_n$ are i.i.d. with distribution F, which has a finite third moment, then there exists a constant C (independent of F) such that for all x,*

$$(2.4.10) \qquad |G_n(x) - \Phi(x)| \le \frac{C}{\sqrt{n}} \frac{E|X_1 - \xi|^3}{\sigma^3}.$$

For a proof, see, for example, Feller (Vol. 2) (1966) or Chung (1974).

Note: The important aspect of (2.4.10) is not the value of C but the fact that C is independent of F. The smallest value of C for which (2.4.10) holds is not known, but Theorem 2.4.2 does hold with $C = .7975$ and does not when $C < .4097$ (van Beek (1972)).

The bound (2.4.10), unlike most of our earlier results, is not a limit statement; it does not even assume that $n \to \infty$, but is an exact statement valid for any F, n, and x. The result has, however, important asymptotic implications for the following situation.

Suppose that $X_1, \ldots, X_n$ are i.i.d. according to a distribution F, but that this common distribution depends on n. In such a case, Theorem 2.4.1 is not applicable since it assumes F to be fixed. The following example shows that, in fact, the CLT then no longer holds without some additional restrictions.

Example 2.4.7 Counterexample. Let $X_1, \ldots, X_n$ be i.i.d. according to the Poisson distribution $P(\lambda)$ with $\lambda = 1/n$. Then $\sum_{i=1}^{n} X_i$ is distributed as $P(1)$, and hence with $\xi = \sigma^2 = 1/n$

$$Z = \frac{\sqrt{n}(\bar{X} - \xi)}{\sigma} = \frac{\sum X_i - n\xi}{\sqrt{n}\,\sigma} = \sum X_i - 1,$$

has the distribution of $Y - 1$, where Y is distributed as $P(1)$, and therefore Z is not asymptotically normal. □

A sufficient condition for the CLT to remain valid when F is allowed to depend on n can be read from Theorem 2.4.2.

Corollary 2.4.1 *Under the assumptions of Theorem 2.4.2,*

$$(2.4.11) \qquad G_n(x) \to \Phi(x) \text{ as } n \to \infty$$

for any sequence F_n with mean ξ_n and variance σ_n^2 for which

$$(2.4.12) \qquad \frac{E_n|X_1 - \xi_n|^3}{\sigma_n^3} = o\left(\sqrt{n}\right),$$

and thus in particular if (2.4.12) is bounded. Here E_n denotes the expectation under F_n.

Proof. Immediate consequence of Theorem 2.4.2.   ∎

Example 2.4.8 Binomial with $p_n \to 0$. If S_n has the binomial distribution $b(p, n)$, it was seen in Example 2.4.1 that

$$\frac{S_n - np}{\sqrt{npq}} \overset{L}{\to} N(0, 1).$$

Let us now consider the case that $p = p_n$ depends on n. Then it follows from the representation of S_n as a sum of i.i.d. random variables X_i given by (2.1.8), and by Corollary 2.4.1, that

$$(2.4.13) \qquad \frac{S_n - np_n}{\sqrt{np_nq_n}} \overset{L}{\to} N(0, 1),$$

provided the standardized absolute third moment γ_n satisfies

$$(2.4.14) \qquad \gamma_n = \frac{E|X_1 - p_n|^3}{(p_nq_n)^{3/2}} = o\left(\sqrt{n}\right).$$

Now

$$E|X_1 - p_n|^3 = p_n(1 - p_n)^3 + q_np_n^3 = p_nq_n\left(p_n^2 + q_n^2\right)$$

and hence

$$\gamma_n = \frac{p_n^2 + q_n^2}{\sqrt{p_nq_n}}.$$

If $p_n \to p\ (0 < p < 1)$, then $\gamma_n \to \dfrac{p^2 + q^2}{\sqrt{pq}}$ and hence satisfies (2.4.14). However, (2.4.14) holds also when $p_n \to 0$, provided this tendency is sufficiently slow. For if $p_n \to 0$, then

$$\gamma_n \sim \frac{1}{\sqrt{p_n}}.$$

A sufficient condition for (2.4.13) is therefore $\frac{1}{\sqrt{p_n}} = o(\sqrt{n})$ or

$$(2.4.15) \qquad\qquad \frac{1}{n} = o(p_n).$$

Thus in particular if $p_n \asymp \frac{1}{n^\alpha}$, (2.4.13) holds for any $0 < \alpha < 1$. For $\alpha = 1$, we are, of course, in the case of the Poisson limit (1.2.7). $\qquad\qquad\square$

Example 2.4.9 Sample median. Let $X_1, \ldots, X_n$ be i.i.d. according to the distribution

$$P(X \leq x) = F(x - \theta).$$

Suppose that $F(0) = 1/2$ so that θ is a median of the distribution of X and that n is odd: $n = 2m - 1$, say. If $X_{(1)} \leq \cdots \leq X_{(n)}$ denotes the ordered sample, the median of the X's is then $\tilde{X}_n = X_{(m)}$. Let us now find the limit distribution of $\sqrt{n}(\tilde{X}_n - \theta)$ under the assumption that F has a density f with $F'(0) = f(0) > 0$.

Since $\tilde{X}_n - \theta$ is the median of the variables $X_1 - \theta, \ldots, X_n - \theta$, its distribution is independent of θ, and

$$(2.4.16) \quad P_\theta \left[\sqrt{n}\,(\tilde{X}_n - \theta) \leq a \right] = P_0 \left[\sqrt{n}\,\tilde{X}_n \leq a \right] = P_0 \left[X_{(m)} \leq a/\sqrt{n} \right].$$

Let S_n be the number of X's exceeding $a/\sqrt{n}$. Then

$$X_{(m)} \leq \frac{a}{\sqrt{n}} \quad \text{if and only if} \quad S_n \leq m - 1 = \frac{1}{2}(n - 1).$$

Since S_n has the binomial distribution $b(p_n, n)$ with

$$(2.4.17) \qquad\qquad p_n = 1 - F\left(a/\sqrt{n}\right),$$

the probability (2.4.16) is equal to

$$P_0 \left[S_n \leq \frac{n-1}{2} \right] = P_0 \left[\frac{S_n - np_n}{\sqrt{np_n q_n}} \leq \frac{\frac{1}{2}(n-1) - np_n}{\sqrt{np_n q_n}} \right].$$

Since $p_n \to 1 - F(0) = 1/2$ as $n \to \infty$, it follows from the previous example and Theorem 2.4.2 that

$$P_0 \left(S_n \leq \frac{n-1}{2} \right) - \Phi \left[\frac{\frac{1}{2}(n-1) - np_n}{\sqrt{np_n q_n}} \right] \to 0 \text{ as } n \to \infty.$$

The argument of Φ is equal to

$$\begin{aligned}
x_n &= \frac{\sqrt{n}\left(\frac{1}{2} - p_n\right) - \frac{1}{2\sqrt{n}}}{\sqrt{p_n q_n}} \sim 2\sqrt{n}\left(\frac{1}{2} - p_n\right) \\
&= 2a\frac{F(a/\sqrt{n}) - F(0)}{a/\sqrt{n}} \to 2af(0).
\end{aligned}$$

Therefore, finally,

$$(2.4.18) \qquad P\left[\sqrt{n}\,(\tilde{X}_n - \theta) \le a\right] \to \Phi(2f(0)a),$$

which shows that (Problem 4.5(i))

$$(2.4.19) \qquad \sqrt{n}\,(\tilde{X}_n - \theta) \xrightarrow{L} N(0, 1/4f^2(0)).$$

So far, it has been assumed that n is odd. For even n, $n = 2m$ say, the sample median is defined as $\tilde{X}_n = \frac{1}{2}\left[X_{(m)} + X_{(m+1)}\right]$, and (2.4.16) can be shown to remain valid (Problem 4.5(ii)).   □

The error bound (2.4.10) holds for all F. For any particular F, the error can therefore be expected to be smaller. Is it even of order $1/\sqrt{n}$? This question is answered by the following result.

Theorem 2.4.3 *Let $X_1, \ldots, X_n$ be i.i.d. with a distribution F which is not a lattice distribution (defined in Section 1.6) and has finite third moment. Then*

$$(2.4.20) \qquad G_n(x) = \Phi(x) + \frac{\mu_3}{6\sigma^3\sqrt{n}}(1 - x^2)\phi(x) + o\left(\frac{1}{\sqrt{n}}\right),$$

where $\mu_k = E(X - \xi)^k$ denotes the kth central moment of the distribution F. The second term in (2.4.20) is the first Edgeworth correction to the normal approximation.

For a proof, see, for example, Feller (Vol. 2) (1966) or Gnedenko and Kolmogorov (1954, §42).

Note: Under the assumptions of Theorem 2.4.3 the following stronger statement holds. The remainder term $R_n(x)$ which in (2.4.20) is denoted by $o(1/\sqrt{n})$ satisfies $\sqrt{n}\,R_n(x) \to 0$ not only for each fixed x but uniformly in x; i.e., for any $\epsilon > 0$, there exists $n_0 = n_0(\epsilon)$ independent of x such that $|\sqrt{n}\,R_n(x)| < \epsilon$ for all $n > n_0$ and all x. (For a more detailed discussion of uniform convergence, see Section 2.6).

Under the stated assumptions, the error made in approximating $G_n(x)$ by $\Phi(x)$ is of order $1/\sqrt{n}$ when $\mu_3 \ne 0$. It is of smaller order when $\mu_3 = 0$ and therefore in particular when the distribution of the X's is symmetric (about ξ). Roughly speaking, the second term on the right side of (2.4.20) corrects for the skewness of G_n when F is asymmetric, bringing G_n closer to the symmetry of the limit distribution Φ.

Example 2.4.10 Chi squared (continued). As an illustration of the accuracy of the normal approximation, consider the approximation for χ^2 given in (2.4.2) and its refinement (2.4.20). These approximations are shown

TABLE 2.4.2. Accuracy of three χ^2 approximations

$1 - G_n(X)$	.5	.1	.05	.01	$F = \chi_1^2$	$F = \chi_2^2$	$F = \chi_5^2$	$F = \chi_{10}^2$
$1 - \Phi(X)$	.590	.091	.026	.001	$n = 4$	$n = 2$		
First Edg. Corr.	.503	.121	.066	.005				
Second Edg. Corr.	.500	.090	.049	.014				
$1 - \Phi(X)$	.575	.090	.029	.001	$n = 6$	$n = 3$		
First Edg. Corr.	.502	.115	.061	.006				
Second Edg. Corr.	.500	.094	.049	.014				
$1 - \Phi(X)$	.559	.090	.032	.002	$n = 10$	$n = 5$	$n = 2$	$n = 1$
First Edg. Corr.	.501	.110	.058	.007				
Second Edg. Corr.	.500	.097	.049	.012				
$1 - \Phi(X)$	.542	.092	.036	.003	$n = 20$	$n = 10$	$n = 4$	$n = 2$
First Edg. Corr.	.500	.105	.054	.009				
Second Edg. Corr.	.500	.099	.050	.011				
$1 - \Phi(X)$	.527	.094	.040	.004	$n = 50$	$n = 25$	$n = 10$	$n = 5$
First Edg. Corr.	.500	.102	.052	.010				
Second Edg. Corr.	.500	.100	.050	.010				
$1 - \Phi(X)$	.519	.095	.043	.006	$n = 100$	$n = 50$	$n = 20$	$n = 10$
First Edg. Corr.	.500	.101	.051	.010				
Second Edg. Corr.	.500	.100	.050	.010				

in the first column of Table 2.4.2 when F is the χ^2-distribution with one degree of freedom for sample sizes $n = 4, 6, 10, 20, 50$, and 100.

The first row for each sample size gives the normal approximation $1 - \Phi(x)$ for the exact probability

$$1 - G_n(x) = P\left[\frac{\chi_n^2 - n}{\sqrt{2n}} > x\right].$$

(Note that in this table, the values $1 - G_n(x)$ are fixed. Thus x and hence $1 - \Phi(x)$ depend on n.)

It is seen from this table that for $1 - G_n(x) = .05$ or $.01$, the relative error of the normal approximation is still quite large, even at $n = 100$. This poor performance is explained by the extreme skewness of the χ_1^2-distribution shown in Figure 2.4.1. The improvement as the skewness decreases can be seen by considering instead the performance of the approximation for $F = \chi_2^2, \chi_5^2$, and χ_{10}^2, which are consecutively less skewed. Since χ_k^2 is the sum of k independent χ^2 variables, the same tabular values correspond to χ^2 variables with successively larger degrees of freedom and proportionally smaller sample sizes, as shown in the last three columns of Table 2.4.2. The values for sample size $n = 10$ are pulled together in Table 2.4.3, which illustrates the improvement of the normal approximation as F gets closer to the normal shape.

TABLE 2.4.3. Accuracy of the normal approximation for $n = 10$

F	χ_1^2	χ_2^2	χ_5^2	χ_{10}^2	
	.559	.542	.527	.519	.5
	.090	.092	.094	.095	.1
	.032	.036	.040	.043	.05
	.002	.003	.004	.006	.01
		$1 - \Phi(x)$			$1 - G_n(x)$

The second row in Table 2.4.2 for each sample size shows the first Edgeworth correction (2.4.20). In the present case, with $F = \chi_1^2$, we have $\mu_3 = 8$ and $\sigma^3 = 2\sqrt{2}$ so that (2.4.20) becomes

$$(2.4.21) \qquad G_n(x) = \Phi(x) + \frac{\sqrt{2}}{3\sqrt{n}}[1 - x^2]\phi(x) + o\left(\frac{1}{\sqrt{n}}\right).$$

In all cases covered by the table, the correction is in the right direction but, for small n, it sometimes overcorrects so much that the resulting error exceeds that of the original (normal) approximation. However, generally adding the correction term provides a considerable improvement.

The $1/\sqrt{n}$ term on the right side of (2.4.20) is the first term of the so-called Edgeworth expansion. This expansion under mild assumptions on F permits successive corrections to the approximation of $G_n(x)$ by $\Phi(x)$ of order $1/\sqrt{n}$, $1/n$, $1/n\sqrt{n}$, ..., with the remainder at each stage being of smaller order than the last term. A detailed treatment of the expansion, which is beyond the scope of this book, can be found, for example, in Feller (Vol. 2) (1971) or in Gnedenko and Kolmogorov (1954). (See also McCune and Gray (1982), Barndorff-Nielsen and Cox (1989), and Hall (1992)). The process of successive approximations may be illustrated by the second step, which subtracts from the right side of (2.4.20) the term

$$(2.4.22) \qquad \frac{\phi(x)}{n\sigma^4}\left[\frac{\mu_4 - 3\mu_2^2}{24}(x^3 - 3x) + \frac{\mu_3^2}{72\sigma^2}(x^5 - 10x^3 + 15x)\right]$$

with the remaining error then being $o(1/n)$. The results for χ^2 are given in the third row of Table 2.4.2, which shows a clear improvement over the second row. Alternative more accurate approximations for χ^2 are discussed in Chapter 18 of Johnson, Kotz, and Balakrishnan (1994).

Theorem 2.4.3 assumes that the distribution F is not a lattice distribution. When it is, (2.4.20) is not correct since another term of order $1/\sqrt{n}$ must be taken into account, the so-called *continuity correction*. Suppose that the difference between successive values of X is h. Then the difference between successive values of $Y = \sqrt{n}(\bar{X} - \xi)/\sigma$ is $h/\sigma\sqrt{n}$. Consider now the histogram of the random variable Y. The bar of this histogram corresponding to the value y of Y extends from $y - \dfrac{h}{2\sigma\sqrt{n}}$ to $y + \dfrac{h}{2\sigma\sqrt{n}}$, and

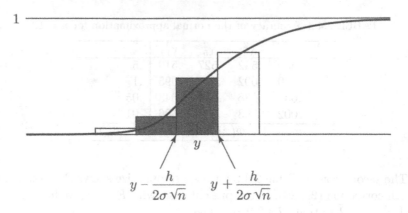

FIGURE 2.4.4. Continuity correction

the probability $P(Y \leq y)$ is therefore the area of the histogram to the left of $y + \dfrac{h}{2\sigma\sqrt{n}}$. When this area is approximated by the corresponding area under the normal curve, it is natural to approximate it by $\Phi\left(y + \dfrac{h}{2\sigma\sqrt{n}}\right)$ rather than by $\Phi(y)$. This change is the *continuity correction*. [†] It follows from Taylor's theorem (to be discussed in the next section) that

$$\Phi\left(y + \frac{h}{2\sigma\sqrt{n}}\right) = \Phi(y) + \frac{h}{2\sigma\sqrt{n}}\phi(y) + o(\frac{1}{\sqrt{n}}).$$

This explains the version of Theorem 2.4.3 for lattice distributions F, which states that for any lattice point x of F,

$$(2.4.23) \qquad G_n(x) = \Phi(x) + \frac{\phi(x)}{\sqrt{n}}\left[\frac{\mu_3}{6\sigma^3}(1 - x^2) + \frac{h}{2\sigma}\right] + o\left(\frac{1}{\sqrt{n}}\right).$$

For a proof, see Gnedenko and Kolmogorov (1954, §43). □

The normal approximation plays a central role in large-sample statistics and its Edgeworth corrections are useful for more delicate investigations. The importance of these approximations stems from their very general applicability, but this advantage also carries a drawback: better approximations in any particular case can usually be attained by taking into account the special features of a given distribution. For many standard distributions, such approximations are discussed, for example, in Johnson, Kotz, and Kemp (1992) and Johnson, Kotz, and Balakrishnan(1994,1995). In addition, extensive tables are available for most standard distributions, or needed values can be obtained on a computer.

[†]Continuity corrections are discussed in more detail and with references to the literature in Maxwell (1982).

Summary

1. If $X_1, X_2, \ldots$ are i.i.d. with mean ξ and finite variance σ^2, then the CLT states that

$$\sqrt{n}(\bar{X} - \xi) \xrightarrow{L} N(0, \sigma^2) \quad \text{and} \quad \frac{\sqrt{n}(\bar{X} - \xi)}{\sigma} \xrightarrow{L} N(0, 1).$$

2. If G_n denotes the cdf of $\sqrt{n}(\bar{X} - \xi)/\sigma$ and Φ that of $N(0,1)$, then the Berry-Esseen theorem provides a bound for the error $|G_n(x) - \Phi(x)|$ in the normal approximation, which holds simultaneously for all x and all distributions F of the X_i that have finite third moment.

3. A more accurate approximation than that stated in 1 is given by the first Edgeworth correction, which adds a $1/\sqrt{n}$ correction term to the approximation $\Phi(x)$ for $G_n(x)$, and which holds for any distribution F that is not a lattice distribution. For a lattice distribution, an additional $1/\sqrt{n}$ term is required which corresponds to the continuity correction.

2.5 Taylor's theorem and the delta method

A central result of the calculus is Taylor's theorem concerning the expansion of a sufficiently smooth function about a point.

Theorem 2.5.1
(i) Suppose that $f(x)$ has r derivatives at the point a. Then

$$(2.5.1) \qquad f(a + \Delta) = f(a) + \Delta f'(a) + \cdots + \frac{\Delta^r}{r!} f^{(r)}(a) + o(\Delta^r),$$

where the last term can also be written as

$$\frac{\Delta^r}{r!} \left[f^{(r)}(a) + o(1) \right].$$

(ii) If, in addition, the $(r+1)$st derivative of f exists in a neighborhood of a, the remainder $o(\Delta^r)$ in (2.5.1) can be written as

$$R_r = \frac{\Delta^{r+1}}{(r+1)!} f^{(r+1)}(\xi),$$

where ξ is a point between a and $a + \Delta$.

(For a proof, see Hardy (1992), Sec. 151).

An easy consequence is the following result which greatly extends the usefulness of the central limit theorem.

Theorem 2.5.2 *If*

(2.5.2) $$\sqrt{n}(T_n - \theta) \overset{L}{\to} N(0, \tau^2),$$

then

(2.5.3) $$\sqrt{n}\,[f(T_n) - f(\theta)] \overset{L}{\to} N\left(0, \tau^2[f'(\theta)]^2\right),$$

provided $f'(\theta)$ exists and is not zero.

Proof. By Taylor's theorem, with $a = \theta$ and $\Delta = T_n - \theta$,

$$f(T_n) = f(\theta) + (T_n - \theta)f'(\theta) + o_p(T_n - \theta)$$

and hence

(2.5.4) $$\sqrt{n}\,[f(T_n) - f(\theta)] = \sqrt{n}(T_n - \theta)f'(\theta) + o_p[\sqrt{n}\,(T_n - \theta)]\,.$$

The first term on the right side tends in law to $N\left(0, \tau^2\,[f'(\theta)]^2\right)$. On the other hand, it follows from (2.5.2) and Theorem 2.3.2 that $\sqrt{n}\,(T_n - \theta)$ is bounded in probability and hence that the remainder tends to zero in probability (Problem 5.1). The result now follows from Corollary 2.3.1. ∎

This theorem may seem surprising, since if X is normally distributed, the distribution of $f(X)$, for example, $1/X$, $\log X$, or e^X, will typically be non-normal. The explanation for this apparent paradox is found in the proof. Since $T_n \overset{P}{\to} \theta$, we are nearly certain that when n is large, T_n is very close to θ; however, in a small neighborhood, a differentiable function is nearly linear, and a linear function of a normal variable is again normal. The process of approximating the difference $f(T_n) - f(\theta)$ by the linear function $(T_n - \theta)f'(\theta)$ and the resulting limit result (2.5.3) is called the *delta method.*

Example 2.5.1 For estimating p^2, suppose that we have the choice between

(a) n binomial trials with probability p^2 of success

or

(b) n binomial trials with probability p of success,

and that as estimators of p^2 in the two cases, we would use respectively X/n and $(Y/n)^2$, where X and Y denote the number of successes in cases (a) and (b), respectively. Then we have

$$\sqrt{n}\left(\frac{X}{n} - p^2\right) \overset{L}{\to} N\left(0, p^2\left(1 - p^2\right)\right)$$

and

$$\sqrt{n}\left(\left(\frac{Y}{n}\right)^2 - p^2\right) \xrightarrow{L} N\left(0, pq \cdot 4p^2\right).$$

At least for large n, X/n will thus be more accurate than $(Y/n)^2$, provided

$$p^2\left(1 - p^2\right) < pq \cdot 4p^2.$$

On dividing both sides by $p^2(1 - p)$, it is seen that

$$\frac{X}{n} \text{ or } \frac{Y^2}{n^2} \text{ is preferable as } p > \frac{1}{3} \text{ or } p < \frac{1}{3}.$$

□

Theorem 2.5.2 provides the basis for deriving variance-stabilizing transformations, that is, transformations leading to an asymptotic variance that is independent of the parameter. Suppose, for example, that $X_1, \ldots, X_n$ are i.i.d. Poisson variables with expectation λ. The variance of the X's is then also λ and it follows from the central limit theorem that

$$(2.5.5) \qquad \sqrt{n}\left(\bar{X} - \lambda\right) \xrightarrow{L} N\left(0, \lambda\right).$$

For inference problems concerning λ, it is often inconvenient that λ occurs not only in the expectation but also in the variance of the limit distribution. It is therefore of interest to look for a function f for which $\sqrt{n}\left[f\left(\bar{X}\right) - f\left(\lambda\right)\right]$ tends in law to $N(0, c^2)$, where c^2 does not depend on λ.

Suppose more generally that

$$(2.5.6) \qquad \sqrt{n}\left(T_n - \theta\right) \xrightarrow{L} N\left(0, \tau^2\left(\theta\right)\right).$$

Then by Theorem 2.5.2,

$$\sqrt{n}\left[f\left(T_n\right) - f\left(\theta\right)\right] \xrightarrow{L} N\left(0, \tau^2\left(\theta\right)\left(f'\right)^2\left(\theta\right)\right),$$

provided the derivative of f' of f exists at θ and is $\neq 0$. The limit distribution on the right side will therefore have constant variance c^2 if

$$(2.5.7) \qquad f'(\theta) = \frac{c}{\tau(\theta)}.$$

The resulting transformation f is said to be *variance stabilizing*.

The extensive literature on variance-stabilizing transformations and transformations to approximate normality is reviewed in Hoyle (1973). Two later references are Efron (1982) and Bar-Lev and Enis (1990).

Example 2.5.2 Poisson. In the Poisson case, one has $\theta = \lambda$, $\tau(\theta) = \sqrt{\lambda}$, and (2.5.7) reduces to

$$f'(\lambda) = \frac{c}{\sqrt{\lambda}} \text{ or } f(\lambda) = 2c\sqrt{\lambda}.$$

Putting $c = 1$, we see that in fact

(2.5.8) $$2\sqrt{n}\left(\sqrt{\bar{X}} - \sqrt{\lambda}\right) \xrightarrow{L} N(0, 1).$$

□

Example 2.5.3 Chi squared. Let $Y_i = X_i^2$, where the X's are independent $N(0, \sigma^2)$. Then $E(Y_i) = \sigma^2$ and $\text{Var}(Y_i) = 2\sigma^4$, and (2.5.6) holds with $T_n = \bar{Y}$, $\theta = \sigma^2$, and $\tau^2(\theta) = 2\theta^2$. Equation (2.5.7) thus becomes

$$f'(\theta) = \frac{c}{\sqrt{2}\theta} \text{ or } f(\theta) = \frac{c}{\sqrt{2}} \log \theta.$$

With $c = 1$, we see that

(2.5.9) $$\sqrt{\frac{n}{2}} \log\left(\frac{\bar{Y}}{\sigma^2}\right) \xrightarrow{L} N(0, 1).$$

□

To illustrate the usefulness of such transformations, consider the problem of finding (approximate) confidence intervals for the Poisson parameter based on a large sample.

Example 2.5.4 Poisson confidence intervals. It follows from (2.5.8) that for any $\lambda > 0$,

$$P\left(\left|\sqrt{\bar{X}} - \sqrt{\lambda}\right| < \frac{u_{\alpha/2}}{2\sqrt{n}}\right) \to 1 - \alpha,$$

where $u_{\alpha/2}$ is the point for which $1 - \Phi(u_{\alpha/2}) = \alpha/2$. This provides for $\sqrt{\lambda}$ the intervals

(2.5.10) $$\sqrt{\bar{X}} - \frac{u_{\alpha/2}}{2\sqrt{n}} < \sqrt{\lambda} < \sqrt{\bar{X}} + \frac{u_{\alpha/2}}{2\sqrt{n}}$$

at approximate confidence level $1 - \alpha$. The lower end point can be negative since $\bar{X}$ can be arbitrarily close to zero. However, for any positive λ, we have that $\bar{X} \xrightarrow{P} \lambda$ and $u_{\alpha/2}/\sqrt{n} \to 0$, so that the probability of a negative end point tends to zero as $n \to \infty$. When this unlikely event does occur, one would replace the negative end point by zero without changing the probability of the resulting statement. From the so modified intervals for

$\sqrt{\lambda}$, one obtains the corresponding intervals for λ at the same level by squaring. This leads to the approximate confidence intervals $\underline{\lambda} < \lambda < \overline{\lambda}$, where

(2.5.11)

$$\underline{\lambda} = \begin{cases} \left(\sqrt{\bar{X}} - \dfrac{u}{2\sqrt{n}} \right)^2 & \text{if } \dfrac{u}{2\sqrt{n}} < \sqrt{\bar{X}} \\ 0 & \text{otherwise} \end{cases} \quad \text{and } \overline{\lambda} = \left(\sqrt{\bar{X}} + \dfrac{u}{2\sqrt{n}} \right)^2$$

Here and in what follows, we write u for $u_{\alpha/2}$.

There is an alternative way of deriving approximate confidence intervals for λ. From (2.5.5), it follows that

$$\frac{\sqrt{n}\,(\bar{X} - \lambda)}{\sqrt{\lambda}} \xrightarrow{L} N(0, 1)$$

and this limit result remains correct if in the denominator λ is replaced by a consistent estimator, say $\hat{\lambda} = \bar{X}$. The probability therefore tends to $1 - \alpha$ that λ lies in the interval

(2.5.12) $$\bar{X} - \frac{u}{\sqrt{n}}\sqrt{\bar{X}} < \lambda < \bar{X} + \frac{u}{\sqrt{n}}\sqrt{\bar{X}},$$

where the lower limit can again be replaced by 0 when it is negative.

The upper limits

$$\left(\sqrt{\bar{X}} + \frac{u}{2\sqrt{n}} \right)^2 \quad \text{and } \bar{X} + \frac{u}{\sqrt{n}}\sqrt{\bar{X}}$$

at first sight look rather different. However, the fact that

$$\left(\sqrt{\bar{X}} + \frac{u}{2\sqrt{n}} \right)^2 = \bar{X} + \frac{u}{\sqrt{n}}\sqrt{\bar{X}} + \frac{u^2}{4n}$$

shows that they differ only by the term $u^2/4n$, which is of smaller order than the first two terms and can therefore be expected to be small in relation to them. The corresponding remark applies to the lower limit. □

In Theorem 2.5.2, it was assumed that $f'(\theta) \neq 0$. Let us now consider what happens when $f'(\theta) = 0$ but $f''(\theta) \neq 0$. Since the leading term in the expansion (2.5.4) then drops out, it is natural to carry the expansion one step further. By Theorem 2.5.1, this gives

$$f(T_n) = f(\theta) + (T_n - \theta)f'(\theta) + \frac{1}{2}(T_n - \theta)^2 [f''(\theta) + R_n]$$

and since $f'(\theta) = 0$,

(2.5.13) $$k_n [f(T_n) - f(\theta)] = \frac{k_n}{2}(T_n - \theta)^2 f''(\theta) + o_p [k_n(T_n - \theta)^2].$$

Now it follows from (2.5.2) that

$$\frac{n(T_n - \theta)^2}{\tau^2(\theta)} \overset{L}{\to} \chi_1^2 \text{ or } n(T_n - \theta)^2 \overset{L}{\to} \tau^2(\theta)\chi_1^2,$$

where χ_1^2 denotes the χ^2-distribution with 1 degree of freedom. In order to get a non-degenerate limit distribution for (2.5.13), it is therefore necessary to take $k_n = n$ rather than $\sqrt{n}$. With this choice, the proof of Theorem 2.5.2 then shows that

$$(2.5.14) \qquad n\left[f(T_n) - f(\theta)\right] \overset{L}{\to} \frac{1}{2}\tau^2(\theta)f''(\theta)\chi_1^2.$$

It follows from this result that when $f'(\theta) = 0$ and $f''(\theta) \neq 0$, the convergence of $f(T_n)$ to $f(\theta)$ is faster than when $f'(\theta) \neq 0$ so that (for large n) $f(T_n)$ does a better job of approximating $f(\theta)$ when $f'(\theta) = 0$ than when $f'(\theta) \neq 0$. The reason for the faster convergence is qualitatively easy to see from Figure 2.5.1. The vanishing of f' at θ means that f changes very little in a small neighborhood of θ. Thus if T_n is close to θ, $f(T_n)$ will be very close to $f(\theta)$ and thus provide an excellent approximation.

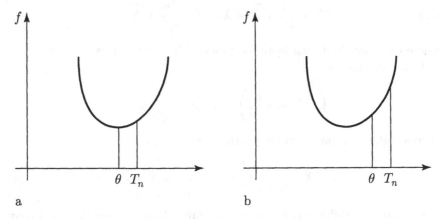

FIGURE 2.5.1. Derivative and rate of change

Example 2.5.5 Binomial variance. Suppose X has the binomial distribution $b(p, n)$ corresponding to n trials and success probability p so that

$$\sqrt{n}\left(\frac{X}{n} - p\right) \overset{L}{\to} N(0, pq).$$

The maximum likelihood estimator of the variance pq is $\delta_n = \frac{X}{n}\left(1 - \frac{X}{n}\right)$. With $\theta = p$ and $f(\theta) = pq$, it is seen from Theorem 2.5.2 that

$$\sqrt{n}\left(\delta_n - pq\right) \overset{L}{\to} N\left(0, pq(1 - 2p)^2\right) \text{ if } p \neq 1/2.$$

Since $f'(p) = 1 - 2p$, it follows that at $p = 1/2$, f' is zero and $f'' = -2$, and by (2.5.14),

$$n(\delta_n - pq) \overset{L}{\rightarrow} -\frac{1}{4}\chi_1^2 \text{ if } p = 1/2.$$

$\square$

Let us finally consider an example in which $f'(\theta)$ does not exist.

Example 2.5.6 Absolute value. Suppose T_n is a sequence of statistics satisfying (2.5.2) and that we are interested in the limiting behavior of $|T_n|$. Since $f(\theta) = |\theta|$ is differentiable with derivative $f'(\theta) = \pm 1$ at all values of $\theta \neq 0$, it follows from Theorem 2.5.2 that

$$\sqrt{n}\,(|T_n| - |\theta|) \overset{L}{\rightarrow} N(0, \tau^2) \text{ for all } \theta \neq 0.$$

When $\theta = 0$, Theorem 2.5.2 does not apply, but it is easy to determine the limit behavior of $|T_n|$ directly. With $|T_n| - |\theta| = |T_n|$, we then have

$$
\begin{aligned}
P\left[\sqrt{n}\,|T_n| < a\right] &= P[-a < \sqrt{n}\,T_n < a] \\
&\rightarrow \Phi\left(\frac{a}{\tau}\right) - \Phi\left(-\frac{a}{\tau}\right) = P(\tau\chi_1 < a),
\end{aligned}
$$

where $\chi_1 = \sqrt{\chi_1^2}$ is the distribution of the absolute value of a standard normal variable. The convergence rate of $\delta_n = |T_n|$ therefore continues to be $1/\sqrt{n}$, but the form of the limit distribution is χ_1 rather than normal. $\square$

When the function f in (2.5.1) has derivatives of all orders, it is tempting to let $r \to \infty$ and obtain the expansion, *Taylor's series*,

$$(2.5.15) \qquad f(a + \Delta) = \sum_{i=0}^{\infty} \frac{f^{(i)}(a)}{i!}\Delta^i.$$

If the $o(\Delta^r)$ remainder in (2.5.1) is denoted by R_r, the expansion (2.5.15) of $f(a + \Delta)$ as a power series in Δ is legitimate only if

$$(2.5.16) \qquad R_r \to 0 \text{ as } r \to \infty.$$

In particular, for $a = 0$, we get the expansion of $f(x)$ as a power series in x,

$$(2.5.17) \qquad f(x) = \sum_{i=0}^{\infty} \frac{f^{(i)}(0)}{i!}x^i,$$

provided

$$(2.5.18) \qquad R_r = f(x) - \sum_{i=0}^{r} \frac{f^{(i)}(0)}{i!}x^i \to 0 \text{ as } r \to \infty.$$

For the values of x for which (2.5.18) holds, (2.5.17) is *Mac Laurin's* series for $f(x)$.

Example 2.5.7 The geometric series. An example in which condition (2.5.18) is easy to check is the case

$$f(x) = \frac{1}{1-x}.$$

Then

$$f'(x) = \frac{1}{(1-x)^2}, \ f''(x) = \frac{2}{(1-x)^3}, \dots, f^{(r)}(x) = \frac{r!}{(1-x)^{r+1}},$$

so that (2.5.1) with $a = 0$ and $\Delta = x$ becomes

$$\frac{1}{1-x} = 1 + x + +x^2 + \cdots + x^r + o(x^r).$$

The remainder R_r is then

$$\begin{aligned} R_r &= \frac{1}{1-x} - (1 + x + \cdots + x^r) = \frac{1}{1-x}[1 - (1-x)(1+x+\cdots+x^r)] \\ &= x^{r+1}/(1-x). \end{aligned}$$

Since this tends to 0 as $r \to \infty$ provided $|x| < 1$, we get the expansion

(2.5.19) $$\frac{1}{1-x} = \sum_{i=0}^{\infty} x^i \text{ for all } |x| < 1.$$

The right side of (2.5.19) is the *geometric series* considered in Example 1.3.3. When $|x| > 1$, (2.5.19) is no longer valid since R_r does not tend to 0. For example, if $x = 2$,

$$R_r = \frac{x^{r+1}}{1-x} = -2^{r+1} \to -\infty;$$

the left side of (2.5.19) is -1 and the right side is $+\infty$, so that (2.5.19) does not hold.   □

Other examples of (2.5.17) are given in Examples 1.3.4–1.3.6 of Section 1.3 (Problem 5.11).

Summary

1. Taylor's theorem approximates any sufficiently smooth function by a polynomial—in the simplest and most important case by a linear function—in the neighborhood of a given point.

2. The delta method shows that if T_n is approximately normal with mean 0 and variance $\tau^2(\theta)/n$, then for any differentiable function f with $f'(\theta) \neq 0$, $f(T_n)$ is approximately normal with mean $f(\theta)$ and variance $\tau^2(\theta)[f'(\theta)]^2/n$.

3 In a number of important cases, the delta method permits the calculation of a variance-stabilizing transformation by determining f so that $\tau^2(\theta)[f'(\theta)]^2$ is independent of θ.

4. When the assumptions of 2 hold except that $f'(\theta) = 0$ but $f''(\theta) \neq 0$, the delta method shows that $n[f(T_n) - f(\theta)]$ is approximately distributed as $\frac{1}{2}\tau^2 f''(\theta)\chi_1^2$.

5. By letting the number of terms in Taylor's theorem tend to infinity, one obtains Taylor's series, and for $a = 0$ and $\Delta = x$ Mac Laurin's series for a function $f(x)$ all of whose derivatives exist, provided the remainder tends to zero.

2.6   Uniform convergence

Consideration of approximate confidence intervals $(\underline{\theta}, \bar{\theta})$ for a parameter θ such as those derived in Example 2.5.4 leads to an important distinction.

Suppose first that the intervals are exact confidence intervals with confidence coefficient $1 - \alpha$. This can be expressed by saying that

(2.6.1)   for every θ: $P_\theta\left(\underline{\theta} < \theta < \bar{\theta}\right) = 1 - \alpha$.

An obvious consequence is that

(2.6.2) $\inf_\theta P_\theta\left(\underline{\theta} < \theta < \bar{\theta}\right) = 1 - \alpha$.

Suppose now that $(\underline{\theta}_n, \bar{\theta}_n)$ are only approximate confidence intervals for large n in the sense of the Poisson intervals of the last section. Then (2.6.1) becomes

(2.6.3)   for every θ: $P_\theta\left(\underline{\theta}_n < \theta < \bar{\theta}_n\right) \to 1 - \alpha$ as $n \to \infty$.

However, (2.6.3) no longer implies

(2.6.4) $\inf_\theta P_\theta\left[\underline{\theta}_n < \theta < \bar{\theta}_n\right] \to 1 - \alpha$.

To see why this is so, consider several sequences of numbers $\{a_n(\theta), n = 1, 2, \ldots\}$ (in our case, the probabilities (2.6.3) for varying θ) converging to a common limit c. We are concerned with the convergence of $\inf_\theta a_n(\theta)$, $n = 1, 2, \ldots$

If there are only two such sequences, say $\{a_n\}$ and $\{b_n\}$ with $a_n \to c$ and $b_n \to c$ as $n \to \infty$, then also $\min(a_n, b_n) \to c$ as $n \to \infty$.

Proof. Given $\epsilon > 0$, there exist n_1 and n_2 such that $|a_n - c| < \epsilon$ for $n > n_1$ and $|b_n - c| < \epsilon$ for $n > n_2$. Then if $n > n_0 = \max(n_1, n_2)$, it follows that both $|a_n - c|$ and $|b_n - c|$ are $< \epsilon$ and hence also that $|\min(a_n, b_n) - c| < \epsilon$.

This argument easily extends to any finite number of sequences (Problem 6.1) but breaks down when the number of sequences is infinite.

Example 2.6.1 Counterexample. Consider the following sequences

(2.6.5)

$$
\begin{array}{llll}
1^{\text{st}} \text{ sequence}: & 0\ 1\ 1\ 1\ 1\ \cdots \\
2^{\text{nd}} \text{ sequence}: & 0\ 0\ 1\ 1\ 1\ \cdots \\
3^{\text{rd}} \text{ sequence}: & 0\ 0\ 0\ 1\ 1\ \cdots
\end{array}
$$

$$\cdots \qquad \cdots$$

If we denote the ith sequence by $\{a_n^{(i)}, n = 1, 2, \ldots\}$, then clearly, for every i,

$$a_n^{(i)} \to 1 \text{ as } n \to \infty.$$

On the other hand,

$$\min_{i=1,2,\ldots} a_n^{(i)} = 0 \text{ for every } n$$

since every column of (2.6.5) contains a zero. It follows that also

$$\lim_{n \to \infty} \left[\min_{i=1,2,\ldots} a_n^{(i)} \right] = 0.$$

$\square$

Let us now apply this consideration to the upper end of the intervals (2.5.12) for λ in the Poisson case, which is

(2.6.6)

$$\bar{X} + \frac{u_{\alpha/2}}{\sqrt{n}} \sqrt{\bar{X}}.$$

When $X_1 = \cdots = X_n = 0$, the upper limit (2.6.6) is zero and the Poisson interval (2.5.12) therefore does not cover λ. How frequently this event occurs depends on both λ and n. In fact,

$$P(X_1 = \cdots = X_n = 0) = e^{-n\lambda}.$$

For any fixed n, this probability tends to 1 as $\lambda \to 0$. It follows that for any fixed n,

$$\inf_{\lambda} P\left[\bar{X} - \frac{u}{\sqrt{n}} \sqrt{\bar{X}} < \lambda < \bar{X} + \frac{u}{\sqrt{n}} \sqrt{\bar{X}} \right] = 0.$$

To get a better understanding of this phenomenon, consider the more general situation of a sequence of events A_n whose probability $P_n(\theta) = P_\theta(A_n)$ depends on n and a parameter θ. In our example, $\theta = \lambda$ and the events are

$$\bar{X} - \frac{u}{\sqrt{n}} \sqrt{\bar{X}} < \lambda < \bar{X} + \frac{u}{\sqrt{n}} \sqrt{\bar{X}}.$$

Suppose that for each fixed θ, $P_n(\theta) \to 1 - \alpha$ or, more generally, that for each fixed θ, $P_n(\theta)$ tends to a limit $P(\theta)$. Still more generally, consider a sequence of functions f_n (not necessarily probabilities) converging *pointwise* to a limit function f, i.e., such that

(2.6.7) $$f_n(x) \to f(x) \text{ for all } x.$$

Then for each x and each $\epsilon > 0$, there exists an integer $n_0(\epsilon, x)$ such that

$$|f_n(x) - f(x)| < \epsilon \text{ if } n > n_0(\epsilon, x).$$

Definition 2.6.1 The sequence $f_n(x)$ is said to converge to $f(x)$ *uniformly* (in x) if for each $\epsilon > 0$, there exists $n_0(\epsilon)$ independent of x such that

(2.6.8) $$|f_n(x) - f(x)| < \epsilon \text{ for all } x \text{ if } n > n_0(\epsilon).$$

Example 2.6.2

(i) Let $f_n(x) = x + 1/n$. Then for each x, $f_n(x) \to f(x) = x$ as $n \to \infty$. Since $|f_n(x) - f(x)| = 1/n$, the convergence of f_n to f is clearly uniform.

(ii) Let $f_n(x) = x/n$. For each x, $f_n(x) \to f(x) = 0$ as $n \to \infty$.

Now

(2.6.9) $$|f_n(x) - f(x)| = x/n < \epsilon$$

if $n > x/\epsilon$. The smallest n for which (2.6.9) holds is the smallest integer $> x/\epsilon$, and this tends to ∞ as $x \to \infty$. There is thus no $n_0(\epsilon)$ for which (2.6.9) holds simultaneously for all x. $\square$

Graphically, uniform convergence means that for any $\epsilon > 0$, $f_n(x)$ will lie entirely in the band $f(x) \pm \epsilon$ for all sufficiently large n. This is illustrated in Figure 2.6.1(a). In contrast, Figure 2.6.1(b) shows the case $f_n(x) = x^n$ ($0 \le x \le 1$), in which the convergence is not uniform (Problem 6.3).

In the light of Definition 2.6.1, the difficulty with the sequences (2.6.5) is seen to be that these sequences converge to 1 for every i but not uniformly in i. Similarly, the coverage probability of the intervals (2.5.12) tends to $1 - \alpha$ pointwise but not uniformly. Confidence intervals are discussed more systematically in Section 4.1.

Convergence may be required not for all x but only for all x in some set X of x-values. Suppose for example that in (ii) of Example 2.6.2, we are only concerned with the set X of x-values between 0 and 1. Clearly, (2.6.8) will hold for all $0 < x < 1$ if $n > 1/\epsilon$. Hence in this case, $f_n \to f$ uniformly on X.

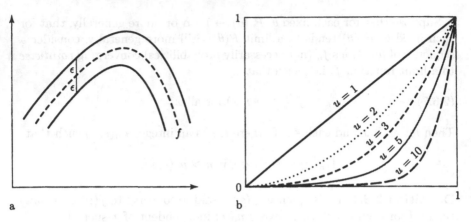

FIGURE 2.6.1. (a) Uniform convergence; (b) non-uniform convergence $f_n(x) = x^n$

Lemma 2.6.1 *The sequence $\{f_n\}$ converges to f uniformly on X if and only if*

(2.6.10) $$M_n = \sup_{x \in X} |f_n(x) - f(x)| \to 0 \text{ as } n \to \infty.$$

Proof. This is clear from the fact that (2.6.10) holds if and only if for each $\epsilon > 0$, there exists $n_0(\epsilon)$ such that

$$\sup |f_n(x) - f(x)| < \epsilon \text{ when } n > n_0(\epsilon),$$

and that this statement is equivalent to (2.6.8).

Another way of characterizing uniform convergence is provided by the next lemma.

Lemma 2.6.2 *The sequence $\{f_n\}$ converges to f uniformly on X if and only if*

(2.6.11) $$f_n(x_n) - f(x_n) \to 0 \text{ as } n \to \infty$$

for every sequence $\{x_n\}$ of points in X.

Proof. If $f_n \to f$ uniformly, (2.6.11) is an immediate consequence of Lemma 2.6.1. For a proof of the converse, see, for example, Knopp (1990, p. 334).

Theorem 2.6.1 Polya. *If a sequence of cumulative distribution functions H_n tends to a continuous cdf H, then $H_n(x)$ converges to $H(x)$ uniformly in x.*

For a proof, see Parzen (1992, p. 438).

This result is illustrated by Theorem 2.4.2, where (2.4.10) shows that the convergence of $G_n(x)$ to $\Phi(x)$ is uniform in x and that the difference

$[G_n(x) - \Phi(x)]$ divided by the standardized third moment $E|X_1 - \xi|^3/\sigma^3$ tends to zero uniformly in F on the family of distributions with finite third moment.

Still another example of uniform convergence can be found in the Note following Theorem 2.4.3.

Summary

1. A sequence of functions f_n defined over a space X converges to a limit function f *pointwise* if $f_n(x) \to f(x)$ for all $x \in X$; it converges to f *uniformly* if the maximum (i.e., sup) difference between f_n and f tends to 0. This stronger mode of convergence is equivalent to the requirement that

$$f_n(x_n) - f(x_n) \to 0 \text{ for every sequence of points } x_n \in X.$$

2. Polya's theorem states that if a sequence of cdf's H_n tends to a continuous cdf H, then it does so uniformly.

3. The distinction between pointwise and uniform convergence implies a corresponding distinction for the coverage probability $P_\theta\left(\underline{\theta}_n < \theta < \bar{\theta}_n\right)$ of large-sample confidence intervals.

2.7   The CLT for independent non-identical random variables

The classical CLT refers to the case that the X's are i.i.d. The following result provides an extension to sums of independent random variables that are not necessarily identically distributed.

Theorem 2.7.1 Liapounov. *Let X_i $(i = 1, \ldots, n)$ be independently distributed with means $E(X_i) = \xi_i$ and variances σ_i^2, and with finite third moments. If*

$$(2.7.1) \qquad Y_n = \frac{\bar{X} - E(\bar{X})}{\sqrt{\text{Var}(\bar{X})}} = \frac{\sqrt{n}(\bar{X} - \bar{\xi})}{\sqrt{(\sigma_1^2 + \cdots + \sigma_n^2)/n}},$$

then

$$(2.7.2) \qquad Y_n \overset{L}{\to} N(0,1),$$

provided

$$(2.7.3) \qquad \left[E\left(\sum|X_i - \xi_i|^3\right)\right]^2 = o\left[\left(\sum\sigma_i^2\right)^3\right].$$

This is proved, for example, in Feller (Vol. 1) (1957) and Feller (Vol. 2) (1966).

Note: To see that (2.7.2) requires some condition, consider the case that each of the variables $X_2, \ldots, X_n$ is a constant, say 0. Then the left side of (2.7.2) reduces to $(X_1 - \xi_1)/\sigma_1$ which has mean 0 and variance 1 but otherwise can have any distribution whatever.

Condition (2.7.3) holds in particular when the X's are i.i.d. with a finite third moment since then the left side is of order n^2 and the right side of order n^3. However, it does not cover Theorem 2.4.1, which has no third moment requirement. A more general condition which is essentially both necessary and sufficient is given in Section A.1 of the Appendix.

Corollary 2.7.1 *Let X_i $(i = 1, \ldots, n)$ be independently distributed with means ξ_i and variances σ_i^2 and suppose that the X's are uniformly bounded; i.e., there exists a constant A such that*

(2.7.4) $$|X_i| \leq A \quad \text{for all } i.$$

Then (2.7.2) holds, provided

(2.7.5) $$s_n^2 = \sum_{i=1}^{n} \sigma_i^2 \to \infty.$$

Proof. We have

$$\sum |X_i - \xi_i|^3 \leq 2A \sum (X_i - \xi_i)^2$$

and hence

$$E \sum |X_i - \xi_i|^3 \leq 2A s_n^2.$$

The left side of (2.7.3) is therefore $\leq 4A^2 s_n^4$, which is $o\left[(s_n^2)^3\right]$ when $s_n^2 \to \infty$. This proves (2.7.3) and hence (2.7.2). ∎

Example 2.7.1 Poisson binomial. Consider a sequence of so-called Poisson-binomial trials (i.e., binomial trials with varying p) with success probabilities $p_1, p_2, \ldots$. Let $X_i = 1$ or 0 as the ith trial is a success or failure and let $X = \sum X_i$, so that $X/n = \bar{X}$. Then the X's are uniformly bounded and hence

(2.7.6) $$\frac{\sqrt{n}\left(\dfrac{X}{n} - \bar{p}\right)}{\sqrt{\sum p_i q_i / n}} \xrightarrow{L} N(0, 1),$$

provided

(2.7.7)
$$s_n^2 = \sum_{i=1}^{n} p_i q_i \to \infty \text{ as } n \to \infty.$$

Condition (2.7.7) will hold if the p's are bounded away from 0 and 1, that is, if there exists a constant $a > 0$ such that

$$a < p_i < 1 - a \text{ for all } i.$$

Since then both p_i and q_i are $> a$, we have $p_i q_i > a^2$ and hence $s_n^2 > na^2 \to \infty$. However, (2.7.7) can also hold when $p_i \to 0$, provided it tends to 0 sufficiently slowly. Suppose, for example, $p_i = 1/i$. Then

$$\sum p_i q_i = \sum \frac{1}{i} - \sum \frac{1}{i^2}$$

and it follows from Examples 1.3.1 and 1.3.2 that $\sum_{i=1}^{n} \frac{1}{i} \to \infty$ and $\sum_{i=1}^{n} \frac{1}{i^2}$ tends to a finite limit, and hence that (2.7.7) holds. On the other hand, the same example shows that (2.7.7) does not hold when $p_i = \frac{1}{i^k}$ with $k > 1$.□

The limit (2.7.6) generalizes the normal limit (2.3.10) from binomial to Poisson binomial trials. Corresponding generalizations are also available for the Poisson limit (1.2.8) of the binomial in the case of rare events. We shall consider such a result at the end of the section.

Example 2.7.2 Binomial sampling: Wilcoxon statistic. Let K_i be a sequence of independent random variables taking on the values 1 and 0 with probabilities p and $q = 1 - p$ respectively, and let $X_i = a_i K_i$ and $S_n = \sum_{i=1}^{n} X_i$. This corresponds to the process of binomial sampling where the ith element of a population of size n has the "value" a_i attached to it and where a sample is drawn by including or excluding each of the n elements with probability p or q, respectively. Thus S_n is the sum of the a-values of the elements included in the sample. For the case $a_i = i$ and $p = 1/2$, S_n (with change of notation) is the signed-rank Wilcoxon statistic (3.2.27).

In the present case,

$$\xi_i = a_i p, \sigma_i^2 = a_i^2 pq, \text{ and } E|X_i - \xi_i|^3 = a_i^3 pq \left(p^2 + q^2\right).$$

Thus condition (2.7.3) reduces to

$$\left(\sum a_i^3\right)^2 = o\left[\left(\sum a_i^2\right)^3\right].$$

When $a_i = i$, it follows from Example 1.1.3 that

$$\sum a_i^2 \sim cn^3 \text{ and } \sum a_i^3 \sim c'n^4,$$

and condition (2.7.3) is seen to be satisfied. In particular, this proves asymptotic normality of the one-sample Wilcoxon statistic.   □

Note: Condition (2.7.3), although *sufficient* for many applications, is not *necessary* for (2.7.2). For weaker conditions, see Section A.1. A Berry–Esseen type version of Theorem 2.7.1 is given in Petrov (1995, p. 111) and Stroock (1993, p. 69).

The versions of the CLT considered so far have been associated with the following two situations:

(A) $X_1, X_2, \ldots$ are independent according to a common distribution F (Theorem 2.4.1).

(B) $X_1, X_2, \ldots$ are independent according to distributions $F_1, F_2, \ldots$ (Theorem 2.7.1).

In Corollary 2.4.1, we had to deal with a generalization of (A), namely that for each fixed n, the variables $X_1^{(n)}, \ldots, X_n^{(n)}$ are i.i.d. but that their common distribution F_n may depend on n. With a different notation, a generalization of this model is described by a double array of random variables:

(A′)
$$
\begin{array}{ll}
X_{11} & \text{with distribution } F_1 \\
X_{21}, X_{22} & \text{independent, each with distribution } F_2 \\
\ldots & \\
X_{n1}, X_{n2}, \ldots, X_{nn} & \text{independent, each with distribution } F_n.
\end{array}
$$

A CLT for (A′) can be obtained from Corollary 2.4.1 as a consequence of the Berry–Esseen theorem. A more general result will be given in Theorem 2.7.2.

Even models (B) and (A′) are not general enough to cover many important situations.

Example 2.7.3 Common mean. In Example 2.2.2, we considered the consistency of the estimator $\delta = \sum w_i X_i$ with

(2.7.8)
$$w_i = \frac{1/\sigma_i^2}{\displaystyle\sum_{j=1}^{n} 1/\sigma_j^2}$$

of the common mean ξ of n variables X_i independently distributed with variance σ_i^2. Consistency was proved for the case that

(2.7.9)
$$\sum_{j=1}^{n} 1/\sigma_j^2 \to \infty \text{ as } n \to \infty.$$

When the estimator is consistent, one would like to know its asymptotic distribution. Since $\delta = \sum Y_i$ with $Y_i = w_i X_i$, it seems at first glance that one is dealing with a situation of type (B) so that Theorem 2.7.1 is applicable. Note, however, that the notation is misleading. The weights w_i depend not only on i but also on n and should have been denoted by w_{ni}. □

Such situations are covered by the following generalization (B′) of (A), (B) and (A′), called a triangular array:

(B′)
$$\begin{array}{ll} X_{11} & \text{with distribution } F_{11} \\ X_{21}, X_{22} & \text{independent, with distribution } F_{21}, F_{22} \\ \ldots & \\ X_{n1}, \ldots, X_{nn} & \text{independent, with distributions } F_{n1}, \ldots, F_{nn}. \end{array}$$

Example 2.7.4 Simple linear regression. As in Example 2.2.3 let

(2.7.10) $$X_i = \alpha + \beta v_i + E_i,$$

where the E's are i.i.d. with mean 0, and the v's are given constants. The standard (least squares) estimators of β and α are

(2.7.11) $$\hat{\beta} = \frac{\displaystyle\sum_{i=1}^{n}\left(X_i - \bar{X}\right)\left(v_i - \bar{v}\right)}{\displaystyle\sum_{j=1}^{n}(v_j - \bar{v})^2} = \sum_{i=1}^{n}\frac{v_i - \bar{v}}{\sum(v_j - \bar{v})^2}X_i,$$

$$\hat{\alpha} = \bar{X} - \hat{\beta}\bar{v}.$$

Thus $\hat{\alpha}$ and $\hat{\beta}$ are again linear functions of the X's in which the coefficients depend on both i and n. □

The following result extends Theorem 2.7.1 from model (B) to model (B′).

Theorem 2.7.2 (Liapounov) *Let the X_{ij} be distributed as in (B′) and let*

(2.7.12) $E(X_{ij}) = \xi_{ij}$, $\mathrm{Var}(X_{ij}) = \sigma_{ij}^2 < \infty$, *and* $s_n^2 = \sigma_{n1}^2 + \cdots + \sigma_{nn}^2$.

Then

(2.7.13) $$\sum_{j=1}^{n}\frac{(X_{nj} - \xi_{nj})}{s_n} = \frac{\bar{X}_n - \bar{\xi}_n}{\sqrt{\mathrm{Var}\,\bar{X}_n}} \overset{L}{\to} N(0,1) \quad \text{as } n \to \infty,$$

provided

(2.7.14) $$\left[\sum_{j=1}^{n}E\,|X_{nj} - \xi_{nj}|^3\right]^2 = o\left[(s_n^2)^3\right].$$

Note: In (2.7.14), the exponent 3 can be replaced by $2 + \delta$ for any $\delta > 0$. This is proved, for example, in Billingsley (1986) and in Dudley (1989).

As an important application, we shall now give conditions under which a linear function of i.i.d. random variables is asymptotically normal. In particular, this will cover both Examples 2.7.3 and 2.7.4.

Theorem 2.7.3 *Let* $Y_1, Y_2, \ldots$ *be i.i.d. with* $E(Y_i) = 0$, *Var*$(Y_i) = \sigma^2 > 0$, *and* $E|Y_i^3| = \gamma < \infty$. *Then*

(2.7.15)
$$\frac{\sum_{i=1}^{n} d_{ni} Y_i}{\sigma \sqrt{\sum_{i=1}^{n} d_{ni}^2}} \xrightarrow{L} N(0,1),$$

provided

(2.7.16)
$$\left(\sum_{i=1}^{n} |d_{ni}|^3 \right)^2 = o\left[\left(\sum_{i=1}^{n} d_{ni}^2 \right)^3 \right].$$

Proof. With $X_{ni} = d_{ni} Y_i$, the left side of (2.7.15) reduces to (2.7.13). Also, $\xi_{ni} = 0$,

$$\sigma_{ni}^2 = \sigma^2 d_{ni}^2 \quad \text{and hence} \quad s_n^2 = \sigma^2 \sum_{i=1}^{n} d_{ni}^2,$$

and

$$E|X_{ni} - \xi_{ni}|^3 = |d_{ni}|^3 \gamma.$$

Condition (2.7.14) thus becomes (2.7.16).   ∎

Condition (2.7.16), in turn, is equivalent to another which typically is simpler to apply and to interpret.

Theorem 2.7.4 *The sufficient condition (2.7.16) is equivalent to*

(2.7.17)
$$\max_{i=1,\ldots,n} (d_{ni}^2) = o\left(\sum d_{ni}^2 \right).$$

Proof. If (2.7.16) holds, then

$$\left(\max (d_{ni})^2 \right)^3 = \left(\max |d_{ni}|^3 \right)^2 \le \left(\sum |d_{ni}|^3 \right)^2 = o\left[\left(\sum d_{ni}^2 \right)^3 \right],$$

which implies (2.7.17). Conversely, if (2.7.17) holds, then

$$\sum |d_{ni}^3| \leq \max |d_{ni}| \sum d_{ni}^2 = o\left(\sqrt{\sum d_{ni}^2} \sum d_{ni}^2\right) = o\left[\left(\sum d_{ni}^2\right)^{3/2}\right],$$

which implies (2.7.16). ∎

The meaning of condition (2.7.17) is clear. It requires the individual weights d_{ni}^2 to become negligible compared to their sum, and thereby prevents any one of the variables $d_{ni}Y_i$ to have an undue influence and thereby the kind of possibility mentioned in the note following Theorem 2.7.1.

As illustrations of Theorem 2.7.3 with condition (2.7.17), we shall now determine conditions for asymptotic normality for the estimators of Examples 2.7.3 and 2.7.4.

Example 2.7.5 Common mean (continued). Consider the asymptotic normality of the estimator δ of Example 2.7.3. Suppose that $E|X_i|^3 < \infty$ for all i and let

$$(2.7.18) \qquad Y_i = (X_i - \xi)/\sigma_i.$$

Then the assumptions of Theorem 2.7.3 are satisfied with $\sigma^2 = 1$ and we have

$$(2.7.19) \qquad \delta - \xi = \sum w_i(X_i - \xi) = \sum d_{ni}Y_i$$

with

$$(2.7.20) \qquad d_{ni} = \frac{1/\sigma_i}{\sum\limits_{j=1}^{n} 1/\sigma_j^2}$$

and hence

$$\frac{\sum d_{ni}Y_i}{\sqrt{\sum d_{ni}^2}} = (\delta - \xi)/\sqrt{\sum 1/\sigma_j^2}.$$

It follows that

$$(2.7.21) \qquad (\delta - \xi)/\sqrt{\sum 1/\sigma_j^2} \xrightarrow{L} N(0,1),$$

provided (2.7.17) holds, which in the present case reduces to

$$(2.7.22) \qquad \max_{i=1,\dots,n} (1/\sigma_i^2) = o\left(\sum \frac{1}{\sigma_j^2}\right).$$

This condition is satisfied, for example, when the σ_j^2 are equally spaced, say $\sigma_j^2 = j\Delta$ (Problem 7.2). □

Example 2.7.6 Simple linear regression (continued). In the model of Example 2.7.4, suppose that $E(E_i) = 0, \mathrm{Var}(E_i) = \sigma^2$, and $E(|E_i|^3) < \infty$. By (2.7.11),

$$\hat{\beta} - \beta = \sum d_{ni} (X_i - \xi_i),$$

where $\xi_i = E(X_i)$ and $d_{ni} = (v_i - \bar{v})/\sum_{j=1}^{n} (v_j - \bar{v})^2$. It follows that

$$(2.7.23) \qquad \left(\hat{\beta} - \beta\right) \sqrt{\sum (v_i - \bar{v})^2}/\sigma \xrightarrow{L} N(0, 1),$$

provided

$$(2.7.24) \qquad \max_{i=1,\ldots,n} (v_i - \bar{v})^2 = o\left(\sum (v_j - \bar{v})^2\right).$$

This same condition also ensures the asymptotic normality of $(\hat{\alpha} - \alpha)$ (Problem 7.11). □

Example 2.7.7 ‡Poisson binomial trials with rare events. In Section 1.2, we found that the Poisson distribution is the limit as $n \to \infty$ of the binomial distribution $b(p, n)$ for rare events, more specifically when $p = \lambda/n$. To generalize this result to the case of varying p, consider a triangular array of Poisson binomial trials with success probabilities

$$p_{11}; p_{21}, p_{22}; \cdots ; p_{n1}, \cdots, p_{nn}.$$

In generalization of the assumption $np = \lambda$, we shall assume that

$$(2.7.25) \qquad \sum_{i=1}^{n} p_{ni} \to \lambda, \quad 0 < \lambda < \infty.$$

Although (2.7.25) implies that the p's are small on the average when n is large, it does not completely capture the idea of "rare events" in that individual p's can remain large. To avoid this possibility, we add to (2.7.25) the condition

$$(2.7.26) \qquad \max(p_{n1}, \ldots, p_{nn}) \to 0 \quad \text{as} \quad n \to \infty.$$

A different argument leads to another condition. If the distribution of X is to tend to the Poisson distribution $P(\lambda)$ which has variance λ, we would expect that

$$\mathrm{Var}(X) = \sum_{i=1}^{n} p_{ni} q_{ni} = \sum_{i=1}^{n} p_{ni} - \sum_{i=1}^{n} p_{ni}^2 \to \lambda.$$

---

‡This somewhat more difficult example is not required in the remainder of the book.

In view of (2.7.25), this would require

$$(2.7.27) \qquad \sum_{i=1}^{n} p_{ni}^2 \to 0.$$

Actually, it is easy to see that when (2.7.25) holds, conditions (2.7.26) and (2.7.27) are equivalent. That (2.7.27) implies (2.7.26) is obvious. Conversely, the inequality

$$\sum_{i=1}^{n} p_{ni}^2 \le \max\left(p_{n1}, \cdots, p_{nn}\right) \sum_{j=1}^{n} p_{nj}$$

shows that (2.7.25) and (2.7.26) together imply (7.27).

Theorem 2.7.5 *If X is the number of successes in n Poisson binomial trials satisfying (2.7.25) and (2.7.27), then the distribution of X has the Poisson limit $P(\lambda)$.*

Proof. The result will be proved[§] by constructing suitable joint distributions for variables (X_i, Y_i) such that

$$(2.7.28) \qquad P\left(X_i = 1\right) = p_i, \ P\left(X_i = 0\right) = q_i$$

and that the Y_i are independent with Poisson distribution $P(p_i)$. Then $Y = \sum_{i=1}^{n} Y_i$ has the Poisson distribution $P\left(\sum_{i=1}^{n} p_i\right)$ and hence by (2.7.25) converges in law to the Poisson distribution $P(\lambda)$. If $X = \sum X_i$, we shall show that

$$(2.7.29) \qquad P\left(X = Y\right) \to 1 \ \text{as} \ n \to \infty,$$

and hence that X also tends in law to $P(\lambda)$.

Let $P\left(X_i = a, Y_i = b\right) = p_{a,b}^{(i)}$ for $a = 0, 1$ and $b = 0, 1, 2, \ldots$ and let

$$p_{00}^{(i)} = e^{-p_i} - p_i\left(1 - e^{-p_i}\right), \ p_{01}^{(i)} = 0, \ p_{0y}^{(i)} = p_i^y e^{-p_i}/y! \ \text{for} \ y \ge 2,$$

$$p_{10}^{(i)} = p_i\left(1 - e^{-p_i}\right), \ p_{11}^{(i)} = p_i e^{-p_i}, \ p_{1y}^{(i)} = 0 \ \text{for} \ y \ge 2,$$

with the Y's being independent of each other, and with $p_i \le .8$ so that $p_{00}^{(i)} \ge 0$. Then it is easily verified (Problem 7.9) that (2.7.28) holds and that the distribution of Y_i is $P(p_i)$. To show (2.7.29), note that (Problem 7.10) $e^{-p_i} \ge 1 - p_i$ and hence that

$$P\left(X_i \ne Y_i\right) = 1 + p_i - (1 + 2p_i) e^{-p_i} \le 1 + p_i - (1 + 2p_i)(1 - p_i) = 2p_i^2.$$

[§]This proof is due to Hodges and Le Cam (1960). It is an early example of the coupling method. For a detailed treatment of this method, see, for example, Lindvall (1992).

It follows that

$$P(X \neq Y) \leq P(X_i \neq Y_i \text{ for some } i) \leq \sum P(X_i \neq Y_i) \leq 2 \sum p_i^2$$

and hence by (2.7.27) that

$$P(X \neq Y) \to 0 \text{ as } n \to \infty.$$

This concludes the proof.   ∎

For additional results concerning this limit process, see Arratia et al. (1990) and Barbour et al. (1992), Chapter 1, which is also a good source for Poisson approximations in the case of dependent trials.   □

Summary

1. The central limit theorem is generalized from the i.i.d. case in two directions. Of the resulting situations, (A) and (B) can be represented by a single sequence of random variables; on the other hand, (A') and (B'), because of the dependence on n, require a triangular array.

2. A central limit theorem for (B') is Lyapounov's Theorem 2.7.2 (and its special case, Theorem 2.7.1, for (B)) which assumes the existence of third moments and states that the standardized average of the nth row of the array tends in law to the standard normal distribution, provided the sum of the third absolute central moments is not too large.

3. An application of Lyapounov's theorem leads to sufficient conditions for the asymptotic normality of weighted sums of i.i.d. random variables. Slightly weaker conditions can be obtained from the Lindeberg version of the CLT, which is discussed in Section A.1 of the Appendix.

4. The Lyapounov and Lindeberg extensions of the classical CLT provide conditions under which the standardized number of successes in a sequence of Poisson binomial trials has a normal limit distribution. Under different conditions corresponding to the case of "rare events," the number of successes has a Poisson limit.

2.8 Central limit theorem for dependent variables

In the preceding section, we considered the asymptotic behavior of the average $\bar{X}$ when $X_1, X_2, \ldots$ are a sequence of independent but not identically distributed random variables. The assumption of independence tends to be made rather casually, even though often it is not appropriate (see, for example, Kruskal (1988) and Beran (1992)). One reason for this neglect may be the difficulty of specifying appropriate models incorporating dependence. Most of the work involving dependent sequences is beyond the scope of this book (the appropriate literature is that dealing with stochastic processes and time series). We shall in this and the next section mainly restrict attention to the consideration of a few important examples.

If $X_1, \ldots, X_n$ are n dependent variables with $E(X_i) = \theta$ and $\text{Var}(X_i) = \sigma^2$, it will frequently continue to be true that $\sqrt{n}(\bar{X} - \theta)$ tends to a normal distribution with mean 0 and variance

$$(2.8.1) \qquad \tau^2 = \lim \text{Var}\left(\sqrt{n}\left(\bar{X} - \theta\right)\right)$$

so that

$$(2.8.2) \qquad \frac{\sqrt{n}\left(\bar{X} - \theta\right)}{\tau} \xrightarrow{L} N\left(0, 1\right).$$

Now

$$(2.8.3) \qquad \text{Var}\left[\sqrt{n}\left(\bar{X} - \theta\right)\right] = \sigma^2 + \frac{1}{n}\sum\sum_{i \neq j}\text{Cov}\left(X_i, X_j\right),$$

which will tend to a finite limit provided the second term tends to a finite limit γ,

$$(2.8.4) \qquad \frac{1}{n}\sum\sum_{i \neq j}\text{Cov}\left(X_i, X_j\right) \to \gamma.$$

Note that the sum in (2.8.4) contains $n(n - 1) = O(n^2)$ terms. A simple way in which (2.8.4) can occur is that $\text{Cov}(X_i, X_j) = 0$ for all but $O(n)$ number of pairs (i, j). Alternatively, (2.8.4) may hold if $\text{Cov}(X_i, X_j)$ tends to 0 sufficiently fast as $|j - i| \to \infty$. If (2.8.2) and (2.8.4) hold, the asymptotic distribution of $\sqrt{n}\left(\bar{X} - \theta\right)$ will differ from what it is in the independent case only, in that the asymptotic variance is no longer σ^2 but $\tau^2 = \sigma^2 + \gamma$, which may be either less than or greater than σ^2.

Of particular interest is the case in which the sequence $X_1, X_2, \ldots$ is *stationary* (Definition 2.2.1). The variance (2.8.3) then reduces to

$$(2.8.5) \qquad \text{Var}\left[\sqrt{n}\left(\bar{X} - \theta\right)\right] = \sigma^2 + \frac{2}{n}\sum_{k=1}^{n-1}(n - k)\gamma_k$$

where $\gamma_k = \text{Cov}\left(X_i, X_{i+k}\right)$, and (2.8.4) becomes

$$\frac{2}{n}\sum_{k=1}^{n-1}(n - k)\gamma_k \to \gamma.$$

The limit γ will be finite, for example, when γ_k tends to zero at the rate a^k with $0 < a < 1$, that is, when the covariances γ_k decay at an exponential rate (Problem 8.1).

Example 2.8.1 m-dependence. Recall from Example 2.2.7 that the sequence $X_1, X_2, \ldots$ is m-dependent if $(X_1, \ldots, X_i)$ and $(X_j, X_{j+1}, \ldots, X_n)$ are independent whenever $j - i > m$. An important class of examples of m-dependent sequences is obtained from a sequence of independent variables $Z_1, Z_2, \ldots$ by letting

$$(2.8.6) \qquad X_i = \phi(Z_i, \ldots, Z_{i+m}).$$

If the sequence of Z's is i.i.d., the X's are not only m-dependent but also stationary (Problem 8.2).

As pointed out in Example 2.2.7, a special case of m-dependence is a sequence of *moving averages*, given by

$$(2.8.7) \qquad \phi(Z_i, \ldots, Z_{i+m}) = (Z_i + \cdots + Z_{i+m}) / (m+1).$$

In the simplest case, $m = 1$ and $X_i = (Z_i + Z_{i+1})/2$, and hence

$$(2.8.8) \qquad \sum_{i=1}^{n} X_i = \frac{1}{2} Z_1 + (Z_2 + \cdots + Z_n) + \frac{1}{2} Z_{n+1}.$$

If $E(Z_i) = \theta$, then $\sqrt{n}(\bar{X} - \theta)$ clearly has the same limit distribution as $\sqrt{n}(\bar{Z} - \theta)$, which is $N(0, \sigma^2)$ with $\sigma^2 = \mathrm{Var}(Z_i)$. Thus (2.8.2) holds with $\tau^2 = \sigma^2$ and this argument easily extends to general m (Problem 8.3).

In an m-dependent sequence, each X_i can be dependent on at most $2m$ of the variables X_j with $j \neq i$, so that the total number of pairs (X_i, X_j) with $\mathrm{Cov}(X_i, X_j) \neq 0$ cannot exceed $2m \cdot n = O(n)$. An m-dependent sequence therefore provides an example of the first possibility mentioned after (2.8.4).

The central limit theorem extends to the general case of a stationary m-dependent sequence. □

Theorem 2.8.1 *Let $X_1, X_2, \ldots$ be a stationary m-dependent sequence of random variables. Then (2.8.2) will hold with*

$$(2.8.9) \qquad \tau^2 = \sigma^2 + 2 \sum_{i=2}^{m+1} \mathrm{Cov}(X_1, X_i)$$

provided $0 < \sigma^2 = \mathrm{Var}(X_i) < \infty$.

Proof. Without loss of generality, let $E(X_i) = 0$. To prove the result divide the sequence of X's into alternate blocks of lengths N and m, where m is fixed but N will tend to infinity. Assume for the moment that n is a multiple of $N + m$, say $n = r(N + m)$. Then

$$(2.8.10) \qquad \sum_{i=1}^{n} X_i = \sum_{i=1}^{r} A_i + \sum_{i=1}^{r} B_i$$

with

$$A_1 = X_1 + \cdots + X_N, B_1 = X_{N+1} + \cdots + X_{N+m},$$
$$A_2 = X_{N+m+1} + \cdots + X_{2N+m,\ldots} \ .$$

The crucial aspect of this decomposition of $\sum X_i$ is that the A's are independent since the intervening B's separate them by m X's.

We have

(2.8.11) $$\sqrt{n}\bar{X} = \frac{1}{\sqrt{r(N+m)}} \sum_{i=1}^{r} A_i + \frac{1}{\sqrt{r(N+m)}} \sum_{i=1}^{r} B_i.$$

Since the A's are i.i.d., it follows from the central limit theorem that for any fixed N, the first term tends to $N\left(0, \mathrm{Var}\left[A_1/\sqrt{N+m}\right]\right)$ as $r \to \infty$. On the other hand,

(2.8.12)

$$E\left[\frac{1}{\sqrt{r(N+m)}} \sum_{i=1}^{r} B_i\right]^2 = \frac{1}{r(N+m)} \mathrm{Var}\left(\sum_{i=1}^{r} B_i\right) = \frac{1}{N+m} \mathrm{Var}\left(B_1\right)$$
$$= \frac{1}{N+m} E\left(X_1 + \cdots + X_m\right)^2 \leq \frac{m^2}{N+m}\sigma^2.$$

Since the right side tends to zero as $N \to \infty$, this appears to complete the proof of asymptotic normality by Theorem 2.3.3.

However, this argument overlooks the fact that the convergence of the first term of (2.8.11) was asserted for N fixed, while convergence of the second term requires N to tend to infinity. For the closing of this gap see, for example, Anderson (1971, Section 7.7) or Ferguson (1996, Section 11).

So far, we have restricted n to multiples of $N + m$. If, instead, $n = r(N+m) + k$ with $k < N+m$, then $X_1 + \cdots + X_n$ is given by the right side of (2.8.10) plus a remainder term consisting of fewer than $N+m$ X's. This remainder term tends in probability to 0 uniformly in r as $N \to \infty$ (Problem 8.7). ∎

We shall next consider an example in which all the covariances $\mathrm{Cov}\left(X_i, X_j\right)$ are different from zero.

Example 2.8.2 Stationary first order autoregressive process. Let the X's be defined successively by

(2.8.13) $$X_{i+1} = \theta + \beta\left(X_i - \theta\right) + U_{i+1}, \ |\beta| < 1, i = 1, 2, \ldots$$

with the U_i independent $N(0, \sigma_0^2)$. From (2.8.13), we see that

(2.8.14) $$X_{i+k} - \theta = \beta^k\left(X_i - \theta\right) + \beta^{k-1}U_{i+1} + \cdots + \beta U_{i+k-1} + U_{i+k}.$$

The variances of the X's by (2.8.13) satisfy

(2.8.15) $$\mathrm{Var}\left(X_{i+1}\right) = \beta^2\mathrm{Var}\left(X_i\right) + \sigma_0^2.$$

If the sequence of the X's is to be stationary, we must have

$$\text{Var}\,(X_{i+1}) = \text{Var}\,(X_i) \text{ for all } i = 1, 2, \dots \,,$$

and hence

$$\sigma^2 = \sigma_0^2 / \left(1 - \beta^2\right),$$

where σ^2 denotes the common variance of the X's. Suppose that the sequence is started with an X_1 which is $N(\theta, \sigma^2)$. Then this is the distribution of all the X_i, and it follows from the next paragraph that the sequence is stationary.

Establishing (2.8.2) requires the proof of two facts: asymptotic normality and that the variance of the limit distribution agrees with the limit (2.8.1) of the variance. In the present case, the joint distribution of the X's is multivariate normal and hence $\sqrt{n}\,(\bar{X} - \theta)$ is normal for all n (see Chapter 5, Sec. 4), say $N\left(0, \sigma_n^2\right)$. If $\sigma_n \to \tau$, it then follows from Example 2.3.4 that (2.8.2) holds. Since by (2.8.14)

$$(2.8.16) \qquad\qquad \gamma_j = \text{Cov}\,(X_1, X_j) = \beta^{j-1}\sigma^2,$$

it is seen that (Problem 8.5)

$$(2.8.17) \qquad\qquad \text{Var}\left[\sqrt{n}\,(\bar{X} - \theta)\right] \to \sigma^2 \frac{1+\beta}{1-\beta}$$

which is therefore the variance of the limiting distribution. $\square$

The following definition illustrates one way in which one can characterize dependence, the influence of which becomes small when the variables are sufficiently far apart.

Definition 2.8.1 (α-mixing.) The sequence $X_1, X_2, \dots$ is said to be α-mixing if

$$(2.8.18) \qquad \begin{aligned} &|P\left[(X_1, \dots, X_i) \in A \text{ and } (X_{i+m}, X_{i+m+1,\dots}) \in B\right] \\ &-P\left[(X_1, \dots, X_i) \in A\right] P\left[(X_{i+m}, X_{i+m+1,\dots}) \in B\right]| \leq \alpha_m \end{aligned}$$

for all i, m, A, and B, with

$$(2.8.19) \qquad\qquad \alpha_m \to 0 \text{ as } m \to \infty.$$

It can be shown that a stationary sequence $X_1, X_2, \dots$ which is α-mixing satisfies (2.8.2) if $\alpha_m \to 0$ sufficiently fast and under some mild additional assumptions. This result is proved, for example, in Billingsley (1986) under the assumptions that $\alpha_m = O\left(m^{-5}\right)$ and $E|X_i^{12}| < \infty$.

Example 2.8.3 Two-state Markov chains. Construct a sequence of random variables $X_1, X_2, \ldots$ each taking on the values 1 and 0, as follows. The starting variable X_1 takes on the values 1 and 0 with probabilities

$$(2.8.20) \qquad P(X_1 = 1) = p_1 \text{ and } P(X_1 = 0) = 1 - p_1.$$

Given the ith variable X_i, the probability that the next variable X_{i+1} is 1 is

$$(2.8.21) \qquad P(X_{i+1} = 1 | X_i = 0) = \pi_0, \ P(X_{i+1} = 1 | X_i = 1) = \pi_1.$$

(Note that these *transition probabilities* are assumed to be independent of i.) Such a sequence is a *Markov chain*, i.e., the probability of success on the $(i+1)$st trial, given the outcomes of all the previous trials depends only on the outcome of the ith trial.

If $p_k = P(X_k = 1)$, it follows from (2.8.21) that

$$(2.8.22) \qquad p_{k+1} = p_k (\pi_1 - \pi_0) + \pi_0.$$

In general, $p_{k+1} \neq p_k$, and the sequence will therefore not be stationary. A necessary condition for stationarity is $p_{k+1} = p_k$ for all k, which implies

$$(2.8.23) \qquad p_k = \frac{\pi_0}{1 - \pi_1 + \pi_0}$$

and hence in particular

$$(2.8.24) \qquad p_1 = \frac{\pi_0}{1 - \pi_1 + \pi_0} = p, \text{ say.}$$

It is easy to see that conditions (2.8.20) and (2.8.23) are not only necessary but also sufficient for stationarity since any segment $(X_{i+1}, \ldots, X_{i+k})$ then has the same starting probability (2.8.23) and transition probabilities (2.8.21) as the segment $(X_1, \ldots, X_k)$. Since $E(\bar{X}) = E(X_1) = p$, the statistic $\bar{X}$ in the stationary case is an unbiased estimate of p. It will therefore be a consistent estimator of p, provided $\text{Var}(\bar{X}) \to 0$ as $n \to \infty$.

To evaluate $\text{Var}(\bar{X})$, let us determine the covariance

$$\gamma_k = \text{Cov}(X_i, X_{i+k})$$

which because of stationarity depends only on k. We can establish a simple relationship between γ_{k-1} and γ_k as follows. Using the fact that

$$E(X_i X_{i+k}) = \gamma_k + p^2,$$

we find

$$\begin{aligned}
E(X_i X_{i+k}) &= P(X_i = X_{i+k} = 1) \\
&= P(X_i = X_{i+k-1} = X_{i+k} = 1) + P(X_i = 1, X_{i+k-1} = 0, X_{i+k} = 1) \\
&= \pi_1 \left(\gamma_{k-1} + p^2 \right) + \pi_0 \left(p - \gamma_{k-1} - p^2 \right) \\
&= \pi_1 p^2 + \pi_0 pq + (\pi_1 - \pi_0) \gamma_{k-1}.
\end{aligned}$$

Now (Problem 8.10)

$$(2.8.25) \qquad \pi_1 p^2 + \pi_0 pq = p^2$$

so that finally

$$(2.8.26) \qquad \gamma_k = (\pi_1 - \pi_0)\, \gamma_{k-1}.$$

Since

$$\gamma_1 = E\left(X_i X_{i+1}\right) - p^2 = p\left(\pi_1 - p\right),$$

this leads to

$$(2.8.27) \qquad \gamma_k = p\left(\pi_1 - p\right)\left(\pi_1 - \pi_0\right)^{k-1}.$$

An easy calculation shows that (Problem 8.10)

$$(2.8.28) \qquad \left(\pi_1 - p\right) = q\left(\pi_1 - \pi_0\right),$$

which gives the final result

$$(2.8.29) \qquad \gamma_k = \left(\pi_1 - \pi_0\right)^k pq.$$

It follows from the last statement preceding Example 2.8.1 that Var $\bar{X} \to 0$ and hence that $\bar{X}$ is a consistent estimate of p. It further can be shown that (Problem 8.11)

$$(2.8.30) \qquad \mathrm{Var}\left[\sqrt{n}\left(\bar{X} - p\right)\right] \to pq\frac{1 + \pi_1 - \pi_0}{1 - \pi_1 + \pi_0}.$$

That the limit result (2.8.2) holds in the present situation can be proved by using the fact that the X's are α-mixing with α_m tending to 0 exponentially (see Billingsley (1986)) and in other ways. $\qquad \square$

Examples 2.8.1–2.8.3 illustrate situations in which the dependence becomes weak as the variables move further apart. This is obviously so for m-dependent variables, where the variables in fact become independent once their distance exceeds m; in Examples 2.8.2 and 2.8.3, it is reflected by the fact that the covariances and correlations decrease to 0 exponentially as the distance between the variables tends to infinity.

Such short-term dependence is, however, not the only possibility and, in fact, in many areas of application long-term dependence is not uncommon (see, for example, Cox (1984), Hampel et al. (1986), and Beran (1992)). Formally, a stationary sequence is said by some authors to exhibit short- or long-term dependence as $\sum_{k=1}^{\infty} \gamma_k$ converges or diverges. (Convergence of

$\sum_{k=1}^{\infty} \gamma_k$ implies convergence to the same limit of $\dfrac{1}{n} \sum_{k=1}^{n} (n-k)\,\gamma_k$, i.e., the condition following (2.8.5); see Anderson (1971, Lemma 8.3.1).

In particular situations, which have been much studied but which are beyond the scope of this book, the γ's turn out to be given by (see for example Beran (1989))

$$(2.8.31) \qquad \gamma_k = \frac{\sigma^2}{2}\left[(k+1)^d - 2k^d + (k-1)^d\right], \; k = 1, 2, \ldots,$$

where $\sigma^2 = \operatorname{Var}(X_i)$ and $1 < d < 2$. (For $d = 1$, the X's are uncorrelated.) Then (Problem 8.12)

$$(2.8.32) \qquad\qquad\qquad \gamma_k \sim \frac{d(d-1)}{k^{2-d}},$$

so that $\gamma_k \to 0$ very slowly (for $d = 3/2$, for example, at the rate $1/\sqrt{k}$), $\sum \gamma_k$ diverges, and $\operatorname{Var}(\bar{X}_n)$ is of order $1/n^{2-d}$. Since $2 - d < 1$, (2.8.2) cannot be expected to hold.

Example 2.8.4 Sampling from a 2-valued population. In Example 2.2.4, we considered a population Π of N items of which D are defective and $N - D$ are not, and a sample of n is drawn at random, i.e., in such a way that all $\binom{N}{n}$ possible samples of size n are equally likely. As before, let $X_i = 1$ or 0 as the ith item drawn is or is not defective. Then for each i,

$$(2.8.33) \qquad\qquad\qquad P(X_i = 1) = D/N$$

so that the X's are identically distributed with mean and variance

$$(2.8.34) \qquad E(X_i) = \frac{D}{N} \text{ and } \operatorname{Var}(X_i) = \frac{D}{N}\left(1 - \frac{D}{N}\right).$$

The X's are dependent with covariances

$$(2.8.35) \qquad\qquad \operatorname{Cov}(X_i, X_j) = -\frac{D(N-D)}{N^2(N-1)}$$

The distribution of $X = \sum_{i=1}^{n} X_i$, the total number of defectives in the sample, is the hypergeometric distribution given in Table 1.6.2. In Example 2.2.4, we were concerned with the consistency of $\bar{X} = X/n$ as an estimator of the proportion D/N of defectives; we shall now consider the asymptotic distribution of this estimator as $n \to \infty$.

Since $n \leq N$, the sample size n cannot tend to infinity without the population size N tending to infinity also. It is therefore necessary to consider not a single population Π with fixed N, but a sequence of populations Π_N with $N \to \infty$ in which to embed the given situation. This leads to a double sequence of the kind described in (A') of the preceding section with n replaced by N, where, however, now the variables

$$(2.8.36) \qquad\qquad X_{N1}, \ldots, X_{Nn}$$

are no longer independent. Furthermore, unlike the situations in Examples 2.8.1–2.8.3, the order of the variables in (8.35) has no significance, since by the assumption of random sampling the joint distribution of these variables is completely symmetric, i.e., unchanged under permutation of the variables.

Two natural embedding sequences are

(i) Fix n, let $N \to \infty$, and let $D = D_N$ be such that

$$(2.8.37) \qquad\qquad \frac{D_N}{N} \to p, \quad 0 < p < 1.$$

This will provide a reasonable approximation when the sample size n is small compared to the population size N, while D is of the same order of magnitude as N.

(ii) The numbers n and $N - n$ of both types of items tend to infinity and D_N satisfies (2.8.37).

In the first case, the n draws become less and less dependent as the size of the population increases and the (hypergeometric) distribution of X tends to the binomial distribution $b(p, n)$ as $N \to \infty$.[¶] A formal proof of this result, which is a slight generalization of Problem 2.6 of Chapter 1, is easily obtained by writing

$$(2.8.38)$$
$$P(x) = \binom{n}{x} \frac{D_N!}{(D_N - x)!} \frac{(N - D_N)!}{(N - D_N - n + x)!} \frac{(N - n)!}{N!}$$
$$\sim \binom{n}{x} \frac{D_N!}{(D_N - x)! N^x} \frac{(N - D_N)!}{(N - D_N - n + x)! N^{n-x}} \to \binom{n}{x} p^x q^{n-x}.$$

Note that this result concerns a limit not as the sample size n tends to infinity but for fixed n as the degree of dependence decreases.

Let us next consider the case, in which n and N both tend to ∞. We saw in Example 2.2.4 that then $\bar{X}$ is a consistent estimator of p. It, furthermore,

---

[¶] For a detailed discussion of the binomial and other approximations to the hypergeometric distribution, see Johnson, Kotz, and Kemp (1992, p. 256).

turns out as a special case of Theorem 2.8.2 that

$$(2.8.39) \qquad \frac{\left(\dfrac{X}{n} - \dfrac{D}{N}\right)}{\sqrt{\mathrm{Var}\,(X/n)}} \xrightarrow{L} N\,(0,1)$$

with

$$(2.8.40) \qquad \mathrm{Var}\left(\frac{X}{n}\right) = \frac{1}{n}\frac{N-n}{N-1}\frac{D}{N}\left(1-\frac{D}{N}\right).$$

Suppose in particular that $D/N \to p$, so that

$$\mathrm{Var}\left(\frac{X}{n}\right) \sim \frac{pq}{n}\frac{N-n}{N-1} \sim \frac{1}{n}\mathrm{Var}\,(X_i)\frac{N-n}{N-1}.$$

If $n = o\,(N)$, the third factor on the right side $\to 1$, and with $\dfrac{X}{n} = \bar{X}$, it follows from (2.8.39) that

$$\sqrt{n}\left[\bar{X} - E\,(X_i)\right]/\sqrt{\mathrm{Var}\,(X_i)} \xrightarrow{L} N\,(0,1)$$

as it would if the X's were independent. On the other hand, if n is of the same order as N, $n \sim \lambda N$ say, $\sqrt{n}\left[\bar{X} - E\,(X_i)\right]/\sqrt{\mathrm{Var}\,(X_i)}$ will continue to have a normal limit distribution but the asymptotic variance will be $1 - \lambda$ rather than 1.

These results reflect the dependence structure of the X's. The correlation of any two X's is $\rho = -1/\,(N-1)$. If $n = o\,(N)$, then $\rho = o\left(\dfrac{1}{n}\right)$, so that the dependence decreases rapidly with the sample size and becomes asymptotically negligible. On the other hand, if $n \sim \lambda N$, it follows that $\rho \sim -\lambda/n$ is of order $1/n$ and the dependence is strong enough to affect the limit distribution.

Note: In the earlier examples, dependence was discussed in terms of the covariances γ_k rather than the corresponding correlation coefficients. There the two were in fact equivalent, since the variance σ^2 was independent of n. This is no longer the case in the present situation where $\mathrm{Var}\,(X_i)$ depends on N and where correlation is then the more meaningful measure of dependence. $\qquad\qquad\qquad\qquad\qquad\qquad\qquad\qquad\square$

Example 2.8.4 is a special case of the following more general situation.

Example 2.8.5 Sampling from a finite population. Let the population Π consist of N elements, to each of which is attached a numerical value, $v_1, v_2, \dots, v_N$. From this population, a random sample of n is drawn; let the v-values in the sample be denoted by $X_1, \dots, X_n$. Example 2.8.4 is the special case in which D of the v's are equal to 1; the remaining $N - D$ equal to zero. As in this special case, the X's are identically distributed (with each X taking on the values $v_1, \ldots, v_N$, each with probability $1/N$), but they are not independent.

In order to obtain a limit theory for $\sqrt{n}\left[\bar{X} - E\left(X_i\right)\right]$ as $n \to \infty$, we must have not only n but also N tending to infinity and therefore, as before, embed the given population Π in a sequence of populations $\Pi_1, \Pi_2, \ldots$ of increasing size. Let us suppose that

Π_1 consists of a single element with value v_{11};

Π_2 consists of two elements with values v_{21}, v_{22};

$$\vdots \qquad \vdots$$

Π_N consists of N elements with values $v_{N1}, \ldots, v_{NN}$.

$\square$

Theorem 2.8.2 *Let the n sample values from Π_N be denoted by $X_{N1}, \ldots, X_{Nn}$ and their average by $\bar{X}_n$. Then*

$$(2.8.41) \qquad \frac{\bar{X}_n - E\left(\bar{X}_n\right)}{\sqrt{\operatorname{Var}\left(\bar{X}_n\right)}} \to N\left(0, 1\right),$$

provided

$$(2.8.42) \qquad n \text{ and } N - n \text{ both tend to infinity}$$

and either of the following conditions is satisfied:

(a) $\dfrac{n}{N}$ *is bounded away from 0 and 1 as $N \to \infty$ and*

$$(2.8.43) \qquad \frac{\max\left(v_{Ni} - v_{N\cdot}\right)^2}{\sum\left(v_{Nj} - v_{N\cdot}\right)^2} \to 0$$

or

(b)

$$(2.8.44) \qquad \frac{\max\left(v_{Ni} - v_{N\cdot}\right)^2}{\sum\left(v_{Nj} - v_{N\cdot}\right)^2/N} \quad \text{remains bounded as } N \to \infty.$$

Before considering the proof of this result, note that its implementation requires evaluating $E\left(\bar{X}_n\right)$ and $\operatorname{Var}\left(\bar{X}_n\right)$. Since each X_i takes on the values $v_{N1}, \ldots, v_{NN}$ with probability $1/N$, we have

$$(2.8.45) \qquad E\left(\bar{X}_n\right) = v_{N\cdot}.$$

The variance of $\bar{X}_n$ satisfies

(2.8.46)
$$\frac{1}{n^2} \text{Var}\,(X_{N1} + \cdots + X_{Nn}) = \frac{1}{n}\text{Var}\,(X_{N1}) + \frac{n-1}{n}\text{Cov}\,(X_{N1}, X_{N2}).$$

The first term on the right side is determined by

(2.8.47) $\qquad \text{Var}\,(X_{N1}) = E\,(X_{N1} - v_{N\cdot})^2 = \frac{1}{N}\sum_{i=1}^{N}(v_{Ni} - v_{N\cdot})^2 .$

The covariance term can be evaluated in the same way (Problem 8.14) or more simply by noting that for $n = N$, we have $\text{Var}\,(X_{N1} + \cdots + X_{NN}) = 0$ since $X_{N1} + \cdots + X_{NN} = v_{n1} + \cdots + v_{NN}$ is a constant. Setting $n = N$ in (2.8.46), we therefore see that

(2.8.48) $\qquad \text{Cov}\,(X_{N1}, X_{N2}) = \frac{-1}{(N-1)}\text{Var}\,(X_{N1}),$

and, on combining (2.8.46) and (2.8.47), that

(2.8.49) $\qquad \text{Var}\,(\bar{X}_n) = \frac{N-n}{n\,(N-1)} \cdot \frac{1}{N}\sum_{i=1}^{N}(v_{Ni} - v_{N\cdot})^2 .$

In the situation of Example 2.8.4 in which D_N of the v's are equal to 1 and the remaining $N - D_N$ are equal to 0, (2.8.49) reduces to (2.8.40) (Problem 8.15).

Condition (2.8.43) is reminiscent of condition (2.7.17) with $d_{Ni} = v_{Ni} - v_N$. It is in fact possible to construct a sequence of *independent* variables which in a suitable sense is asymptotically equivalent to the sequence of the X's and which satisfies (2.7.17). This equivalence can be used to prove Theorem 2.8.2. An indication of this proof, which is due to Hajek (1961), will be given in Section 4.4.

In Example 2.8.4(ii), $\sqrt{n}$ is still a suitable normalizing factor for $[\bar{X} - E(X_i)]$ even though the limiting variance is no longer guaranteed to be pq. In (2.8.41), the needed normalizing factor may be quite different (Problem 8.16).

Example 2.8.6 Wilcoxon rank-sum statistic. Let $v_{Ni} = i\,(i = 1, \ldots, N)$ so that

(2.8.50) $\qquad\qquad W_s = \sum_{i=1}^{n} X_{Ni}$

is the sum of n numbers drawn at random from the integers $1, \ldots, N$. Then W_s is the Wilcoxon rank-sum statistic, which will be discussed in Section 3.2.

Since $W_s = n\bar{X}_n$ and $v_N. = \dfrac{1}{2}(N+1)$, it follows from (2.8.45) that

(2.8.51) $$E(W_s) = \frac{1}{2}n(N+1).$$

Similarly,

$$\frac{1}{N}\sum_{i=1}^{N}(v_{Ni} - v_N.)^2 = \frac{1}{N}\sum v_{Ni}^2 - v_N^2.$$
$$= \frac{(N+1)(2N+1)}{6} - \frac{(N+1)^2}{4} = \frac{N^2-1}{12}.$$

and hence

(2.8.52) $$\mathrm{Var}(W_s) = n(N-n)(N+1)/12.$$

To check condition (2.8.44), note that

$$\max|v_{Ni} - v_N.| = \left|N - \frac{1}{2}(N+1)\right| = \frac{1}{2}(N-1).$$

Thus

$$\frac{\max(v_{Ni} - v_N.)^2}{\sum(v_{Nj} - v_N.)^2/N} = \frac{(N-1)^2/4}{(N^2-1)/12},$$

which is clearly bounded as $N \to \infty$. It therefore follows from Theorem 2.8.2 that

(2.8.53) $$\frac{W_s - \frac{1}{2}n(N+1)}{\sqrt{mn(N+1)/12}} \xrightarrow{L} N(0,1)$$

as n and $m = N - n$ tend to infinity.   □

Summary

1. The CLT continues to hold for a sequence of dependent variables if the dependence is sufficiently weak. This is illustrated on a number of stationary sequences, in particular for m-dependent variables, for the first order autoregressive process and for 2-state Markov chains.

2. A different kind of dependence structure holds for simple random sampling from a finite population. Conditions are given under which the CLT holds in this case.

2.9   Problems

Section 1

1.1 Prove (2.1.10) and (2.1.11).

[Hint for (2.1.10):

$$|A_n B_n - ab| = |A_n (B_n - b) + b (A_n - a)|$$
$$\leq |A_n (B_n - b)| + |b (A_n - a)| .]$$

1.2 In Example 2.1.4, show that

(i) $Y_n \xrightarrow{P} 1$ if $p_n \to 0$;

(ii) $E (Y_n - 1)^2 \to \infty$ if $p_n = 1/n$.

1.3 Prove Theorem 2.1.4

[Hint: Given any $a > 0$, there exists b such that $|f(y) - f(c)| < a$, provided $|y - c| < b$. It then follows that

$$P [|f (Y_n) - f (c)| < a] \geq P [|Y_n - c| < b] .]$$

1.4 In generalization of Lemma 2.1.2, show that if

$$P(E_n) \to p \text{ and } P(F_n) \to 1,$$

then

$$P(E_n \text{ and } F_n) \to p.$$

1.5 (i) If Y_n is a sequence of random variables for which

(2.9.1) $$E |Y_n - c| \to 0,$$

then Y_n converges in probability to c.

(ii) Find an example in which (2.9.1) is satisfied but (2.1.6) is not.

1.6 Let c_n be a sequence of numbers converging to c, and let X_n be a random variable assigning probability 1 to c_n. Then $X_n \xrightarrow{P} c$.

1.7 Let $X_1, \dots, X_n$ be i.i.d., each with probability density f, and let $X_{(1)}$ be the smallest of the X's. Show that $X_{(1)}$ is consistent for estimating ξ when

(i) $f(x) = 1$ for $\xi < x < \xi + 1$, $f(x) = 0$ elsewhere;

(ii) $f(x) = 2(x - \xi)$ for $\xi < x < \xi + 1$, $f(x) = 0$ elsewhere;

(iii) $f(x) = e^{-(x-\xi)}$ for $\xi < x$, $f(x) = 0$ elsewhere;

(iv) Suppose that $f(x) = 0$ for all $x < \xi$. Give a simple condition on f for $X_{(1)}$ to be consistent for estimating ξ which contains (i)–(iii) as special cases.

[**Hint:** $P\left[X_{(1)} \geq a\right] = \left[P\left(X_1 \geq a\right)\right]^n.$]

1.8 If there exists M such that

$$P\left(|Y_n - c| < M\right) = 1 \text{ for all } n,$$

and if $Y_n \xrightarrow{P} c$, then

(i) $E\left(Y_n\right) \to c$; (ii) $E\left(Y_n - c\right)^2 \to 0$.

1.9 For any positive integer k, if $E|X|^k < \infty$, then $E|X - a|^k < \infty$ for any a.

[**Hint:** Expand $(X - a)^k.$]

1.10 (i) Prove (2.1.18) and (2.1.19).

(ii) Generalize the remaining properties of o stated in Lemma 1.4.1 to o_p.

1.11 Work the following problems of Chapter 1 with the indicated modifications:

(i) Problem 4.16 with O replaced by O_p;

(ii) Problem 4.17 with o and O replaced by o_p and O_p, respectively;

(iii) Problem 4.18 with o and O replaced by o_p and O_p, respectively.

Section 2

2.1 In Example 2.2.1, if $X_i = \pm 3^{i-1}$ with probability $1/2$ each, show that for any $M > 0$ there exists N such that for $n > N$ both $P(\bar{X} > M)$ and $P(\bar{X} < -M)$ are exactly $1/2$.

[**Hint:** Evaluate the smallest value that $(X_1 + \cdots + X_n)/n$ can take on when $X_n = 3^{n-1}.$]

2.2 Verify (2.2.9).

2.3 An alternative expression for $\hat{\beta}$ given by (2.2.13) is

$$\sum (v_i - \bar{v})(X_i - \bar{X}) / \sum (v_j - \bar{v})^2.$$

2.4 Verify

(i) (2.2.16);

(ii) (2.2.18);

(iii) (2.2.19).

2.5 If $X_1, \ldots, X_m$ and $Y_1, \ldots, Y_n$ are independent with common mean ξ and variances σ^2 and τ^2, respectively, use (2.2.1) to show that the grand mean $(X_1 + \cdots + X_m + Y_1 + \cdots + Y_n)/(m+n)$ is a consistent estimator of ξ provided $m + n \to \infty$.

2.6 Verify (2.2.24).

2.7 (i) Show that if $v_n \to \infty$, then also $\bar{v}_n = (v_1 + \cdots + v_n)/n \to \infty$.

(ii) Use (i) to show that if $v_n \to \infty$, then $\sum (v_i - \bar{v})^2 \to \infty$ and hence $\hat{\beta}$ given by (2.2.13) is a consistent estimator of β.

[**Hint**: (i) Given any M, let k be so large that $v_i > M$ for all $i > k$. Then for any $n > k$,

$$\bar{v}_n > \frac{v_1 + \cdots + v_k}{n} + \frac{(n-k)}{n}M.$$

Holding M and k fixed, let $n \to \infty$.

(ii) Note that $\sum (v_i - \bar{v})^2 > (v_1 - \bar{v})^2$.]

2.8 (i) Generalize part (i) of the preceding problem to show that if v_n tends to any limit ℓ, then also $\bar{v}_n \to \ell$.

(ii) Give an example of a sequence of numbers v_n such that $\bar{v}_n$ tends to a limit, but v_n does not.

(iii) Show that (2.2.35) implies (2.2.31).

2.9 In Example 2.2.6, show that as $n \to \infty$ the estimator δ has a limit distribution which assigns probability p and q to the values 1 and 0, respectively.

2.10 Let $Y_1, Y_2, \ldots$ be i.i.d. with mean ξ and variance $\sigma^2 < \infty$, and let

$$X_1 = \frac{Y_1 + \cdots + Y_{m_1}}{m_1}; \quad X_2 = \frac{Y_{m_1+1} + \cdots + Y_{m_1+m_2}}{m_2}, \ldots$$

so that $\mathrm{Var}(X_i) = \sigma_i^2 = \sigma^2/m_i$, $i = 1, \ldots, n$.

(i) The estimator δ_n of ξ given by (2.2.5) and (2.2.8) is consistent for estimating ξ, provided $\sum_{i=1}^{n} m_i \to \infty$ and hence in particular if either (a) $n \to \infty$ or (b) n is finite but at least one of $m_1, \ldots, m_n$ tends to ∞.

(ii) The estimator $\bar{X}$ is consistent if either (c) $n \to \infty$ or (d) n is finite and $m_i \to \infty$ for all $i = 1, \ldots, n$, but not necessarily if only (b) holds.

2.11 In the standard two-way random effects model, it is assumed that

$$X_{ij} = A_i + U_{ij} \ (j = 1, \ldots, m; i = 1, \ldots, s)$$

with the A's and U's all independently, normally distributed with

$$E(A_i) = \xi, \ \text{Var}(A_i) = \sigma_A^2, \ E(U_{ij}) = 0, \ \text{Var}(U_{ij}) = \sigma^2.$$

Show that $\bar{X} = \sum \sum X_{ij}/sm$ is

(i) not consistent for ξ if $m \to \infty$ and s remains fixed,

(ii) consistent for ξ if $s \to \infty$ and m remains fixed.

Section 3

3.1 If $\{H_n, n = 1, 2, \ldots\}$ is a sequence of cdf's and H is a function such that $H_n(x) \to H(x)$ for all x, then

(i) $0 \leq H(x) \leq 1$ for all x;

(ii) H is non-decreasing.

3.2 (i) If $Y_n \overset{P}{\to} c$ (c finite), then Y_n is bounded in probability.

(ii) Prove Lemma 2.3.1.

3.3 (i) If Y_n is bounded in probability and if K_n is any sequence of constants tending to ∞, then

(2.9.2) $P(|Y_n| < K_n) \to 1$ as $n \to \infty$.

(ii) If Y_n is bounded in probability and c_n is any sequence of constants tending to 0, then $c_n Y_n \overset{P}{\to} 0$.

3.4 In Example 2.3.5, give an example each of a sequence of random variables Y_n with cdf H_n which converges to c in probability such that $H_n(c)$ does or does not converge to $H(c) = 1$.

3.5 Any cdf H has arbitrarily large continuity points.

[**Hint:** The points of discontinuity of any cdf are countable. This follows from the fact that any collection of mutually exclusive intervals is countable since each interval contains a rational number and the rationals are countable. Apply this fact to the intervals $[H(x-), H(x)]$ at points x of discontinuity of H.]

3.6 If $k_n(Y_n - c) \overset{L}{\to} H$ and $k_n \to \infty$, then $Y_n \overset{P}{\to} c$.

3.7 Give an example in which $k_n(Y_n - c)$ tends in law to random variable Y with cdf, but where $\text{Var}[k_n(Y_n - c)]$ does not tend to $v^2 = \text{Var}(Y)$.

3.8 In Example 2.3.7, verify (2.3.11).

3.9 (i) If $X_1, \ldots, X_n$ are i.i.d. according to the uniform distribution $U(0, \theta)$, obtain the suitably normalized limit distribution for

(a) $\dfrac{n+1}{n} X_{(n)}$,

(b) $\dfrac{n}{n-1} X_{(n)}$.

(ii) For large n, which of the three statistics (a), (b), or $X_{(n)}$ would you prefer as an estimator for θ.

3.10 (i) If $X_1, \ldots, X_n$ are i.i.d. according to the exponential density e^{-x}, $x > 0$, show that

(2.9.3) $P\left[X_{(n)} - \log n < y\right] \to e^{-e^{-y}}, \quad -\infty < y < \infty.$

(ii) Show that the right side of (2.9.3) is a cumulative distribution function. (The distribution with this cdf is called the *extreme value distribution*.)

(iii) Graph the cdf of $X_{(n)} - \log n$ for $n = 1, 2, 5$ together with the limit $e^{-e^{-y}}$.

(iv) Graph the densities corresponding to the cdf's of (iii).

3.11 Suppose that $k_n (T_n - g(\theta)) \xrightarrow{L} H$, with H continuous. What can be said about the limiting behavior of $k'_n (T_n - g(\theta))$ when

(i) $\dfrac{k'_n}{k_n} \to d \neq 0$;

(ii) $\dfrac{k'_n}{k_n} \to 0$;

(iii) $\dfrac{k'_n}{k_n} \to \infty$?

3.12 Let $X_1, \ldots, X_n$ be i.i.d. according to $U(0, 1)$. Find the limit distribution of $n\left[1 - X_{(n-1)}\right]$.

3.13 Show that if H_n is the normal distribution with mean ξ_n and variance σ_n^2, then H_n tends to the normal distribution H with mean ξ and variance $\sigma^2 > 0$ if and only if

$$\xi_n \to \xi \text{ and } \sigma_n^2 \to \sigma^2.$$

3.14 In Example 2.3.8, show that $Y_n \xrightarrow{L} Y$ if $p_n \to 0$.

3.15 If Y_n is bounded in probability and $R_n = o_p(Y_n)$, then $R_n \xrightarrow{P} 0$.

3.16 If $H_n \to H$, then

    (i) $H_n(a_n) \to 1$ if $a_n \to \infty$,

      $H_n(a_n) \to 0$ if $a_n \to -\infty$.

    (ii) $H_n(a_n) \to H(a)$ if $a_n \to a$ and a is a continuity point of H.

    (iii) If $V_n \xrightarrow{L} H$ and $W_n \xrightarrow{P} 0$ and if a is a continuity point of H, then

$$P[a - W_n \le V_n < a] \to 0.$$

3.17 Prove Theorem 2.3.4.

Section 4

4.1 Replace (2.4.2) by an equivalent statement the right side of which is $N(0,1)$ instead of $N(0,2)$.

4.2 Let X_n have distribution F_n satisfying

(2.9.4) $\hspace{3cm} P(0 \le X_n \le 1) = 1.$

    (i) Then $E(X_n) = \xi_n$, $\mathrm{Var}(X_n) = \sigma_n^2$, and $E|X_n - \xi_n|^3$ are all between 0 and 1.

    (ii) Determine a sequence of distributions F_n satisfying (2.9.4) but for which the standardized third moment

(2.9.5) $\hspace{3cm} E|X_n - \xi_n|^3 / \sigma_n^3$

is not $o(\sqrt{n})$.

4.3 (i) Determine the expectation and variance of the distribution

$$F(x) = (1 - \epsilon) \Phi(x) + \epsilon \Phi\left(\frac{x - \eta}{\tau}\right).$$

    (ii) By simulation, determine

(2.9.6) $\hspace{2cm} p = P_F\left(\frac{\bar{X} - E(\bar{X})}{\sqrt{\mathrm{Var}(\bar{X})}} \le x\right),$

where $X_1, \ldots, X_n$ are i.i.d. according to F, for

$$x = 0, 1, 2, \ \epsilon = .1, .2, .3, \ n = 20, 50, 100$$

and the following combinations of η and τ:

    (a) $\eta = 0$, $\tau = 1, 2, 3$;

    (b) $\tau = 1$, $\eta = .5, 1, 2$;

    (c) $\eta = \tau = .5, 1, 2$,

and compare your results to the normal approximation $\Phi(x)$.

4.4 Show that (2.4.7) is the probability density of $X = 1/Y^2$ when Y is distributed as $N(0,1)$.

4.5 (i) Show that (2.4.19) follows from (2.4.18).

(ii) Show that (2.4.19) remains valid when n is even.

[**Hint for (ii)**: Show that (2.4.19) remains valid when $n = 2m$ and $\tilde{X}_n$ is replaced by either $X_{(m+1)}$ or $X_{(m)}$, and hence also holds for their average.]

4.6 Determine the limit distribution of $\sqrt{n}\left(\tilde{X}_n - \theta\right)$ for the following distributions [||]:

(i) $F = N\left(0, \sigma^2\right)$,

(ii) $F = U(-a, a)$,

(iii) $F =$ double exponential,

(iv) logistic with $F(x) = 1/(1 + e^{-x})$.

4.7 (i) For each of the distributions (i)–(iv) of the preceding problem, compare the limit distribution of $\sqrt{n}\left(\tilde{X} - \theta\right)$ with that of $\sqrt{n}\left(\bar{X} - \theta\right)$, and in each case, determine whether for large n you would prefer $\bar{X}$ or $\tilde{X}$ as an estimator of θ.

(ii) Use simulation to graph the cdf of $\sqrt{n}\left(\tilde{X}_n - \theta\right)$ for $n = 1, 3, 5$, and compare it with that of the limit distribution for the distributions (ii) and (iii) of Problem 4.6.

4.8 Let $X_1, \ldots, X_n$ be i.i.d. with cdf F and density f. For any $0 < p < 1$, let ξ_p be such that

$$(2.9.7) \qquad\qquad F(\xi_p) = p,$$

and suppose that $f(\xi_p) > 0$. If

$$(2.9.8) \qquad\qquad \frac{n_1}{n} = p + o\left(\frac{1}{\sqrt{n}}\right),$$

then

$$(2.9.9) \qquad \sqrt{n}\left(X_{(n_1)} - \xi_p\right) \xrightarrow{L} N\left[0, \frac{pq}{[f(\xi_p)]^2}\right].$$

4.9 If $n = 2m$, find the limit distribution of $\sqrt{n}\left(X_{(m+d)} - \theta\right)$ as $m \to \infty$ and d remains fixed, under the assumptions of Example 2.4.9.

[||] The densities are given in Table 1.6.1 of Chapter 1.

4.10 Note that the first correction term in (2.4.20) drops out when $x = 1$. To see the effect of this phenomenon, make a table analogous to Table 2.4.2, but with the top row reading $x = .5, 1, 2$ instead of $1 - G_n(x) = .5, .1, .05, .01$.

4.11 When F is symmetric, the correction given by the first correction term of (2.4.20) drops out. Make a table like Table 2.4.2 using the correction term (2.4.22) instead of (2.4.20) for the following distributions:

(i) logistic,

(ii) double exponential,

(iii) uniform $U(0, 1)$.

4.12 If Y is the negative binomial variable defined in Table 1.6.2 of Chapter 1, then

$$E(Y) = \frac{mq}{p} \text{ and Var}(Y) = \frac{mq}{p^2}.$$

Show that

$$[Y - E(Y)]/\sqrt{\text{Var}Y} \xrightarrow{L} N(0, 1) \text{ as } m \to \infty.$$

[**Hint:** Represent Y as the sum of m i.i.d. random variables and apply the central limit theorem.]

4.13 In Example 2.4.7, show that the sequence of distributions $F_n = P(1/n)$ does not satisfy condition (2.4.12).

[**Hint:** If X is distributed as $P(\lambda)$, then $E(X - \lambda)^2 = E(X - \lambda)^3 = \lambda$.]

Section 5

5.1 If Y_n is bounded in probability and $X_n = o_p(Y_n)$, then $X_n \xrightarrow{P} 0$.

5.2 Let $X_1, \ldots, X_n$ be i.i.d. as $N(\theta, \sigma^2)$, $\sigma = $ known, and let $\delta = \Phi\left(\frac{\bar{X} - a}{\sigma}\right)$ denote the maximum likelihood estimator of

$$p = P(X_i > a) = 1 - \Phi\left(\frac{a - \theta}{\sigma}\right) = \Phi\left(\frac{\theta - a}{\sigma}\right).$$

Determine the limit distribution of $\sqrt{n}(\delta - p)$.

5.3 Let $X_1, \ldots, X_n$ be i.i.d. according to the logistic distribution with cdf $F(x) = 1/(1 + e^{\theta - x})$. Construct an estimator $\delta(\bar{X})$ of $F(0)$ and state the limit distribution of $\sqrt{n}[\delta(\bar{X}) - F(0)]$.

5.4 If X has the binomial distribution $b(p, n)$, show that

(2.9.10) $$\sqrt{n}\left(\arcsin\sqrt{\frac{X}{n}} - \arcsin\sqrt{p}\right)$$

tends to a normal distribution with variance independent of p.

5.5 Let $X_1, \ldots, X_n$ be i.i.d. $N(\theta, \sigma^2)$ and consider the estimation of θ^2. The maximum likelihood estimator is $\delta_{1n} = \bar{X}^2$; an alternative (unbiased) estimator is $\delta_{2n} = \bar{X}^2 - \dfrac{1}{n(n-1)}S^2$, where $S^2 = \sum(X_i - \bar{X})^2$. Obtain the limit distribution of

(i) $\sqrt{n}\left(\delta_{in} - \theta^2\right)$, $i = 1, 2$ when $\theta \neq 0$;

(ii) a suitably normalized $\left(\delta_{in} - \theta^2\right)$ when $\theta = 0$.

5.6 For any positive integer k, determine, under the assumptions of the preceding problem, the limit distribution of a suitably normalized $\left(\bar{X}^k - \theta^k\right)$.

[**Hint:** For $\theta = 0$, generalize the argument leading to (2.5.14).]

5.7 If X has the binomial distribution $b(p, n)$, determine the limit distribution of $f\left(\dfrac{X}{n}\right) - f(p)$, suitably normalized, where $f(p) = \min(p, q)$.

[**Hint:** Consider separately the cases $p < q$, $p > q$, and $p = q$, and follow the approach of Example 2.5.6.]

5.8 Suppose that $\sqrt{n}(T_n - \theta) \xrightarrow{L} N(0, \tau^2)$, and let

$$f(t) = \begin{cases} -at & \text{if } t < 0 \\ bt & \text{if } t > 0. \end{cases}$$

Find the limit distribution of $f(T_n) - f(\theta)$, suitably normalized.

5.9 (i) If X is distributed as $b(p, n)$, use (2.3.9) and the fact that $\dfrac{X}{n}\left(1 - \dfrac{X}{n}\right)$ is a consistent estimator of pq to obtain approximate confidence intervals for p.

(ii) Obtain alternative approximate confidence intervals for p by using the result of Problem 5.4.

(iii) Show that the two lower confidence limits of (i) and (ii) can be negative and the upper ones greater than 1 but that for any fixed $0 < p < 1$, the probability of either of these events tends to 0 as $n \to \infty$.

[**Hint for (ii):** Use the fact that $\sin(\arcsin x) = x$.]

5.10 If $X_1, \ldots, X_n$ are i.i.d. according to $U(0, \theta)$ and $T_n = X_{(n)}$, the limit distribution of $n(\theta - X_{(n)})$ is given in Example 2.3.7. Use this result to determine the limit distribution of

(i) $n[f(\theta) - f(T_n)]$, where f is any function with $f'(\theta) \neq 0$;

(ii) $[f(\theta) - f(T_n)]$ is suitably normalized if $f'(\theta) = 0$ but $f''(\theta) \neq 0$.

5.11 For the functions $f(x) = e^x, (1+x)^\alpha$, and $-\log(1-x)$ respectively, show that the coefficients $f^{(i)}(0)/i!$ of x^i in (2.5.17) reduce to those given in

(i) (1.3.9) of Chapter 1,

(ii) (1.3.10) of Chapter 1,

(iii) (1.3.14) of Chapter 1.

5.12 Graph the approximations

$$f_r(x) = \sum_{i=0}^{r} \frac{f^{(i)}(0)}{i!} x^i$$

for $i = 1, 2, 3, 5$ together with $f(x)$ over the interval of convergence for the functions of (1.3.9) and (1.3.14).

5.13 If $f(x) = 1/(1-x)$, evaluate the Taylor series (2.5.15) for $f(2 + \Delta)$ and use (2.5.19) to show that it gives a valid expansion for $\dfrac{1}{1 - (2 + \Delta)}$

$$= \frac{-1}{1 + \Delta}.$$

Section 6

6.1 Let $\left\{a_n^{(i)}, n = 1, 2, \cdots\right\}, i = 1, \ldots, k$, be k sequences satisfying

$$a_n^{(i)} \to c \text{ as } n \to \infty \text{ for every } i = 1, \ldots, k.$$

Then $\min_{i=1,\ldots,k} a_n^{(i)} \to c$ as $n \to \infty$.

[**Hint**: Given ϵ, for each i there exists $n_0^{(i)}$ such that $\left|a_n^{(i)} - c\right| < \epsilon$ for $n > n_0^{(i)}$.]

6.2 Explain where the argument of the preceding problem breaks down if the finite number of sequences is replaced by an infinite number $\left\{a_n^{(i)}, n = 1, 2, \cdots\right\}, i = 1, 2, \ldots$.

6.3 If $f_n(x) = x^n$, then $f_n(x) \to 0$ for every $0 < x < 1$.

(i) Show that for any $0 < a < 1$, the convergence of $f_n(x)$ to 0 is uniform for $0 < x < a$, but that it is not uniform for $0 < x < 1$.

(ii) Construct a sequence of values x_n such that $f_n(x_n)$ tends to a limit different from 0.

(iii) If M_n denotes the sup of $f_n(x)$ in $0 < x < 1$, evaluate M_n and show that it does not tend to zero.

6.4 Solve the three parts of the preceding problem when $f_n(x) = \dfrac{x}{x+n}$ and with the intervals $0 < x < a$ and $0 < x < 1$ replaced by $m < x < M$ and $-\infty < x < \infty$, respectively.

6.5 Use the result of Problem 4.12 to derive approximate confidence intervals for p based on a random variable Y with negative binomial distribution $Nb(p, m)$.

6.6 (i) For the approximate confidence intervals $\underline{p} < p < \bar{p}$ of Problem 5.9 show that

$$P\left[\underline{p} < p < \bar{p} \text{ for all } 0 < p < 1\right] = 0.$$

(ii) Solve the corresponding problem for the intervals of Problem 6.5.

Section 7

7.1 Verify that $\hat{\alpha}$ and $\hat{\beta}$ given by (2.7.11) are the least squares estimators of α and β, i.e., minimize $\sum (X_i - \alpha - \beta v_i)^2$.

7.2 Show that (2.7.22) is satisfied for $\sigma_j^2 = j\Delta$.

7.3 Show that (2.7.24) holds when
(i) $v_i = a + i\Delta$,
(ii) $v_i = a + \Delta i^k$ for any positive integer k,
(iii) but not when $v_i = 2^i$.

7.4 Give an example of a sequence of positive numbers v_i which is strictly decreasing and tends to zero for which (2.7.24) does not hold.

7.5 If $\left(\sum_{i=1}^{n} |d_{ni}|^k\right)^2 = o\left(\sum_{i=1}^{n} d_{ni}^2\right)^k$, then the same inequality holds when k is replaced by $k+1$.

[**Hint:** Apply the Schwarz inequality to $\left[\sum |d_{ni}|^{k+1}\right]^2 = \left[\sum |d_{ni}|^k \cdot |d_{n_i}|\right]^2$.

7.6 (i) Suppose that $k_n(T_n - \theta)$ tends in law to a limit distribution H which does not assign probability 1 to any single point. It was shown in Theorem 2.3.4 that if $k_n \to \infty$, then $T_n \xrightarrow{P} \theta$. Show that if $k_n \to 0$, then T_n does not tend to θ in probability.

(ii) For what kind of variables would you expect k_n to tend to 0?.

[**Hint for (i)**: Find the limit of $P\left(|T_n - \theta| < a\right)$ as $n \to \infty$.]

7.7 The following provides an example of the situation considered in the preceding problem. Let X_j $(j = 1, \dots, n)$ be independent and take on the values $\pm j$ with probability $1/2$ each. Show that

$$\sqrt{\frac{3}{n}}\bar{X} \xrightarrow{L} N(0, 1),$$

but that $\bar{X} \to \pm\infty$ with probability $1/2$ each.

[**Hint**: $P\left(\bar{X} > M\right) \to 0$ for any $M > 0$.]

Note: For a generalization of this problem, see Feller (Vol. 1) (1957), pp. 239–240.

7.8 Give an example showing that (2.7.26) no longer implies (2.7.27) when (2.7.25) does not hold.

7.9 If the pair of variables (X_i, Y_i) has the joint distribution stipulated in the proof of Theorem 2.7.5, then X_i has the distribution (2.7.28) and Y_i has the Poisson distribution $P(p_i)$.

7.10 Show that $e^{-x} \geq 1 - x$ for all x.

[**Hint**: Use Taylor's theorem 2.5.1(ii).]

7.11 (i) Let $\{u_i\}, \{v_i\}$, $i = 1, \dots, n$, be two sequences of numbers, each of which satisfies (2.7.24) and for which in addition $\sum u_i v_i = 0$. Then the sequence $\{u_i + v_i\}$ also satisfies (2.7.24).

(ii) In Example 2.7.6 with $\hat{\alpha}$ given by (2.7.11), we have $\sqrt{n}(\hat{\alpha} - \alpha)/\sqrt{\text{Var}\,\hat{\alpha}} \to N(0, 1)$ (with $\text{Var}(\hat{\alpha})$ given by (2.2.19)), provided the v's satisfy (2.7.24).

[**Hint for (i)**: Use the facts that $(u_i + v_i)^2 \leq 2(u_i^2 + v_i^2)$ and that in the present case $\sum(u_i + v_i)^2 = \sum u_i^2 + \sum v_i^2$.]

Section 8

8.1 Show that $\displaystyle\sum_{k=1}^{n}(n - k)\,|\gamma_k|$ tends to a finite limit as $n \to \infty$ if γ_k/a^k tends to a finite limit for some $0 < a < 1$.

8.2 Show that the sequence $X_1, X_2, \dots$ defined by (2.8.6) is stationary if the Z's are i.i.d.

8.3 For the sequence of moving averages given by (2.8.6) and (2.8.7), use the representation corresponding to (2.8.8) for general m to show that (2.8.2) holds with $\tau^2 = \text{Var}(Z_i)$.

8.4 If $Z_1, Z_2, \ldots$ are i.i.d. with $E\left(Z_i\right) = 0$ and $\text{Var}\left(Z_i\right) = \sigma^2$, determine the limit distribution of $\sum_{i=1}^{n} Z_i Z_{i+1}/\sqrt{n}$.

8.5 For the autoregressive process (2.8.13), verify the limit (2.8.17).

8.6 Determine the limit distribution of $\sqrt{n}\left(\bar{X} - \theta\right)$ when the X's are given by (2.8.13) without assuming the normality of X_1 and the U's.

[**Hint**: Express the sum $X_1 + \cdots + X_n$ as a linear function of X_1 and the U's and apply (2.7.17).]

8.7 In Theorem 2.8.1, suppose that n is not restricted to be a multiple of $N + m$, so that $n = r\left(N + m\right) + k$ with $k < N + m$. Then a sum

$$C = X_{r(N+m)+1} + \cdots + X_{r(N+m)+k}$$

has to be added to the right side of (2.8.10). Prove the analog of (2.8.11) for this case by showing that

$$E\left[\frac{1}{\sqrt{r\left(N+m\right)+k}} \left(\sum_{i=1}^{r} B_i + C\right)\right]^2 \to 0 \text{ as } N \to \infty$$

uniformly in r and k.

8.8 Consider the simplest ARMA (autoregressive moving average) process, (an autoregressive process in which the U's, instead of being i.i.d., constitute a moving average process) given by

$$X_{i+1} = \xi + \beta\left(X_i - \xi\right) + \left(Z_{i-1} + Z_i\right),$$

where the Z's are i.i.d. $N\left(0, \sigma_1^2\right)$ and X_1 is $N\left(0, \sigma^2\right)$.

(i) State conditions under which this process is stationary.

(ii) Show that the resulting stationary process satisfies $\sqrt{n}\left(\bar{X} - \xi\right) \to N\left(0, \tau^2\right)$ and evaluate τ^2.

8.9 In the preceding problem, determine the limit distribution when X_1 and the Z's are no longer assumed to be normal.

8.10 Under the assumptions of Example 2.8.3, show that

(i) $\pi_1 p^2 + \pi_0 pq = p^2$,

(ii) $\pi_1 - p = q\left(\pi_1 - \pi_0\right)$.

8.11 Verify (2.8.30).

8.12 Verify equation (2.8.32).

[**Hint**: For large k, multiply and divide the right side of (2.8.31) by k^d and expand $\left(1 + \frac{1}{k}\right)^d$ and $\left(1 - \frac{1}{k}\right)^d$ by (1.3.10) to as many terms as necessary.]

8.13 For $1 \leq n < N$, show that $(N - n) / (N - 1)$ can tend to any value between 0 and 1, depending on the limit of n/N.

8.14 Use the direct method leading to (2.8.47) to evaluate the covariance term in (2.8.46).

8.15 Show that (2.8.49) reduces to (2.8.40) when D_N of the v's are equal to 1 and the remaining $N - D_N$ are equal to 0.

8.16 (i) If $v_{Ni} = i^k$ for some positive integer k, show that

$$\sum_{i=1}^{N} (v_{Ni} - v_N.)^2 \sim \sum_{i=1}^{N} v_{Ni}^2 - c_k N^{2k+1}$$

for some $c_k > 0$.

(ii) In Example 2.8.5, use (i) and Theorem 2.8.2 to show that with n proportional to N and v_{Ni} given in (i), $k_n \left[\bar{X}_n - E(X_i) \right]$ may be asymptotically normally distributed with k_n not proportional to $\sqrt{n}$.

8.17 It can be shown (Hajek (1960)) that (2.8.41) is valid if n and $N - n$ both tend to ∞ and

(2.9.11)
$$\frac{\max (v_{Ni} - v_N.)^2}{\sum_{j=1}^{N} (v_{Nj} - v_N.)^2} \max \left(\frac{N - n}{n}, \frac{n}{N - n} \right) \to 0 \text{ as } N \to \infty.$$

(i) Show that this result implies the validity of (2.8.41) both under (a) and under (b) of Theorem 2.8.2.

(ii) Construct an example in which (2.9.11) holds but neither (a) nor (b) are satisfied.

Bibliographic Notes

More detailed discussions (with proofs) of convergence in probability, convergence in law, and the central limit theorem can be found in standard texts on probability theory such as Feller (Vol. 1 (1968); Vol. 2 (1971)) and Billingsley (1995). Monographs dealing more specifically with these topics are Gnedenko and Kolmogorov (1954) and Petrov (1995). Statistically oriented treatments are provided by Cramér (1946) and Serfling (1980). For the early history of the central limit theorem, see Adams (1974) and Stigler (1986).

3
Performance of Statistical Tests

Preview

The basic quantities needed in hypothesis testing, once a test statistic has been specified, are (i) the critical value that provides the desired level α, (ii) the power of the test, and (iii) the sample size(s) required to achieve a given power β. In addition, one will wish to know (iv) whether the test is robust against departures from the assumptions under which it was derived and (v) its efficiency relative to alternative tests.

For statistics that are asymptotically normal, simple approximate formulas for (i)–(iii) are obtained in Sections 3.1–3.3. The same limit theory permits investigating the robustness of the level of the test to departures from assumptions such as in the i.i.d. one-sample case, the assumed shape of the distribution, or the independence of the observations. To compare different tests of the same hypothesis, their asymptotic relative efficiency (ARE) is defined and a simple formula for it is obtained. As an application, it is shown, for example, that certain non-parametric tests offer unexpected efficiency advantages over their parametric competitors.

3.1 Critical values

The classical approach to hypothesis testing assumes that the random observables $X = (X_1, \ldots, X_n)$ have a distribution depending on some unknown parameters. In the simplest case, it depends only on one real-valued parameter θ, and the problem is that of testing the hypothesis

(3.1.1) $$H : \quad \theta = \theta_0$$

against the one-sided alternatives

(3.1.2) $$K : \quad \theta > \theta_0.$$

Suppose that H is rejected in favor of K when a suitable test statistic $T_n = T(X_1, \ldots , X_n)$ is too large, say, when

(3.1.3) $$T_n \geq C_n,$$

where C_n is determined by

(3.1.4) $$P_{\theta_0}(T_n \geq C_n) = \alpha.$$

Here α is the preassigned level of significance which controls the probability of falsely rejecting the hypothesis when it is true.

In this section, we shall consider the problem of obtaining an approximate value for C_n when n is large and for this purpose replace (3.1.4) by the weaker requirement of an *asymptotic level* α satisfying

(3.1.5) $$P_{\theta_0}(T_n \geq C_n) \to \alpha \text{ as } n \to \infty.$$

In the rest of the section, we shall suppress the subscript θ_0 in (3.1.4) and (3.1.5) since all calculations will be performed under this distribution.

In many situations, it turns out that

(3.1.6) $$\sqrt{n}(T_n - \theta_0) \xrightarrow{L} N(0, \tau^2(\theta_0))$$

so that

(3.1.7) $$\frac{\sqrt{n}}{\tau(\theta_0)}(T_n - \theta_0) \xrightarrow{L} N(0,1).$$

The left side of (3.1.5) will then tend to α, provided

(3.1.8) $$\frac{\sqrt{n}}{\tau(\theta_0)}(C_n - \theta_0) \to u_\alpha,$$

where

(3.1.9) $$1 - \Phi(u_\alpha) = \alpha.$$

Here Φ is the cdf of a standard normal variable. The test (3.1.3) will therefore have asymptotic level α when

(3.1.10) $$C_n = \theta_0 + \frac{\tau(\theta_0) u_\alpha}{\sqrt{n}} + o\left(\frac{1}{\sqrt{n}}\right)$$

and therefore in particular when

$$(3.1.11) \qquad C_n = \theta_0 + \frac{\tau(\theta_0) u_\alpha}{\sqrt{n}}.$$

For this critical value, the test can then be written as

$$(3.1.12) \qquad \frac{\sqrt{n}(T_n - \theta_0)}{\tau(\theta_0)} \geq u_\alpha.$$

Example 3.1.1 Poisson. If $X_1, \ldots, X_n$ are i.i.d. according to the Poisson distribution $P(\lambda)$, the hypothesis $H : \lambda = \lambda_0$ is rejected in favor of $K : \lambda > \lambda_0$ if $\bar{X} \geq C_n$. Since (3.1.6) holds for $T_n = \bar{X}$ with $\tau^2 = \lambda_0$, the critical value (3.1.11) becomes

$$(3.1.13) \qquad C_n = \lambda_0 + u_\alpha \sqrt{\frac{\lambda_0}{n}}$$

and the rejection region (3.1.12) reduces to

$$(3.1.14) \qquad \frac{\sqrt{n}(\bar{X} - \lambda_0)}{\sqrt{\lambda_0}} \geq u_\alpha.$$

$\square$

Note: Here and in other situations in which the distribution of T_n is lattice valued, (3.1.12) is typically improved by applying a continuity correction. In the present context this means replacing C_n by $C_n - \frac{1}{2n}$. A key reference to other refinements is Peizer and Pratt (1968) and Pratt (1968). For later references, see Johnson, Kotz, and Kemp (1992, Section 4.5).

The above argument easily extends to situations other than (3.1.6).

Example 3.1.2 Uniform. If $X_1, \ldots, X_n$ are i.i.d. according to the uniform distribution $U(0, \theta)$, the hypothesis $H : \theta = \theta_0$ is rejected against $\theta > \theta_0$ if $X_{(n)} \geq C_n$, where $X_{(n)}$ is the largest of the n X's. We saw in Example 2.3.7 that

$$n\left(\theta - X_{(n)}\right) \overset{L}{\to} E(0, \theta),$$

where $E(0, \theta)$ denotes the exponential distribution with density $\frac{1}{\theta} e^{-x/\theta}$ when $x > 0$. It follows that

$$\frac{n}{\theta}\left(\theta - X_{(n)}\right) \overset{L}{\to} E(0, 1).$$

Since $X_{(n)} \geq C_n$ is equivalent to

$$\frac{n\left(\theta_0 - X_{(n)}\right)}{\theta_0} \leq \frac{n\left(\theta_0 - C_n\right)}{\theta_0},$$

(3.1.5) will hold, provided

$$n\left(1 - \frac{C_n}{\theta_0}\right) \to l_\alpha$$

and hence in particular when

(3.1.15) $$C_n = \theta_0 - \frac{\theta_0 l_\alpha}{n},$$

where l_α is the lower α-point of the distribution $E(0,1)$. Since the cdf of the latter is $1 - e^{-x}, x > 0$, we have

(3.1.16) $$l_\alpha = -\log(1 - \alpha).$$

An approximate rejection region is therefore given by

(3.1.17) $$\frac{n(\theta_0 - X_{(n)})}{\theta_0} \le l_\alpha.$$

$\square$

In most applications, the distribution of the X's depends not only on the parameter θ being tested but also on certain additional *nuisance parameters* ϑ. For the sake of simplicity, suppose again that T_n satisfies (3.1.6) and (3.1.7), where, however, the asymptotic variance $\tau^2(\theta_0, \vartheta)$ may now depend on ϑ. Since ϑ is unknown, (3.1.11) or (3.1.12) cannot be used. However, often there is a simple remedy. If $\hat{\tau}_n$ is any consistent estimator of $\tau(\theta_0, \vartheta)$, it follows from (3.1.7) with $\tau(\theta_0) = \tau(\theta_0, \vartheta)$ and Slutsky's theorem that

(3.1.18) $$\frac{\sqrt{n}(T_n - \theta_0)}{\hat{\tau}_n} \xrightarrow{L} N(0, 1).$$

The process of dividing $\sqrt{n}(T_n - \theta_0)$ by its estimated standard error in order to produce a test with approximate level α is called *studentization*. The test with rejection region

(3.1.19) $$\frac{\sqrt{n}(T_n - \theta_0)}{\hat{\tau}_n} \ge u_\alpha$$

has asymptotic level α and we can put

(3.1.20) $$C_n = \theta_0 + \frac{\hat{\tau}_n}{\sqrt{n}} u_\alpha.$$

Note: The term "studentization" is a misnomer. The idea of replacing τ by $\hat{\tau}_n$ was already used by Laplace. Student's contribution was to work out the *exact* distribution of (3.1.18) in the following situation.

Example 3.1.3 One-sample t-test. Let $X_1, \dots, X_n$ be i.i.d. as $N\left(\xi, \sigma^2\right)$, both ξ and σ unknown, and consider testing

$$(3.1.21) \qquad\qquad H: \quad \xi = \xi_0$$

against the alternatives $\xi > \xi_0$. Since $\bar{X}$ is the natural estimator of ξ, one would want to reject H when $\bar{X}$ is sufficiently large. Now

$$(3.1.22) \qquad\qquad \frac{\sqrt{n}\left(\bar{X} - \xi\right)}{\sigma} \xrightarrow{L} N\left(0, 1\right),$$

so it is necessary to estimate σ. The standard estimator of σ^2 is

$$(3.1.23) \qquad\qquad \hat{\sigma}^2 = \frac{1}{n-1}\sum\left(X_i - \bar{X}\right)^2,$$

which is consistent by Example 2.1.3 and Theorem 2.1.3. On replacing σ by $\hat{\sigma}$, the rejection region (3.1.19) becomes

$$(3.1.24) \qquad\qquad t_n = \frac{\sqrt{n}\left(\bar{X} - \xi_0\right)}{\sqrt{\dfrac{1}{n-1}\sum\left(X_i - \bar{X}\right)^2}} \geq u_\alpha.$$

As in all asymptotic work, the question arises as to how accurate the approximation is. In the present case this means how close the probability of (3.1.24) is to the intended α. The actual level of (3.1.24) is shown for $\alpha = .1, .05$, and $.01$ in Table 3.1.1. The table makes it clear that the approximation is inadequate when $n = 5$ and is still very rough when $n = 10$. It becomes more acceptable when $n = 20$ and is reasonably satisfactory for most purposes when $n \geq 30$. The last of these statements does not follow from the fact that the probability of (3.1.24) tends to α combined with the values of Table 3.1.1 but stems from the belief—founded on experience, not on mathematically established results—that in a situation as regular as that of t_n, the smooth improvement exhibited by Table 3.1.1 will continue to obtain also for larger values of n. (For reference on the accuracy of the normal approximation and on further refinements, see Johnson, Kotz, and Balakrishnan (1994) and Stuart and Ord (1987).

TABLE 3.1.1. Level of approximate t-test (3.1.24)

	$n = 5$	10	20	30	50	80	100
$\alpha = .1$	.134	.116	.108	.105	.103	.102	.100
$= .05$	.088	.067	.058	.056	.054	.052	.050
$= .01$	.040	.023	.016	.013	.012	.011	.010

Note: The exact distribution of t_n, found by Student (1908), is now known as the Student t-distribution (with $n-1$ degrees of freedom). This distribution, which has been extensively tabled and which is widely available in statistical software, provides the exact value of C_n for which $P_{\xi_0}(t_n \geq C_n) = \alpha$. It follows from Example 2.4.3 that $C_n \to u_\alpha$ as $n \to \infty$. $\qquad\square$

Example 3.1.4 Normal variance. Under the assumptions of the preceding example, consider testing the hypothesis $H: \quad \sigma = \sigma_0$ against the alternatives $\sigma > \sigma_0$. The hypothesis is rejected when $S^2 = \sum (X_i - \bar{X})^2/n$ is sufficiently large. The unknown mean ξ is now a nuisance parameter; however, the distribution of S^2 depends only on σ, not on ξ. Despite the presence of a nuisance parameter, we are in the situation described at the beginning of the section: Once the data have been reduced to S^2, only a single parameter is involved. Example 2.4.4 found that

$$(3.1.25) \qquad \sqrt{n}\left(\frac{\sum (X_i - \bar{X})^2}{n} - \sigma^2\right) \to N\left(0, \tau^2\right),$$

where $\tau^2 = \text{Var}\left(X_i^2\right)$. In the present case, which assumes normality, $\tau^2 = 2\sigma^4$ and the rejection region (3.1.12) becomes

$$(3.1.26) \qquad \sqrt{n}\left(\frac{\sum (X_i - \bar{X})^2}{n} - \sigma_0^2\right) \geq u_\alpha \sqrt{2}\sigma_0^2$$

or

$$(3.1.27) \qquad \frac{1}{n}\sum (X_i - \bar{X})^2 \geq \sigma_0^2 + \sqrt{\frac{2}{n}}u_\alpha\sigma_0^2.$$

Note: The asymptotic level of this test is not affected if the factor $1/n$ of $\sum (X_i - \bar{X})^2$ is replaced by $1/(n-1)$ (Problem 1.7). $\qquad\square$

We next consider some two-sample problems. For this purpose, the following fact is useful, which in Chapter 5 will be seen to be a special case of a very general result (Theorem 5.1.5).

Lemma 3.1.1 *If U_N and V_N are independent, and if U, V are independent variables for which*

$$U_N \overset{L}{\to} U, \quad V_N \overset{L}{\to} V,$$

then

$$U_N \pm V_N \overset{L}{\to} U \pm V.$$

Suppose we have independent samples $X_1, \ldots, X_m$ and $Y_1, \ldots, Y_n$ with distributions depending on real-valued parameters ξ and η, respectively. Suppose that the hypothesis

$$H: \quad \eta = \xi$$

is rejected in favor of

$$K: \quad \eta > \xi$$

when the difference $V_n - U_m$ is sufficiently large, where $V_n = V(Y_1, \dots, Y_n)$ and $U_m = U(X_1, \dots, X_m)$. As in the one-sample case, we suppose that

(3.1.28)
$$\sqrt{m}(U_m - \xi) \xrightarrow{L} N(0, \sigma^2) \quad \text{as } m \to \infty,$$
$$\sqrt{n}(V_n - \eta) \xrightarrow{L} N(0, \tau^2) \quad \text{as } n \to \infty,$$

where σ^2 and τ^2 may depend on ξ and η, respectively. Let $N = m + n$ and suppose that for some $0 < \rho < 1$,

(3.1.29)
$$\frac{m}{N} \to \rho, \quad \frac{n}{N} \to 1 - \rho \text{ as } m \text{ and } n \to \infty,$$

so that

(3.1.30)
$$\sqrt{N}(U_m - \xi) \xrightarrow{L} N(0, \sigma^2/\rho),$$
$$\sqrt{N}(V_n - \eta) \xrightarrow{L} N(0, \tau^2/(1-\rho)).$$

It then follows from Lemma 3.1.1 that

$$\sqrt{N}[(V_n - \eta) - (U_m - \xi)] \xrightarrow{L} N\left(0, \frac{\sigma^2}{\rho} + \frac{\tau^2}{1-\rho}\right)$$

or, equivalently, that

(3.1.31)
$$\frac{[(V_n - U_m) - (\eta - \xi)]}{\sqrt{\dfrac{\sigma^2}{m} + \dfrac{\tau^2}{n}}} \xrightarrow{L} N(0,1).$$

Note: The assumption (3.1.29) can be avoided by applying the argument to subsequences for which m/N converges. More useful perhaps is the realization that in practice we are interested in (3.1.31) primarily as a basis for approximating the situation at hand. For this purpose, we embed the situation with given sample sizes m and n in a sequence of situations (with m and n tending to infinity) which can be chosen at will and so, in particular, by keeping m/N fixed. If, for example, $m = 10$ and $n = 20$, we could embed it in the sequence $\{m = K, n = 2K, K = 1, 2, \dots\}$.

If σ^2 and τ^2 are known, it follows from (3.1.31) that the rejection region

(3.1.32)
$$\frac{V_n - U_m}{\sqrt{\dfrac{\sigma^2}{m} + \dfrac{\tau^2}{n}}} \geq u_\alpha$$

provides a test of H against K at asymptotic level α. If σ^2 and τ^2 are unknown, let $\hat{\sigma}^2$ and $\hat{\tau}^2$ be consistent estimators. Then we shall show that

(3.1.32) remains valid when σ^2 and τ^2 are replaced by $\hat{\sigma}^2$ and $\hat{\tau}^2$ and hence that the test with rejection region

$$(3.1.33) \qquad \frac{V_n - U_m}{\sqrt{\dfrac{\hat{\sigma}^2}{m} + \dfrac{\hat{\tau}^2}{n}}} \geq u_\alpha$$

also has asymptotic level α.

To see that this substitution is legitimate, write the left side of (3.1.32) as

$$\frac{(V_n - U_m) - (\mu - \xi)}{\sqrt{\dfrac{\hat{\sigma}^2}{m} + \dfrac{\hat{\tau}^2}{n}}} \cdot \sqrt{\frac{\dfrac{\hat{\sigma}^2}{m} + \dfrac{\hat{\tau}^2}{n}}{\dfrac{\sigma^2}{m} + \dfrac{\tau^2}{n}}} \cdot$$

Now

$$\frac{N\left(\dfrac{\hat{\sigma}^2}{m} + \dfrac{\hat{\tau}^2}{n}\right)}{\dfrac{\sigma^2}{\rho} + \dfrac{\tau^2}{1-\rho}} \xrightarrow{P} 1 \text{ and } \frac{N\left(\dfrac{\sigma^2}{m} + \dfrac{\tau^2}{n}\right)}{\dfrac{\sigma^2}{\rho} + \dfrac{\tau^2}{1-\rho}} \to 1$$

and hence by Slutsky's theorem,

$$\frac{(V_n - U_m) - (\mu - \xi)}{\sqrt{\dfrac{\hat{\sigma}^2}{m} + \dfrac{\hat{\tau}^2}{n}}} \xrightarrow{L} N(0,1),$$

as was to be proved.

Example 3.1.5 Poisson and binomial two-sample problem. (i) Let $X_1, \ldots, X_m$ and $Y_1, \ldots, Y_n$ be i.i.d. according to Poisson distributions $P(\lambda)$ and $P(\mu)$, respectively. Then (3.1.28) applies with $U_m = \bar{X}$, $V_n = \bar{Y}$, $\xi = \lambda$, $\eta = \mu$, $\sigma^2 = \lambda$, and $\tau^2 = \mu$. Since $\bar{X}$ and $\bar{Y}$ are consistent estimators of λ and μ, the test (3.1.33) of $H : \lambda = \mu$ becomes

$$(3.1.34) \qquad \frac{\bar{Y} - \bar{X}}{\sqrt{\dfrac{\bar{X}}{m} + \dfrac{\bar{Y}}{n}}} \geq u_\alpha.$$

(ii) Let X and Y be independent with binomial distributions $b(p, m)$ and $b(\pi, n)$, respectively, representing the numbers of successes in binomial trials performed under two different conditions A and $\bar{A}$. Then (3.1.28) applies with $U_m = X/m$, $V_n = Y/n$, $\xi = p$, $\eta = \pi$, $\sigma^2 = p(1-p)$, and

$\tau^2 = \pi (1 - \pi)$. Since X/m and Y/n are consistent estimators of p and π, respectively, the test (3.1.33) of $H : \pi = p$ against $\pi < p$ becomes

$$(3.1.35) \qquad \frac{\dfrac{X}{m} - \dfrac{Y}{n}}{\sqrt{\dfrac{X}{m^2}\left(1 - \dfrac{X}{m}\right) + \dfrac{Y}{n^2}\left(1 - \dfrac{Y}{n}\right)}} \geq u_\alpha.$$

Note: The data of Part (ii) of this example can be displayed in a 2×2 contingency table showing the counts in the four possible categories:

	Success	Failure
A	X	$m - X$
$\bar{A}$	Y	$n - Y$

For the large literature concerning refinements of (3.1.35) and some of the issues involved, see Agresti (1990, p. 68). □

Example 3.1.6 Comparing two normal means (Behrens-Fisher problem). Let $X_1, \ldots, X_m$ and $Y_1, \ldots, Y_n$ be samples from normal distributions $N(\xi, \sigma^2)$ and $N(\eta, \tau^2)$, respectively, where all four parameters are unknown, and consider the problem of testing $H : \eta = \xi$ against the alternatives $K : \eta > \xi$. Since $\bar{Y} - \bar{X}$ is a natural estimator of $\eta - \xi$, let us reject H when $\bar{Y} - \bar{X}$ is too large and, more specifically, when (3.1.33) holds with $U_m = \bar{X}$, $V_n = \bar{Y}$,

$$\hat{\sigma}^2 = S_X^2 = \frac{1}{m-1} \sum (X_i - \bar{X})^2 \text{ and } \hat{\tau}^2 = S_Y^2 = \frac{1}{n-1} \sum (Y_j - \bar{Y})^2.$$

The exact probability $\alpha_{m,n}(\gamma)$ of (3.1.33) in this case is complicated and depends on $\gamma = \sigma^2/\tau^2$. However, as m and n tend to infinity, $\alpha_{m,n}(\gamma)$ tends to α regardless of the value of γ. (For the extensive literature concerning refinements of this test see Robinson (1982), Lehmann (1986, p. 304), and Stuart and Ord (1991, p. 772).)

An assumption that is frequently made in this example is that $\sigma^2 = \tau^2$. Then the natural test statistic is

$$(3.1.36) \qquad \frac{(\bar{Y} - \bar{X})/\sqrt{\dfrac{1}{m} + \dfrac{1}{n}}}{\sqrt{\left[\sum (X_i - \bar{X})^2 + \sum (Y_j - \bar{Y})^2\right]/(m+n-2)}}$$

and the situation is essentially that of Example 3.1.3 with the null distribution of (3.1.36) being the t-distribution with $m+n-2$ degrees of freedom. However, the gain in simplicity and a trivial gain in power must be balanced against the effect which a violation of the assumption $\sigma^2 = \tau^2$ may have on the level. (See Example 3.1.6 (continued).) □

Instead of testing the hypothesis (3.1.1) at a fixed level α by means of the test (3.1.3), one often prefers to state the associated p-value, i.e., the probability

$$(3.1.37) \qquad \hat{\alpha}(t) = P(T_n \geq t),$$

where t is the observed value of the test statistic T_n. If T_n satisfies the asymptotic normality assumption (3.1.6), the p-value $\hat{\alpha}(t)$ can be approximated by

$$(3.1.38) \qquad P_{\theta_0}\left[\frac{\sqrt{n}(T_n - \theta_0)}{\tau(\theta_0)} \geq \frac{\sqrt{n}(t - \theta_0)}{\tau(\theta_0)}\right] \doteq 1 - \Phi\left[\frac{\sqrt{n}(t - \theta_0)}{\tau(\theta_0)}\right].$$

It is seen from Examples 3.1.1–3.1.5 how limit theorems can be used to derive tests with approximately known signficance levels and to obtain approximate p-values. However, it is important to be clear about the limitations of this approach. The method provides a rather crude approximation which is intended to give a simple ballpark figure but not a value of reliably high accuracy. High accuracy can often be obtained from tables or computer programs, but typically at a price. The results lose their simplicity and the insights that the simple asymptotic formulas provide, and they often require specific distributional assumptions which cannot be expected to be exactly satisfied in practice.

The accuracy of an asymptotic method derived from a limit theorem depends on the speed of the convergence to the limit. When applying such a method to a new situation, it is good practice to spot-check the speed of convergence through simulation.

To conclude this section, we shall consider two issues that cannot be answered by the crude approach employed so far. We shall illustrate each of them with a typical example.

Consider once more the test (3.1.19) for testing $H : \theta = \theta_0$ against $\theta > \theta_0$ in the presence of a nuisance parameter ϑ. The probability of the rejection region (3.1.19) under the hypothesis will in general depend on ϑ; let us denote it by $\alpha_n(\vartheta)$. There we saw that under (3.1.18)

$$(3.1.39) \qquad \alpha_n(\vartheta) \to \alpha \text{ for all } \vartheta,$$

which was summarized by saying that the asymptotic level of the test (3.1.19) is α. This statement is justified in the sense that for any given values of ϑ and $\epsilon > 0$, there exists $n_0(\vartheta)$ such that

$$(3.1.40) \qquad |\alpha_n(\vartheta) - \alpha| < \epsilon \text{ for all } n > n_0(\vartheta).$$

There is, however, an alternative way of looking at the concept of asymptotic level. For any given n, the actual (finite sample) level of (3.1.19) is

$$(3.1.41) \qquad \alpha_n^* = \sup_{\vartheta} \alpha_n(\vartheta),$$

the maximum probability of rejection when H is true. One may therefore wish to require that

(3.1.42) $$\alpha_n^* \to \alpha \text{ as } n \to \infty.$$

Unfortunately, (3.1.39) is not sufficient to guarantee (3.1.42) (Problem 1.13). What is required instead (see Lemma 2.6.1) is that the convergence (3.1.39) is uniform in ϑ.

Example 3.1.6 Comparing two normal means (continued). In Example 3.1.6, the probability $\alpha_{m,n}(\gamma)$ of (3.1.33) under H depends on $\gamma = \sigma^2/\tau^2$. The function $\alpha_{m,n}(\gamma)$ has been carefully investigated (see, for example, Scheffé (1970)) and it has been shown that

(3.1.43) $$\alpha_{\nu_1} \le \alpha_{m,n}(\gamma) \le \alpha_{\nu_0} \text{ for all } \gamma,$$

where α_ν is the tail probability of the t-distribution with ν degrees of freedom

$$\alpha_\nu = P(t_\nu \ge u_\alpha)$$

and

(3.1.44) $$\nu_0 = \min(m-1, n-1), \quad \nu_1 = m+n-2.$$

Since, by Example 2.4.3, $\alpha_\nu \to \alpha$ as $\nu \to \infty$, it is seen that both the lower and upper bounds of $\alpha_{m,n}(\gamma)$ in (3.1.43) tend to α and hence that $\alpha_{m,n}(\gamma) \to \alpha$ uniformly in γ and $\sup_\gamma \alpha_{m,n}(\gamma) \to \alpha$ as m and n both tend to infinity. Table 3.1.2 gives the range (3.1.43) of $\alpha_{m,n}(\gamma)$ for some values of α and of m and n.

TABLE 3.1.2. Range of $\alpha_{m,n}(\gamma)$

	$\alpha = .01$	$\alpha = .05$	$\alpha = .1$
$m = n = 5$	(.024, .040)	(.069, .088)	(.118, .135)
$m = 5, n = 10$	(.018, .040)	(.062, .088)	(.111, .135)
$m = 5, n = 20$	(.015, .040)	(.057, .088)	(.110, .135)
$m = n = 10$	(.016, .022)	(.059, .067)	(.108, .116)
$m = 10, n = 20$	(.014, .022)	(.056, .067)	(.105, .116)
$m = 15, n = 20$	(.012, .018)	(.055, .061)	(.104, .110)

There exist relatively simple refinements of (3.1.33) such as the Welch approximate t-test and the Welch-Aspin test for which $\alpha_{m,n}(\gamma)$ differs only slightly from α over the whole range of γ when $\nu_0 \ge 4$. For references, see Scheffé (1970), Wang (1971), and Lehmann (1986).   □

An aspect of the large-sample approach which can cause confusion and uncertainty is non uniqueness i.e., the existence of several different large-sample tests all of which seem equally natural. It will often turn out that these tests are asymptotically equivalent (at least under the hypothesis) in the following sense.

Definition 3.1.1 Two sequences of tests with rejection regions R_n and R'_n respectively are *asymptotically equivalent* if under the hypothesis the probability of their leading to the same conclusion tends to 1 as $n \to \infty$.

If the rejection regions are given respectively by

$$(3.1.45) \qquad\qquad V_n \geq u_\alpha \text{ and } V'_n \geq u_\alpha,$$

then the tests will be asymptotically equivalent provided as $n \to \infty$,

$$(3.1.46) \qquad P\left[V_n \geq u_\alpha, V'_n < u_\alpha\right] + P\left[V_n < u_\alpha, V'_n \geq u_\alpha\right] \to 0.$$

Example 3.1.5(i) Poisson two-sample problem (continued).   In Example 3.1.5(i), the difference $\bar{Y} - \bar{X}$ of the sample means was studentized by dividing it by a suitable estimate of its standard deviation $\sqrt{\dfrac{\lambda}{m} + \dfrac{\mu}{n}}$. Since $\bar{X}$ and $\bar{Y}$ are consistent estimators of λ and μ, respectively, the denominator $\sqrt{\dfrac{\bar{X}}{m} + \dfrac{\bar{Y}}{n}}$ was chosen in (3.1.34). However, since the level of the test is calculated under the hypothesis $\mu = \lambda$, we could instead write the denominator in (3.1.32) as $\sqrt{\lambda}\sqrt{\dfrac{1}{m} + \dfrac{1}{n}}$ and estimate it by $\sqrt{\dfrac{m\bar{X} + n\bar{Y}}{m+n}}$ $\times \sqrt{\dfrac{1}{m} + \dfrac{1}{n}}$, which leads to the rejection region

$$(3.1.47) \qquad\qquad \frac{(\bar{Y} - \bar{X})\sqrt{mn}}{\sqrt{m\bar{X} + n\bar{Y}}} \geq u_\alpha.$$

Let us begin by comparing the test statistic of (3.1.34)

$$(3.1.48) \qquad\qquad V_N = \frac{\bar{Y} - \bar{X}}{\sqrt{\dfrac{\bar{X}}{m} + \dfrac{\bar{Y}}{n}}}$$

with that of (3.1.47)

$$(3.1.49) \qquad\qquad V'_N = \frac{(\bar{Y} - \bar{X})\sqrt{mn}}{\sqrt{m\bar{X} + n\bar{Y}}}.$$

An easy calculation shows that

$$(3.1.50) \qquad\qquad \frac{V'_N}{V_N} = \sqrt{\frac{n\bar{X} + m\bar{Y}}{m\bar{X} + n\bar{Y}}}$$

and hence that

$$(3.1.51) \qquad V_N' < V_N \iff (n - m)(\bar{Y} - \bar{X}) > 0.$$

Suppose now first that $m < n$. Then if H is rejected by V_N' at level $\alpha < 1/2$, it follows that H is also rejected by V_N. If R_N and R_N' denote the rejection regions of the test (3.1.34) and (3.1.47) respectively, we thus have

$$(3.1.52) \qquad R_N' \subset R_N.$$

It may seem surprising that of two tests with the same asymptotic level, one has a larger rejection region than the other. This could of course not happen if they had exactly the same level, but it causes no problem asymptotically where it implies that the probability of the set $R_N - R_N'$ tends to zero as $N \to \infty$. Since by (3.1.52), the tests (3.1.34) and (3.1.47) can lead to different conclusions only if the sample point falls into R_N but not into R_N', it follows from the remark just made that (3.1.46) holds in the present case and the two tests are therefore asymptotically equivalent.   □

The argument in the case $m > n$ is analogous.

In this example, the two tests differ only in the denominator used for studentizing the test statistic $\bar{Y} - \bar{X}$. The following theorem provides a very general class of situations in which two tests differing only in this way are asymptotically equivalent.

Theorem 3.1.1 *Consider two tests of $H : \theta = \theta_0$, one of them given by (3.1.19), the other by the rejection region*

$$(3.1.53) \qquad \frac{\sqrt{n}\,(T_n - \theta_0)}{\hat{\tau}_n'} \geq u_\alpha,$$

where $\hat{\tau}_n'$ is another consistent estimator of $\tau(\theta_0, \vartheta)$. Then the tests (3.1.19) and (3.1.53) are asymptotically equivalent.

Proof. If V_n and V_n' denote the left sides of (3.1.19) and (3.1.53), respectively, then the first term of (3.1.46) is equal to

$$(3.1.54) \qquad P\left[u_\alpha \hat{\tau}_n \leq \sqrt{n}\,(T_n - \theta_0) < u_\alpha \hat{\tau}_n'\right].$$

Since the middle term tends in law to a normal distribution and the first and third terms both tend in probability to $u_\alpha \tau(\theta_0, \vartheta)$, it follows from Theorem 2.3.3 that (3.1.54) tends to 0. By symmetry, the same is true for the second term of (3.1.46).   ∎

Summary

1. When a sequence of test statistics is asymptotically normal, this fact provides an approximation to the critical value when the hypothesis is simple. In the case of a composite hypothesis, the asymptotic variance of the test statistic can be estimated and an approximate critical value then follows from Slutsky's theorem. The calculations are illustrated on a number of parametric situations. Numerical examples show how the accuracy of the approximation improves with increasing sample size.

2. Two sequences of tests are asymptotically equivalent if the probability of their reaching opposite conclusions tends to 0 as the sample size(s) tend to infinity. To the order of approximation considered here, the theory is unable to distinguish between asymptotically equivalent tests.

3.2   Comparing two treatments

A frequently occurring question is whether a Treatment B is better than another Treatment A (for example, a standard treatment or a placebo). To test the hypothesis of no difference, Treatments A and B are applied respectively to m and n subjects drawn at random from a large population. Let $X_1, \ldots, X_m$ and $Y_1, \ldots, Y_n$ denote their responses and denote by F and G the distributions of the responses over the population if A or B were applied to all of its members. If the population is large enough for the $m+n$ subjects to be essentially independent, the situation can be represented by the *population model*

$$(3.2.1) \qquad \begin{array}{ll} X_1, \ldots, X_m : & \text{i.i.d. } F, \\ Y_1, \ldots, Y_n : & \text{i.i.d. } G \end{array}$$

with the two samples $X_1, \ldots, X_m$ and $Y_1, \ldots, Y_n$ also being independent of each other.

Note: The model (3.2.1) is appropriate also for the comparison of two populations Π and Π' on the basis of samples of m and n subjects drawn at random from these two populations.

In this section, we shall be concerned with testing the hypothesis

$$(3.2.2) \qquad\qquad H : F = G \text{ (continuous)}$$

under various assumptions.

When F and G are normal with common variances $F = N\left(\xi, \sigma^2\right)$ and $G = N\left(\mu, \sigma^2\right)$, the hypothesis reduces to $H : \mu = \xi$, and the two-sample t-test based on (3.1.36) is appropriate. When the distributions instead are

$F \sim N\left(\zeta, \sigma^2\right)$ and $G - N\left(\mu, \tau^2\right)$, we are dealing with the Behrens-Fisher problem (defined in Example 3.1.6) We shall now discuss the problem of testing (3.2.2) without the assumption of normality against alternatives which may be loosely described by saying that the Y's tend to be larger than the X's. (More specific alternatives will be considered in the next section.) Of course, the normal-theory tests are asymptotically valid without the assumption of normality, but their exact level depends on the common distribution F. We shall now describe a test for which even the exact level is independent of F.

Example 3.2.1 The Wilcoxon two-sample test. Let the responses of the $N = m + n$ subjects $(X_1, \ldots, X_m; Y_1, \ldots, Y_n)$ be ranked, with the smallest observation having rank 1, the next smallest rank 2, and so on. (The assumption of continuity in (3.2.2) ensures that the probability of any ties is 0. For a discussion of the problem without this assumption, see, for example, Lehmann (1998) or Pratt and Gibbons (1981).)

If $S_1 < \cdots < S_n$ denote the ranks of the Y's among the set of all N variables, then it follows from symmetry considerations that under the hypothesis (3.2.2)

$$(3.2.3) \qquad P_H\left(S_1 = s_1, \ldots, S_n = s_n\right) = \frac{1}{\binom{N}{n}}$$

for all $\binom{N}{n}$ possible choices of $1 \leq s_1 < s_2 < \cdots < s_n \leq N$ (Problem 2.1). The hypothesis is rejected by the Wilcoxon test when the rank-sum

$$W_s = \sum_{i=1}^{n} S_i$$

is sufficiently large. It follows from (3.2.3) that under H, the distribution of W_s is independent of the common underlying (continuous) distribution F of the X's and Y's. The null distribution of the Wilcoxon rank-sum statistics W_s has been extensively tabled and is available in statitistical packages.

By (3.2.3), this distribution is the same as that of a random sample of n from the set $\{1, \ldots, N\}$ of all ranks. This is exactly the situation of Example 2.8.6 where it was shown that

$$(3.2.4) \qquad \frac{W_s - \frac{1}{2}n\left(N+1\right)}{\sqrt{mn\left(N+1\right)/12}} \xrightarrow{L} N\left(0, 1\right).$$

The rejection region

$$(3.2.5) \qquad \frac{W_s - \frac{1}{2}n\left(N+1\right)}{\sqrt{mn\left(N+1\right)/12}} \geq u_\alpha$$

therefore has asymptotic level α as m and n tend to ∞.

The subjects used for the comparison of two treatments are often not obtained as a random sample from some population but may be the patients in the hospital or the students in the classroom available at the time. It is therefore fortunate that the Wilcoxon test can also be applied when the $N = n + m$ subjects are simply those at hand, *provided*

(3.2.6)   the assignment of the $m + n$ subjects, m to A and n to B, is made at random, i.e., in such a way that all possible $\binom{N}{n}$ assignments are equally likely.

Under this *randomization model*, the distribution of the B-ranks $S_1, \ldots, S_n$ continues to be given by (3.2.3). To see this, note that under H, the response of each subject is the same regardless to which treatment it is assigned. Thus the response, and therefore also its rank, can be considered to be attached to the subject even before the assignment is made. The selection of a random sample of subjects to be assigned to Treatment B therefore implies the random selection of their ranks and hence proves (3.2.3).

As a consequence, the rejection region (3.2.5) has asymptotic level α not only in the *population model* (3.2.1) but also in the *randomization model* (3.2.6) as m and n tend to ∞.

An alternative representation of W_s, which will be useful later in this chapter, can be obtained in terms of the Mann-Whitney statistic

$$(3.2.7) \qquad W_{XY} = \text{Number of pairs } (i, j) \text{ for which } X_i < Y_j.$$

It is, in fact, easy to show (Problem 2.2) that

$$(3.2.8) \qquad W_s = W_{XY} + \frac{1}{2}n(n+1).$$

It follows from (3.2.4) that

$$(3.2.9) \qquad \frac{W_{XY} - \frac{1}{2}mn}{\sqrt{mn(N+1)/12}} \xrightarrow{L} N(0,1).$$

The asymptotic normality of W_{XY} is proved by a very different argument (which extends to the case $F \neq G$) in Section 6.1. Tables for the Wilcoxon statistic are usually given in terms of W_{XY} rather than W_s.

If w denotes the observed value of W_{XY}, the p-value of the Wilcoxon test is

$$\hat{\alpha}(w) = P_H(W_{XY} \geq w),$$

which can be approximated by

$$(3.2.10) \qquad \hat{\alpha}(w) \doteq 1 - \Phi\left[\frac{w - \frac{1}{2}mn}{\sqrt{mn(N+1)/12}}\right].$$

The following table shows for three combinations of sample sizes and several values of w, in the region of greatest interest, the value of $\hat{\alpha}(w)$, its normal approximation without continuity correction (3.2.10), and the corresponding approximation with continuity correction given by

$$(3.2.11) \qquad \hat{\alpha}(w) \doteq 1 - \Phi\left[\frac{w - \frac{1}{2} - \frac{1}{2}mn}{\sqrt{mn(N+1)/12}}\right].$$

TABLE 3.2.1. Normal approximation of p-value of Wilcoxon test*

w	18	17	16	15	14		
Exact	.012	.024	.048	.083	.131		
without	.010	.019	.035	.061	.098		$m = 3, n = 6$
with	.014	.026	.047	.078	.123		
w	45	43	38	35	33		
Exact	.004	.010	.052	.106	.158		
without	.005	.011	.045	.091	.138		$m = 4, n = 12$
with	.006	.012	.051	.102	.151		
w	48	46	44	40	36	34	
Exact	.005	.010	.019	.052	.117	.164	
without	.006	.010	.018	.047	.104	.147	$m = n = 8$
with	.007	.012	.020	.052	.114	.159	

The table suggests that even with fairly small sample sizes, the normal approximation with continuity correction will be adequate for most purposes. It is also seen that in most cases the continuity correction improves the approximation but that this is not so for all cases, particularly not for very small probabilities. (For more accurate approximations, see Hodges, Ramsey, and Wechsler (1990).)   □

Example 3.2.2 The randomization t-test. For testing $H : F = G$ in the population model when F and G are normal with common variance, the appropriate test is the two-sample t-test, which rejects when

$$(3.2.12) \qquad t = \frac{(\bar{Y} - \bar{X})/\sqrt{\frac{1}{m} + \frac{1}{n}}}{\sqrt{\left[\sum(X_i - \bar{X}) + \sum(Y_j - \bar{Y})^2\right]/(m+n-2)}}$$

exceeds the critical value $t_{m,n}(\alpha)$ of the t-distribution with $m + n - 2$ degrees of freedom. However, the exact level of this t-test—unlike that of

---

*Adapted from Table 1.1 of Lehmann (1998).

the Wilcoxon test—is no longer α when F is not normal, although it tends to α as m and $n \to \infty$ (and hence $t_{m,n}(\alpha)$ to u_α).

We shall now discuss an alternative, also based on t, which is free of this defect. For this purpose, consider the null distribution of (3.2.12) in the randomization model (3.2.6). In this model, the subjects are fixed and, as was discussed in the proof of (3.2.3) following (3.2.6), the $N = m + n$ responses, say $a_1, \ldots, a_N$, of the N subjects can then also be regarded as fixed: The only random feature is the assignment of n of these subjects to receive Treatment B and their a-values thus becoming Y's. The proof of (3.2.3) therefore also shows that

$$(3.2.13) \qquad P\left(Y_1 = a_{i_1}, \ldots, Y_n = a_{i_n}\right) = \frac{1}{\binom{N}{n}}$$

for each choice of the n subscripts

$$1 \le i_1 < i_2 < \cdots < i_n \le N.$$

It follows that in the randomization model, the t-statistic (3.1.12) can take on $\binom{m+n}{n}$ values, each of which has probability $1/\binom{m+n}{n}$. This randomization distribution of t can be used as the basis for a test of H. If $r/\binom{N}{n} = \alpha$, the test rejects H for the r largest among the $\binom{N}{n}$ values of t, say, when

$$(3.2.14) \qquad t \ge K\left(a_1, \ldots, a_N\right).$$

Unfortunately, the evaluation of $K\left(a_1, \ldots, a_N\right)$ is prohibitive except for small values of m and n.

Let us therefore next consider the asymptotic behavior of the critical value K. For this purpose, we must embed the given vector $(a_1, \ldots, a_N)$ in a sequence of such vectors, say $\{(a_{N1}, \ldots, a_{NN}), N = 1, 2, \ldots\}$. We shall show below that for such a sequence

$$(3.2.15) \qquad K\left(a_{N1}, \ldots, a_{NN}\right) \to u_\alpha,$$

provided

$$(3.2.16) \qquad \frac{n}{N} \text{ is bounded away from 0 and 1}$$

and the a's satisfy

$$(3.2.17) \qquad \frac{\max\left(a_{Ni} - a_{N\cdot}\right)^2}{\sum\left(a_{Nj} - a_{N\cdot}\right)^2} \to 0 \text{ as } N \to \infty,$$

where

(3.2.18) $$a_{N.} = (a_{N1} + \cdots + a_{NN})/N.$$

Under these conditions, it follows from Theorem 2.8.2 and formulas (2.8.45) and (2.8.49) that for every u,

(3.2.19) $$P\left[\frac{\bar{Y} - a_{N.}}{\sqrt{\dfrac{m}{nN(N-1)}\sum(a_{Nj} - a_{N.})^2}} \le u\right] \to 1 - \Phi(u).$$

After some manipulation (Problem 2.4), it can be shown that the inequality in square brackets holds if and only if

(3.2.20) $$t \le \sqrt{\frac{N-2}{N-1}}\frac{u}{\sqrt{1 - \dfrac{1}{N-1}u^2}}.$$

Since the right side of (3.2.20) tends to u as $N \to \infty$, it follows that in the randomization model (3.2.6),

(3.2.21) $$P(t \le u) \to \Phi(u),$$

i.e., the randomization distribution of t tends to the standard normal distribution. This finally proves (3.2.15).

The limit result (3.2.21) suggests that if m and n are large and comparable in size, and if $\max(a_i - \bar{a})^2/\sum(a_j - \bar{a})^2$ is small, the critical value u_α of the normal distribution provides a reasonable approximation to the exact critical value of the randomization t-test. In this way, the usual t-test with either the critical value of the t-distribution (or the asymptotically equivalent u_α which equals α only asymptotically) can be viewed as an approximation to the randomization t-test. The latter has a level that is independent of F and can therefore be made exactly equal to α. (For the literature on randomization tests see, for example, Romano (1990) and Welch (1990).)

Let us finally consider the randomization t-test (3.2.14) under the assumptions of the population model (3.2.1). The ordered constants $a_{(1)} < \cdots < a_{(N)}$ then become the values taken on by the order statistics of the combined sample

$$(Z_1, \ldots, Z_{m+n}) = (X_1, \ldots, X_m; Y_1, \ldots, Y_n),$$

when $F = G$, and (3.2.13) becomes the conditional distribution of the Y's given the N values of the Z's. (That (3.2.13) is the correct formula of this conditional distribution follows from the fact that the joint distribution of the Z's is symmetric in its $m+n$ variables.) Thus the randomization t-test (3.2.14) has conditional level α for each set of values of the Z's and hence also unconditionally.

Furthermore, it can be shown that if F has finite fourth moment, then condition (3.2.17) (with the Z's in place of the a's) holds with probability tending to 1. It follows that the convergence (3.2.15) holds in probability (Problems 3.2.10 and 3.2.11). Thus the usual t-test approximates the randomization t-test not only in the randomization model but also in the population model.   □

For the two-sample problem, Examples 3.2.1 and 3.2.2 considered the t- and Wilcoxon tests for both the population and the randomization model. Corresponding possibilities are available for the one-sample problem. Again, there will be two models, but we shall this time consider three tests: the Wilcoxon, t-, and sign test.

A principal application of the two-sample tests was the comparison of two Treatments A and B on the basis of m and n subjects receiving A and B, respectively. Unless these subjects are fairly homogeneous, a treatment difference, when it exists, may be masked by the variability of the responses. It is then often more efficient to carry out the comparison on matched pairs of subjects. In the population case, we assume a population of matched pairs. Examples are a population of twins, or situations in which both treatments (e.g., two headache remedies) can be applied to the same subject. In the randomization case, there exists an additional possibility: to form pairs by matching on variables such as age, gender, severity of disease, and so forth.

Example 3.2.3 One-sample t-test (population model).   Consider a population of pairs from which a sample of N pairs is drawn, with one member from each pair receiving Treatment A and the other Treatment B. Let the N pairs of responses be (X_i, Y_i), $i = 1, \dots, N$. If we assume that they are i.i.d. according to a bivariate normal distribution, the differences $Z_i = Y_i - X_i$ will be i.i.d. univariate normal (see Section 5.2), say $N\left(\xi, \sigma^2\right)$, and the hypothesis of no treatment differences $H : \xi = 0$ is tested by the t-test of Example 3.1.3 with Z_i, instead of X_i and N instead of n.

If we drop the assumption of normality and only assume that the pairs (X_i, Y_i) are i.i.d. according to some unknown bivariate distribution, then under the hypothesis of no treatment difference, the variable $Z_i = Y_i - X_i$ has the same distribution as $-Z_i = X_i - Y_i$, i.e. the distribution of Z_i is symmetric about 0. If the Z's have finite variance σ^2, it is seen that (3.1.22) and (3.1.23) (with Z_i in place of X_i and N instead of n) remain valid and the t-test (3.1.24) therefore continues to have asymptotic level α.   □

Example 3.2.4 Sign test.   If the distribution of the Z's in the preceding example is not normal, the level of the t-test is only asymptotically equal to

α. A very simple test of the hypothesis H of symmetry about 0, which has an exact level is the sign test based on the number N_+ of positive Z's. This variable has a binomial distribution $b(p, N)$ with $p = P(Z_i > 0)$. Under H, we have $p = 1/2$, provided F is continuous. Under the alternatives that Y_i tends to be larger than X_i, and Z_i is therefore shifted to the right, H is rejected by the sign test when N_+ is large, with the critical value calculated from the binomial distribution. Since, under H,

$$(3.2.22) \qquad \left(\frac{N_+}{N} - \frac{1}{2}\right) / \sqrt{1/4} \overset{L}{\to} N(0,1) \quad \text{as } N \to \infty,$$

the exact rejection region can for large N be approximated by

$$(3.2.23) \qquad 2\left(\frac{N_+}{N} - \frac{1}{2}\right) \geq u_\alpha$$

which has asymptotic level α. $\qquad\qquad\qquad\qquad\qquad\qquad\qquad\qquad$ $\square$

Example 3.2.5 One-sample Wilcoxon test. The sign test uses from each Z_i only its sign, i.e., whether it is positive or negative. Suppose now that $N = 3$ and consider the two possible samples

$$-1, 4, 5 \quad \text{and} \quad -5, 1, 4.$$

They have the same value $N_+ = 2$, yet one feels that the first gives a stronger indication of a shift to the right. This suggests looking not only at the signs of the Z's but also at their absolute values. To obtain an analog of the two-sample Wilcoxon test, let us rank $|Z_1|, \ldots, |Z_N|$ and denote by $R_1 < \cdots < R_m$ and $S_1 < \cdots < S_n$ the ranks of the absolute values of the negative and positive Z's, where $n = N_+$ and $m = N - N_+$. For the two cases with $N = 3$ displayed above, we have $m = 1$ and $n = 2$, and, respectively,

$$R_1 = 1, \; S_1 = 2, \; S_3 = 3 \quad \text{and} \quad R_1 = 3, \; S_1 = 1, \; S_2 = 2.$$

The Wilcoxon signed-rank test rejects the hypothesis H of symmetry with respect to 0 when

$$(3.2.24) \qquad\qquad V_s = S_1 + \cdots + S_n > C.$$

Here the rank sum $S_1 + \cdots + S_n$ reflects both the number of positive Z's and their absolute values.

The rejection region (3.2.24) is formally the same as that based on $W_s = S_1 + \cdots + S_n$ in the two-sample case. However, in the earlier case, n was fixed; now it is the value of N_+ which is a random variable. This difference is reflected in the null distribution of the S's. In the two-sample case, it was given by (3.2.3). In the present case, we shall show that instead

$$(3.2.25) \qquad P_H(N_+ = n; S_1 = s_1, \ldots, S_n = s_n) = \frac{1}{2^N}$$

for each possible set $(n; S_1, \ldots, s_n)$.

When the assumption that the pairs (X_i, Y_i) consitute a random sample from a large population is not satisfied, the Wilcoxon test (3.2.24) can be applied if

(3.2.26)   the assignment of the two subjects within each pair to the two treatments is made at random, i.e., with probability $1/2$ each and independently for each of the N pairs.

In this *randomization model*, the absolute values $|Z_1|, \ldots, |Z_N|$ of the N responses, and hence their ranks, are fixed and only their signs are random. By (3.2.26), each of the 2^N possible sign combinations of the *signed ranks* $\pm 1, \ldots, \pm N$ is equally likely and has probability $1/2^N$. This proves (3.2.25) for the randomization model.

To obtain the asymptotic distribution of V_s, let $Z^{(1)}, \ldots, Z^{(N)}$ denote the Z's ordered according to their absolute values and let

$$I_j = \begin{cases} 1 & \text{if } Z^{(j)} > 0 \\ 0 & \text{if } Z^{(j)} \leq 0. \end{cases}$$

Then

(3.2.27) $$V_s = \sum_{j=1}^{N} j I_j$$

and, by (3.2.26), the I_j's are independent and $P(I_j = 1) = 1/2$. It follows that

(3.2.28) $$E(V_s) = \frac{1}{2} \sum_{j=1}^{N} j = \frac{1}{4} N(N+1)$$

and

(3.2.29) $$\text{Var}(V_s) = \frac{1}{4} \sum_{i=1}^{N} j^2 = \frac{1}{24} N(N+1)(2N+1).$$

Furthermore, it follows from Theorem 2.7.3 with $Y_j = I_j$ and $d_{N_j} = j$ that

(3.2.30) $$\frac{V_s - \frac{1}{4} N(N+1)}{\sqrt{N(N+1)(2N+1)/24}} \to N(0,1).$$

To prove (3.2.30), it is only necessary to check condition (2.7.17) (Problem 2.6). For a comparison of the normal approximation with the exact values, see Ramsey, Hodges, and Shaffer (1993).

So far we have considered the distribution of V_s for the randomization model. Suppose now that we are instead concerned with the population model in which the variables $Z_1, \dots, Z_N$ are i.i.d. according to a distribution F which is symmetric with respect to 0. In addition, we assume that F is continuous, so as to rule out (with probability 1) the possibility of zeros and ties. Then given $|Z_1|, \dots, |Z_N|$, (3.2.25) will continue to hold since each Z_j is as likely to be positive as negative. It follows from the earlier argument for which (3.2.25) was the starting point that (3.2.30) continues to hold in the present population model.

An alternative proof of (3.2.30) can be based on a representation of V_s which is analogous to the representation (3.2.8) for W_s. For each $1 \le i \le j \le N$, let $U_{ij} = 1$ when $(Z_i + Z_j)/2 > 0$ and $U_{ij} = 0$ otherwise. Then it can be shown that

$$(3.2.31) \qquad \sum\sum_{1 \le i \le j \le N} U_{ij} = V_s.$$

(For a proof see, for example, Lehmann (1998 p. 128).) A proof of (3.2.30) based on (3.2.31) will be given in Section 6.1.   □

Example 3.2.6 Randomization t-test (one-sample).   Both the signed Wilcoxon test statistic V_s and the sign test statistic N_+ have the same null distribution under the population and randomization model. However, as discussed in Example 3.2.3, this equality no longer holds for the t-test, which rejects when

$$(3.2.32) \qquad t_N = \frac{\sqrt{N}\bar{Z}}{\sqrt{\dfrac{1}{N-1}\sum(Z_i - \bar{Z})^2}}$$

exceeds the critical value $t_N(\alpha)$ of the t-distribution with $N - 1$ degrees of freedom. In the population model, if the common distribution F of the Z's satisfies the hypothesis

$$(3.2.33) \qquad H : F \text{ is symmetric about } 0$$

and if F has finite variance, the level of the t-test tends to α as N tend to infinity.

In analogy to the randomization t-test of Example 3.2.2, there exists a test based on t_N with exact level α in the present one-sample case. Consider the randomization model under which the N absolute values of the Z's are fixed, say $a_1, \dots, a_N$ and

$$(3.2.34) \qquad \begin{array}{l} Z_i = \pm a_i \text{ with probability } 1/2 \text{ each,} \\ \text{independently for } i = 1, \dots, N. \end{array}$$

This model obtains in particular under the assumption (3.2.26). Then the 2^N possible observation vectors $(Z_1, \dots, Z_N) = (\pm a_1, \dots, \pm a_N)$ are equally likely, each having probability $1/2^N$, and the same is true for the resulting 2^N values of t_N. If $r/2^N = \alpha$, the test that rejects for the largest r value of t_N, say, when

$$(3.2.35) \qquad t_N \geq K(a_1, \dots, a_N),$$

then will have exact level α under the model (3.2.34).

To obtain an approximation for $K = K(a_1, \dots, a_N)$, we proceed as in Example 3.2.2 and embed the given vector $(a_1, \dots, a_N)$ in a sequence $\{(a_{N1}, \dots, a_{NN}), N = 1, 2, \dots\}$ and shall show that then

$$(3.2.36) \qquad K(a_{N1}, \dots, a_{NN}) \to u_\alpha \text{ as } N \to \infty,$$

provided

$$(3.2.37) \qquad \frac{\max a_{Ni}^2}{\sum_{j=1}^N a_{Nj}^2} \to 0.$$

To see this, write Z_i as

$$(3.2.38) \qquad Z_i = a_{Ni} J_i,$$

where

$$J_i = 1 \text{ or } -1 \text{ as } Z_i > 0 \text{ or } < 0,$$

so that in particular

$$E(J_i) = 0 \text{ and } \mathrm{Var}(J_i) = 1.$$

Then it follows from Theorems 2.7.3 and 2.7.4 that

$$(3.2.39) \qquad \frac{\sum Z_i}{\sqrt{\sum Z_j^2}} = \frac{\sum Z_i}{\sqrt{\sum a_{Nj}^2}} \overset{L}{\to} N(0,1),$$

provided (3.2.37) holds. Thus (3.2.39) implies that

$$(3.2.40) \qquad P\left[\frac{\sum Z_i}{\sqrt{\sum a_{Nj}^2}} \geq u_\alpha \right] \to \alpha \text{ as } N \to \infty.$$

It is easy to see (Problem 2.8) that the inequality in square brackets holds if and only if

$$(3.2.41) \qquad t_N \geq \frac{u_\alpha}{\sqrt{1 - \dfrac{u_\alpha^2}{N}}} \cdot \sqrt{\frac{N}{N-1}}$$

and hence that

(3.2.42) $P\left[t_N \geq u_\alpha\right] \to \alpha,$

which implies (3.2.36).

The randomization t-test (3.2.35) has level α not only in the randomization model but also in the population model of Example 3.2.3, since then the conditional distribution of the Z's given $|Z_i| = a_i\,(i = 1, \ldots, N)$ is given by (3.2.34)—and hence its conditional level is α—for each vector $(a_1, \ldots, a_N)$. In addition, it follows from (3.2.39) and (3.2.37) that the conditional critical value $K\left(a_1, \ldots, a_N\right)$ can be expected to be close to the critical point u_α of the normal or the critical value $t_N\left(\alpha\right)$ of the t-distribution when N is large and $\max a_i^2 / \sum_{j=1}^{N} a_j^2$ is small. Whether u_α or $t_N\left(\alpha\right)$ provides the better approximation is investigated by Diaconis and Holmes (1994). They show that up to second order terms, and with certain additional conditions on the a's, the normal approximation is more accurate than the t-approximation if and only if $K \leq 3/2$ where

$$K = N \sum d_i^4 / \left(\sum d_i^2\right)^2.$$

Their paper also develops an algorithm which makes it practicable to evaluate $K\left(a_1, \ldots, a_N\right)$ exactly for $N \leq 30$.   □

Summary

1. For the comparison of two treatments, two types of models are proposed: population models in which the subjects are assumed to be drawn at random from some large population, and randomization models in which the subjects are fixed but their assignment to the two treatments is random. Both the sampling and the assignment are carried out according to some specified design. Two designs are considered: (a) completely random drawing or assignment and (b) paired comparisons.

2. The Wilcoxon tests for (a) and (b) and the sign tests for (b) are distribution-free in the population model and are the same in the population and the randomzation model.

3. The exact level of the t-tests for (a) and (b) in the population model depends on the underlying distribution and (except in the normal case) attains the nominal level only asymptotically. These tests can be viewed as approximations to the randomization version of the t-test, for which the nominal level is exact.

3.3   Power and sample size

When testing $H : \theta = \theta_0$ against $\theta > \theta_0$, one important aspect of the test (3.1.3) is the level α which controls the probability of rejecting H when in fact $\theta = \theta_0$. The calculations necessary to achieve the assigned level at least approximately were discussed in the preceding sections. Another equally important aspect which we shall now take up is the *power* of the test against an alternative θ, i.e., the probability

$$(3.3.1) \qquad \beta_n(\theta) = P_\theta(T_n \geq C_n)$$

of rejecting H when in fact an alternative θ is the true parameter value. If the hypothesis being tested is one of "no effect," the power is the probability of detecting an effect when it exists. Since it is desirable to reject the hypothesis when it is false, one would like $\beta_n(\theta)$ to be large when $\theta > \theta_0$. As the sample size n increases, one would expect $\beta_n(\theta)$ to increase, hopefully to 1, as $n \to \infty$.

Definition 3.3.1 The sequence of tests (3.1.3) is said to be *consistent* against the alternative θ if

$$(3.3.2) \qquad \beta_n(\theta) \to 1 \text{ as } n \to \infty.$$

Theorem 3.3.1 *Suppose that not only (3.1.6) holds but also that*

$$(3.3.3) \qquad \sqrt{n}(T_n - \theta) \overset{L}{\to} N\left(0, \tau^2(\theta)\right) \text{ for all } \theta > \theta_0.$$

Then the test (3.1.12) is consistent against all alternatives $\theta > \theta_0$.

Proof. The power of the test with rejection region (3.1.12) can be written as

$$(3.3.4) \qquad \beta_n(\theta) = P\left[\sqrt{n}(T_n - \theta) \geq u_\alpha \tau(\theta_0) - \sqrt{n}(\theta - \theta_0)\right];$$

since $u_\alpha \tau(\theta_0) - \sqrt{n}(\theta - \theta_0) \to -\infty$ when $\theta_0 < \theta$, it follows from (3.3.3) and Problem 2.3.16 that $\beta_n(\theta) \to 1$ as $n \to \infty$. ∎

This result establishes consistency in Examples 3.1.1 and 3.1.4 and a slight extension also shows it for Example 3.1.2 (Problem 3.1).

Note: The same argument also shows that $\beta_n(\theta) \to 0$ for $\theta < \theta_0$ (assuming (3.3.3) to hold also for all $\theta < \theta_0$).

Let us next consider the situation with nuisance parameters.

Theorem 3.3.2 *Suppose that*

$$(3.3.5) \qquad \sqrt{n}(T_n - \theta) \overset{L}{\to} N\left(0, \tau^2(\theta, \vartheta)\right) \text{ for all } \theta > \theta_0 \text{ and all } \vartheta$$

and that $\hat{\tau}_n^2$ is a consistent estmator of $\tau^2(\theta, \vartheta)$. Then the test (3.1.19) is consistent against all alternatives (θ, ϑ) with $\theta > \theta_0$.

Proof. The power of the test (3.1.19) against (θ, ϑ) is

$$\beta_n(\theta, \vartheta) = P\left[\sqrt{n}(T_n - \theta) - u_\alpha \hat{\tau}_n \geq -\sqrt{n}(\theta - \theta_0)\right].$$

By Theorem 2.3.3 of Chapter 2,

$$\sqrt{n}(T_n - \theta) - u_\alpha \hat{\tau}_n \xrightarrow{L} N\left(-u_\alpha \tau(\theta, \vartheta), \tau^2(\theta, \vartheta)\right)$$

and the result then follows from the fact that $\sqrt{n}(\theta - \theta_0) \to \infty$. ■

Theorem 3.3.2 establishes consistency of the tests in Examples 3.1.3, 3.1.5, and 3.1.6 (Problem 3.2).

While consistency is reassuring, it provides little information concerning the power of a test for a given sample size and alternative. What is needed for this purpose is not that the limiting power is 1 but some limit value which depends on the alternative and is strictly between α and 1. Such a result can be obtained by embedding the actual situation with fixed n and θ in a suitable sequence (n, θ_n). Consistency obtains when the information provided by the data increases with increasing sample size, so that eventually near-perfect discrimination between θ_0 and any fixed alternative θ becomes possible. To keep the power away from 1, it is then necessary to make the discrimination between θ_0 and the alternative more difficult as n increases, and this is achieved by considering alternatives θ_n which get closer to θ_0 as n increases and in the limit tend to θ_0.

The approach seems contrived but becomes quite reasonable when seen in context. The aim is to calculate the value of the power against a *fixed* alternative θ for a *given* sample size n_0. We expect this power to have some intermediate value β, larger than α but smaller than 1. A limit argument requires embedding the actual situation (θ, n_0) in a sequence (θ_n, n), $n = 1, 2, \ldots$, and one may hope for a good approximation if $\lim \beta_n(\theta_n)$ is close to β. An approximate value of $\beta_n(\theta_0)$ is then obtained by identifying n with n_0 and θ_n with θ in this limiting power. How to implement this identification will be shown following the proof of Theorem 3.3.3.

To see how to choose θ_n, consider the situation assumed in Theorem 3.3.1. Replacing θ by θ_n in (3.3.4) gives

$$\beta_n(\theta_n) = P\left[\sqrt{n}(T_n - \theta_n) \geq u_\alpha \tau(\theta_0) - \sqrt{n}(\theta_n - \theta_0)\right],$$

and this will tend to a limit strictly between α and 1, provided $\sqrt{n}(\theta_n - \theta_0)$ tends to a finite positive limit Δ. These considerations motivate the following result.

Theorem 3.3.3 *Suppose that (3.3.3) holds in some neighborhood of θ_0 and that, in addition,*

(3.3.6)
$$\frac{\sqrt{n}(T_n - \theta_n)}{\tau(\theta_n)} \xrightarrow{L} N(0, 1) \text{ when } \theta_n \to \theta_0$$

and that $\tau^2(\theta)$ is a continuous function of θ. Then for

(3.3.7)
$$\theta_n = \theta_0 + \frac{\Delta}{\sqrt{n}} + o\left(\frac{1}{\sqrt{n}}\right),$$

the power function $\beta_n(\theta_n)$ of the test (3.1.3) against the alternatives θ_n has the limit

(3.3.8)
$$\beta_n(\theta_n) \to \Phi\left(\frac{\Delta}{\tau(\theta_0)} - u_\alpha\right).$$

Proof. Since $\tau(\theta_n) \to \tau(\theta_0)$ as $\theta_n \to \theta_0$, it follows from (3.3.6) and Theorem 2.3.3 that

(3.3.9)
$$\frac{\sqrt{n}(T_n - \theta_n)}{\tau(\theta_0)} \xrightarrow{L} N(0,1)$$

when θ_n is the true value. Now, by (3.3.4),

$$\beta_n(\theta_n) = P\left[\frac{\sqrt{n}(T_n - \theta_n)}{\tau(\theta_0)} \geq u_\alpha - \frac{\sqrt{n}(\theta_n - \theta_0)}{\tau(\theta_0)}\right]$$

and this tends to the right side of (3.3.8) by Theorem 2.3.3.   ∎

We shall refer to the right side of (3.3.8) as the *asymptotic power* of the test (3.1.3) against the alternatives (3.3.7) and when considered as a function of Δ, as the *asymptotic power function* of (3.1.3).

This limit result can be used to obtain an approximation to the power of the test (3.1.3) for a given alternative θ and sample size n by identifying θ with $\theta_n = \theta_0 + \dfrac{\Delta}{\sqrt{n}}$ and solving for Δ to get

(3.3.10)
$$\Delta = \sqrt{n}(\theta_n - \theta_0).$$

Substitution in (3.3.8) gives for $\beta_n(\theta)$ the approximation

(3.3.11)
$$\beta_n(\theta) \doteq \Phi\left[\frac{\sqrt{n}(\theta - \theta_0)}{\tau(\theta_0)} - u_\alpha\right].$$

A different approximation to $\beta_n(\theta)$ is obtained by starting from (3.3.6) instead of (3.3.9). One then has

$$\beta_n(\theta_n) = P\left[\frac{\sqrt{n}(T_n - \theta_n)}{\tau(\theta_n)} \geq \frac{\tau(\theta_0)}{\tau(\theta_n)}u_\alpha - \frac{\sqrt{n}(\theta_n - \theta_0)}{\tau(\theta_n)}\right],$$

which leads, instead of (3.3.11), to the approximation

(3.3.12)
$$\beta_n(\theta) \doteq \Phi\left[\frac{\sqrt{n}(\theta - \theta_0)}{\tau(\theta)} - u_\alpha\frac{\tau(\theta_0)}{\tau(\theta)}\right].$$

One would expect (3.3.12) typically to be slightly more accurate than (3.3.11) since the latter replaces the true value $\tau(\theta)$ by an approximation. However, since other approximations are also involved, this will not always be the case. In certain applications, the variance of T_n is much easier to obtain under the hypothesis than under the alternatives; in such situations, (3.3.11) has the advantage of not requiring the latter calculation. An example comparing the accuracy of the two approximations is given in Table 3.3.1.

Lack of uniqueness such as that just encountered is a central feature of large-sample theory which attains its results by embedding the given situation in a sequence with an infinite variety of different embedding sequences available. As was pointed out in Section 1.2, the choice is typically guided by the desire for both simplicity and accuracy. Simplicity is an important virtue since it helps to highlight the most important features of a procedure, thereby providing a powerful aid to intuitive understanding.

On the other hand, simplicity and highlighting of essentials often are achieved by ignoring many of the finer details and may blur important distinctions. In addition, such a broad-gauged approach cannot be expected to always achieve high accuracy. The resulting approximations can typically be improved, and some of the distinctions restored by adding second (or higher) order terms. Although we shall occasionally discuss such refinements, the emphasis in this book will be on the unrefined first order approximations obtainable by the convergence in law of a sequence of distributions.

Example 3.3.1 Binomial. If X has the binomial distribution $b(p, n)$, then (3.3.3) holds with $T_n = X/n$, $\theta = p$, and $\tau^2(\theta) = pq$. Clearly, $\tau^2(\theta)$ is continuous, and (3.3.6) follows for any $0 < p_0 < 1$ from Example 2.4.8.

Theorem 3.3.3 thus applies, and for the problem of testing $H : p = p_0$ against $p > p_0$ at asymptotic level α it shows that the power against the sequence of alternatives

$$(3.3.13) \qquad p_n = p_0 + \frac{\Delta}{\sqrt{n}}$$

has the limit

$$(3.3.14) \qquad \beta_n(p_n) \to \Phi\left(\frac{\Delta}{\sqrt{p_0 q_0}} - u_\alpha\right).$$

For the power $\beta_n(p)$ against a fixed alternative p, we obtain from (3.3.11) and (3.3.12) the approximations

$$(3.3.15) \qquad \beta_n(p) \doteq \Phi\left(\frac{\sqrt{n}(p - p_0)}{\sqrt{p_0 q_0}} - u_\alpha\right)$$

and

$$(3.3.16) \qquad \beta_n(p) \doteq \Phi \left[\frac{\sqrt{n}(p - p_0)}{\sqrt{pq}} - u_\alpha \sqrt{\frac{p_0 q_0}{pq}} \right].$$

TABLE 3.3.1. Approximations to level and power in the binomial case; $n = 100$.

		Level		Power at $p = .1$		
		Exact	Approx.	Exact	Approx. (3.3.15)	Approx. (3.3.16)
	.1	.040	.023	.872	.954	.897
$p_0 = .3$	.053	.041	.693	.704	.692
	.5	.044	.036	.623	.639	.641

Table 3.3.1 shows for $n = 100$ and $p_0 = .1, .3, .5$ the exact level closest to $\alpha = .05$, the corresponding asymptotic level of (3.1.12), the exact power against the alternative $p = p_0 + .1$, and its approximations (3.3.15) and (3.3.16). The table shows how rough the approximation (3.3.15) is even when $n = 100$.[†] Improvement is possible by use of the continuity correction (Problem 3.3). The table shows clearly the decrease of power for a fixed value of $p - p_0$ as $p_0 q_0$ increases, i.e., as p_0 moves toward $1/2$.   □

The principal difficulty in the application of Theorem 3.3.3 is the checking of condition (3.3.6), which requires a certain amount of uniformity in the convergence of T_n. Let us consider some examples which illustrate the argument that can be used to establish (3.3.6).

Example 3.3.2 Poisson. Let $X_1, \ldots, X_n$ be i.i.d. according to the Poisson distribution $P(\lambda)$. To validate Theorem 3.3.3 and thereby obtain an approximation for the power of the test (3.1.14) of Example 3.1.1, we need to show that

$$(3.3.17) \qquad \frac{\sqrt{n}(\bar{X} - \lambda_n)}{\sqrt{\lambda_n}} \xrightarrow{L} N(0, 1)$$

as λ_n tends to the hypothetical value λ_0. This will follow from the Berry-Esseen theorem, provided the standardized third moment

$$\frac{E|X_1 - \lambda_n|^3}{\lambda_n^{3/2}}$$

[†]For a careful discussion of the accuracy of the normal approximation to the level of the binomial test with and without continuity correction, see Ramsey and Ramsey (1988) and Johnson, Kotz, and Kemp (1992, Section 6.1).

can be shown to remain bounded as $\lambda_n \to \lambda_0$. This third moment condition is an immediate consequence of the facts that

$$E|X_1 - \lambda_n|^3 \to E|X_1 - \lambda_0|^3 < \infty$$

and

$$\lambda_n^{3/2} \to \lambda_0^{3/2} > 0.$$

What saves the day in this and similar cases is that boundedness of the standardized third moment is not required as λ_n tends to zero or infinity but only in the neighborhood of some fixed $0 < \lambda_0 < \infty$.

It now follows from Theorem 3.3.3 that the power $\beta_n(\lambda_n)$ of the test (3.1.14) has the limit

$$(3.3.18) \qquad \beta_n(\lambda_n) \to \Phi\left(\frac{\Delta}{\sqrt{\lambda_0}} - u_\alpha\right)$$

against the alternatives $\lambda_n = \lambda_0 + \dfrac{\Delta}{\sqrt{n}}$. An approximation to the power $\beta_n(\lambda)$ against a given alternative λ is obtained from (3.3.11) as

$$(3.3.19) \qquad \Phi\left(\frac{\sqrt{n}(\lambda - \lambda_0)}{\sqrt{\lambda_0}} - u_\alpha\right).$$

$\square$

Theorem 3.3.3 extends easily to problems involving nuisance parameters in addition to the parameter θ being tested.

Theorem 3.3.4 *Under the assumptions of Theorem 3.3.3 and the additional assumptions that for a sequence of distributions (θ_n, ϑ) with $\theta_n \to \theta_0$, the estimator $\hat{\tau}_n^2$ is a consistent estimator of $\tau^2(\theta_0, \vartheta)$ and that $\tau^2(\theta, \vartheta)$ is a continuous function of θ for each ϑ, formula (3.3.8) remains valid for the power of the test (3.1.19), where $\beta_n(\theta_n)$ and $\tau^2(\theta_0)$ have to be replaced by $\beta_n(\theta_n, \vartheta)$ and $\tau_n^2(\theta_0, \vartheta)$, respectively.*

Theorem 3.1.1 showed that tests obtained by studentizing $\sqrt{n}(T_n - \theta_0)$ by means of different consistent denominators are asymptotically equivalent. From formula (3.3.8), we now see in addition that these different asymptotically equivalent tests all have the same asymptotic power function. Note, however, that consistency of $\hat{\tau}_n$ as the estimator of $\tau(\theta_0, \vartheta)$ is now required not only under the distribution $P_{\theta_0, \vartheta}$ but also under the sequence $P_{\theta_n, \vartheta}$ where $\theta_n \to \theta_0$. For an example showing that the latter does not follow automatically, see Problem 3.4.

Example 3.3.3 One-sample t-test. Under the assumptions of Example 3.1.3, it follows from the Berry-Esseen theorem that for a sequence $\xi_n \to \xi_0$ and fixed σ,

$$\frac{\sqrt{n}\,(\bar{X} - \xi_n)}{\sigma} \overset{L}{\to} N(0,1)$$

and hence that also

$$(3.3.20) \qquad \frac{\sqrt{n}\,(\bar{X} - \xi_n)}{\hat{\sigma}} \overset{L}{\to} N(0,1),$$

where $\hat{\sigma}^2$ given by (3.1.23) is a consistent estimator of σ^2 regardless of the value of ξ. This verifies condition (3.3.6) and by Theorem 3.3.4 shows that the power of the t-test against a sequence of alternatives

$$\xi_n = \xi_0 + \frac{\Delta}{\sqrt{n}}$$

and fixed σ has the limit

$$(3.3.21) \qquad \beta_n(\xi_n) \to \Phi\left(\frac{\Delta}{\sigma} - u_\alpha\right).$$

By the usual argument, this leads to the approximation for the power against a fixed alternative (ξ, σ),

$$(3.3.22) \qquad \beta_n(\xi) \doteq \Phi\left(\frac{\sqrt{n}\,(\xi - \xi_0)}{\sigma} - u_\alpha\right).$$

□

Let us next apply this approach to the two-sample tests considered in Section 3.1 following Lemma 3.1.1. Under assumptions (3.1.28) and (3.1.29), it was seen that (3.1.33) provides a test of asymptotic level α. Suppose now that the test statistics U_m and V_n satisfy not only (3.1.28) but in analogy with (3.3.6),

$$(3.3.23) \qquad \begin{aligned} \sqrt{N}\,(U_m - \xi_m) &\overset{L}{\to} N(0, \sigma^2/\rho), \\ \sqrt{N}\,(V_n - \eta_n) &\overset{L}{\to} N(0, \tau^2/(1-\rho)), \end{aligned}$$

where $\xi_m \to \xi, \eta_n \to \eta$. Then for

$$(3.3.24) \qquad \xi_m = \xi, \quad \eta_n = \xi + \frac{\Delta}{\sqrt{N}},$$

the power function of the test (3.1.32) for testing $H : \eta = \xi$ satisfies

$$(3.3.25) \qquad \beta_N(\xi_m, \eta_n) \to \Phi\left[\frac{\Delta}{\sqrt{\dfrac{\sigma^2}{\rho} + \dfrac{\tau^2}{1-\rho}}} - u_\alpha\right].$$

Suppose, finally, that $\hat{\sigma}_N^2$ and $\hat{\tau}_N^2$ are consistent estimators of σ^2 and τ^2 not only when $\eta = \xi$ but also under the alternatives (3.3.24). Then (3.3.25) is valid for the power not only of (3.1.32) but also of (3.1.33).

The limit result (3.3.25) can be used like that of Theorem 3.3.3 to obtain an approximation to the power of (3.1.33) against a given alternative (ξ, η) and sample sizes m and n. Identifying η with η_n, we find $\Delta = \sqrt{N}\,(\eta - \xi)$. Substitution in (3.3.25) with $m = \rho N$ and $n = (1 - \rho)\,N$ then leads to

$$(3.3.26) \qquad \beta_N\,(\xi, \eta) \doteq \Phi\left[\frac{\eta - \xi}{\sqrt{\dfrac{\sigma^2}{m} + \dfrac{\tau^2}{n}}} - u_\alpha\right].$$

Example 3.3.4 Poisson two-sample problem. As an illustration, consider the power function of (3.1.34). Here $\xi = \lambda, \eta = \mu, \sigma^2 = \lambda$, and $\tau^2 = \mu$. Condition (3.3.23) was checked in Example 3.3.2. It therefore remains to check the conditions on the variances $\hat{\sigma}_m^2 = \bar{X}$ and $\hat{\tau}_n^2 = \bar{Y}$. To see that $\bar{Y} \to \mu$ in probability when $Y_1, \ldots, Y_n$ are i.i.d. $P\,(\mu_n)$ with $\mu_n \to \mu$, recall Theorem 2.1.1. This shows that $\bar{Y} - \mu_n \overset{P}{\to} 0$, provided $\mathrm{Var}\,(\bar{Y}) = \mu_n / n \to 0$, which clearly is the case. Formula (3.3.25) thus shows that the power of (3.1.34) against a sequence of alternatives $\mu_n = \lambda + \frac{\Delta}{\sqrt{N}}$ satisfies

$$(3.3.27) \qquad \beta_N\,(\lambda, \mu_n) \to \Phi\left[\frac{\Delta}{\sqrt{\lambda}}\sqrt{\rho\,(1 - \rho)} - u_\alpha\right],$$

which by (3.3.26) leads to the approximate formula

$$(3.3.28) \qquad \beta_N\,(\lambda, \mu) \doteq \Phi\left[\frac{\mu - \lambda}{\sqrt{\dfrac{\lambda}{m} + \dfrac{\mu}{n}}} - u_\alpha\right].$$

$\square$

In Theorem 3.3.3, the parameter θ plays a double role. It both labels the distribution and indicates the location of the statistic T_n. For later applications, it is convenient to allow for the separation of these two roles and to replace (3.3.3) by

$$(3.3.29) \qquad \sqrt{n}\,[T_n - \mu\,(\theta)] \overset{L}{\to} N\,(0, \tau^2\,(\theta)).$$

This modification leads to the following generalization of Theorem 3.3.3.

Theorem 3.3.5
(i) Suppose that (3.3.29) holds for $\theta = \theta_0$. Then the test of $H : \theta = \theta_0$, which rejects when

$$(3.3.30) \qquad \frac{\sqrt{n}\,[T_n - \mu\,(\theta_0)]}{\tau\,(\theta_0)} \geq u_\alpha$$

has asymptotic level α as $n \to \infty$.
(ii) Suppose, in addition, that

(3.3.31)
$$\frac{\sqrt{n}\,[T_n - \mu(\theta_n)]}{\tau(\theta_0)} \xrightarrow{L} N(0,1) \quad \text{when } \theta_n \to \theta_0$$

and that the function μ is differentiable at θ_0 with

(3.3.32)
$$\mu'(\theta_0) > 0.$$

Then the power $\beta_n(\theta_n)$ of the test (3.3.30) against the alternatives (3.3.7) has the limit

(3.3.33)
$$\beta_n(\theta_n) \to \Phi\left[\frac{\Delta\mu'(\theta_0)}{\tau(\theta_0)} - u_\alpha\right].$$

Proof. We have

$$\beta_n(\theta_n) = P\left\{\frac{\sqrt{n}\,[T_n - \mu(\theta_n)]}{\tau(\theta_0)} \geq u_\alpha - \frac{\sqrt{n}\,[\mu(\theta_n) - \mu(\theta_0)]}{\tau(\theta_0)}\right\}.$$

Now, by Theorem 2.5.1 of Chapter 2,

$$\mu(\theta_n) - \mu(\theta_0) = (\theta_n - \theta_0)\,\mu'(\theta_0) + o(\theta_n - \theta_0)$$
$$= \frac{\Delta}{\sqrt{n}}\mu'(\theta_0) + o\left(\frac{1}{\sqrt{n}}\right)$$

and substitution in $\beta_n(\theta_n)$ completes the proof.   ∎

For a fixed alternative θ, one finds, in analogy with (3.3.11), the approximation

(3.3.34)
$$\beta_n(\theta) \doteq \Phi\left[\frac{\sqrt{n}\,(\theta - \theta_0)\,\mu'(\theta_0)}{\tau(\theta_0)} - u_\alpha\right].$$

As an illustration, consider the following version of Example 3.2.4.

Example 3.3.5 Sign test for center of symmetry. For comparing two treatments in a paired comparisons design, Example 3.2.4 discussed the sign test based on the number N_+ of positive differences $Z_i = Y_i - X_i$ $(i = 1, \ldots, N)$. As was pointed out in the discussion preceding Example 3.2.4, under the hypothesis H of no treatment difference, the distribution of the Z's is symmetric about 0. In order to discuss the power of the test, it is necessary to state the alternatives to H more specifically than was done in Example 3.2.4. A natural such specification is obtained by assuming that superiority of Treatment B over Treatment A results in shifting the distribution of the Z's to the right by a fixed amount θ, so that under the alternatives, this distribution is symmetric about some $\theta > 0$. The problem then becomes that of testing

(3.3.35)
$$H : \theta = 0 \quad \text{against} \quad K : \theta > 0,$$

on the basis of observations $Z_1, \ldots, Z_N$, which are i.i.d. according to a distribution $F(z - \theta)$ where F is symmetric about 0.

The sign test rejects H when the number N_+ of positive Z's is sufficiently large. If $T_N = N_+/N$ and

$$p = P_\theta (Z_i > 0) = 1 - F(-\theta) = F(\theta),$$

H is equivalent to $p = 1/2$ and K to $p > 1/2$, and we are therefore in the situation of Example 3.3.1. However, we are now interested in the power of the test not against an alternative value of p but of θ. We could, of course, translate (3.3.14) into an equivalent result concerning θ (Problem 3.17). More conveniently, Theorem 3.3.5 gives the result directly. If we put

$$(3.3.36) \qquad \mu(\theta) = p = F(\theta)$$

and if F has density f, the assumptions of Theorem 3.3.5 hold, and the power of the test against the alternatives (3.3.7) satisfies

$$(3.3.37) \qquad \beta_n(\theta_n) \to \Phi(2\Delta f(0) - u_\alpha).$$

Note 1: The test and its asymptotic power (3.3.37) remain valid if the assumption of symmetry about 0 is replaced by the broader assumption that 0 is the median of F so that θ is the median of the distribution of the Z's.

Note 2: Instead of the alternatives $F(z - \theta)$, one might want to consider the more general alternatives $F\left(\dfrac{z - \theta}{\tau(\theta)}\right)$. Let $\theta_n = \Delta/\sqrt{n}$ as before and suppose that $\tau(\theta)$ is continuous so that

$$\tau(\theta_n) \to \tau(0) = 1.$$

Then under the assumptions of Example 3.3.5, formula (3.3.37) continues to hold (Problem 3.18). $\square$

Example 3.3.6 Wilcoxon test for center of symmetry. Under the assumptions of the preceding example, an alternative test of H is the one-sample Wilcoxon test discussed in Example 3.2.5. It follows from (3.2.30) that the asymptotic form of this test has the rejection region

$$(3.3.38) \qquad \frac{V_s - \frac{1}{4}N(N+1)}{\sqrt{N(N+1)(2N+1)/24}} \geq u_\alpha.$$

Here

$$V_s = \sum\sum_{1 \leq i \leq j \leq N} U_{ij},$$

with U_{ij} given above (3.2.31). The test thus has the form (3.3.30) with

(3.3.39) $$T_N = \frac{V_s}{\binom{N}{2}}, \quad \mu(\theta_0) = \frac{1}{2}, \quad \text{and } \tau_N(\theta_0) = \frac{1}{\sqrt{3}}.$$

It will be shown in Section 6.1 that, more generally, T_N satisfies (3.3.31) with

(3.3.40) $$\mu(\theta) = P_\theta\left[(Z_1 + Z_2) > 0\right].$$

To evaluate the asymptotic power (3.3.33), it is only necessary to determine $\mu'(\theta_0)$. Let us write

(3.3.41) $$\mu(\theta) = P\left[(Z_1 - \theta) + (Z_2 - \theta) > -2\theta\right].$$

The conditional probability of $Z_2 - \theta > -(z_1 - \theta) - 2\theta$ given z_1 is $1 - F\left[-(z_1 - \theta) - 2\theta\right]$ since $Z_2 - \theta$ is distributed according to F. The unconditional probability (3.3.41) is obtained by integrating the conditional probability with respect to the distribution of $(z_1 - \theta)$, which is again F, so that finally

(3.3.42) $$\mu(\theta) = \int\left[1 - F(z - 2\theta)\right] f(z)\, dz = \int F(-z + 2\theta) f(z)\, dz.$$

We thus find $\mu'(\theta) = 2\int f(-z + 2\theta) f(z)\, dz$ and hence, with $\theta_0 = 0$,

(3.3.43) $$\mu'(\theta_0) = 2\int f(-z) f(z)\, dz = 2\int f^2(z)\, dz.$$

This result requires that the integral on the right side of (3.3.42) can be obtained by differentiating under the integral sign. It can be shown (although we shall not do so here) that this is permissible, provided the integral in (3.3.43) is finite. For references, see Lehmann (1998, p. 373).

It follows from (3.3.39) and (3.3.43) that the asymptotic power of the Wilcoxon test (3.3.38) is given by

(3.3.44) $$\beta_N(\theta_N) \to \Phi\left[\sqrt{12}\Delta \int f^2(z)\, dz - u_\alpha\right].$$

$\square$

Example 3.3.7 Wilcoxon two-sample test. In parallel to Example 3.3.6, let us consider the power of the two-sample Wilcoxon test of Example 3.2.1 against the class of slight alternatives

(3.3.45) $$G(y) = F(y - \theta).$$

Under the hypothesis $H : \theta = 0$, the asymptotic behavior of W_{XY} is given by (3.2.9).

Let us now embed the given situation in a sequence with sample sizes (m_k, n_k), $k = 1, 2, \ldots$, satisfying

$$(3.3.46) \qquad \frac{m_k}{N_k} \to \rho \, (0 < \rho < 1) \,, \; N_k = m_k + n_k,$$

and a sequence of alternatives

$$(3.3.47) \qquad \theta_k = \frac{\Delta}{\sqrt{N_k}}.$$

We wish to apply Theorem 3.3.5 to $T_k = W_{XY}/m_k n_k$. With n replaced by N_k, we shall see in Section 6.5 that (3.3.29) and (3.3.31) hold with

$$(3.3.48) \qquad \mu(\theta) = p(\theta) = P(X < Y) = \int \left[1 - F\left(x - \theta\right)\right] f(x) dx$$

and

$$(3.3.49) \qquad \tau\left(\theta_0\right) = \tau(0) = \lim \sqrt{\frac{N_k\left(N_k + 1\right)}{12 m_k n_k}} = \frac{1}{\sqrt{12\rho(1 - \rho)}}.$$

Theorem 3.3.5 thus applies and it only remains to evaluate $\mu'(0)$. In analogy with (3.3.43), we find

$$(3.3.50) \qquad \mu'(0) = \int f^2(x) dx$$

and therefore for the asymptotic power of the two-sample Wilcoxon test

$$(3.3.51) \qquad \beta\left(\theta_k\right) \to \Phi\left[\sqrt{12\rho\left(1 - \rho\right)} \, \Delta \int f^2(x) dx - u_\alpha\right].$$

To complete the evaluation of the asymptotic power for any particular F, it is now only necessary to calculate $\int f^2$ (Problem 3.23).

Note: The formal similarity of formulas (3.3.44) and (3.3.51) masks an important difference: In the context of comparing two treatments, F refers to the distribution of the X's and Y's in (3.3.51) but to the distribution of the differences $Y_i - X_i$ in (3.3.44).

As usual, the limit result (3.3.51) also provides an approximation. Replacement of Δ by $\sqrt{N}\theta$ and ρ by m/N leads to the approximate power against a fixed alternative θ

$$(3.3.52) \qquad \beta(\theta) \doteq \Phi\left[\sqrt{\frac{12mn}{N}}\theta \int f^2(x) dx - u_\alpha\right].$$

TABLE 3.3.2. Power of the Wilcoxon test against normal shift alternatives[‡]

Δ/σ	0	.2	.4	.6	.8	1.0	1.5	2.0
Exact	.049	.094	.165	.264	.386	.520	.815	.958
(3.3.52)	.049	.098	.177	.287	.423	.568	.861	.977

Table 3.3.2 illustrates the accuracy of this approximation for the case $m = n = 7$, $\alpha = .049$, and F normal. Since the relative error throughout the range of the table is less than 10%, the approximation is fairly satisfactory even for these small sample sizes. □

The power approximations obtained in this section can be used to determine approximately the sample size required to achieve a desired power β against a given alternative θ.

Under the assumptions of Theorem 3.3.3, the power against θ of the test (3.1.12) based on n observations is approximately

$$\Phi\left[\frac{\sqrt{n}\,(\theta - \theta_0)}{\tau\,(\theta_0)} - u_\alpha\right].$$

If we wish the power to be equal to β, we must solve the equation

$$\Phi\left[\frac{\sqrt{n}\,(\theta - \theta_0)}{\tau\,(\theta_0)} - u_\alpha\right] = \beta$$

for n. Since by definition of u_β

(3.3.53) $\Phi\,(-u_\beta) = \beta,$

comparison of the last two displayed equations yields

$$\frac{\sqrt{n}\,(\theta - \theta_0)}{\tau\,(\theta_0)} = u_\alpha - u_\beta$$

and hence

(3.3.54) $$n \doteq \frac{(u_\alpha - u_\beta)^2}{(\theta - \theta_0)^2}\tau^2\,(\theta_0)\,.$$

The same formula, with the obvious trivial changes, applies to the situation covered by Theorem 3.3.4.

Example 3.3.8 Binomial sample size. The test is given by (3.1.12) with $T_n = X/n, \theta = p$ and $\tau^2\,(\theta) = pq$ so that (3.3.54) becomes

(3.3.55) $$n = \frac{(u_\alpha - u_\beta)^2}{(p - p_0)^2}p_0 q_0.$$

[‡]From Table 2.1 of Lehmann (1998).

Table 3.3.3 shows the actual values of α and β for $p = p_0 + .1$ when n is determined by (3.3.55) with nominal values $\alpha = .05$ and $\beta = .9$. As is clear from (3.3.55), the required sample size decreases as p_0 moves away from $1/2$.

TABLE 3.3.3. Approximate sample sizes in the binomial case

p_0	.1	.2	.3	.4	.5	.6	.7	.8
n	78	138	180	206	215	206	180	138
α	.045	.050	.046	.058	.051	.044	.042	.038
β	.808	.864	.873	.907	.906	.906	.916	.942

It is seen that the crude formula (3.3.55) gives fairly satisfactory results unless p_0 is close to 0 or 1. There are two reasons for this exception. For one, the normal approximation to the binomial is less satisfactory for p close to 0 and 1. In addition, the fact that the variance pq under the alternative was replaced by the null variance $p_0 q_0$ matters relatively little when p is close to $1/2$ since pq as a function of y is fairly flat near $1/2$ but has a more important effect for p near 0 and 1 where pq becomes quite steep. This is seen clearly from Table 3.3.4. □

TABLE 3.3.4. The function pq

p	.1	.2	.3	.4	.5	.6	.7	.8
pq	.09	.16	.21	.24	.25	.24	.21	.16

Example 3.3.9 Sample size for Student's t. As another illustration of (3.3.54), consider Student's t-test with the rejection region given by (3.1.19) with $T_n = \bar{X}$, $\theta = \xi$, and $\tau^2 = \sigma^2$, so that (3.3.54) becomes

$$(3.3.56) \qquad n \doteq \frac{(u_\alpha - u_\beta)^2}{(\xi - \xi_0)^2 / \sigma^2}.$$

Table 3.3.5 compares the approximate sample size obtained from (3.3.56) with the exact sample size required to achieve power β against an alternative ξ at $\alpha = .05$ for different values of β and $(\xi - \xi_0)/\sigma^2$. The agreement is remarkably good over the whole range considered. □

Example 3.3.10 Sample size for Poisson two-sample problem. Setting the approximate power (3.3.28) equal to β, and letting $m = \rho N$ and $n = (1 - \rho)N$, we find for N the approximate formula

$$(3.3.57) \qquad N \doteq \left(\frac{u_\alpha - u_\beta}{\mu - \lambda} \right)^2 \left(\frac{\lambda}{\rho} + \frac{\mu}{1 - \rho} \right).$$

TABLE 3.3.5. Sample size required by one-sided t-test

$(\xi - \xi_0)/\sigma$	Appr.	Exact	Appr.	Exact	Appr.	Exact
	.2		.5		.8	
$\alpha = .05, \beta = .8$	155	157	25	27	10	12
$\alpha = .05, \beta = .9$	215	217	35	36	14	15
$\alpha = .05, \beta = .95$	271	274	44	45	17	19

This formula also shows how to distribute a total sample of size N among the two groups so as to minimize (approximately) the value of N required to achieve power β against a given alternative (λ, μ). The value of ρ minimizing the second factor in (3.3.57) is given by (Problem 3.20).

$$(3.3.58) \qquad \frac{1}{\rho} = 1 + \sqrt{\frac{\mu}{\lambda}}.$$

In particular, the sample size is approximately minimized for close-by alternatives (μ close to λ) by putting $\rho = 1 - \rho = 1/2$, i.e., splitting the total sample equally between the two groups. $\qquad \square$

The argument leading to (3.3.54) can also be applied to the more general situation covered by Theorem 3.3.5. Solving for n from (3.3.34) instead of (3.3.11), one obtains, instead of (3.3.54), the approximate sample size

$$(3.3.59) \qquad n \doteq \frac{(u_\alpha - u_\beta)^2}{(\theta - \theta_0)^2} \Big/ \left[\frac{\mu'(\theta_0)}{\tau(\theta_0)}\right]^2.$$

Here the first factor depends on α, β, and the alternative θ but is independent of the particular test being used. On the other hand, the second factor is a characteristic of the test statistic and is independent of α, β, and θ. The larger the (positive) quantity

$$(3.3.60) \qquad \frac{\mu'(\theta_0)}{\tau(\theta_0)},$$

the smaller is the (approximate) sample size required to achieve power β against the alternative θ at level α for any α, β, and θ. For this reason, the quantity (3.3.60) is called the *efficacy*[§] of the test sequence $\{T_n\}$ for testing $H: \theta = \theta_0$. If

$$(3.3.61) \qquad \beta_n(\theta) \to \Phi(c\Delta - u_\alpha), \quad c > 0,$$

it follows from Theorem 3.3.5 that the efficacy is equal to c.

[§]Different authors use different variants of this definition.

Example 3.3.11 Sample sizes for two-sample Wilcoxon test. As an illustration of (3.3.59), consider the situation of Example 3.3.7 where the assumptions of Theorem 3.3.5 are satisfied with $N = m + n$ in place of n, $\tau(\theta_0) = 1/\sqrt{12\rho(1 - \rho)}$, $\theta_0 = 0$, and

$$(3.3.62) \qquad \mu'(\theta_0) = \int f^2(x)\,dx.$$

Therefore

$$(3.3.63) \qquad N \doteq \frac{(u_\alpha - u_\beta)^2}{\theta^2} \Bigg/ 12\rho(1 - \rho)\left(\int f^2(x)\,dx\right)^2$$

is the total (approximate) sample size needed to achieve power β when testing $H: \theta = 0$ against the alternative (3.3.45) at level α, with $m = \rho N$ and $n = (1 - \rho)N$ being the sample sizes of the two groups. The required sample size is minimized by maximizing $\rho(1 - \rho)$, i.e., by setting $m = n = N/2$. $\qquad\square$

Summary

1. Standard tests are typically consistent in the sense that their power against any fixed alternative tends to 1 as the sample size(s) tend to infinity.

2. To obtain a useful approximation to the power of a test, one therefore considers the power not against a fixed alternative but against a sequence of alternatives tending to the hypothesis at a suitable rate.

3. By equating the asymptotic power to a preassigned value, one obtains a simple formula for the sample size required to attain a given power.

3.4 Comparison of tests: Relative efficiency

The power of a test, in addition to providing a measure of performance, also serves as a basis for the comparison of different tests.

Example 3.4.1 Testing a point of symmetry. Let $Z_1, \ldots, Z_N$ be i.i.d. with a distribution $F(z - \theta)$ that is symmetric about a point θ. Three tests with asymptotic level α for testing $H: \theta = 0$ against $\theta > 0$ were discussed in Examples 2.3–2.5. Their rejection regions are respectively

$$(3.4.1) \qquad \sqrt{N}\bar{Z}/\hat{\sigma} \geq u_\alpha \quad (t\text{-test}),$$

where $\hat{\sigma}$ is the usual denominator of the t-statistic (3.1.24) (with Z_i in place of X_i), and

(3.4.2) $$2\sqrt{N}\left[\frac{N_+}{N}-\frac{1}{2}\right]\geq u_\alpha \quad \text{(sign test)},$$

where N_+ is the number of positive Z's, and

(3.4.3) $$\frac{V_S-\frac{1}{4}N(N+1)}{\sqrt{N(N+1)(2N+1)/24}}\geq u_\alpha \quad \text{(Wilcoxon test)},$$

where V_S is the one-sample Wilcoxon statistic defined in Example 3.2.5.

Actually, the tests (3.4.1)–(3.4.3) are not the exact versions of the t-, sign, and Wilcoxon tests but the large-sample approximations in which the exact critical value has been replaced by its limit u_α (as $N \to \infty$). The results of the present discussion are not affected by this replacement, nor by replacing $\hat{\sigma}$ in (3.4.1) by any other consistent estimator of the standard deviation σ of the Z's.

The asymptotic power of these three tests against the alternatives $\theta_N = \Delta/\sqrt{N}$ was shown in Section 3.3 to be given respectively by

(3.4.4) $$\beta(\theta_N) \to \Phi\left(\frac{\Delta}{\sigma}-u_\alpha\right), \quad t\text{–test, (3.3.21)},$$

(3.4.5) $$\beta'(\theta_N) \to \Phi(2\Delta f(0)-u_\alpha), \quad \text{sign test (3.3.37)},$$

and

(3.4.6)
$$\beta''(\theta_N) \to \Phi\left(\Delta\sqrt{12}\int f^2(z)\,dz - u_\alpha\right), \quad \text{Wilcoxon test (3.3.44)}.$$

Which of these three tests is best (i.e., most powerful) is determined by which of them has the largest coefficient of Δ, i.e., which of the three quantities

(3.4.7) $$c_1=\frac{1}{\sigma}, \quad c_2=2f(0), \quad c_3=\sqrt{12}\int f^2(z)\,dz$$

is largest. (It is seen from (3.3.61) that c_i is the efficacy of the associated test.) This question has no universal answer, rather the answer depends on the distribution F of the Z's. In particular, for example, the best of the three tests is the t-test when F is normal, the sign test when F is double exponential, and the Wilcoxon test when F is logistic (Problem 4.1). $\quad\square$

Consider now more generally two tests with power functions satisfying

(3.4.8) $$\beta_i(\theta_N) \to \Phi(c_i\Delta - u_\alpha), \quad i = 1, 2,$$

against the alternatives $\theta_N = \theta_0 + \Delta/\sqrt{N}$. A natural asymptotic measure of the superiority of test 2 over test 1 against a particular alternative $\Delta > 0$ is

(3.4.9) $$\lim[\beta_2(\theta_N) - \beta_1(\theta_N)] = \Phi(c_2\Delta - u_\alpha) - \Phi(c_1\Delta - u_\alpha).$$

The sign of this difference is independent of Δ; however, its value is not.

To obtain a comparison that is independent of Δ, consider the situation with which one is confronted when planning the study. One would then wish to determine the sample size required by each test to achieve the same asymptotic power β ($\alpha < \beta < 1$) against the same alternative at the same asymptotic level α. Under the assumptions of Theorem 3.3.5, this sample size was seen to be given approximately by (3.3.59), i.e., by

(3.4.10) $$N \doteq \frac{(u_\alpha - u_\beta)^2}{(\theta - \theta_0)^2} \cdot \frac{1}{c^2},$$

where $|c|$ is the efficacy of the given test sequence,

(3.4.11) $$c = \mu'(\theta_0)/\tau(\theta_0) \text{ with } \mu \text{ and } \tau \text{ defined in (3.3.29).}$$

By (3.3.33), the asymptotic power of this test is

$$\beta(\theta_N) \to \Phi(c\Delta - u_\alpha)$$

against the alternatives

$$\theta_N = \theta_0 + \frac{\Delta}{\sqrt{N}}.$$

If N_1 and N_2 are the sample sizes required by two different tests with efficacies c_1 and c_2 to achieve the same power β against the same alternatives at the same level α, we have, by (3.4.10),

(3.4.12) $$\frac{N_1}{N_2} \doteq \frac{c_2^2}{c_1^2} = e_{2,1}.$$

The left side is the *relative efficiency* of test 2 with respect to test 1. If, for example, it equals $1/2$, then approximately $N_1 = \frac{1}{2}N_2$ or $N_2 = 2N_1$; the second test is half as efficient as test 1 since it requires twice as large a sample to achieve the same end. The right side of (3.4.12) is the *asymptotic relative efficiency* (ARE). Under suitable regularity assumptions, the right side is actually the limit of the left side as the sample sizes tend to ∞, as will be shown in Theorem 3.4.1. Note that the ARE is independent of α and β.

Example 3.4.2 Testing a point of symmetry (continued). Let us return now to the three tests of Example 3.4.1. From (3.4.4)–(3.4.6), we can read off the relative efficiencies of the sign, Wilcoxon, and t-tests to each other, as

$$(3.4.13) \qquad e_{S,t}(F) = 4\sigma^2 f^2(0),$$

$$(3.4.14) \qquad e_{W,t}(F) = 12\sigma^2 \left(\int f^2(z)dz \right)^2,$$

and

$$(3.4.15) \qquad e_{S,W}(F) = f^2(0)/3 \left(\int f^2(z)dz \right)^2.$$

These formulas provide a basis for comparing the three tests for different distributions F. Before doing so, we note that the efficiencies are independent of scale; they are also independent of location since the distribution of the Z's was taken to be $F(z - \theta)$ (Problem 4.2). The efficiencies (3.4.13)–(3.4.15) therefore depend only on the shape of the distribution F, which is assumed to be symmetric about 0.

Consider first the case that F is the standard normal distribution, $F = \Phi$. It is easily seen (Problem 4.3) that

$$(3.4.16) \qquad e_{S,t}(\Phi) = \frac{2}{\pi} \sim .637, \quad e_{W,t}(\Phi) = \frac{3}{\pi} \sim .955.$$

When the assumption of normality holds, the efficiency of the sign to the t-test is therefore less than $2/3$, which means that the sign test requires in excess of 50% more observations than the t-test to achieve the same power.

The situation is much more favorable for the Wilcoxon test, which has an efficiency loss of less than 5% relative to the t-test. For this reason, the Wilcoxon test is a serious competitor of the t-test, and it is of interest to see how their efficiencies compare when the assumption of normality does not hold. From (3.4.14), one sees that $e_{W,t}(F)$ is very sensitive to small disturbances in the tail of the distribution. Moving small masses out toward $\pm\infty$ will make σ^2 tend to infinity while causing only little change in $\int f^2$. This shows that the Wilcoxon test can be infinitely more efficient than the t-test for distributions differing only slightly from the normal. On the other hand, it turns out that the Wilcoxon test can never be very much less efficient than the t-test, that, in effect,

$$(3.4.17) \qquad e_{W,t}(F) \geq .864 \text{ for all } F.$$

(For a proof, see Lehmann (1998).)   □

Example 3.4.3 Two-sample Wilcoxon vs. t. A two-sample analog of Example 3.4.1 is concerned with samples $X_1, \ldots, X_m$ and $Y_1, \ldots, Y_n$ from distributions $F(x) = P(X \leq x)$ and $G(y) = P(Y \leq y) = F(y - \theta)$ respectively. Two tests for testing $H : \theta = 0$ against $\theta > 0$ are

  (a) the t-test based on (3.1.36)

and

  (b) the Wilcoxon test based on (3.2.9).

The approximate sample size $N = m + n$ required by the Wilcoxon test to achieve power β against an alternative value of θ was seen in Example 3.3.11 to be

$$(3.4.18) \qquad N \doteq \frac{(u_\alpha - u_\beta)^2}{\theta^2} / 12\rho (1 - \rho) \left(\int f^2 \right)^2,$$

where $\rho = m/N$. The corresponding sample size N' for the t-test is (Problem 4.8)

$$(3.4.19) \qquad N' \doteq \frac{(u_\alpha - u_\beta)^2}{\theta^2} \frac{\sigma^2}{\rho (1 - \rho)}.$$

It follows that the efficiency of the Wilcoxon test relative to the two-sample t-test continues to be given by the corresponding one-sample efficiency (3.4.14), where, however, F is no longer restricted to be symmetric. In particular, when F is normal, the efficiency is again equal to $3/\pi \sim .955$. Also, (3.4.17) continues to hold for all F, symmetric or not.

TABLE 3.4.1. Exact efficiency $e_{w,t}(F)$ for $F = $ normal $m = n = 5$, $\alpha = 4/126$

θ	.5	1.0	1.5	2.0	2.5	3.0	3.5
β	.072	.210	.431	.674	.858	.953	.988
e	.968	.978	.961	.956	.960	.960	.964

Source: Table 2.3 of Lehmann (1998).

The efficiency (3.4.12) is the ratio of two approximate sample sizes, and one must be concerned about its accuracy. Such a concern seems particularly appropriate since this approximation is independent of α, β, and θ while the actual ratio of sample sizes required to achieve power β at level α against an alternative θ will vary with these parameters. Table 3.4.1 shows the exact efficiency in the normal case, i.e., the ratio $(m' + n') / (m + n)$, where it is assumed that $m' = n'$. Here $m' + n'$ is the total sample size required by the t-test to equal the power of the Wilcoxon test with $m = n = 5$ at level $\alpha = 4/126$ over a range of values of θ and β. [Because of the discrete nature of n', no value of n' will give exactly the desired power. To overcome this difficulty, n' is calculated by randomizing between the two sample sizes n'_0 and $n'_0 + 1$, the powers β' and β'' of which bracket the desired power β, i.e., by $n' = pn'_0 + (1 - p)(n'_0 + 1)$, where p satisfies $p\beta' + (1 - p)\beta'' = \beta$].

The most striking feature of the table is the near-constancy of e. As β varies from .072 to .988, the range of e is .956 to .978, all values being slightly above the large-sample approximate value of .955. $\square$

Example 3.4.4 Choice of design in simple linear regression. Let X_i be independent $N\left(\xi_i, \sigma^2\right)$, where

$$(3.4.20) \qquad \xi_i = \alpha + \beta v_i \quad (i = 1, \dots, N).$$

Then the test which rejects $H : \beta = \beta_0$ when

$$(3.4.21) \qquad \frac{\left(\hat{\beta} - \beta_0\right)\sqrt{\sum (v_i - \bar{v})^2}}{\sqrt{\sum \left(X_i - \hat{\alpha} - \hat{\beta}v_i\right)^2 / (N - 2)}} \geq u_\alpha,$$

with $\hat{\alpha}$ and $\hat{\beta}$ given by (2.7.11), has asymptotic level α. This follows from the fact that the exact null distribution of the test statistic (3.4.21) is the t-distribution with $N - 2$ degrees of freedom. If

$$(3.4.22) \qquad \frac{1}{N} \sum (v_i - \bar{v})^2 \to d,$$

the power of this test against the alternatives

$$(3.4.23) \qquad \beta_n = \beta_0 + \frac{\Delta}{\sqrt{N}}$$

tends to (Problem 4.9)

$$(3.4.24) \qquad \Phi\left(\frac{\Delta\sqrt{d}}{\sigma} - u_\alpha\right)$$

and the efficacy of the test is therefore $c = \sqrt{d}/\sigma$.

The values of the v's can be chosen by the experimenter and constitute the *design* of the experiment. (Often the v's depend on N and should be denoted by v_{Ni}, but for simplicity, we shall retain the one-subscript notation.) The most efficient design is obtained by maximizing d and hence $\sum (v_i - \bar{v})^2$. This sum can be made arbitrarily large by choosing the v's sufficiently far apart. However, in practice, the v's will typically have to lie in some finite interval which without loss of generality we may take to be $[0,1]$. Subject to this condition, it is easy to see that for maximum efficiency, the v's must all lie on the boundary of the interval (i.e., take on only the values 0 and 1) and that the maximum value of $\sum (v_i - \bar{v})^2$ satisfies (Problem 4.10(i))

$$\sum (v_i - \bar{v})^2 \sim \frac{N}{4} \text{ as } N \to \infty.$$

(The value $N/4$ is attained exactly when N is even and half of the observations are placed at each end point.)

Unfortunately, a design which places all observations at either 0 or 1 has a great drawback: It does not permit checking the assumption that the regression is linear. For this reason, other less efficient designs are often preferred. If we consider two designs $(v_1, \ldots, v_N)$ and $(v_1', \ldots, v_N')$, the efficiency of the test based on the primed design to that of the unprimed one is

$$(3.4.25) \qquad\qquad e = \frac{\sum (v_i' - \bar{v}')^2}{\sum (v_i - \bar{v})^2}.$$

As an example, consider the design which places the v's at the points 0, $\frac{1}{N-1}, \frac{2}{N-1}, \ldots, 1$. Then (Problem 4.10(ii))

$$\sum (v_i - \bar{v})^2 \sim \frac{N}{12},$$

so that the efficiency of this uniform design to the earlier 2-point design is 1/3.

Intermediate designs of course exist which make it possible to obtain a check on linearity with a less drastic loss of efficiency. If, for example, we put $N/4$ observations each at 0, 1/3, 2/3, and 1, the efficiency of this 4-point design to the earlier 2-point design is 5/9 (Problem 4.11).

When condition (3.4.22) holds with $d > 0$, as it does for the three designs mentioned above, and if the v's are bounded, then (2.7.24) of Chapter 2 is satisfied since $\max (v_i - \bar{v})^2 / N \to 0$ while $\sum (v_i - \bar{v})^2 / N \to d > 0$. It follows that (3.4.24) and the efficiencies derived from it are valid without the assumption of normality for the more general model (2.7.10).   □

Relative efficiency was defined in (3.4.12) as the ratio of the approximate sample sizes required by two tests to achieve the same performance. We shall now give a somewhat more careful derivation of (3.4.12) as a limit result, and at the same time provide a slight generalization which will be useful below.

Theorem 3.4.1 *Let* $T^{(i)} = \left\{ T_N^{(i)}, N = 1, 2, \ldots ; i = 1, 2 \right\}$ *be two sequences of test statistics for testing* $H : \theta = \theta_0$ *based on* N *observations, and suppose that the power of* $T_N^{(i)}$ *against the alternatives*

$$(3.4.26) \qquad\qquad \theta_N = \theta_0 + \frac{\Delta}{N^{\gamma/2}}$$

satisfies

(3.4.27) $$\beta_i(\theta_N) \to \Phi(c_i\Delta - u_\alpha).$$

Suppose now that $N_k^{(i)}$, $k = 1, 2, \ldots$, are two sequences of sample sizes and Δ_1, Δ_2 two positive constants such that the power of the test $T_{N_k^{(i)}}^{(i)}$ against the same sequence of alternatives

(3.4.28) $$\theta_k^{(i)} = \theta_0 + \frac{\Delta_i}{\left[N_k^{(i)}\right]^{\gamma/2}} \quad \text{and} \quad \theta_k^{(1)} = \theta_k^{(2)}$$

has the same limit

(3.4.29) $$\Phi(c_1\Delta_1 - u_\alpha) = \Phi(c_2\Delta_2 - u_\alpha) = \beta.$$

Then the ARE of test 2 with respect to test 1 is given by

(3.4.30) $$e_{2,1} = \lim \frac{N_k^{(1)}}{N_k^{(2)}} = \left(\frac{c_2}{c_1}\right)^{2/\gamma}.$$

Note: The sequences $\left\{N_k^{(i)}\right\}$ are only required to satisfy (3.4.26) and hence are not unique. However, it follows from (3.4.30) that the ARE $e_{2,1}$ is independent of the particular sequences chosen to satisfy (3.4.27) and that the proof of (3.4.30) therefore at the same time establishes the existence of the ARE.

Proof. From (3.4.28), we have

$$\frac{\Delta_1}{\left[N_k^{(1)}\right]^{\gamma/2}} = \frac{\Delta_2}{\left[N_k^{(2)}\right]^{\gamma/2}}.$$

Therefore

$$\Delta_2 = \Delta_1\left[\frac{N_k^{(2)}}{N_k^{(1)}}\right]^{\gamma/2}.$$

It follows that the limit of $N_k^{(1)}/N_k^{(2)}$ exists and equals

(3.4.31) $$\lim_{k\to\infty} \frac{N_k^{(1)}}{N_k^{(2)}} = \left(\frac{\Delta_1}{\Delta_2}\right)^{2/\gamma}.$$

On the other hand, (3.4.29) implies

(3.4.32) $$c_1\Delta_1 = c_2\Delta_2$$

and hence (3.4.30), as was to be proved.   ■

For $\gamma = 1$, conditions under which (3.4.27) holds are given in Theorem 3.3.5. The constants c_1 and c_2 in (3.4.30) are then just the efficacies of $T^{(1)}$ and $T^{(2)}$, respectively.

Example 3.4.5 Paired comparisons. In the paired comparison situation of Example 3.2.3 (with slight changes in notation), suppose that

$$E\left(X_i\right) = \xi, \; E\left(Y_i\right) = \eta, \; \text{Var}\left(X_i\right) = \text{Var}\left(Y_i\right) = \sigma^2,$$

that the correlation between X_i and Y_i is ρ, and that $Z_i = Y_i - X_i (i = 1, \ldots , n)$ are independent normal. If $E\left(Z_i\right) = \zeta$, the hypothesis $H : \zeta = 0$ is being tested against $K : \zeta > 0$ by the t-test with rejection region

$$(3.4.33) \qquad\qquad \sqrt{n}\bar{Z}/\hat{\tau} \geq k,$$

where $\hat{\tau}^2 = \sum \left(Z_i - \bar{Z}\right)^2 / (n - 1)$ is a consistent estimator of

$$\tau^2 = \text{Var}\left(Z_i\right) = 2\sigma^2 \left(1 - \rho\right)$$

and k is the critical value of the t_{n-1}-distribution.

Suppose we are interested in two different methods of forming the pairs for which the means ξ and η and the variance σ^2 are the same but which lead to different correlation coefficients ρ_1 and ρ_2. Then it follows from Theorem 3.4.1 that the ARE of the second to the first method is

$$(3.4.34) \qquad\qquad e_{2,1} = \frac{1 - \rho_1}{1 - \rho_2}.$$

The higher the correlation, the greater is the efficiency of the associated method of pairing.   □

Example 3.4.6 Simple linear regression (continued). In Example 3.4.4, the regression model (3.4.20) was discussed in the context of a designed experiment with the v's restricted to a fixed range. But this model arises also in situations such as time series in which the v's may tend to ∞ as $N \to \infty$. As an example, suppose that we are dealing with a linear trend

$$(3.4.35) \qquad\qquad v_i = ia.$$

Then (Problem 4.14)

$$(3.4.36) \qquad\qquad \sum \left(v_i - \bar{v}\right)^2 = a^2 \left(N^3 - N\right)/12,$$

which violates (3.4.22). Such possibilities can often be accommodated by replacing (3.4.22) by

$$(3.4.37) \qquad \frac{1}{N^\gamma} \sum (v_i - \bar{v})^2 \to d.$$

Under the assumptions of Example 3.4.4, the power of the test (3.4.21) against the alternatives

$$(3.4.38) \qquad \beta_N = \beta_0 + \frac{\Delta}{N^{\gamma/2}}$$

then continues to tend to (3.4.24) when (2.7.24) of Chapter 2 holds (Problem 4.15), and the efficacy of the test is therefore $c = \sqrt{d}/\sigma$, as before.

For $v_i = ai$, this condition is easily checked (Problem 4.16), and (3.4.36) shows that $\gamma = 3$ and $d = a^2/12$, so that the efficacy of (3.4.21) with $v_i = ia$ is $c = a/\sigma\sqrt{12}$. It follows that the ARE of the spacing $v_i = ai$ to that with $v_i = i$ is $a^{2/3}$. □

Example 3.4.7 Tests of randomness. Let us next consider a nonparametric version of the preceding example. If $X_1, \dots, X_N$ are independent random variables with continuous distributions $F_1, \dots, F_N$, a standard test of the hypothesis $H : F_1 = \cdots = F_N$ against the alternatives of an upward trend (i.e., that the X_i tend to give rise to larger values with increasing i) rejects when

$$(3.4.39) \qquad \sum iR_i \geq c_N,$$

where $R_1, \dots, R_N$ are the ranks of $X_1, \dots, X_N$. Since, under H,

$$(3.4.40) \qquad P(R_1 = r_1, \dots, R_N = r_N) = 1/N!$$

for each permutation $(r_1, \dots, r_N)$ of $(1, \dots, N)$, the null distributions of $\sum iR_i$ is independent of the common null distribution $F_1 = \cdots = F_N = F$ of the X's, so that the test (3.4.39) is distribution-free.

The expectation and variance of $\sum iR_i$ under H are given by (Problem 3.4.17)

$$(3.4.41) \qquad E_H\left(\sum iR_i\right) = \frac{N(N+1)^2}{4}$$

and

$$(3.4.42) \qquad \mathrm{Var}_H\left(\sum iR_i\right) = \frac{N^2(N+1)^2(N-1)}{144}.$$

Furthermore, it can be shown that

$$\left[\sum iR_i - N(N+1)^2/4\right] / \sqrt{\mathrm{Var}\left(\sum iR_i\right)}$$

tends in law to $N(0,1)$ (for details, see, for example, Hajek and Sidak (1967) or Lehmann (1998). The rejection region

(3.4.43)
$$\frac{\sum iR_i - \frac{N(N+1)^2}{4}}{N(N+1)\sqrt{N-1/12}} \geq u_\alpha$$

therefore has asymptotic level α.

Let us consider the ARE of the nonparametric trend test (3.4.43) with respect to the corresponding normal theory test (3.4.21) with $v_i = i$, from which it differs mainly in its replacement of the original observations X_i by their ranks R_i. The asymptotic power $\beta_N^{(1)}$ of the latter test against the alternatives (3.4.28) with $\gamma = 3$ continues to be given by (Problem 4.18)

(3.4.44)
$$\beta_N^{(1)} \to \Phi\left(\frac{\Delta}{\sigma\sqrt{12}} - u_\alpha\right).$$

The asymptotic distribution of $\sum iR_i$ under the alternatives is more difficult. It is discussed in Hajek and Sidak (1967) as a special case of the theory of simple linear rank statistics, i.e., statistics of the form $\sum_{i=1}^{N} b_i h(R_i)$, and more specifically in Aiyar, Guillier, and Albers (1979) and in the papers cited there. We shall not develop this theory here, but without proof state that under the alternatives (3.4.38) with $\gamma = 3$,

(3.4.45)
$$\frac{\sum iR_i - \frac{N(N+1)^2}{4} - \frac{\Delta N^{5/2}}{12}\int f^2}{N^{5/2}/12} \to N(0,1).$$

It follows from (3.4.45) that the power of the trend test (3.4.43) against the alternatives (3.4.38) with $\gamma = 3$ satisfies

(3.4.46)
$$\beta_N^{(2)} \to \Phi\left(\Delta \int f^2 - u_\alpha\right),$$

and hence from Theorem 3.4.1 that the ARE of the trend test (3.4.43) with respect to (3.4.21) with $v_i = i$ is

(3.4.47)
$$e_{2,1} = \left[12\sigma^2\left(\int f^2\right)^2\right]^{1/3}.$$

This is equal to

(3.4.48)
$$e_{2,1}(F) = [e_{W,t}(F)]^{1/3}$$

where $e_{W,t}(F)$ is the ARE of the Wilcoxon to the t-test obtained in Examples 3.4.1 and 3.4.2.

From (3.4.48), it follows that

$$e_{W,t}(F) < e_{2,1}(F) < 1 \text{ when } e_{W,t}(F) < 1$$

and

$$e_{W,t}(F) > e_{2,1}(F) > 1 \text{ when } e_{W,t}(F) > 1.$$

In particular, when F is normal, $e_{2,1} = .98$, and the lower bound (3.4.17) is replaced by

(3.4.49) $e_{2,1}(F) \geq .95$ for all F.

Under the assumptions of Theorem 3.4.1, the asymptotic relative efficiency is independent of both α and β. That this is not the case in all circumstances is shown by the following example.   □

Example 3.4.8 One-sided vs. two-sided test. The one-sided test (3.1.12) of $H : \theta = \theta_0$ provides good power only against alternatives $\theta > \theta_0$. If it is desired to reject H also when $\theta < \theta_0$, one may want to use the two-sided rejection rule

(3.4.50) $$\frac{\sqrt{n}\,|T_n - \theta_0|}{\tau(\theta_0)} \geq u_{\alpha/2},$$

which also has asymptotic level α (Problem 3.7(i)). Since this test must guard against a larger class of alternatives, one would expect it to be less powerful against any $\theta > \theta_0$ than the one-sided test which concentrates its power on these alternatives. The limiting power of the one-sided test against the alternatives

(3.4.51) $$\theta_n = \theta_0 + \frac{\Delta_1}{\sqrt{n}}$$

is given by (3.3.8); that of the test (3.4.50) against the alternatives

(3.4.52) $$\theta_n = \theta_0 + \frac{\Delta_2}{\sqrt{n}}$$

is (Problem 3.7(ii))

$$\lim \beta_n^{(2)}(\theta_n) = 1 - \Phi\left[u_{\alpha/2} - \frac{\Delta_2}{\tau(\theta_0)}\right] + \Phi\left[-u_{\alpha/2} - \frac{\Delta_2}{\tau(\theta_0)}\right].$$

The proof of Theorem 3.4.1 shows that if the alternatives (3.4.51) and (3.4.52) coincide and if the one-sided test based on n_1 and the two-sided test on n_2 observations achieve the same power against these alternatives, then

(3.4.53) $$e_{2,1} = \lim \frac{n_1}{n_2} = \frac{\Delta_1^2}{\Delta_2^2}.$$

The ratio Δ_1^2/Δ_2^2 and hence the efficiency $c_{2,1}$ is determined by the fact that the limiting power of both tests is to be equal to some given value β. The ARE $e_{2,1}$ is therefore obtained by solving the equations

$$\Phi\left(u_\alpha - \frac{\Delta_1}{\tau(\theta_0)}\right) = \Phi\left[u_{\alpha/2} - \frac{\Delta_2}{\tau(\theta_0)}\right] - \Phi\left[-u_{\alpha/2} - \frac{\Delta_2}{\tau(\theta_0)}\right] = 1 - \beta$$

for Δ_1 and Δ_2. The resulting efficiency depends heavily on α and β, as is shown in Table 3.4.2.   □

TABLE 3.4.2. Efficiency of two-sided to one-sided test

α	β .2	.5	.8	.95
.01	.73	.82	.86	.89
.05	.52	.70	.79	.83
.10	.32	.61	.73	.79

Example 3.4.9 Uniform. If $X_1, \ldots, X_n$ are i.i.d. according to the uniform distribution $U(0, \theta)$ it was seen in Example 3.1.2 that the test (3.1.17) has asymptotic level α for testing $H : \theta = \theta_0$ against $\theta > \theta_0$. If the maximum of the X's is not considered a safe basis for a test—perhaps because of fear of gross errors—one might want to consider as an alternative the rejection region

$$(3.4.54) \qquad \frac{\sqrt{n}\,(2\bar{X} - \theta_0)}{\theta_0/\sqrt{3}} \geq u_\alpha,$$

which also has asymptotic level α (Problem 4.19(i)). The power of this test against a sequence of alternatives $\theta_n \to \theta_0$ is

$$(3.4.55) \qquad \beta_1(\theta_n) = P\left[\frac{\sqrt{n}\,(2\bar{X} - \theta_n)}{\theta_0/\sqrt{3}} \geq u_\alpha - \frac{\sqrt{n}\,(\theta_n - \theta_0)}{\theta_0/\sqrt{3}}\right]$$

and that of the test (3.1.17)

$$(3.4.56) \qquad \beta_2(\theta_n) = P\left[\frac{n\,(\theta_n - X_{(n)})}{\theta_0} \leq l_\alpha + \frac{n\,(\theta_n - \theta_0)}{\theta_0}\right].$$

The left side of the inequality in square brackets in both these expressions tends in law to a non-degenerate limit distribution. For $\beta_1(\theta_n)$ and $\beta_2(\theta_n)$ to tend to a common limit β ($\alpha < \beta < 1$) therefore requires that

$$\theta_n - \theta_0 \sim \frac{\Delta_1}{\sqrt{n}} \quad \text{and} \quad \theta_n - \theta_0 \sim \frac{\Delta_2}{n}$$

for some positive Δ_1 and Δ_2. It follows that the ARE of the test based on the mean to that based on the maximum is zero (Problem 4.19(ii)).

Since this is unsatisfactory, let us consider as another alternative to (3.1.17) a third rejection region of the form

$$(3.4.57) \qquad \frac{n\left(\theta_0 - X_{(n-1)}\right)}{\theta_0} \leq w_\alpha.$$

Now the limit distribution of $Y = n\left(\theta_0 - X_{(n-1)}\right)/\theta_0$ has density xe^{-x}, $x > 0$, and cdf (Problem 3.12 of Chapter 2)

$$P\left(Y \leq y\right) = 1 - \left(1 + y\right)e^{-y}, \quad 0 < y.$$

Hence w_α is determined by

$$(3.4.58) \qquad \left(1 + w_\alpha\right)e^{-w_\alpha} = 1 - \alpha.$$

The power of this third test against the alternatives $\theta_n \sim \theta_0 + \dfrac{\Delta_3}{n}$ is

$$P_{\theta_n}\left[\frac{n\left(\theta_0 - X_{(n-1)}\right)}{\theta_0} \leq w_\alpha\right] = P\left[\frac{n\left(\theta_n - X_{(n-1)}\right)}{\theta_0} \leq w_\alpha + \frac{n\left(\theta_n - \theta_0\right)}{\theta_0}\right]$$

$$= P\left[\frac{n\left(\theta_n - X_{(n-1)}\right)}{\theta_0} \leq w_\alpha + \frac{\Delta_3}{\theta_0}\right],$$

which tends to (Problem 3.4.20(i))

$$(3.4.59) \qquad \beta_3\left(\theta_n\right) \to 1 - \left(1 + w_\alpha + \frac{\Delta_3}{\theta_0}\right)e^{-w_\alpha - \Delta_3/\theta_0}.$$

The asymptotic efficiency of the test based on $X_{(n-1)}$ relative to that based on $X_{(n)}$ is therefore

$$e_{3,2} = \Delta_2/\Delta_3,$$

where the right side has to be calculated from the equation

$$(3.4.60) \qquad e^{-l_\alpha - \Delta_2/\theta_0} = \left(1 + w_\alpha + \frac{\Delta_3}{\theta_0}\right)e^{-w_\alpha - \Delta_3/\theta_0}.$$

The solution will depend on both α and β (Problem 4.20(ii)). $\square$

Summary

1. The criterion proposed for comparing two tests is their asymptotic relative efficiency (ARE), the ratio of the sample sizes required to achieve the same asymptotic power at the same asymptotic level α.

2. Under assumptions that are frequently satisfied, the ARE is shown to be independent of α and β, and to be equal to the ratio of the efficacies of the two tests.

3. Among the examples in which the assumptions of 2 hold is the comparison of three tests of $H : \theta = \theta_0$, where θ is the center of a symmetric distribution and the choice of design in linear regression.

4. Two examples are given in which the ARE varies with α and β.

3.5 Robustness

Many of the tests discussed in the preceding sections of this chapter deal with hypotheses concerning parameters θ of a parametric model. The distribution of the observable random variables is completely specified except for the value of θ and possibly same nuisance parameters. Such models are based on assumptions that often are not reliable. One such assumption which is pervasive throughout the discussion of the preceding sections is the independence of the observations $X_1, \ldots, X_n$. Another is the assumption of identity of their distributions. Finally, there is the assumption that the form of the distribution is known, for example, to be normal or exponential. We shall refer to these assumptions as *independence, identity,* and *distribution*, respectively.

When the assumptions underlying the calculations of the preceding sections are not valid, neither, of course, are the resulting values of the significance level, power, or sample size. In the present context, we are not primarily concerned with exact validity of these quantities but rather with asymptotic validity as $n \to \infty$. As we shall see, asymptotic validity may continue to hold under violation of some model assumptions or it may be seriously affected. These asymptotic results, reinforced by spot checks for finite n, provide a useful guide to the reliability of standard tests and confidence intervals.

Let us consider some asymptotically normal situations of the kind treated in Sections 3.1–3.3.

Theorem 3.5.1 *Let T_n be a sequence of test statistics which under the postulated model satisfies*

$$\frac{\sqrt{n}\,[T_n - \mu\,(\theta_0)]}{\tau\,(\theta_0)} \xrightarrow{L} N\,(0,1)$$

so that the test (3.3.30) has nominal asymptotic level α. We shall assume α to be $< 1/2$, so that $u_\alpha > 0$. Suppose that the postulated model is wrong *and that under the true model,*

$$(3.5.1) \qquad \frac{\sqrt{n} \left[T_n - \mu \left(\theta_0 \right) \right)}{\tau' \left(\theta_0 \right)} \xrightarrow{L} N \left(0, 1 \right).$$

(i) *If α_n' denotes the actual level of (3.3.30), then $\alpha' = \lim \alpha_n'$ exists and is given by*

$$(3.5.2) \qquad \alpha' = 1 - \Phi \left[u_\alpha \frac{\tau \left(\theta_0 \right)}{\tau' \left(\theta_0 \right)} \right].$$

Therefore

$$\alpha' \lesseqgtr \alpha \text{ as } \tau' \left(\theta_0 \right) \lesseqgtr \tau \left(\theta_0 \right).$$

(ii) *If $\tau \left(\theta_0 \right)$ depends on nuisance parameters and the test (3.3.30) is replaced by*

$$\frac{\sqrt{n} \left[T_n - \mu \left(\theta_0 \right) \right]}{\hat{\tau}_n} \geq u_\alpha,$$

where $\hat{\tau}_n$ is a consistent estimator of $\tau \left(\theta_0 \right)$, the conclusion (3.5.2) remains valid.

Proof.

(i) We have

$$\alpha_n' = P \left[\frac{\sqrt{n} \left(T_n - \mu \left(\theta_0 \right) \right)}{\tau' \left(\theta_0 \right)} \geq u_\alpha \frac{\tau \left(\theta_0 \right)}{\tau' \left(\theta_0 \right)} \right] \rightarrow 1 - \Phi \left(u_\alpha \frac{\tau \left(\theta_0 \right)}{\tau' \left(\theta_0 \right)} \right) = \alpha'.$$

If $\tau' \left(\theta_0 \right) \lesseqgtr \tau \left(\theta_0 \right)$, then $\alpha' \lesseqgtr 1 - \Phi \left(u_\alpha \right) = \alpha$, as was to be proved.

The proof for (ii) is completely analogous. ∎

Table 3.5.1, which shows α' as a function of $r = \tau \left(\theta_0 \right) / \tau' \left(\theta_0 \right)$, gives an idea of how much the actual α' of (3.5.2) differs from the nominal α when $\alpha = .05$. Thus, for $r = .75$ when the actual $\tau' \left(\theta_0 \right)$ is $\frac{4}{3} \tau \left(\theta_0 \right)$, α' is about twice the nominal level.

TABLE 3.5.1. Actual level α' given by (3.5.2) when nominal level α is .05

r	.25	.50	.75	1.0	1.1	1.2	1.5	2.0
α'	.340	.205	.109	.050	.035	.024	.007	.0005

Let us denote by $\mathcal{F}$ the original model in which we are testing the hypothesis $H : \theta = \theta_0$. For example, $\mathcal{F}$ might specify that $X_1, \ldots, X_n$ is a sample from $N \left(\theta, \sigma^2 \right)$. If we do not trust the normality assumption, we might wish to investigate the behavior of the test of $H : \theta = \theta_0$ under a broader model $\mathcal{F}'$ such as

(i) the X's are a sample from $F(x - \theta)$, where $\theta = E(X_i)$ and where F is an unspecified distribution with finite variance

or

(ii) the X's are a sample from a symmetric distribution with θ as its center of symmetry.

More generally, we shall consider the behavior of a test of H, which was originally derived under some $\mathcal{F}$, under a broader model $\mathcal{F}'$ for which θ has a meaningful extension. Such extensions are illustrated by the models $\mathcal{F}'$ given by (i) and (ii). We shall then be interested in the asymptotic rejection probability $\alpha'(F)$ of the test when $F \in \mathcal{F}'$ and $\theta = \theta_0$. A test which has asymptotic level α under $\mathcal{F}$ (the *nominal* level) is said to be *conservative* under $\mathcal{F}'$ if

$$\alpha'(F) \leq \alpha \text{ for all } F \epsilon \mathcal{F}' \text{ with } \theta = \theta_0,$$

liberal if

$$\alpha'(F) \geq \alpha \text{ for all } F \epsilon \mathcal{F}' \text{ with } \theta = \theta_0,$$

and *robust* if

$$\alpha'(F) = \alpha \text{ for all } F \epsilon \mathcal{F}' \text{ with } \theta = \theta_0.$$

While conservative tests provide satisfactory control of the probability of false rejection of $H' : \theta = \theta_0$ in $\mathcal{F}'$, a level α' unnecessarily smaller than α is a mixed blessing since such a test is apt to be less powerful than it might be at the permitted level α.

Example 3.5.1 Normal variance. If $X_1, \dots, X_n$ are assumed to be i.i.d. according to the normal distribution $N(\xi, \sigma^2)$, we saw in Example 3.1.4 that (3.1.26) is a rejection region for testing $H : \sigma = \sigma_0$ against $\sigma > \sigma_0$ with asymptotic level α. Suppose now that in fact the assumption of normality is in error and that instead the X's are i.i.d. according to some other distribution F with variance $\text{Var}(X_i) = \sigma^2$ and with $0 < \text{Var}(X_i^2) = \lambda^2 < \infty$. Let us denote the resulting true probability of (3.1.26) by $\alpha'_n(F)$. It follows from Example 2.4.4 that

$$(3.5.3) \qquad \frac{\sqrt{n}\left[\frac{1}{n}\sum (X_i - \bar{X})^2 - \sigma_0^2\right]}{\lambda} \xrightarrow{L} N(0, 1).$$

The conditions of Theorem 3.5.1(i) are therefore satisfied with $\tau^2 = 2\sigma_0^4$ and $(\tau')^2 = \lambda^2$, the variances of X_i^2 when X_i is normal or has distribution F, respectively.

Since λ can take on any value between 0 and ∞, $\alpha' = \lim \alpha'_n$ can take on any value between 0 and $1/2$ regardless of the nominal level α, and the test (3.5.3) is therefore very non-robust against violations of the normality assumption. What is particularly unfortunate is that a very small disturbance of the normal distribution which would be impossible to detect could cause an enormous increase in $\text{Var}\left(X_i^2\right)$ and therefore in α'. (In this connection, see the note "a crucial distinction" following Example 2.1.4.)

To illustrate this non-robustness result for some actual distributions, suppose that F, instead of being normal, is a t-distribution with ν degrees of freedom. Table 3.5.2 shows the value of $\alpha' = \alpha'(F)$ when $\alpha = .05$ as a function of ν. As $\nu \to \infty$, the t-distribution tends to the normal and $\alpha'(F)$ to $\alpha = .05$. The tables shows dramatically how very misleading the

TABLE 3.5.2. $\alpha'(F)$ for $F = t$ with ν degrees of freedom

ν	5	6	7	10	20	30	50	100	∞
$\alpha'(F)$	.205	.149	.122	.0896	.0656	.0597	.0555	.0526	.0500

nominal level α can be when the assumption of normality is violated.

It is important to be clear about the distinction among three different uses that are being made of t in this chapter.

(a) There is the t-statistic t_n defined by (3.1.24). This is defined for any random variables $X_1, \ldots, X_n$ regardless of the distribution of the X's.

(b) There is the classical result, discovered by Student, that t_n is distributed according to the t-distribution with $n-1$ degrees of freedom when the X's are i.i.d. according to a normal distribution $N\left(0, \sigma^2\right)$.

(c) Finally, in Table 3.5.2, the level of the χ^2-test for variance is considered when the data distribution F of the X's is a t-distribution (rather than normal) for various degrees of freedom.

In connection with both (b) and (c), it is useful to realize that the t-distributions are more heavy tailed than the normal distribution and that they tend to the normal distribution as the number ν of degrees of freedom tends to infinity. Already for $\nu = 10$, the density of t_ν visually closely resembles that of the standard normal density, and this resemblance increases with increasing ν (Problem 5.5). The choice of a t-distribution for Table 3.5.2 is unconnected with the result (b). Its common use as a possible data distribution is motivated primarily by its familiarity and simplicity.

For an example in which F is lighter tailed than the normal distribution, see Problem 5.3.

Since in practice normality can never be guaranteed, it seems clear that (3.1.26) is not a practicable test, However, in the present situation, an easy remedy is available. All that is required is studentization of (3.5.3) in the sense of (3.1.18), i.e., replacing the true λ by a consistent estimator $\hat{\lambda}_n$. Since λ^2 is the variance of X_i^2, a consistent estimator of λ^2 is $\sum (Y_i - \bar{Y})^2 / (n-1)$ where $Y_i = X_i^2$. A rejection region with asymptotic level α is therefore given by

(3.5.4)
$$\frac{\sqrt{n} \left[\frac{1}{n} \sum (X_i - \bar{X})^2 - \sigma_0^2 \right]}{\sqrt{\sum \left(X_i^2 - \overline{X^2} \right)^2 / (n-1)}} \geq u_\alpha,$$

where $\overline{X^2} = \sum X_i^2 / n$. Unfortunately, the convergence to α in this case tends to be very slow. This is illustrated in Table 3.5.3, which shows the rejection probability α' of the test (3.5.4) as a function of n when the nominal level is $\alpha = .05$ and when the distribution of the X's is $F = N(0,1)$. □

TABLE 3.5.3. True level α' of (3.5.4) when $\alpha = .05$ and $F = N(0,1)$

n	10	50	100	500	1,000	5,000	10,000
α'	.0082	.0173	.0235	.0364	.0396	.0456	.0466

Example 3.5.2 One-sample t-test. In the situation of the preceding example, suppose we wish to test the mean rather than the variance. Believing the X's to be i.i.d. normal, we reject $H : \xi = \xi_0$ in favor of $\xi > \xi_0$ when (3.1.24) holds. Consider now the situation in which the X's are i.i.d., but instead of being normally distributed, they are a sample from some other distribution F with mean ξ and variance σ^2. Let $\alpha'_n(F)$ denote the true rejection probability when $\xi = \xi_0$. It follows immediately from the central limit theorem and Theorem 2.3.3 that

(3.5.5) $\alpha'_n(F) \to \alpha$ as $n \to \infty$

for any distribution F with finite variance. The level of the t-test is therefore asymptotically robust against non-normality.

Before trusting this robustness result, we need to know how close $\alpha'_n(F)$ can be expected to be to α in practice, that is, for various distributions F and sample sizes n. Some indications are provided by the following three examples.

(i) F is the t-distribution with ν degrees of freedom. Here and in case (ii), F is symmetric and the level of the one-sided test is just half of that of the two-sided test given in Tables 3.5.4 (and 3.5.5). The table shows clearly how $\alpha'_n(F)$ gets closer to α as ν increases and hence t_ν gets closer to normality.

TABLE 3.5.4. $\alpha'_n(F)$ for two-sided t-test when F is a t-distribution with ν degrees of freedom

n	α	$\nu = 3$	$\nu = 5$	$\nu = 9$
	.01	.00624	.00754	.00851
5	.05	.03743	.04241	.04553
	.10	.08375	.09085	.09465
	.01	.00593	.00751	.00864
10	.05	.03935	.04445	.04720
	.10	.08876	.09494	.09755
	.01	.00671	.00818	.00915
20	.05	.04313	.04686	.04868
	.10	.09441	.09804	.09965

Source: Yuen and Murthy (1974).

(ii) F is the standard normal distribution, truncated as $\pm a$, so that its density is

$$(3.5.6) \qquad f(x) = \begin{cases} \phi(x)/[\Phi(a) - \Phi(-a)] & \text{if } |x| < a \\ 0 & \text{elsewhere.} \end{cases}$$

The uniform distribution with mean 0 is included as the limiting case corresponding to $a \to 0$. The test is the two-sided t-test.

TABLE 3.5.5. $\alpha'_n(F)$ for two-sided t-test; F = standard normal truncated at $\pm a$

n	α	$a = 3.0$	$a = 2.0$	$a = 1.0$	Uniform
5		.0102	.0118	.0168	.0200
10	.01	.0102	.0114	.0134	.0142
21		.0102	.0108	.0114	.0116
31		.0102	.0104	.0108	.0106
5		.0504	.0544	.0622	.0658
10	.05	.0504	.0522	.0538	.0542
21		.0502	.0508	.0512	.0514
31		.0502	.0504	.0508	.0508

Source: Scott and Saleh (1975).

On the whole, these two tables are reassuring for the t-test when F is symmetric. Even for distributions as far from normal as t_3 (heavy-tailed) and the uniform distribution (short-tailed), the values of $\alpha'_n(F)$ are relatively close to the nominal α once n is as large as 20 or 30, although they can be far off for small values of n such as 5 or even 10.

(iii) To give an idea of how skewness of F affects $\alpha'_n(F)$, Table 3.5.6 shows $\alpha'_n(F)$ for four distributions from the family of Pearson curves. These distributions can be indexed by their values of $\beta_1 = \mu_3/\sigma^3$ and $\beta_2 = \mu_4/\sigma^4$, which are standard measures of skewness and kurtosis, respectively. (For the normal distribution, $\beta_2 = 3$.) The table is taken from a large Monte Carlo study of Posten (1979). Since this study shows that for these distributions, $\alpha'_n(F)$ varies much less with β_2 than with β_1, Table 3.5.6 gives $\alpha'_n(F)$ for only one value of β_2 ($\beta_2 = 4.2$) and four values of β_1. The Pearson curves with $\beta_1 = 0$ are all symmetric, and the case $\beta_1 = 0$, $\beta_2 = 4.2$ corresponds to $F = t_9$. The distributions corresponding to $\beta_1 = .4$ and $.8$ are skewed to the right and their ranges are $(-\infty, \infty)$ and $(0, \infty)$ respectively. Finally, the distribution corresponding to $\beta_1 = 1.6$, $\beta_2 = 4.2$ is a distribution which is so strongly skewed to the right that it is J-shaped. (For more details about the Pearson curves, see Johnson, Kotz, and Balakrishnan (1994, Chapter 12) and Ord (1985), and the references listed there.)

TABLE 3.5.6. $\alpha'_n(F)$ for t-test at $\alpha = .05$ when F is a Pearson curve with $\beta_2 = 4.2$ and indicated β_1

β_1	Lower tail			Upper tail			Double tail		
n	10	20	30	10	20	30	10	20	30
0.0	.048	.052	.050	.049	.049	.050	.047	.050	.050
0.4	.068	.067	.063	.035	.037	.040	.052	.054	.052
0.8	.081	.076	.070	.029	.032	.036	.060	.059	.055
1.6	.109	.091	.082	.020	.025	.030	.083	.070	.063

Source: Posten (1979).

The table shows the strong effect of skewness on the level. As is typically the case when F is skewed to the right, $\alpha'_n(F) > \alpha$ for the lower-tail rejection region and $< \alpha$ for the upper tail. The two effects balance to some extent for the two-sided (double-tailed) test. For values of β_1 as large as $.8$ or 1.6, the table shows that even with $n = 30$ observations the difference between $\alpha'_n(F)$ and α can still be substantial.

It should be pointed out that, in general, the closeness of $\alpha'_n(F)$ to α depends not only on β_1 and β_2 but also on other aspects of F. For a numerical example, see Lee and Gurland (1977).

The discussion of Table 3.5.6 suggests an explanation for the very slow convergence of the level of (3.5.4) in the normal case. If in (3.5.4) the sample mean $\bar{X}$ is replaced by the population mean ξ, the resulting test statistic can be written as

$$\frac{\sqrt{n}\left[\bar{Y} - E\left(Y_i\right)\right]}{\sqrt{\sum \left(Y_i - \bar{Y}\right)^2 / (n-1)}}$$

with $Y_i = \left(X_i - \xi\right)^2$. It therefore becomes a t-statistic with the variables Y_i having a χ^2-distribution with one degree of freedom. This distribution is extemely skewed with $\beta_1 = 2\sqrt{2} = 2.8284$, much more skewed even than the distributions with $\beta_1 = 1.6$ considered in Table 3.5.6. It is therefore not surprising that it requires an enormous sample size for the distribution of this t-statistic to overcome the skewness and take on an approximately normal shape.

Since the asymptotic level of the t-test is asymptotically robust against non-normality in the sense of (3.5.5), it is natural to ask whether the same is true for the asymptotic power of the test. An examination of the limit result (3.3.21) for the power in the normal case shows (Problem 5.6) that it does not use the assumption of normality but applies equally if the X's are a sample from any other distribution with finite third moment. (The third moment assumption is made to ensure applicability of the Berry-Esseen Theorem.) The power of the t-test is therefore robust against non-normality in the sense that even for non-normal F, it is approximately the same as it is when F is normal. This is, however, not as desirable a property as it seems at first sight. For although the t-test has many optimum properties in the normal case, in other cases much higher power can often be achieved by other tests, as we saw in the preceding section.   □

The robustness properties against non-normality of the one-sample t-test extend with only slight modifications to the two-sample problem.

Example 3.5.3 Comparing two means. Consider the normal two-sample test with rejection region

(3.5.7)
$$\frac{\bar{Y} - \bar{X}}{\sqrt{\dfrac{\hat{\sigma}^2}{m} + \dfrac{\hat{\tau}^2}{n}}} \geq u_\alpha$$

discussed in Example 3.1.6, but suppose that the X's and Y's, in fact, are samples from non-normal distributions $F\left(\dfrac{x - \xi}{\sigma}\right)$ and $G\left(\dfrac{y - \eta}{\tau}\right)$, where $\xi = E(X_i)$, $\sigma^2 = \mathrm{Var}(X_i)$, $\eta = E(Y_j)$ and $\tau^2 = \mathrm{Var}(Y_j)$. Then it is seen from Lemma 3.1.1 that the probability of (3.5.7) under the hypothesis $H : \eta = \xi$ continues to tend to α. The asymptotic level of (3.5.7) is therefore robust against non-normality. Note that this result does not even require the distribution of the X's and Y's to have the same form. Analogously, it is easy to see that the asymptotic power of the test (3.5.7) against the alternatives $\eta - \zeta + \dfrac{\Delta}{\sqrt{n}}$ is also robust against non-normality (Problem 5.6). □

Let us next consider the simple linear regression situation of Example 2.7.4.

Example 3.5.4 Simple linear regression. Let $X_i \, (i = 1, \ldots, N)$ be independently distributed as $N\left(\xi_i, \sigma^2\right)$, where $\xi_i = \alpha + \beta v_i$. Then it was seen in Example 3.4.4 that the test of $H : \beta = \beta_0$ with rejection region (3.4.21) has asymptotic level α.

Let us now consider the actual level of this test when in fact

$$(3.5.8) \qquad X_i = \alpha + \beta v_i + E_i,$$

where the E's are i.i.d., not necessarily normal, with zero mean and finite variance σ^2. It was seen in Example 2.7.6 that then the numerator of (3.4.21) satisfies

$$(3.5.9) \qquad \left(\hat{\beta} - \beta\right) \sqrt{\sum (v_i - \bar{v})^2} \xrightarrow{L} N\left(0, \sigma^2\right),$$

provided

$$(3.5.10) \qquad \frac{\max (v_i - \bar{v})^2}{\sum (v_j - \bar{v})^2} \to 0.$$

We shall now show that the square of the denominator of (3.4.21) tends in probability to σ^2. Combining these two facts, we see that under the model (3.5.8), the asymptotic level of the test (3.4.21) continues to be α if (3.5.10) holds, and that then the level of the test (3.4.21) is therefore robust against non-normality.

The distribution of the denominator is independent of α and β (Problem 5.12) and we can therefore assume without loss of generality that $\alpha = \beta = 0$ so that the X's are i.i.d. with mean zero and variance σ^2. From the definition of $\hat{\alpha}$ and $\hat{\beta}$, it is seen that (Problem 5.13)

$$(3.5.11) \qquad \sum \left(X_i - \hat{\alpha} - \hat{\beta} v_i\right)^2 = \sum \left(X_i - \bar{X}\right)^2 - \hat{\beta}^2 \sum (v_i - \bar{v})^2 .$$

Since

$$\sum \left(X_i - \bar{X}\right)^2 / (N - 2) \xrightarrow{P} \sigma^2$$

by Example 2.1.3, it remains only to show that

$$(3.5.12) \qquad \hat{\beta}^2 \sum (v_i - \bar{v})^2 / (N - 2) \xrightarrow{P} 0,$$

which follows immediately from (3.5.9). Now

$$(3.5.13) \qquad \mathrm{Var}\left[\hat{\beta}\sqrt{\sum (v_i - \bar{v})^2}\right] = \sigma^2$$

by (2.7.11), and it therefore follows that the asymptotic level of (3.4.21) continues to be α under model (3.5.8). $\qquad\qquad\square$

The robustness question of Example 3.5.2 can be viewed in another way. If the true distribution F of the X's is not known to be normal, the problem should perhaps be treated non-parametrically, that is, by testing the hypothesis

$$(3.5.14) \qquad H^* : \xi = \xi_0$$

in the model

$$(3.5.15) \qquad X_1, \dots, X_n \text{ i.i.d. according to } F \in \mathcal{F}$$

with $\xi = E(X_i)$ and $\mathcal{F}$ the class of all distributions with finite variance. From this point of view, the level of the t-test of H^* is

$$(3.5.16) \qquad \alpha_n^* = \sup_{F \in \mathcal{F}_0} \alpha_n'(F),$$

where $\mathcal{F}_0$ is the family of distributions F in $\mathcal{F}$ with $\xi = \xi_0$ and the question is whether $\alpha_n^* \to \alpha$ as $n \to \infty$.

Closely related is the question of whether the convergence in (3.5.5) is uniform in F, i.e., whether for any $\epsilon > 0$, there exists n_0 (independent of F) such that

$$|\alpha_n'(F) - \alpha| < \epsilon \text{ for all } n > n_0.$$

It follows in fact from Lemma 2.6.1 that uniform convergence in (3.5.5) implies $\alpha_n^* \to \alpha$. Unfortunately, these desirable properties do not hold. We have instead

Lemma 3.5.1 *For the t-test in its approximate form (3.1.24) or given exactly in the note following Example 3.1.3,*

$$(3.5.17) \qquad \inf_{F \in \mathcal{F}_0} \alpha_n(F) = 0 \quad and \quad \sup_{F \in \mathcal{F}_0} \alpha_n(F) = 1$$

holds for every $n \geq 2$.

Proof. To prove the result for the sup, consider the subclass $\mathcal{F}_1$ of $\mathcal{F}$ consisting of the distributions

$$(3.5.18) \qquad F = \gamma G + (1 - \gamma) H, \quad 0 \leq \gamma \leq 1,$$

where $G = N(\mu_1, 1)$ and $H = N(\mu_2, 1)$ with $\gamma\mu_1 + (1-\gamma)\mu_2 = \xi_0$. Then for any set S in n-space,

$$P_F[(X_1,\ldots,X_n) \in S] = \int \cdots_S \int \Pi[\gamma g(x_i) + (1-\gamma)h(x_i)]\, dx_1 \cdots dx_n$$
$$\geq \gamma^n \int \cdots_S \int g(x_1) \cdots g(x_n)\, dx_1 \cdots dx_n = \gamma^n P_G[(X_1,\ldots,X_n) \in S],$$

where g and h denote the densities of G and H, respectively.

If S is the rejection region of the t-test, then it is easy to see that $P_G[(X_1,\ldots,X_n) \in S]$, which is the power of the t-test against the alternative G, tends to 1 as $\mu_1 \to \infty$ (Problem 5.9(i)). On the other hand, $\gamma^n \to 1$ as $\gamma \to 1$. Thus by choosing μ_1 and γ sufficiently large, the rejection probability of the t-test under $\mathcal{F}_1$ can be made arbitrary close to 1 and this proves the second statement in (3.5.17). The proof of the result for the inf is completely analogous (Problem 5.9(ii)).   ∎

Lemma 3.5.1 thus leads to the disappointing conclusion that there exists no sample size n for which $\alpha_n(F) \leq \alpha + \epsilon$ for all $F \in \mathcal{F}$. No matter how large n is chosen, there will exist distributions $F \in \mathcal{F}$ for which $\alpha_n(F)$ is arbitrary close to 1. While this fact should not be forgotten, Tables 3.5.3–3.5.5 nevertheless suggest that for the distributions commonly encountered, the nominal level of the t-test provides a reasonable approximation to the actual level.   □

Note: The conclusions of Lemma 3.5.1 hold not only for the t-test but also (with the obvious modifications) for the tests of Examples 3.5.3, and 3.5.4.

In the examples considered so far, we examined the robustness of some standard normal theory tests against violation of the normality assumption. We next examine what happens to the level of such a test when the assumption of independence is violated. This assumption is difficult to check but frequently is not valid (for example, because measurements taken close together in time or space often exhibit dependence). (For further discussion of this assumption and the importance of guarding against its violations see, for example, Cochran (1968) and Kruskal (1988).)

Here we give just a few illustrations. In each of these, the test statistic will be a standardized or studentized mean such as in (3.1.22) or (3.1.24) based on identically distributed variables $X_1,\ldots,X_n$ with $E(X_i) = \theta$ and $\mathrm{Var}(X_i) = \sigma^2 < \infty$, for which

(3.5.19) $$\frac{\sqrt{n}(\bar{X} - \theta)}{\sqrt{\mathrm{Var}[\sqrt{n}(\bar{X} - \theta)]}} \to N(0,1).$$

Let us denote the correlation coefficient of X_i, X_j by ρ_{ij} and the covariance by $\gamma_{ij} = \rho_{ij}\sigma^2$. Then the square of the denominator of (3.5.19) is given by

$$(3.5.20) \qquad \mathrm{Var}\left[\sqrt{n}\,(\bar{X} - \theta)\right] = \sigma^2\left[1 + \frac{1}{n}\sum\sum_{i \neq j}\rho_{ij}\right].$$

We are not interested in extreme forms of dependence such as, for example, $X_1 = X_2 = \cdots = X_n$, but, instead, shall restrict attention to cases in which the correlations satisfy

$$(3.5.21) \qquad \frac{1}{n^2}\sum_{i=1}^{n}\sum_{j=1}^{n}\rho_{ij} \to 0.$$

This condition holds in particular when the ρ's satisfy

$$(3.5.22) \qquad \frac{1}{n}\sum\sum_{i \neq j}\rho_{ij} \to \gamma$$

for some finite γ. Examples are given in Section 2.8. Note that γ is not the average of the ρ_{ij} and hence that $|\gamma|$ is not necessarily ≤ 1.

Example 3.5.5 The one-sample t-test. (Effect of dependence). In studying the effect of dependence on the t-test, we shall retain the assumption of normality of the X's and the identity of their distributions. Let us assume then that $(X_1, \ldots, X_n)$ have a joint multivariate normal distribution (see Chapter 5, Section 4) with the marginal distribution of each X_i being $N(\xi, \sigma^2)$. In analogy with (3.5.21), which implies that $\bar{X} \to \xi$ in probability (see (2.2.21)), we shall also assume that

$$(3.5.23) \qquad \frac{1}{n^2}\sum\sum \mathrm{Correlation}\left(X_i^2, X_j^2\right) \to 0$$

so that

$$(3.5.24) \qquad \frac{1}{n}\sum X_i^2 - \bar{X}^2 \xrightarrow{P} \sigma^2.$$

Under these conditions, it turns out that the asymptotic behavior of the t-test for testing the hypothesis $H : \xi = \xi_0$ is determined by the limiting behavior of $\sum\sum \rho_{ij}/n$, that is, by the value of γ. $\qquad \square$

Theorem 3.5.2 Let $(X_1, \ldots, X_n)$ have a joint multivariate normal distribution with $E(X_i) = \xi$, $\mathrm{Var}(X_i) = \sigma^2$, and satisfying (3.5.22) and (3.5.23). Then

  (i) the distribution of the t-statistic (3.1.24) under H tends to the normal distribution $N(0, 1 + \gamma)$;

(ii) the asymtotic level of the t-test (3.1.24) is

(3.5.25)
$$\alpha' = 1 - \Phi\left(\frac{u_\alpha}{\sqrt{1+\gamma}}\right).$$

Proof. Since the X's are jointly normal, the numerator $\sqrt{n}\left(\bar{X} - \xi_0\right)$ of t is also normal, with mean 0 and variance given by (3.5.20); it therefore tends in law to $N\left(0, \sigma^2 (1+\gamma)\right)$. Consider next the denominator of t. Since it is independent of ξ, assume without loss of generality that $\xi = 0$. Then the second term on the right side of

$$\frac{1}{n-1}\sum (X_i - \bar{X})^2 = \frac{1}{n-1}\sum X_i^2 - \frac{n}{n-1}\bar{X}^2$$

tends in probability to 0 while the first term by (3.5.23) tends to σ^2. This completes the proof of (i). Part (ii) is now an obvious consequence. ∎

It follows from this result that the level of the t-test is robust against situations in which the dependence is so weak that

(3.5.26)
$$\frac{1}{n}\sum_{i\neq j}\sum \rho_{ij} \to 0,$$

a condition that will typically not hold unless the ρ_{ij} are allowed to depend on n and to tend to 0 as $n \to \infty$.

To illustrate Theorem 3.5.2, we consider two stationary models with dependent observations, which were formulated in Section 2.8.

Example 3.5.6 Moving averages. Suppose that $Z_1, Z_2, \ldots$ are i.i.d. with distribution $N\left(\xi, \sigma_0^2\right)$ and that

(3.5.27)
$$X_i = \frac{Z_i + \cdots + Z_{i+m}}{m+1}.$$

To see whether the t-test is still applicable, we only need to evaluate γ from (3.5.22) (Problem 5.18). Alternatively, we can use the identity

(3.5.28)
$$\sum_{i=1}^{} X_i = \frac{Z_1 + 2Z_2 + \cdots + mZ_m}{m+1} + (Z_{m+1} + \cdots + Z_n)$$
$$+ \frac{mZ_{n+1} + \cdots + 2Z_{n+m-1} + Z_{n+m}}{m+1}$$

to see that $\sqrt{n}\left(\bar{X} - \xi\right)$ has the same limit distribution as $\sqrt{n}\left(\bar{Z} - \xi\right)$ and hence that

$$\sqrt{n}\left(\bar{X} - \xi\right) \xrightarrow{L} N\left(0, \sigma_0^2\right) = N\left(0, (m+1)\sigma^2\right).$$

It follows that

$$(3.5.29) \qquad \alpha' = 1 - \Phi\left(\frac{u_\alpha}{\sqrt{m+1}}\right) > \alpha.$$

For example, when $m = 3$ and the nominal level is $\alpha = .05$, it is seen from Table 3.5.1 that $\alpha' = .205$. In the present case, it is, of course, easy to correct for this problem by replacing the critical value u_α in (3.1.24) by $u_\alpha \sqrt{m+1}$. □

Example 3.5.7 First order autoregressive process. Instead of (3.5.27), suppose that the X's are given by the stationary process (2.8.13) with $|\beta| < 1$ so that

$$(3.5.30) \qquad \mathrm{Var}\left[\sqrt{n}\left(\bar{X} - \theta\right)\right] \to \sigma^2 \frac{1+\beta}{1-\beta},$$

and hence that

$$(3.5.31) \qquad 1 + \gamma = \frac{1+\beta}{1-\beta}.$$

Then the asymptotic level of the t-test can take on any value between 0 and $1/2$ as β varies from -1 to $+1$. Thus under this model too, the t-test is not robust. □

Example 3.5.8 Markov chains. In a sequence of binomial trials, let $X_i = 1$ or 0 as the ith trial is a success or failure and reject $H : p = p_0$ in favor of $p > p_0$ if

$$(3.5.32) \qquad \frac{\sqrt{n}\left(\bar{X} - p_0\right)}{\sqrt{p_0 q_0}} \geq u_\alpha.$$

To illustrate the behavior of this test under dependence, suppose that the dichotomous trials form a stationary Markov chain. As was discussed in Example 2.8.3, then

$$(3.5.33) \qquad \frac{\sqrt{n}\left(\bar{X} - p\right)}{\sqrt{pq}} \overset{L}{\to} N\left(0, \frac{1+\pi_1-\pi_0}{1-\pi_1+\pi_0}\right),$$

so that in the notation of Theorem 3.5.1,

$$(3.5.34) \qquad \frac{\tau'^2(p_0)}{\tau^2(p_0)} = \frac{1+\pi_1-\pi_0}{1-\pi_1+\pi_0}.$$

We must now distinguish two cases.

(a) If

(3.5.35) $\pi_0 < \pi_1,$

then (Problem 5.21)

(3.5.36) $\pi_0 < p < \pi_1,$

that is, the trials are positively dependent in the sense that success on the $(i - 1)$st trial increases the probability of success on the ith trial. In this case, $\tau'(p_0) > \tau(p_0)$ and the level of the test (3.5.32) is liberal. For any fixed $p = p_0$, the ratio $\tau'(p_0)/\tau(p_0)$ can be made arbitrarily large by letting $\pi_0 \to 0$ and $\pi_1 \to 1$ (Problem 3.5.22), in which case, the level of the test by (3.5.2) tends to 1/2.

(b) On the other hand, if

(3.5.37) $\pi_1 < \pi_0,$

we have

(3.5.38) $\pi_1 < p < \pi_0$

so that $\tau'(p_0) < \tau(p_0)$ and the test (3.5.35) is conservative. By letting $\pi_1 \to 0$ and $\pi_0 \to 1$, we get $\tau'(p_0)/\tau(p_0) \to 0$, in which case, the level of the test tends to 0.

We see that under the Markov model, the test (3.5.35) is robust (i.e., $\tau'(p_0)/\tau(p_0) = 1$) if and only if $\pi_1 = \pi_0$, which implies $\pi_1 = \pi_0 = p$ and hence independence of the X's. Weak dependence can, of course, be modeled by

(3.5.39) $\pi_1 = \pi_{1n} = p + \epsilon_n$ with $\epsilon_n \to 0,$

in which case

(3.5.40) $\pi_0 = \pi_{0n} = p - \dfrac{p}{q}\epsilon_n$

by (2.8.23). Then the ratio (3.5.34) tends to 1, and the level of the test tends to α.

If it were known that the only kind of dependence possible was that considered in this example, one could studentize (3.5.33) by replacing $(1 + \pi_1 - \pi_0) / (1 - \pi_1 + \pi_0)$ by $(1 + \hat{\pi}_1 - \hat{\pi}_0) / (1 - \hat{\pi}_1 + \hat{\pi}_0)$, where $\hat{\pi}_0$ and $\hat{\pi}_1$ are consistent estimator of π_0 and π_1, respectively. Unfortunately, it is only rarely the case that this simple Markov dependence is the only kind possible.   □

Summary

1. If the assumed model is wrong, use of the critical values determined in Section 3.1 may lead to an asymptotic level α' different from the nominal level α. If $\alpha' = \alpha$, the level of the test is said to be *robust* against the deviation from the assumed model under consideration.

2. The normal theory tests of a mean, the difference of two means, and of regression coefficients are robust against non-normality. On the other hand, the normal tests of a variance or the ratio of two varicances is non-robust and, in fact, is extremely sensitive to the assumption of normality.

3. Although for any fixed distribution F with mean ξ_0 and finite variance, the level $\alpha_n(F)$ of the one-sample t-test of $H : \xi = \xi_0$ tends to α as $n \to \infty$, the maximum level $\sup_F \alpha_n(F)$ is 1 for every fixed $n \geq 2$. Thus, despite the eventual robustness of the level, for each n_0 there exist distributions F for which the true level $\alpha_n(F)$ is far from the nominal α.

4. The problem of robustness arises not only for non-normality but also for other possible departures from the assumed model, for example, from independence. In particular Student's t-test is typically non-robust against dependence unless that dependence is very weak.

3.6   Problems

Section 1

1.1 Refine (3.1.13) and (3.1.14) by using the continuity correction in (3.1.12).

1.2 Make a table of the exact value of (a) the probability of (3.1.14) and (b) the corresponding probability for the test of Problem 1.1, for $\alpha = .01, .05, .1$ and a number of values of λ_0 and n.

1.3 Let X be binomial $b(p, n)$ and suppose the hypothesis $H : p = p_0$ is rejected in favor of $K : p > p_0$ when $X \geq C_n$. Derive formulas analogous to (3.1.13) and (3.1.14), with and without continuity correction.

1.4 Make a table for the situation of Problem 1.3 analogous to that asked for in Problem 1.2.

1.5 If Y has the negative binomial distribution (defined in Table 1.6.2), then by Problem 4.12 of Chapter 2,

(3.6.1)
$$\frac{\left(Y - \frac{mq}{p}\right)p}{\sqrt{mq}} \xrightarrow{L} N(0,1).$$

Use this fact to obtain a test of $H : p - p_0$ vs. $K : p > p_0$ which for large m has the approximate level α. (For a review of more accurate approximations, see Johnson, Kotz, and Kemp (1992, Sect. 5.6).)

1.6 (i) Make a table giving the exact level of the test (3.1.17) for $\theta_0 = 1$ and $\alpha = .01, .05, .1$ and various values of n.

(ii) Make a table comparing the exact probability of (3.1.27) with the nominal level for $\sigma_0 = 1$ and $\alpha = .01, .05, .1$ and various values of n. Determine how large n has to be before the approximation becomes satisfactory.

1.7 Show that the asymptotic level of the test (3.1.27) remains α if the factor $1/n$ on the left side is replaced by $1/(n - 1)$.

1.8 Let $X_1, \ldots, X_n$ be i.i.d. according to the exponential distribution $E(\xi, a)$.

(i) Determine a test of $H : a = 1$ against $a > 1$ with asymptotic level α, based on $\sum [X_i - X_{(1)}]$, analogous to the test (3.1.27) in the normal case.

(ii) Determine a test of $H : \xi = \xi_0$ against $\xi > \xi_0$ with asymptotic level α, based on $(X_{(1)} - \xi_0) / \frac{1}{n} \sum [X_i - X_{(1)}]$, analogous to (3.1.24).

1.9 If $T_n / T'_n \xrightarrow{P} 1$, and T'_n converges in probability to a finite limit $\neq 0$, then $T_n - T'_n \xrightarrow{P} 0$.

1.10 In Example 3.1.1, show that

(i) the rejection region

$$(3.6.2) \qquad 2\sqrt{n} \left(\sqrt{\bar{X}} - \sqrt{\lambda_0} \right) \geq u_\alpha$$

for testing $H : \lambda = \lambda_0$ against $\lambda > \lambda_0$ has asymptotic level α;

(ii) the tests (3.1.14) and (3.6.2) are asymptotically equivalent.

[**Hint (i)**: Use Example 2.5.2.]

1.11 In generalization of the preceding problem, suppose that T_n is a test statistic satisfying (3.1.7) and that f is a real-valued function for which $f'(\theta_0)$ exists and is $\neq 0$.

(i) The rejection region

$$(3.6.3) \qquad \frac{\sqrt{n} [f(T_n) - f(\theta_0)]}{\tau(\theta_0) f'(\theta_0)} \geq u_\alpha$$

has asymptotic level α for testing $H : \theta = \theta_0$.

(ii) The tests (3.1.12) and (3.6.3) are asymptotically equivalent.

[**Hint**: Use equation (2.5.4).]

1.12 If the sequence of rejection regions R_n is asymptotically equivalent to R'_n, and R'_n to R''_n, show that R_n is also asymptotically equivalent to R''_n.

1.13 Use the following (or some other) example to show that (3.1.39) is not enough to guarantee (3.1.42).

Example. Let $X_1, \ldots, X_n$ be i.i.d. $N\left(\xi, \sigma^2\right)$ and let

$$T_n = \frac{\sqrt{n}\bar{X}}{\sqrt{\dfrac{1}{n-1}\sum\left(X_i - \bar{X}\right)^2}} + \frac{1}{n^a}\sum\left(X_i - \bar{X}\right)^2, \ a > 1.$$

Consider the test of $H : \xi = 0$, which rejects when $T_n \geq u_\alpha$.

1.14 In Example 3.1.5(i) (continued) show that

(i) the test with rejection region

(3.6.4)
$$\frac{\sqrt{\bar{Y}} - \sqrt{\bar{X}}}{\dfrac{1}{2}\sqrt{\dfrac{1}{m} + \dfrac{1}{n}}} \geq u_\alpha$$

has asymptotic level α;

(ii) (3.6.4) is asymptotically equivalent to (3.1.34) and (3.1.47).

1.15 The exact probability of (3.1.34), (3.1.47), and (3.6.4) under $H : \lambda = \mu$ depends on μ. Use simulation to make a table of this value and graph it as a function of μ for each of these tests when (i) $m = 5$, $n = 10$; (ii) $m = n = 10$.

1.16 In Example (3.1.5(i)) (continued), when $\mu = \lambda$ the conditional distribution of $\sum Y_j$ given $\sum X_i + \sum Y_j = t$ is the binomial distribution $b(p, t)$ with $p = n/(m+n)$. A small-sample level α test rejects when $\sum Y_i > C\left(\sum X_i + \sum Y_j\right)$, where $C(t)$ is the smallest value for which $P\left[Y \geq C(t)\right] \leq \alpha$ when Y has the distribution $b(p, t)$. As m and $n \to \infty$, $\sum X_i + \sum Y_j \overset{P}{\to} \infty$ and an approximate large-sample test is obtained by replacing $b(p, t)$ by its normal approximation. Show that the resulting test coincides with (3.1.47).

1.17 Let $X_1, \ldots, X_m$ and $Y_1, \ldots, Y_n$ be i.i.d. according to exponential distributions $E\left(\xi, a\right)$ and $E\left(\eta, b\right)$, respectively. Find a two-sample test of $H : \eta = \xi$ against $\eta > \xi$ analogous to the one-sample test of Problem 1.8(ii), which has asymptotic level α.

1.18 Let X and Y be independent binomial $b(p, m)$ and $b(\pi, n)$, respectively.

(i) Using the normal approximation to the binomial, find a test of $H : p = \pi$ against $p < \pi$ analogous to (3.1.47).

(ii) Show that the resulting test is asymptotically equivalent to (3.1.35).

Note: Detailed comparisons of the two tests are given by Cressie (1978) and Andrés et al (1992).

1.19 In generalization of Theorem 3.1.1, one might conjecture that if the tests (3.1.45) both have asymptotic level α and if, in addition,

$$(3.6.5) \qquad V_n/V_n' \to 1 \text{ in probability,}$$

then the two tests would be asymptotically equivalent. That this is not so can be seen by letting V_n and V_n' be independently identically distributed with the distribution which assigns probability α and $1-\alpha$ respectively to the points $u_\alpha + \dfrac{1}{n}$ and $u_\alpha - \dfrac{1}{n}$.

1.20 Let $X_1, \ldots, X_n$ be i.i.d. according to the uniform distribution $U(\xi - \tau, \xi + \tau)$.

(i) If ξ is known, let the hypothesis $H : \tau = 1$ be rejected in favor of $\tau > 1$ when

$$(3.6.6) \qquad \max |X_i - \xi| > 1 + \frac{1}{n} \log (1 - \alpha).$$

Show that this test has asymptotic level α.

(ii) If ξ is unknown, replace ξ by $\hat{\xi} = \left[X_{(1)} + X_{(n)} \right]/2$. Show that the test (3.6.6) with ξ replaced by $\hat{\xi}$ no longer has asymtotic level α.

[Hint (ii): Note that

$$\max \left| X_i - \frac{X_{(1)} + X_{(n)}}{2} \right| = \frac{1}{2} \left[X_{(n)} - X_{(1)} \right],$$

and use the fact (to be proved in Chapter 5) that if $Y_1, \ldots, Y_n$ are i.i.d. $U(0, \tau)$, then

$$(3.6.7) \qquad P \left[\frac{a}{n} < Y_{(1)} < Y_{(n)} < \tau - \frac{b}{n} \right] \to e^{-a/\tau} \cdot e^{-b/\tau}.$$

Note. This problem shows that replacement of a parameter by a consistent estimator may change the asymptotic distribution of a test statistic.

1.21 Let $X_1, \ldots, X_n$ be i.i.d., let p_-, p_0, p_+ denote the probability that $X_i < 0, = 0, > 0$, respectively, and consider the problem of testing

$$H : p_- = p_+$$

against the alternatives $p_- < p_+$.

(i) If N_-, N_0, N_+ denote the number of X's $< 0, = 0, > 0$, respectively, then

(3.6.8) $\left(N_+ + \dfrac{1}{2} N_0 - \dfrac{n}{2} \right) \Big/ \sqrt{\dfrac{n}{4}(1 - p_0)} \overset{L}{\to} N(0, 1).$

(ii) Replace the unknown p_0 by a consistent estimator to determine a test of H with asymptotic level α.

[**Hint:** Let $Y_i = 1$, $1/2$, 0 as $X_i > 0$, $= 0$, < 0, respectively, so that $N_+ + \frac{1}{2} N_0 = \sum Y_i$.

Note that under the hypothesis $p_+ + \dfrac{1}{2} p_0 = \dfrac{1}{2}$].

Section 2

2.1 Prove (3.2.3) for the population model that $Z_1, \ldots, Z_N$ are i.i.d. according to some continuous distribution.

[**Hint:** Use the facts that the joint distribution of $(Z_1, \ldots, Z_n)$ is symmetric in its N variables and that the ranks $s_1, \ldots, s_n$ are attained by some subset $(Z_{i_1}, \ldots, Z_{i_n})$ of $(Z_1, \ldots, Z_N)$.]

(Note that the result also follows from the fact that (3.2.3) is valid in the randomization model (3.2.6).)

2.2 Verify the identity (3.2.8).

[**Hint:** The number of X's smaller than the smallest Y is $S_1 - 1$, that smaller than the second smallest Y is $S_2 - 2$, and so on.]

2.3 (i) Determine the range (i.e., the set of possible values) of

(a) W_s,

(b) W_{XY}.

(ii) In the light of (i), why is it preferable to table the distribution of W_{XY} rather than that of W_s.

2.4 For $u > 0$, show that (3.2.20) is equivalent to the inequality in square brackets in (3.2.19).

[**Hint:** Use the identities

$$\bar{Y} - a_N = \frac{m}{n} (\bar{Y} - \bar{X})$$

and

$$\sum (a_{Nj} - a_{N.})^2 = \sum X_i^2 + \sum Y_j^2 - \frac{1}{N}\left(\sum X_i + \sum Y_j\right)^2 .]$$

2.5 In Example 3.2.2, suppose that $m = n = 4$, $\alpha = .1$, and the values $a_1, \ldots, a_N$, when ordered, are

.93, 1.08, 1.56, 1.78, 2.15, 2.34, 3.01, 3.12.

(i) Determine the critical value $K(a_1, \ldots, a_N)$ of (3.2.14) and compare it with u_α.

(ii) Obtain the value of (3.2.17).

2.6 Prove (3.2.30) by showing that $d_{n_i} = i$ $(i = 1, \ldots, n)$ satisfies (2.7.16) or (2.7.17).

2.7 In Example 3.2.6, suppose that $N = 7$, $\alpha = .1$, and the values of $a_1, \ldots, a_N$, when ordered, are .21, .25, .41, .47. .49, .58, .80.

(i) Determine the critical value of (3.2.35) and compare it with u_α.

(ii) Obtain the value of (3.2.37).

(iii) How do the results of (i) and (ii) change when the observation .80 is changed to .60?

2.8 For $u_\alpha > 0$, show that (3.2.41) is equivalent to the inequality in the square bracket of (3.2.40).

2.9 Use (2.8.45), (2.8.49), and the identity (3.2.8) to obtain the expectation and variance of W_{XY}.

2.10 (i) If $Z_1, \ldots, Z_N$ are i.i.d. with $E(Z_i) = \sigma^2$, and $E(Z_i^4) < \infty$, then

$$\frac{\max Z_i^2}{\sum Z_i^2} \to 0 \text{ in probability.}$$

(ii) Under the assumptions of (i), suppose that in addition (3.2.36) holds for every sequence $(a_{N1}, \ldots, a_{NN})$ satisfying (3.2.37). Then

$$K(Z_1, \ldots, Z_N) \to u_\alpha \text{ in probability.}$$

[**Hint** (i) Since $\sum Z_i^2/N \xrightarrow{P} \sigma^2$, it is enough to show that $\max Z_i^2/N \xrightarrow{P} 0$. Now $P[\max Z_i^2/N > \epsilon] = P[Z_i^2 > \epsilon N$ at least for one $i] \leq N P(Z_i^2 > \epsilon N)$, and the result follows from Chebyshev's inequality. (ii) Given $\epsilon > 0$, there exists $\delta > 0$ such that $|K(a_{N1}, \ldots, a_{NN}) - u_\alpha| < \epsilon$ if $\frac{\max a_i^2}{\sum a_i^2} < \delta.$]

2.11 (i) If $Z_1, \ldots, Z_N$ are i.i.d. with $E(Z_i^4) < \infty$, then

(3.6.9) $$\frac{\max (Z_i - \bar{Z})^2}{\sum (Z_i - \bar{Z})^2} \to 0 \text{ in probability.}$$

(ii) Under the assumptions of (i), assume (3.2.16) and suppose that, in addition, (3.2.15) holds for every sequence $(a_{N1}, \ldots, a_{NN})$ satisfying (3.2.17). Then

$$K(Z_1, \ldots, Z_N) \to u_\alpha \text{ in probability.}$$

[**Hint**: Without loss of quantity, assume $E(Z_i) = 0$. Use Problem 2.10(i) and the fact that $\max |Z_i - \bar{Z}| \le 2 \max |Z_i|.$]

Section 3

3.1 Prove consistency of the test (i) (3.1.14); (ii) (3.1.24); (iii) (3.1.17).

3.2 Prove consistency of the tests (i) (3.1.34); (ii) (3.1.35).

3.3 (i) Compare the entries of Table 3.3.1 with the corresponding values when the continuity correction is used.

(ii) Prepare the corresponding table for the case that $n = 400$ instead of 100.

3.4 Let $X_1, \ldots, X_n$ be i.i.d. according to the Poisson distribution $P(\lambda)$. Show that

$$\hat{\tau}_n = \bar{X} + \sqrt[3]{n}(\bar{X} - \lambda_0)$$

tends in probability to λ_0 when $\lambda = \lambda_0$ but not when $\lambda = \lambda_0 + \dfrac{\Delta}{\sqrt[3]{n}}$.

3.5 Make a table comparing the approximation (3.3.19) with the corresponding exact power.

3.6 Make a table comparing (3.3.22) with the corresponding exact power of the t-test.

3.7 In Theorem 3.3.3, replace the one-sided test (3.1.3) of $H : \theta = \theta_0$ by the two-sided test with rejection region

(3.6.10) $$\frac{\sqrt{n}\,|T_n - \theta_0|}{\tau(\theta_0)} \ge u_{\alpha/2}.$$

(i) Show that this test has asymptotic level α.

(ii) Show that its power against the alternatives (3.3.7) satisfies

$$(3.6.11) \quad \beta_n(\theta_n) \to 1 - \Phi\left[u_{\alpha/2} - \frac{\Delta}{\tau(\theta_0)}\right] + \Phi\left[-u_{\alpha/2} - \frac{\Delta}{\tau(\theta_0)}\right].$$

(iii) Determine whether the test is consistent against the alternatives $\theta \neq \theta_0$.

3.8 Make a table showing (3.3.8) and (3.6.11) as functions of $\Delta/\tau(\theta_0)$. This shows the approximate loss of power when it is necessary to use a two-sided rather than a one-sided test.

3.9 (Negative binomial) Obtain formulas for the limiting and the approximate power of the test of Problem 1.5 analogous to (3.3.14) and (3.3.15).

3.10 (Poisson) Obtain the asymptotic power of the test (3.6.2) and compare it with that given in (3.3.18) for the test (3.1.14).

3.11 (Uniform) Determine the limiting power of the test (3.1.17) against the alternatives

(i) $\theta_n = \theta_0 + \dfrac{\Delta}{\sqrt{n}}$ and (ii) $\theta_n = \theta_0 + \dfrac{\Delta}{n}$.

3.12 Assuming that the conditions of Theorem 3.3.3 hold with T_n, θ, and $\tau(\theta)$ replaced by $f(T_n)$, $f(\theta)$, and $\tau(\theta)f'(\theta)$, respectively, determine the limiting power of the test (3.6.3) against the alternatives (3.3.7).

3.13 (Exponential) Determine the asymptotic power of the tests of Problems 1.8.

3.14 (i) Determine the asymptotic power of the test (3.1.32) applied to the normal distributions assumed in Example 3.1.6.

(ii) Assuming $\sigma = \tau$, determine the asymptotic power of the test based on (3.1.36) and compare it with that obtained in (i).

3.15 (Poisson two-sample) Show that (3.3.27) remains valid if (3.1.34) is replaced by

(i) (3.1.47);

(ii) (3.6.4).

3.16 Let X and Y be idependently distributed according to the binomial distributions $b(p, m)$ and $b(\pi, n)$, respectively. Determine the asymptotic power of the two tests of Problem 1.18 against the alternatives $\pi = p + \Delta/\sqrt{N}$.

3.17 Translate (3.3.14) into a result concerning θ rather than p and compare it with (3.3.37).

3.18 Show that (3.3.37) continues to hold under the assumptions of Note 2 for Example 3.3.5.

3.19 Determine sample sizes n such that the following tests based on n observations have approximate power β.

(i) (3.1.14) (Poisson);

(ii) (3.1.17) (Uniform);

(iii) The test of Problem 1.5 (Negative binomial);

(iv) The two tests of Problem 1.8 (Exponential);

(v) The sign test of Example 3.3.5.

[**Hint**: Use the results of (i) Example 3.3.2; (ii) Problem 3.11; (iii) Problem 3.9; (iv) Problem 3.13; (v) Example 3.5.]

3.20 Show that $\dfrac{\lambda}{\rho} + \dfrac{\mu}{1-\rho}$ $(0 < \rho < 1)$ is minimized by (3.3.58).

3.21 Determine approximate total sample sizes $N = m + n$ analogous to (3.3.57) for the following two-sample tests:

(i) The test of Problem 3.14(i) (Behrens-Fisher problem);

(ii) The tests of Problems 1.18 and 3.16 (Binomial).

3.22 Let $X_1, \ldots, X_m$ and $Y_1, \ldots, Y_n$ be independently distributed according to the exponential distributions $E(\xi, a)$ and $E(\eta, b)$, respectively.

(i) Determine a test of $H : \eta = \xi$ with asymptotic level α.

(ii) Obtain the limiting power of the test of (i) against a suitable sequence of alternatives.

[**Hint**: Problems 1.8 and 3.19(iv).]

3.23 Calculate $\displaystyle\int f^2(x)\,dx$ for the following distributions:

(i) Double exponential;

(ii) Logistic;

(iii) Exponential;

(iv) Uniform.

3.24 Use (3.3.52) to obtain an approximate formula for the sample size $m = n$ required by the Wilcoxon test of Examples 3.2.1 and 3.3.8 to achieve power β against the alternative (3.3.45).

3.25 (i) Use (3.3.44) to obtain an approximation to the power of the one-sample Wilcoxon test analogous to (3.3.52).

(ii) Determine a sample size N such that the Wilcoxon test (3.3.38) has approximate power β against the alternative of Example 3.3.5.

3.26 Assuming (3.6.11) to hold, determine the approximate sample size needed to get power β against an alternative θ for the test (3.6.10) when $\alpha = .05$ and

(i) $\beta = .8$,

(ii) $\beta = .9$,

(iii) $\beta = .95$.

[**Hint**: Let $\Psi_\alpha(c)$ denote the area under the normal curve between $-u_{\alpha/2} - c$ and $+u_{\alpha/2} - c$, and solve the equation $\Psi_\alpha(c) = 1 - \beta$ numerically.]

3.27 In Example 3.3.8, calculate the entries for Table 3.3.3 using the formula obtained by specializing (3.3.59) instead of formula (3.3.55).

Section 4

4.1 In Example 3.4.1 Show that of the three quantities (3.4.7), the first, second, and third is largest when F is respectively normal, double exponential, and logistic.

4.2 If X has a distribution F that is symmetric about 0, show that if X is replaced by

(i) aX,

(ii) $X + b$

the efficiencies (3.4.13)–(3.4.15) are unchanged.

4.3 Verify the efficiencies (3.4.16).

4.4 Evaluate the efficiencies (3.4.13)–(3.4.15) when F is

(i) double exponential,

(ii) logistic,

(iii) uniform.

4.5 If $F(x) = (1 - \epsilon)\Phi(x) + \epsilon\Phi(x/\tau)$ with $\epsilon < 1/2$, make tables showing how each of the efficiencies (3.4.13)–(3.4.15) varies as a function of ϵ and τ.

4.6 In the preceding problem, determine what happens to each of the efficiencies (3.4.13)–(3.4.15) as $\tau \to \infty$.

4.7 Evaluate each of the efficiencies (3.4.13)–(3.4.15) when F is the t-distribution with ν degrees of freedom.

[**Hint**: $\text{Var}(t_\nu) = \nu/(\nu - 2)$ if $\nu \geq 2$ and $= \infty$ if $\nu = 1$.]

4.8 Verify (3.4.19).

4.9 (i) Verify (3.4.24).

(ii) Determine the approximate sample size required for the test (3.4.21) to achieve power λ against a fixed alternative of $\beta > \beta_0$.

4.10 (i) If $0 \le v_1, \ldots, v_N \le 1$ show that the maximum value of $\sum (v_i - \bar{v})^2 \sim N/4$.

(ii) If $v_i = i/(N-1)$, $i = 0, \ldots, N-1$, show that $\sum (v_i - \bar{v}^2) \sim \dfrac{N}{12}$.

4.11 Show that the ARE (3.4.25) of the 4-point design with $N/4$ observations at each of 0, 1/3, 2/3, 1 to the 2-point design with $N/2$ observations at each of 0, 1 is 5/9.

4.12 (i) Solve Problem 4.10(i) when the v's are only required to satisfy $|v_i| \le 1$ for all i.

(ii) Solve Problem 4.10(ii) when the v's are equally spaced on $(-1, 1)$ instead of on $(0, 1)$.

4.13 Under the assumption of Example 3.4.4, consider the problem of testing $H : \alpha = \alpha_0$ against $K : \alpha > \alpha_0$.

(i) Show that the rejection region

(3.6.12)
$$\frac{(\hat{\alpha} - \alpha_0) \sqrt{N \sum (v_j - \bar{v})^2 / \sum v_j^2}}{\sqrt{\sum \left(X_i - \hat{\alpha} - \hat{\beta} v_i \right)^2 / (N - 2)}} \ge u_\alpha$$

has asymptotic level α when the $X_i - \alpha - \beta v_i$ are i.i.d. according to any distribution F with 0 mean and a finite variance. (The use of the letter α to denote both the asymptotic level of the test and the regression coefficient in (3.4.20) is unfortunate but should cause no confusion.)

(ii) Solve the two parts of Problem 4.9 for the test (3.6.12), i.e., for α instead of β.

(iii) Obtain the efficacy of the tests (3.6.12).

(iv) If $-1 \le v_i \le 1$ for $i = 1, \ldots, N$, determine for what choices of the v's the test (3.6.12) will have maximum efficacy.

[**Hint:** For (i): Problem 2.7.11 and formula (3.5.11).]

4.14 Verify (3.4.36).

4.15 When the v_i satisfy (3.4.37), show that the power of the test (3.4.21) against the alternatives (3.4.38) tends to (3.4.24).

4.16 Show that (2.7.24) holds when $v_i = ai$.

4.17 The expectation and variance of $\sum iR_i$ under the hypothesis H of Example 3.4.7 are given by (3.4.41) and (3.4.42), respectively.

[**Hint:** Under H, R_i takes on the values $1, \ldots, N$ with probability $1/N$ each.]

4.18 Verify (3.4.44).

4.19 (i) Show that the test (3.4.54) has asymptotic level α.

(ii) Show that the ARE of the test (3.4.54) to that given by (3.1.17) is zero.

4.20 (i) Verify (3.4.59).

(ii) Make a table analogous to Table 3.4.2 for the ARE of (3.4.57) to (3.1.17).

4.21 If $X_1, \ldots, X_n$ are i.i.d. according to the uniform distribution $U(0, \theta)$, determine the distribution of $n\left[\theta - X_{(n-2)}\right]/\theta$ where $X_{(1)} < \cdots < X_{(n)}$ are the ordered X's.

[**Hint:** $P\left[\theta - X_{(n-2)} > y/n\right] = p^n + np^{n-1}q + \binom{n}{2}p^{n-2}q^2$, where $p = P\left[X_i \leq \theta - \dfrac{\theta y}{n}\right]$.

4.22 For testing $H : \theta = \theta_0$ against $\theta > \theta_0$ under the assumptions of Problem 4.21, consider the rejection region

$$(3.6.13) \qquad \frac{n\left(\theta_0 - X_{(n-2)}\right)}{\theta_0} \leq w'_\alpha.$$

(i) In analogy to (3.4.58), find an equation determining w'_α.

(ii) Find the ARE of the test (3.6.13) with respect to (3.4.57) and to (3.1.17).

(iii) Make a table analogous to Table 3.4.2 showing the AREs of part (ii).

4.23 Let $\{a_n\}$ and $\{b_n\}$ be two sequences of positive numbers tending to ∞ for which

$$\log b_n / \log a_n \to A.$$

(i) If $A > 1$, then $b_n/a_n \to \infty$; if $0 < A < 1$, then $b_n/a_n \to 0$.

(ii) If $A = 1$ and b_n/a_n tends to a limit l, then l can have any value $0 \leq l \leq \infty$.

[**Hint (i):** If $b_n = a_n\delta_n$, then

$$\frac{\log b_n}{\log a_n} = 1 + \frac{\delta_n}{\log a_n} \to A.$$

(ii) See what happens when (a) $\delta_n = k$, independent of n, (b) when $\delta_n = \log a_n$.]

4.24 Suppose $\log n \, (T_n - \theta_0) / \tau \, (\theta_0)$ tends to a continuous strictly increasing limit distribution H when $\theta = \theta_0$ and that for any sequence $\theta_n \to \theta_0$, $(\log n) \, (T_n - \theta_n) / \tau \, (\theta_0)$ tends to H when θ_n is the true value of θ. Consider the test of $H : \theta = \theta_0$ which rejects when

$$(3.6.14) \qquad \frac{(\log n) \, (T_n - \theta_0)}{\tau \, (\theta_0)} \geq \nu_\alpha,$$

where $H \, (\nu_\alpha) = 1 - \alpha$. Then the power of (3.6.14) against the alternatives

$$(3.6.15) \qquad \theta_n = \theta_0 + \frac{\Delta}{\log n} + o \left(\frac{1}{\log n} \right)$$

tends to

$$(3.6.16) \qquad 1 - H \left(\nu_\alpha - \frac{\Delta}{\tau \, (\theta_0)} \right).$$

4.25 Let $T^{(i)} = \left\{ T_n^{(i)}, n = 1, 2, \dots \right\}$, $i = 1, 2$, both satisfy the assumptions of the preceding problem. Suppose that the power of the test (3.6.14) with $T_n^{(i)}$ and $\tau_i \, (\theta_0)$ in place of T_n and $\tau \, (\theta_0)$ against the alternatives (3.6.15) satisfies

$$(3.6.17) \qquad \beta_i \, (\theta_n) \to H \, (c_i \Delta - \nu_\alpha).$$

Let $n_k^{(i)}, k = 1, 2, \dots$, be two sequences of sample sizes such that the power of the test based on $T^{(i)}$ with $n_k^{(i)}$ observations against the common alternatives

$$(3.6.18) \qquad \theta_k^{(i)} = \theta_0 + \frac{\Delta_i}{\log n_k^{(i)}} + o \left(\frac{1}{\log \left(n_k^{(i)} \right)} \right), \theta_k^{(1)} = \theta_k^{(2)}$$

both have the limit

$$(3.6.19) \qquad H \, (\nu_\alpha - c_1 \Delta_1) = H \, (\nu_\alpha - c_2 \Delta_2).$$

Then (i)

$$\lim \frac{n_k^{(1)}}{n_k^{(2)}} = \left\{ \begin{array}{l} 0 \\ \infty \end{array} \right. \text{if } \Delta_1 \begin{array}{c} < \\ > \end{array} \Delta_2;$$

and

(ii) if $\Delta_1 = \Delta_2$, $\lim \left[n_k^{(1)} / n_k^{(2)} \right]$ need not exist, and if it does exist, it can take on any value $0 \leq e_{2,1} \leq \infty$.

[**Hint:** Problems 4.23 and 4.24.]

Section 5

5.1 (i) For the test (3.1.26) discussed in Example 3.5.1, obtain a formula for $\lim \alpha'_n(F)$ when

(3.6.20) $$F(x) = \rho \Phi(x) + (1 - \rho) \, \Phi \left(\frac{x}{\tau} \right).$$

(ii) Make a table showing how $\lim \alpha'_n(F)$ varies as a function of ρ and τ.

(iii) Use simulation to obtain the actual level $\alpha'_n(F)$ for $n = 20$ and some of the values of ρ and τ used in the table of part (ii).

5.2 Under the conditions of part (iii) of the preceding problem, use simulation to make a table showing the actual level of the test (3.5.4).

5.3 Determine the asymptotic level of the test (3.1.26) of Example 3.5.1 when F is the uniform distribution $U(0, 1)$.

5.4 Let $X_1, \ldots, X_m$ and $Y_1, \ldots, Y_n$ be independent normal $N\left(\xi, \sigma^2\right)$ and $N\left(\eta, \tau^2\right)$, respectively, and consider the test of $H : \sigma^2 = \tau^2$ against $\sigma^2 < \tau^2$ with rejection region

(3.6.21) $$\sqrt{(m + n) \rho (1 - \rho) / 2} \left[\log S_Y^2 - \log S_X^2\right] \geq u_\alpha,$$

where $\rho = \lim \left(m/ (m + n)\right)$, $S_X^2 = \sum \left(X_i - \bar{X}\right)^2 / (m - 1)$ and $S_Y^2 = \sum \left(Y_j - \bar{Y}\right)^2 / (n - 1)$.

(i) Show that this test has asymptotic level α.

(ii) Show that the test is not robust against non-normality as m and $n \to \infty$.

[**Hint (i)**: Use Theorem 2.5.2 and the results of Example 3.5.1.]

5.5 On a single graph, show the density of Student's t-distribution with ν degrees of freedom for $\nu = 2, 5, 10, \infty$.

5.6 Write out formal proofs of the facts that the power functions

(i) of the one-sample t-test (3.1.24) against the alternatives $\xi_n = \xi_0 + \dfrac{\Delta}{\sqrt{n}}$ and

(ii) of the two-sample test (3.5.7) against the alternatives $\eta = \xi + \dfrac{\Delta}{\sqrt{N}}$ are robust against non-normality.

5.7 Make a table analogous to Table 3.5.5 showing $\alpha'_n(F)$ when F is the normal mixture of Problem 5.1 for $\rho = .1, .2, .3$ and $\tau = 1, 2, 3$.

5.8 Discuss the robustness against non-normality of the test (3.6.12) under the model of Example 3.4.4.

5.9 (i) Under the assumptions of Lemma 3.5.1, show that

$$P_G\left[(X_1, \ldots, X_n) \, \epsilon S\right] \to 1 \text{ as } \mu_1 \to \infty.$$

(ii) Prove the first statement of (3.5.17).

5.10 Sketch the density of the distribution (3.5.18) for large μ_1 and γ close to 1.

5.11 Let $X_i = \xi_i + E_i$, where the E's have mean 0, and for each i, let $\hat{\xi}_i = \sum c_{ij} X_j$ be a linear estimator of ξ_i which is unbiased, i.e., satisfy $E\left(\hat{\xi}_i\right) = \xi_i$. Then the distribution of $\sum \left(X_i - \hat{\xi}_i\right)^2$ is independent of the ξ's.

[**Hint**: Using unbiasedness, show that

$$\sum \left(X_i - \hat{\xi}_i\right)^2 = \sum \left(E_i - \sum c_{ij} E_j\right)^2 .]$$

5.12 In the preceding problem, let $\xi_i = \alpha + \beta v_i$ and $\hat{\xi}_i = \hat{\alpha} + \hat{\beta} v_i$ with $\hat{\alpha}$ and $\hat{\beta}$ given by (2.7.11) of Chapter 2. Show that the distribution of $\sum \left(X_i - \hat{\alpha} - \hat{\beta} v_i\right)^2$ is independent of α and β by showing that $E(\hat{\alpha}) = \alpha, E\left(\hat{\beta}\right) = \beta$.

5.13 Prove the identity (3.5.11).

[**Hint**: Use the fact that $\hat{\alpha} = \bar{X} - \hat{\beta}\hat{v}$.]

5.14 Let $X_i \ (i = 1, \ldots, n)$ be independent $N\left(\xi, \sigma_i^2\right)$. Show that the t-test (3.1.24) of $H : \xi = \xi_0$ has asymptotic level α (and is therefore robust against heterogeneity of variance) if

(3.6.22) $\dfrac{1}{n^2} \sum \sigma_i^4 \to 0$ and $\dfrac{1}{n} \sum \sigma_i^2$ is bounded away from 0

and hence in particular when there exists constants $0 < m < M < \infty$ such that $m \leq \sigma_i^2 \leq M$ for all i.

[**Hint**: Show that (3.6.22) implies

(3.6.23) $\dfrac{1}{n} \sum X_i^2 \Big/ \dfrac{1}{n} \sum \sigma_i^2 \overset{P}{\to} 1 \text{ as } n \to \infty.]$

5.15 Let $X_1, \ldots, X_m$ and $Y_1, \ldots, Y_n$ be independently normally distributed as $N\left(\xi, \sigma^2\right)$ and $N\left(\eta, \tau^2\right)$, respectively. In the belief that $\sigma^2 = \tau^2$, the hypothesis $H : \eta = \xi$ is rejected when the two-sample t-statistic (3.1.36) exceeds u_α. Show that if in fact $\sigma^2 \neq \tau^2$ and $m/(m+n) \to \rho$, the asymptotic level of the test is

$$(3.6.24) \qquad 1 - \Phi\left[u_\alpha \sqrt{\frac{\rho\sigma^2 + (1-\rho)\tau^2}{(1-\rho)\sigma^2 + \rho\tau^2}} \right]$$

and therefore that the asymptotic level of the test is close to α if m/n is close to 1 but not otherwise.

5.16 (i) Given ρ, find the smallest and largest value of the factor of u_α in (3.6.24).

(ii) For nominal level $\alpha = .05$ and $\rho = .1, .2, .3, .4$, determine the smallest and largest asymptotic level of the two-sample t-test under the assumptions of Example 3.5.3 as σ^2/τ^2 varies from 0 to ∞.

5.17 Under the assumptions of Example 3.5.3, use simulation to find the actual level

(i) of the two-sample t-test,

(ii) of the test (3.5.7) when $m = 10$, $n = 30$, $\sigma^2/\tau^2 = 1/4, 1, 4$.

5.18 In Example 3.5.5, determine the value of γ defined by (3.5.22) and use (3.5.25) to check whether the resulting level of the t-test is α.

5.19 Consider the asymptotic level of the t-test under the model

$$X_i = \xi + \beta_0 Z_i + \beta_1 Z_{i+1},$$

where the Z's are independent with mean 0 and variance σ^2.

(i) Show that $\gamma = 2\beta_0\beta_1/(\beta_0^2 + \beta_1^2)$.

(ii) If β_0 and β_1 have the same sign, the asymptotic level is liberal and takes on its maximum value when $\beta_0 = \beta_1$. Determine the maximum value of the asymptotic level in this case when $\alpha = .01, .05, .1$.

(iii) When β_0 and β_1 have opposite signs, the test is conservative; find its minimum asymptotic level.

5.20 Let

$$(3.6.25) \quad X_i = \xi + \beta_0 Z_i + \beta_1 Z_{i+1} + \cdots + \beta_k Z_{i+k}, \quad i = 1, 2, \ldots,$$

where the Z's are independent with mean 0 and variance σ^2. Generalize the results of the preceding problem to this case and show that γ can take an arbitrarily large values when k is sufficiently large.

5.21 In Example 3.5.8, show that (3.5.35) implies (3.5.36).

5.22 (i) In Example 3.5.8(a), show that for any fixed p_0, the ratio $\tau'(p_0)/\tau(p_0)$ can be made arbitrarily large by letting $\pi_0 \to 0$ and $\pi_1 \to 1$.

(ii) In Example 3.5.8(b), show that for any fixed p_0, the ratio $\tau'(p_0)/\tau(p_0) \to 0$ as $\pi_1 \to 0$ and $\pi \to 1$.

5.23 Let

$$(3.6.26) \qquad X_{ij} = A_i + U_{ij}; \quad i = 1,\ldots,m; \quad j = 1,\ldots,s,$$

where the unobservable A's and U's are independent normal $N\left(\xi, \sigma_A^2\right)$ and $N\left(0, \sigma^2\right)$, respectively.

(i) Show that

$$\mathrm{Var}(\bar{X}) = \frac{\sigma^2}{ms}\left[1 + (m-1)\rho'\right],$$

where ρ' is the common value of the ρ_{ij} with $i \neq j$.

(ii) Show that the level of the t-test (or the test (3.1.24)) of $H : \xi = \xi_0$ is not robust against this dependence structure as $m \to \infty$ with s remaining fixed.

(iii) Determine the maximum asymptotic level of the t-test when $\alpha = .05$ and $m = 2, 4, 6$.

(iv) Show that the asymptotic level of the t-test is robust if m is fixed and $s \to \infty$.

5.24 Let $Z_1, Z_2, \cdots$ be i.i.d. $N(0,1)$ and $X_i = \gamma Z_i + (1 - \gamma) Z_{i+1}$, $i = 1,\ldots,n$. Use simulation to find the actual level of the t-test when the nominal level is .05 for $n = 5, 10, 20$ and $\gamma = 1/4, 1/3, 1/2$.

Bibliographic Notes

Discussions of the asymptotic theory of hypothesis testing can be found, for example, in Hampel et al (1986), Lehmann (1986), and Staudte and Sheather (1990), all in the context of robustness. The idea of obtaining a large-sample approximation to the power of a test by considering a sequence of alternatives tending to the hypothesis was introduced by Neyman (1937b). Randomization tests orginate with Fisher (1925, 1935); a good discussion can be found in Pratt and Gibbons (1981) and in the books by Edgington (1987) and Good (1994). The asymptotic theory of rank tests is extensively treated by Hajek and Sidak (1967) and Lehmann (1998). The concept of asymptotic relative efficiency (also called Pitman efficiency) was developed by Pitman (1948) and is discussed, for example, in Cox and Hinkley (1974) and in Stuart and Ord (1987), as well as in most non-parametric books.

4

Estimation

Preview

Estimation can be carried out in terms of confidence intervals or point estimation. Of principal interest in the first case (treated in Section 4.1) is the probability of covering the true value, and in the second case, the bias and variance of the estimator. The latter are considered in Section 4.2, both in terms of the asymptotic distribution of the estimator and of the limit behavior of the finite sample quantities. Robustness of point estimators against gross errors and comparison of competing estimators in terms of their asymptotic relative efficiencies are the topics of Section 4.3. Finally, Section 4.4 treats a special class of estimation problems: estimation of the total or average of a finite population based on various methods of sampling the population.

4.1 Confidence intervals

Rather than testing that a parameter θ of interest has a specified value θ_0 (or falls short of θ_0), one will often want to estimate θ. Two approaches to the estimation problem are (i) estimation by confidence intervals with which we shall be concerned in the present section and (ii) point estimation which will be discussed in Section 2.

Confidence intervals for θ are random intervals

$$(\underline{\theta}(X_1, \ldots, X_n), \bar{\theta}(X_1, \ldots, X_n))$$

which have a guaranteed probability of containing the unknown parameter θ, i.e., for which

$$P_\theta\left[\underline{\theta}\left(X_1,\ldots,X_n\right) \le \theta \le \bar{\theta}\left(X_1,\ldots,X_n\right)\right] \ge \gamma \text{ for all } \theta$$

for some preassigned confidence level γ. Since excessively large coverage probability requires unnecessarily long intervals, we shall instead require that

$$(4.1.1) \qquad \inf_\theta P_\theta\left[\underline{\theta}\left(X_1,\ldots,X_n\right) \le \theta \le \bar{\theta}\left(X_1,\ldots,X_n\right)\right] = \gamma.$$

Here we are concerned with intervals for which (4.1.1) holds approximately when n is large and for this purpose, replace (4.1.1) by the weaker requirement

$$(4.1.2) \quad \inf_\theta P_\theta\left[\underline{\theta}\left(X_1,\ldots,X_n\right) \le \theta \le \bar{\theta}\left(X_1,\ldots,X_n\right)\right] \to \gamma \text{ as } n \to \infty.$$

A still weaker condition is that for every θ,

$$(4.1.3) \qquad P_\theta[\underline{\theta}(X_1,\ldots,X_n) \le \theta \le \bar{\theta}(X_1,\ldots,X_n)] \to \gamma \text{ as } n \to \infty.$$

As was discussed in Section 2.6, the requirement (4.1.3) does not imply (4.1.2). We shall refer to intervals satisfying (4.1.3) and (4.1.2) respectively as confidence intervals and as strong confidence intervals with asymptotic confidence coefficient γ. For confidence intervals with asymptotic confidence coefficient γ, the probability of their containing any given true θ will be arbitrarily close to γ when n is sufficiently large, say differs from γ by less than ϵ when $n > n_\epsilon(\theta)$; when the intervals are strong, it is possible to choose an n_ϵ which will work simultaneously for all θ (and the convergence (4.1.3) will then be uniform in θ as discussed in Section 2.6).

Confidence intervals for θ are closely related to tests of the hypothesis

$$(4.1.4) \qquad\qquad\qquad H(\theta_0) : \theta = \theta_0$$

against the two-sided alternatives $\theta \ne \theta_0$. Suppose that for every θ_0, $A(\theta_0)$ is an acceptance region for $H(\theta_0)$ at asymptotic level α, so that

$$(4.1.5) \qquad P_\theta[(X_1,\ldots,X_n) \in A(\theta)] \to 1 - \alpha \text{ for every } \theta.$$

Then if $S(x_1,\ldots,x_n)$ is the set of all values θ for which $H(\theta)$ is accepted when $X_1 = x_1,\ldots,X_n = x_n$, it follows that

$$(4.1.6)$$
$$P_\theta[\theta \in S(X_1,\ldots,X_n)] = P_\theta[(X_1,\ldots,X_n) \in A(\theta)] \to 1 - \alpha \text{ for all } \theta.$$

Confidence sets for θ at asymptotic level $\gamma = 1-\alpha$ can therefore be obtained by solving the inclusion statement $(X_1,\ldots,X_n) \in A(\theta)$ for θ.

Example 4.1.1 Poisson. As in Example 3.1.1, let $X_1, \ldots, X_n$ be i.i.d. according to the Poisson distribution $P(\lambda)$. Then

$$(4.1.7) \qquad \left| \frac{\sqrt{n}\,(\bar{X} - \lambda_0)}{\sqrt{\lambda_0}} \right| \leq u_{\alpha/2}$$

is the acceptance region of a test of $H : \lambda = \lambda_0$ against $\lambda \neq \lambda_0$ which has asymptotic level α.

Solving for λ_0 by squaring and then completing the square shows (4.1.7), with λ in place of λ_0, to be equivalent to

$$(4.1.8) \qquad \left| \lambda - \bar{X} - \frac{u_{\alpha/2}^2}{2n} \right| \leq \sqrt{\frac{\bar{X} u_{\alpha/2}^2}{n} + \frac{u_{\alpha/2}^4}{4n^2}}$$

and hence to an interval for λ centered at $\bar{X} + \dfrac{1}{2n} u_{\alpha/2}^2$ and of length

$$(4.1.9) \qquad 2\sqrt{\frac{\bar{X} u_{\alpha/2}^2}{n} + \frac{u_{\alpha/2}^4}{4n^2}}.$$

Since the probability of (4.1.7) tends to $1 - \alpha$ for every $\lambda > 0$, this is the case also for (4.1.8). As pointed out in Example 3.1.1, the test (4.1.7) and hence the intervals (4.1.8) are typically improved by applying a continuity correction.

Alternative intervals are obtained by noting that $\bar{X}$ is a consistent estimator of λ and that the acceptance region (4.1.7) therefore continues to define a test with asymptotic level α if it is replaced by

$$(4.1.10) \qquad \left| \frac{\sqrt{n}\,(\bar{X} - \lambda_0)}{\sqrt{\bar{X}}} \right| \leq u_{\alpha/2}.$$

This leads to the intervals

$$(4.1.11) \qquad \bar{X} - \frac{u_{\alpha/2}}{\sqrt{n}} \sqrt{\bar{X}} < \lambda < \bar{X} + \frac{u_{\alpha/2}}{\sqrt{n}} \sqrt{\bar{X}}.$$

A third approach is to begin by subjecting $\bar{X}$ to the variance-stabilizing square root transformation discussed in Example 2.5.2. As was shown in Example 2.5.4, this leads to the intervals

$$(4.1.12) \qquad \sqrt{\bar{X}} - \frac{u_{\alpha/2}}{2\sqrt{n}} < \sqrt{\lambda} < \sqrt{\bar{X}} + \frac{u_{\alpha/2}}{2\sqrt{n}}$$

which can be converted into intervals for λ.

It is easy to see (Problem 1.1) that the three intervals (4.1.8), (4.1.11), and (4.1.12) agree up to terms of order $1/\sqrt{n}$.

The intervals (4.1.11) suffer from the obvious defect that the lower limit can be negative while λ is known to be positive. It is thus natural to replace the lower end point of (4.1.11) by 0 whenever $\bar{X}$ is negative, so that the intervals become

$$(4.1.13) \qquad \max\left(0, \bar{X} - \frac{u_{\alpha/2}}{\sqrt{n}}\sqrt{\bar{X}}\right) < \lambda \leq \bar{X} + \frac{u_{\alpha/2}}{\sqrt{n}}\sqrt{\bar{X}}.$$

This change does not affect the coverage probability of the intervals. It is interesting to note that the intervals (4.1.8) do not suffer from this draw-back: Their left end point is always positive (Problem 1.2). □

Of the intervals (4.1.12) and (4.1.13) it was stated above that the probability of their covering any given value λ, when it is the true value, tends to $\gamma = 1 - \alpha$ as $n \to \infty$. On the other hand, it was seen in Section 2.6 following Example 2.6.1 that

$$(4.1.14) \quad \inf_{0<\lambda} P\left[\bar{X} - \frac{u_{\alpha/2}}{\sqrt{n}}\sqrt{\bar{X}} \leq \lambda \leq \bar{X} + \frac{u_{\alpha/2}}{\sqrt{n}}\sqrt{\bar{X}}\right] = 0 \text{ for every } n,$$

so that the intervals (4.1.11) are not strong.

Example 4.1.1 Poisson (continued). Strong intervals obtain in Example 4.1.1 if, as is often the case, it is possible to rule out values of λ less than some $a > 0$ and so bound λ away from 0. In fact, we shall now show that the left side of (4.1.14) tends to $\gamma = 1 - \alpha$ if the inf is taken not over the whole set $\{0 < \lambda\}$ but instead over the values $\lambda \geq a$ for some $a > 0$. To prove this result, we shall show that

$$(4.1.15) \qquad \sup_{a \leq \lambda} P\left[\bar{X} + \frac{u}{\sqrt{n}}\sqrt{\bar{X}} < \lambda\right] \to \frac{\alpha}{2}$$

and

$$(4.1.16) \qquad \sup_{a \leq \lambda} P\left[\bar{X} - \frac{u}{\sqrt{n}}\sqrt{\bar{X}} > \lambda\right] \to \frac{\alpha}{2},$$

where $u = u_{\alpha/2}$.

Proof of (4.1.15). The probability on the left side of (4.1.15) is equal to

$$P\left[(\bar{X} - \lambda)^2 > \frac{u^2\bar{X}}{n} \text{ and } \bar{X} - \lambda < 0\right]$$

$$= P\left\{\left[(\bar{X} - \lambda) - \frac{u^2}{2n}\right]^2 > \frac{\lambda u^2}{n} + \frac{u^4}{4n^2} \text{ and } \bar{X} - \lambda < 0\right\}.$$

Since $\bar{X} - \lambda < 0$ implies that $\bar{X} - \lambda - \dfrac{u^2}{2n} < 0$, the last probability is

$$\leq P\left[\bar{X} - \lambda - \frac{u^2}{2n} < -\sqrt{\frac{\lambda u^2}{n} + \frac{u^4}{4n^2}}\right]$$

$$= P\left[\frac{(\bar{X} - \lambda)\sqrt{n}}{\sqrt{\lambda}} < \frac{u^2}{2\sqrt{\lambda n}} - \sqrt{u^2 + \frac{u^4}{4\lambda n}}\right].$$

By the Berry-Esseen theorem, this last probability differs from

$$\Phi\left[\frac{u^2}{2\sqrt{\lambda n}} - \sqrt{u^2 + \frac{u^4}{4\lambda n}}\right]$$

by less than

$$\frac{C}{\sqrt{n}}\frac{E|X_1 - \lambda|^3}{\lambda^{3/2}}.$$

Therefore the left side of (4.1.15) does not exceed

$$(4.1.17) \qquad \sup_{a \leq \lambda} \Phi\left[\frac{u^2}{2\sqrt{\lambda n}} - \sqrt{u^2 + \frac{u^4}{4\lambda n}}\right] + \frac{C}{\sqrt{n}}\sup_{a \leq \lambda}\frac{E|X_1 - \lambda|^3}{\lambda^{3/2}}.$$

The first term of (4.1.17) tends to $\alpha/2$ (Problem 1.4). It remains to show that the second term tends to zero, which will be the case, provided

$$(4.1.18) \qquad \frac{E|X_1 - \lambda|^3}{\lambda^{3/2}} \text{ is bounded for } \lambda \geq a.$$

Now by a well-known inequality (see, for example, Cramér (1946, p. 176) or Stuart and Ord (1987, p. 81))

$$(4.1.19) \qquad E|X_1 - \lambda|^3 \leq \left\{E(X_1 - \lambda)^4\right\}^{3/4}.$$

Since (for example, see Stuart and Ord (1987))

$$(4.1.20) \qquad E(X_1 - \lambda)^4 = \lambda + 3\lambda^2,$$

it follows that

$$\frac{E|X_1 - \lambda|^3}{\lambda^{3/2}} \leq \frac{(\lambda + 3\lambda^2)^{3/4}}{\lambda^{3/2}} = \left(\frac{1}{\lambda} + 3\right)^{3/4},$$

which is $\leq \left(\dfrac{1}{a} + 3\right)^{3/4}$ for $\lambda \geq a$. This completes the proof of (4.1.15); that of (4.1.16) is quite analogous (Problem 1.3).   □

Example 4.1.2 Binomial. The estimation of binomial p is so similar to the Poisson case that it will only be sketched here, with the details left to problems. If X is the number of successes in n binomial trials with success probability p, then

(4.1.21)
$$\left| \frac{\sqrt{n}\left(\frac{X}{n} - p_0\right)}{\sqrt{p_0 q_0}} \right| \le u_{\alpha/2}$$

is an acceptance region when $H : p = p_0$ is being tested against the two-sided alternatives $p \neq p_0$.

The inequality (4.1.21) with p in place of p_0 can be solved for p to give approximate confidence intervals for p, which can be simplified without change of asymptotic level by neglecting higher order terms (Problem 1.7).

A simpler approach leading to the same intervals replaces (4.1.21) by the asymptotically equivalent acceptance region

(4.1.22)
$$\left| \frac{\sqrt{n}\left(\frac{X}{n} - p_0\right)}{\sqrt{\frac{X}{n}\left(1 - \frac{X}{n}\right)}} \right| \le u_{\alpha/2},$$

with the intervals then becoming

(4.1.23) $$\frac{X}{n} - \frac{u_{\alpha/2}}{\sqrt{n}}\sqrt{\frac{X}{n}\left(1 - \frac{X}{n}\right)} < p < \frac{X}{n} + \frac{u_{\alpha/2}}{\sqrt{n}}\sqrt{\frac{X}{n}\left(1 - \frac{X}{n}\right)},$$

where the lower limit may be replaced by 0 when it is negative and the upper limit by 1 when it exceeds 1.

A third possibility, as in the Poisson case, is to first use a variance-stabilizing transformation.

The binomial intervals (4.1.23) share with the Poisson intervals (4.1.11) the property that the infimum (over $0 < p < 1$) of their coverage probability is 0 for every n (Problem 1.8). On the other hand, again in analogy with Poisson case, the intervals become strong if p is known to satisfy $a_1 < p < a_2$ for some $0 < a_1 < a_2 < 1$ (Problem 1.9).

For a detailed comparison of various asymptotic confidence intervals for p, see Schader and Schmid (1990) and the literature cited there.

It is frequently of interest to obtain intervals of the form

(4.1.24)
$$\left| \frac{X}{n} - p \right| \le d,$$

where d is a given number, and with probability approximately $1 - \alpha$ of (4.1.24) being correct. From (4.1.21), we can determine the approximate sample size n required to achieve this aim. Since

$$P\left[\left|\frac{X}{n} - p\right| \le \sqrt{\frac{pq}{n}} u_{\alpha/2}\right] \to 1 - \alpha \text{ as } n \to \infty,$$

the desired sample size is

(4.1.25)
$$n = pq u_{\alpha/2}^2 / d^2,$$

which depends on the unknown p. A usable value of n can be determined by replacing p by an estimate obtained from previous experience. Alternatively, a conservative value is

(4.1.26)
$$n = u_{\alpha/2}^2 / 4d^2$$

since

(4.1.27)
$$pq \le 1/4 \text{ for all } 0 \le p \le 1.$$

Instead of controlling the absolute error $\left|\dfrac{X}{n} - p\right|$, one may—particularly if p is expected to be small—prefer to control the relative error $\left|\dfrac{X}{n} - p\right| / p$ and hence replace (4.1.24) by

(4.1.28)
$$\left|\frac{X}{n} - p\right| \le cp$$

for some given value of c. Replacing d by γp in (4.1.25) leads to

(4.1.29)
$$n = \frac{u_{\alpha/2}^2}{c^2} \cdot \frac{q}{p}.$$

□

Example 4.1.3 Normal mean. Let $X_1, \ldots, X_n$ be i.i.d. as $N(\xi, \sigma^2)$. Then

(4.1.30)
$$\left|\frac{\sqrt{n}\left(\bar{X} - \xi_0\right)}{\sqrt{\sum \left(X_i - \bar{X}\right)^2 / (n-1)}}\right| \le u_{\alpha/2}$$

is an acceptance region of a test of $H : \xi = \xi_0$ against $\xi \ne \xi_0$, which has asymptotic level α. Replacing ξ_0 by ξ and solving for ξ leads to the confidence intervals

(4.1.31)
$$\bar{X} - u_{\alpha/2}\sqrt{\frac{\sum \left(X_i - \bar{X}\right)^2}{n(n-1)}} \le \xi \le \bar{X} + u_{\alpha/2}\sqrt{\frac{\sum \left(X_i - \bar{X}\right)^2}{n(n-1)}}.$$

The difficulty encountered in Examples 4.1.1 and 4.1.2 does not arise in the present case since the probability $P_n(\xi, \sigma)$ of (4.1.31) is independent of both ξ and σ, so that (4.1.3) implies (4.1.2).

Consider next the problem of confidence intervals for $\xi = E(X_i)$ when the X's are i.i.d. according to an unknown distribution F with finite variance. Then the coverage probability of the intervals (4.1.31) will continue to tend to $\gamma = 1 - \alpha$ for any fixed F. However, it follows from Lemma 3.5.1 that the coverage probability $\gamma_n(F)$ satisfies

$$(4.1.32) \qquad \inf_F \gamma_n(F) = 0 \text{ for every } n.$$

Thus in this non-parametric setting, the intervals (4.1.31) remain confidence intervals for ξ in the sense of (4.1.3) but no longer in the strong sense of (4.1.2).   □

Example 4.1.4 Difference of normal means. Let $X_1, \ldots, X_m$ and $Y_1, \ldots, Y_n$ be independent i.i.d. as $N(\xi, \sigma^2)$ and $N(\eta, \tau^2)$, respectively. Since $(\bar{Y} - \eta) - (\bar{X} - \xi) = \bar{Y} - \bar{X} - \Delta$ with $\Delta = \eta - \xi$ is distributed as $N\left(0, \dfrac{\sigma^2}{m} + \dfrac{\tau^2}{n}\right)$, it is seen that the intervals for Δ obtained by solving

$$(4.1.33) \qquad \frac{|\bar{Y} - \bar{X} - \Delta|}{\sqrt{\dfrac{\hat{\sigma}^2}{m} + \dfrac{\hat{\tau}^2}{n}}} \leq u_{\alpha/2}$$

consitute confidence intervals in the sense of (4.1.13) at asymptotic level $\gamma = 1 - \alpha$ if $\hat{\sigma}^2$ and $\hat{\tau}^2$ are consistent estimators of σ^2 and τ^2; for example, if

$$(4.1.34) \quad \hat{\sigma}^2 = \sum (X_i - \bar{X})^2 / (m - 1) \text{ and } \hat{\tau}^2 = \sum (Y_j - \bar{Y})^2 / (n - 1).$$

For this choice of $\hat{\sigma}^2$ and $\hat{\tau}^2$ (and many others), it is in fact true that the intervals are strong.

The following argument indicates why this latter result is true. Since $(\bar{Y} - \bar{X} - \Delta) / \sqrt{\dfrac{\sigma^2}{m} + \dfrac{\tau^2}{n}}$ is exactly distributed as $N(0, 1)$, it is enough to show that

$$(4.1.35) \qquad \left(\frac{\hat{\sigma}^2}{m} + \frac{\hat{\tau}^2}{n}\right) \Big/ \left(\frac{\sigma^2}{m} + \frac{\tau^2}{n}\right)$$

tends to 1 uniformly in σ^2 and τ^2, i.e., that given any $c > 0$ and $0 < \epsilon < 1$, there exists m_0 and n_0 independent of σ and τ such that

$$P\left[\left|\frac{\dfrac{\hat{\sigma}^2}{m} + \dfrac{\hat{\tau}^2}{n}}{\dfrac{\sigma^2}{m} + \dfrac{\tau^2}{n}} - 1\right| < c\right] > 1 - \epsilon \text{ when } m > m_0 \text{ and } n > n_0.$$

Now $\hat{\sigma}^2$ is a consistent estimator of σ^2 and the distribution of $\hat{\sigma}^2/\sigma^2$ is independent of σ, and the analogous remark holds for $\hat{\tau}^2$ and τ^2. In addition, $\hat{\sigma}^2$ and $\hat{\tau}^2$ are independent. It follows that given any $0 < \epsilon < 1$, there exist m_0 and n_0 such that

$$P\left\{\left|\frac{\hat{\sigma}^2/m}{\sigma^2/m} - 1\right| < c \text{ and } \left|\frac{\hat{\tau}^2/n}{\tau^2/n} - 1\right| < c\right\} > 1 - \epsilon \text{ for all } m > m_0, \; n > n_0.$$

But

$$1 - c < \frac{\hat{\sigma}^2/m}{\sigma^2/m} < 1 + c \text{ and } 1 - c < \frac{\hat{\tau}^2/n}{\tau^2/n} < 1 + c$$

implies that

$$1 - c < \frac{\dfrac{\hat{\sigma}^2}{m} + \dfrac{\hat{\tau}^2}{n}}{\dfrac{\sigma^2}{m} + \dfrac{\tau^2}{n}} < 1 + c$$

and this completes the proof of the uniform convergence in probability to 1 of (4.1.35). □

Example 4.1.5 Difference of two Poisson means. If $X_1, \ldots, X_m$ and $Y_1, \ldots, Y_n$ are independent Poisson with means $E(X_i) = \lambda$ and $E(Y_j) = \mu$, respectively, then

$$(4.1.36) \qquad \frac{(\bar{Y} - \mu) - (\bar{X} - \lambda)}{\sqrt{\dfrac{\bar{X}}{m} + \dfrac{\bar{Y}}{n}}} \xrightarrow{L} N(0, 1)$$

and hence the probability of the intervals

$$(4.1.37) \quad \bar{Y} - \bar{X} - u_{\alpha/2}\sqrt{\frac{\bar{X}}{m} + \frac{\bar{Y}}{n}} < \mu - \lambda < \bar{Y} - \bar{X} + u_{\alpha/2}\sqrt{\frac{\bar{X}}{m} + \frac{\bar{Y}}{n}}$$

tends to $\gamma = 1 - \alpha$ for every λ and μ as m and n tend to infinity. □

Example 4.1.6 Ratio of two Poisson means.* Since in many applications the parameters λ and μ of the preceding example are the rates at which events are generated under two processes leading to the X's and Y's, respectively, it is of interest to obtain confidence intervals also for the ratio μ/λ. Such intervals can be derived by conditioning on the sum $X + Y = T$, where $X = X_1 + \cdots + X_m$ and $Y = Y_1 + \cdots + Y_n$, and using the fact that

---

*This somewhat more difficult example is not required in the remainder of the book.

the conditional distribution of Y given $T = t$ is the binomial distribution $b(p, t)$ with

$$p = \frac{n\mu}{m\lambda + n\mu} = \frac{\mu}{\rho\lambda + \mu}, \quad \rho = \frac{m}{n}.$$

In this conditional situation, approximate confidence intervals for p are given by (4.1.22) with X and n replaced by Y and t, respectively, i.e., by

(4.1.38)
$$\left| \frac{\sqrt{t}\left(\frac{Y}{t} - p\right)}{\sqrt{\frac{X}{t} \cdot \frac{Y}{t}}} \right| \leq u_{\alpha/2}.$$

The asymptotic validity of these intervals for any fixed λ and μ stems from the fact that the conditional probability of (4.1.38) tends to $1 - \alpha$ as $t \to \infty$. However, in the present context, we should like to know that this is true also for the unconditional probability of (4.1.38) with t replaced by T, as m and $n \to \infty$. We shall show this by using an extension of the central limit theorem, which is of interest in its own right. The theorem is concerned with the limit behavior of a sum of i.i.d. random variables when the number of terms in the sum is random rather than fixed.

Theorem 4.1.1 *Let $Z_1, Z_2, \ldots$ be i.i.d. with mean ζ and variance $\sigma^2 < \infty$. Let $\{\nu_k\}$ be a sequence of positive integer-valued random variables which tend to ∞ in probability as $k \to \infty$. If there exists a sequence of positive constants N_k tending to infinity such that*

(4.1.39)
$$\frac{\nu_k}{N_k} \xrightarrow{P} 1,$$

then

(4.1.40)
$$\sqrt{\nu_k}\left(\bar{Z}_k - \zeta\right) \xrightarrow{L} N\left(0, \sigma^2\right),$$

where $\bar{Z}_k = (Z_1 + \cdots + Z_{\nu_k})/\nu_k$.

The proof is sketched in Billingsley (1986).

If the ν's are constants, (4.1.39) holds with $N_k = \nu_k$, and (4.1.40) is an immediate consequence of the central limit theorem since the convergence in law of

$$\sqrt{N}\left(\frac{Z_1 + \cdots + Z_N}{N} - \zeta\right) \xrightarrow{L} N\left(0, \sigma^2\right)$$

implies that for any subsequence $N_k \to \infty$,

(4.1.41)
$$\sqrt{N_k}\left(\frac{Z_1 + \cdots + Z_{N_k}}{N_k} - \zeta\right) \to N\left(0, \sigma^2\right).$$

What the theorem asserts is that in (4.1.41) the constant sample sizes N_k can be replaced by random sample sizes ν_k satisfying (4.1.40).

To apply this result to the Poisson situation, consider a sequence of sample sizes (m_k, n_k) both tending to infinity and with

$$\frac{m_k}{n_k} = \rho \ (0 < \rho < 1).$$

If $X = X_1 + \cdots + X_{m_k}$, $Y = Y_1 + \cdots + Y_{n_k}$, and $T_k = X + Y$, then $T_k \overset{P}{\to} \infty$ as m_k and $n_k \to \infty$ and (Problem 1.19)

$$(4.1.42) \qquad \frac{T_k}{m_k \lambda + n_k \mu} \overset{P}{\to} 1.$$

The unconditional limit distribution of

$$(4.1.43) \qquad \sqrt{T_k} \left(\frac{Y_k}{T_k} - p_k \right)$$

with

$$p_k = \frac{n_k \mu}{m_k \lambda + n_k \mu} = \frac{\mu}{\rho \lambda + \mu} = p$$

can be obtained by combining the facts that (i) conditionally given $T_k = t$, Y is distributed as $b(p, t)$ and that (ii) T_k has the Poisson distribution with mean $E(T_k) = m_k \lambda + n_k \mu$. By (i), the conditional distribution of Y given t can be represented as the distribution of $Z_1 + \cdots + Z_t$, where the Z's are independent, and $Z_i = 1$ or 0 with probability p and $q = 1 - p$, respectively. The unconditional distribution of (4.1.43) is therefore the same as that of $\sqrt{T_k} \left(\bar{Z}_k - p \right)$. It now follows from Theorem 4.1.1 and (4.1.41) that the unconditional limit distribution of (4.1.43) is $N(0, pq)$.

Example 4.1.7 Population median. Let $Z_1, \ldots, Z_N$ be i.i.d. according to an unknown distribution function F with a unique median θ defined by

$$(4.1.44) \qquad F(\theta) = 1/2.$$

We might try to base confidence intervals for θ on the sample median $\tilde{\theta}_N$. In Example 2.4.9, it was shown (with slightly different notation) that if F has density f with $f(\theta) > 0$, then

$$(4.1.45) \qquad \sqrt{n} \left(\tilde{\theta} - \theta \right) \overset{L}{\to} N \left(0, 1/4f^2(\theta) \right).$$

Confidence intervals based on (4.1.45) therefore would require a consistent estimator of $f(\theta)$. (Some aspects of density estimation will be considered in Section 6.4.)

We shall instead derive confidence intervals for θ from the two-sided tests of $H(\theta_0) : \theta = \theta_0$, the one-sided version of which was treated in Examples 3.2.4 and 3.3.5, but without the (now unnecessary) assumption of symmetry. If $S_N = S_N(\theta_0) \leq N - k$ denotes the number of Z's $> \theta_0$, the hypothesis $H(\theta_0)$ is accepted at level α when

$$(4.1.46) \qquad k \leq S_N(\theta_0) \leq N - k,$$

where k satisfies the condition that the probability of (4.1.46) tends to $1-\alpha$ as $N \to \infty$. Since

$$(4.1.47) \qquad 2\sqrt{N}\left[\frac{S_N(\theta_0)}{N} - \frac{1}{2}\right] \xrightarrow{L} N(0,1)$$

when $\theta = \theta_0$, k can be determined from

(4.1.48)
$$P\left[2\sqrt{N}\left(\frac{k}{N} - \frac{1}{2}\right) \leq 2\sqrt{N}\left(\frac{S_N(\theta_0)}{N} - \frac{1}{2}\right) \leq 2\sqrt{N}\left(\frac{N-k}{N} - \frac{1}{2}\right)\right] \to 1-\alpha$$

and hence by

$$\frac{1}{2} - \frac{k}{N} = \frac{1}{2\sqrt{N}}u_{\alpha/2} + o\left(\frac{1}{\sqrt{N}}\right)$$

or

$$(4.1.49) \qquad k = \frac{N}{2} - \frac{\sqrt{N}}{2}u_{\alpha/2} + o\left(\sqrt{N}\right).$$

The conversion of (4.1.46) into confidence intervals can now be achieved by means of the following lemma.

Lemma 4.1.1 *Let $a_{(1)} < \cdots < a_{(N)}$ denote any ordered set of distinct real numbers $a_1, \ldots, a_N$. Then for any real number θ, the number $S_N(\theta)$ of a's $> \theta$ satisfies*

$$(4.1.50) \qquad S_N(\theta) = N - i \text{ if and only if } a_{(i)} \leq \theta < a_{(i+1)}$$

and hence

$$(4.1.51) \qquad k \leq S_N(\theta) \leq N - k \text{ if and only if } a_{(k)} \leq \theta \leq a_{(N-k)}.$$

The proof is immediate from the definition of S_N.

Application of this lemma to (4.1.46) with k given by (4.1.49) shows that if $[x]$ denotes the greatest integer $\leq x$ and if $u = u_{\alpha/2}$, the coverage probability of the intervals

$$(4.1.52) \qquad Z_{\left(\frac{N}{2} - \frac{\sqrt{N}}{2}u\right)} \leq \theta < Z_{\left(\frac{N}{2} + \frac{\sqrt{N}}{2}u\right)}$$

tends to $1 - \alpha$ as $N \to \infty$. Since the probability of (4.1.51) and hence of (4.1.52) is independent of both θ and F but depends only on N (and α), the intervals (4.1.52) are strong. $\square$

Example 4.1.8 Center of symmetry. If in the preceding example, the Z's can be assumed to be symmetrically distributed about θ, confidence intervals for θ which tend to be more efficient can be based on the one-sample Wilcoxon instead of the sign test. Let $V^*(\theta)$ be the number of averages $(Z_i + Z_j)/2$, $i \leq j$, which are greater than θ, and consider the acceptance region for the hypothesis $H(\theta_0) : \theta = \theta_0$,

$$(4.1.53) \qquad k \leq V^*(\theta_0) \leq M - k,$$

where $M = \binom{N}{2} + N = \dfrac{N(N+1)}{2}$ is the number of pairs $i \leq j$. By (3.2.30), when $\theta = \theta_0$

$$(4.1.54) \qquad \frac{V^*(\theta_0) - \dfrac{N(N+1)}{4}}{\sqrt{N(N+1)(2N+1)/24}} \xrightarrow{L} N(0,1) \text{ when } N \to \infty$$

and the constant k of (4.1.53) can therefore be taken as

$$(4.1.55) \quad k = \frac{N(N+1)}{4} - u_{\alpha/2}\sqrt{N(N+1)(2N+1)/24} + o\left(N^{3/2}\right).$$

If $V_{(1)} < \cdots < V_{(M)}$ denotes the ordered set of M averages $(Z_i + Z_j)/2$, $i \leq j$, Lemma 4.1.1 shows that

$$(4.1.56) \qquad V_{(k)} \leq \theta \leq V_{(M+1-k)},$$

with k given by (4.1.55), constitute strong confidence intervals for θ with asymptotic level $1 - \alpha$ (Problem 1.17). $\qquad \square$

Note: The existence of strong confidence intervals in Examples 4.1.3, 4.1.4, 4.1.7 and 4.1.8 is typical for situations in which there exists a *pivot*, i.e., a function of the observations X and the parameter θ to be estimated which has a *fixed* distribution, independent of θ and all other parameters. In the case of the median (Example 4.1.7), the pivot was

$$S_N(\theta) = \text{Number of } X\text{'s} > \theta,$$

which has the binomial distribution $b\left(\dfrac{1}{2}, n\right)$; when the X's were normal $N(\xi, \sigma^2)$, it was $(\bar{X} - \xi)/\hat{\sigma}$ when estimating ξ and $\hat{\sigma}/\sigma$ when estimating σ. Without a pivot, strong confidence intervals are rare unless some bounds are available for θ, as at the end of Example 4.1.1.

Summary

1. Asymptotic confidence intervals for a parameter θ derived from tests of $H(\theta_0) : \theta = \theta_0$ with asymptotic level α have asymptotic coverage

of θ equal to $1 - \alpha$ for each value of θ. They are said to be strong if they have the additional property that also the inf (with respect to θ) of the coverage probability tends to $1 - \alpha$. Intervals both with and without this property are illustrated on a number of parametric and nonparametric situations.

2. The central limit theorem is extended to certain sums of i.i.d. variables with a random number of terms.

4.2   Accuracy of point estimators

Suppose that δ_n is an estimator of a parametric function $h(\theta)$ for which

$$(4.2.1) \qquad \frac{\sqrt{n}\,[\delta_n - h(\theta)]}{\tau_n} \xrightarrow{L} N(0, 1).$$

Then the intervals

$$(4.2.2) \qquad \left(\delta_n - \frac{k\tau_n}{\sqrt{n}}, \; \delta_n + \frac{k\tau_n}{\sqrt{n}} \right)$$

satisfy

$$(4.2.3) \qquad P\left[\delta_n - \frac{k\tau_n}{\sqrt{n}} \le h(\theta) \le \delta_n + \frac{k\tau_n}{\sqrt{n}} \right] \to 2\Phi(k) - 1 = \gamma.$$

They are therefore asymptotic confidence intervals for $h(\theta)$ in the sense of (4.1.3). In particular, when $k = 2$, we have $\gamma \doteq .95$, so that the estimator δ_n differs from $h(\theta)$ by less than two times its asymptotic standard deviation $\tau_n/\sqrt{n}$ with high probability. In this sense, the intervals (4.2.2) provide an assessment of the accuracy of the estimator δ_n. If τ_n depends on θ and/or some other parameters and therefore is not known, it can be replaced in (4.2.1)–(4.2.3) by a consistent estimator $\hat{\tau}_n$.

An alternative approach to the asymptotic accuracy of δ_n is obtained by replacing the asymptotic variance τ_n^2 by the limit of the actual variance

$$(4.2.4) \qquad \text{Var}\left\{ \sqrt{n}\,[\delta_n - h(\theta)] \right\}.$$

As can be seen from the Note "A crucial distinction" in Section 2.1 and from the discussion at the end of Section 2.3, even for large n these measures need not agree. It can be shown (see, for example, Lehmann and Casella (1998)) that if the limits of both τ_n^2 and of (4.2.4) exist, then always

$$(4.2.5) \qquad \lim \tau_n^2 \le \lim \text{Var}\left\{ \sqrt{n}\,[\delta_n - h(\theta)]^2 \right\}.$$

Although strict inequality may hold, in typical situations the two limits in (4.2.5) will be equal, and one can then assert, for example, that δ_n will differ from $h(\theta)$ by less than two times its standard deviation with probability approximately .95.

If the variance (4.2.4) tends to a finite limit $a^2 > 0$, the variance of δ_n is equal to

$$(4.2.6) \qquad \qquad \operatorname{Var}(\delta_n) = \frac{a^2}{n} + o\left(\frac{1}{n}\right).$$

Another aspect of the accuracy of δ_n is its bias

$$(4.2.7) \qquad \qquad b(\delta_n) = E(\delta_n) - h(\theta),$$

which often also is of order $1/n$, say

$$(4.2.8) \qquad \qquad b(\delta_n) = \frac{b}{n} + o\left(\frac{1}{n}\right).$$

As is seen from the bias-variance decomposition (1.4.6), the expected squared error

$$(4.2.9) \qquad \qquad E\left[\delta_n - h(\theta)\right]^2 = \operatorname{Var}(\delta_n) + b^2(\delta_n)$$

is then also of order $1/n$. Its main contribution comes from the variance term since $b^2(\delta_n)$ is of order $1/n^2$.

In the following example, we shall illustrate a technique for obtaining approximations to the bias and variance of an estimator and compare the latter with the corresponding asymptotic variance, i.e., the variance of the corresponding limit distribution.

Example 4.2.1 Estimating a normal probability. Let $X_1, \dots, X_n$ be i.i.d. as $N(\theta, 1)$ and consider the estimation of

$$(4.2.10) \qquad \qquad p = P(X_i \leq u) = \Phi(u - \theta).$$

The maximum likelihood estimator of p is

$$(4.2.11) \qquad \qquad \delta = \Phi(u - \bar{X}),$$

and we shall attempt to obtain large-sample approximations for the bias and variance of this estimator. Since $\bar{X} - \theta$ is likely to be small, it is natural to write

$$(4.2.12) \qquad \qquad \Phi(u - \bar{X}) = \Phi\left[(u - \theta) - (\bar{X} - \theta)\right]$$

and to expand the right side about $u - \theta$ by Taylor's theorem (Section 2.5) as

(4.2.13)
$$\Phi(u - \bar{X}) = \Phi(u - \theta) - (\bar{X} - \theta)\,\phi(u - \theta) + \tfrac{1}{2}(\bar{X} - \theta)^2\,\phi'(u - \theta)$$
$$- \tfrac{1}{6}(\bar{X} - \theta)^3\,\phi''(u - \theta) + \tfrac{1}{24}(\bar{X} - \theta)^4\,\phi'''(\xi),$$

where ξ is a random quantity that lies between $u-\theta$ and $u-\bar{X}$. To calculate the bias

$$E\left[\Phi\left(u-\bar{X}\right)\right] - \Phi\left(u-\theta\right),$$

we take the expectation of (4.2.13), which yields

$$E\left[\Phi\left(u-\bar{X}\right)\right] = p + \frac{1}{2n}\phi'\left(u-\theta\right) + \frac{1}{24}E\left[\left(\bar{X}-\theta\right)^4\phi'''\left(\xi\right)\right].$$

The derivatives of $\phi(x)$ all are of the form $P(x)\phi(x)$, where $P(x)$ is a polynomial in x and are therefore all bounded (Problem 2.1). It follows that

(4.2.14) $$\left|E\left(\bar{X}-\theta\right)^4\phi'''\left(\xi\right)\right| < ME\left(\bar{X}-\theta\right)^4 = 3M/n^2$$

for some finite M. Using the fact that $\phi'(x) = -x\phi(x)$, we therefore find that

(4.2.15) $$E(\delta) = p - \frac{1}{2n}\left(u-\theta\right)\phi\left(u-\theta\right) + O\left(1/n^2\right),$$

where the error term is uniformly $O\left(1/n^2\right)$ by (4.2.14). The estimator δ therefore has a bias of order $1/n$ which tends to zero as $\theta \to \pm\infty$.

In the same way, one can show that (Problem (2.2(i)))

(4.2.16) $$\mathrm{Var}(\delta) = \frac{1}{n}\phi^2\left(u-\theta\right) + O\left(\frac{1}{n^2}\right)$$

and hence that

(4.2.17) $$\mathrm{Var}\left(\sqrt{n}\delta\right) \to \phi^2\left(u-\theta\right).$$

Since (Problem 2.2(ii))

(4.2.18) $$\sqrt{n}\left[\delta - \Phi\left(u-\theta\right)\right] \xrightarrow{L} N\left(0, \phi^2\left(u-\theta\right)\right),$$

the limit of the variance in this case is equal to the asymptotic variance.

It is interesting to see what happens if the expansion (4.2.13) is carried one step less far. Then

$$E\left[\Phi\left(u-\bar{X}\right)\right] = p + \frac{1}{2n}\phi'\left(u-\theta\right) + \frac{1}{6}E\left[\left(\bar{X}-\theta\right)^3\phi''(\xi)\right].$$

Since the third derivative of ϕ is bounded, the remainder now satisfies

$$\frac{1}{6}E\left|\left(\bar{X}-\theta\right)^3\phi'''\left(\xi\right)\right| < M'E\left|\bar{X}-\theta\right|^3 = O\left(\frac{1}{n^{3/2}}\right).$$

The conclusion is therefore weaker than before.   □

The same argument can be used to prove the following theorem (Problem 2.3).

Theorem 4.2.1 *Let $X_1, \ldots, X_n$ be i.i.d. with $E(X_i) = \theta$, $\mathrm{Var}(X_i) = \sigma^2$ and finite fourth moment, and let h be any function which is four times differentiable. Then*
(i) if the fourth derivative $h^{(iv)}(x)$ is bounded,

$$(4.2.19) \qquad E\left(h\left(\bar{X}\right)\right) = h\left(\theta\right) + \frac{\sigma^2}{2n} h''\left(\theta\right) + O\left(\frac{1}{n^2}\right);$$

(ii) if the fourth derivative of h^2 is also bounded,

$$(4.2.20) \qquad \mathrm{Var}\left[h\left(\bar{X}\right)\right] = \frac{\sigma^2}{n} [h'(\theta)]^2 + O\left(\frac{1}{n^2}\right).$$

If $h'(\theta) \neq 0$, it follows from Theorem 2.5.2 that

$$\sqrt{n}\left[h\left(\bar{X}\right) - h(\theta)\right] \xrightarrow{L} N\left(0, \sigma^2 [h'(\theta)]^2\right).$$

If (4.2.20) holds,

$$\mathrm{Var}\left[\sqrt{n} h\left(\bar{X}\right)\right] \rightarrow \sigma^2 [h'(\theta)]^2,$$

so that again the limiting variance agrees with the asymptotic variance.

It is an attractive feature of Theorem 4.2.1 that assumptions are made essentially only on the function h, not on the distribution of the X's. On the other hand, the assumption of a bounded fourth derivative is very strong and often not satisfied in practice. Fortunately, the approximation formulas (4.2.19) and (4.2.20) continue to hold in many cases in which h does not satisfy the assumptions of Theorem 4.2.1. However, their validity then depends on the particular distribution of the X's being considered.

The following example illustrates an approach that may be applicable in such cases.

Example 4.2.2 Let $X_1, \ldots, X_n$ be i.i.d. $N(\theta, 1)$ and consider the estimation of $h(\theta) = e^\theta$ by its maximum likelihood estimator $\delta = h(\bar{X}) = e^{\bar{X}}$. Since all derivatives of h are unbounded, Theorem 4.2.1 is not applicable. However, we can expand $e^{\bar{X}}$ about e^θ in powers of $\bar{X} - \theta$ to obtain

$$E\left(e^{\bar{X}}\right) = e^\theta E\left(\sum_{k=0}^\infty \frac{(\bar{X} - \theta)^k}{k!}\right) = e^\theta \sqrt{\frac{n}{2\pi}} \int_{-\infty}^\infty \left(\sum_{k=0}^\infty \frac{y^k}{k!}\right) e^{-\frac{n}{2}y^2} dy.$$

Since it is permissible to integrate a convergent power series term by term and since

$$(4.2.21) \qquad \sqrt{\frac{n}{2\pi}} \int y^{2k} e^{-\frac{n}{2}y^2} dy = \frac{1 \cdot 3 \cdots (2k-1)}{n^k} = \frac{(2k)!}{n^k 2^k k!}$$

and the odd moments of $Y = \bar{X} - \theta$ are zero, we see that

$$(4.2.22) \qquad E\left(e^{\bar{X}}\right) = e^{\theta} \sum_{k=0}^{\infty} \frac{1}{n^k} \cdot \frac{1}{2^k k!}.$$

The right side is a power series in $1/n$, and it follows from Theorem 2.5.1 that

$$(4.2.23) \qquad E\left(e^{\bar{X}} - e^{\theta}\right) = \frac{e^{\theta}}{2n} + O\left(\frac{1}{n^2}\right).$$

The bias of δ is therefore again of order $1/n$. Analogously, one finds (Problem 2.10).

$$(4.2.24) \qquad \lim \text{Var}\left[\sqrt{n}\left(\delta - e^{\theta}\right)\right] = e^{2\theta}.$$

This agrees with the asymptotic variance of $e^{\bar{X}}$ (Problem 2.11).

However, unlike the situation of Theorem 4.2.1 where for any h with bounded fourth derivatives this agreement held for *all* distributions with finite fourth moment, the present approach involves the moments of $(\bar{X} - \theta)$ and requires additional assumptions on the underlying distribution of the X's. □

Example 4.2.3 A counterexample. Consider once more the estimator $\delta = e^{\bar{X}}$ of $h(\theta) = e^{\theta}$, but without the assumption that the i.i.d. X's are normal. Note that

$$(4.2.25) \qquad E\left(e^{\bar{X}}\right) = E\left[e^{X_1/n} \cdots e^{X_n/n}\right] = \left[E\left(e^{X_1/n}\right)\right]^n.$$

Thus for any finite n,

$$(4.2.26) \qquad E\left(e^{\bar{X}}\right) = \infty$$

for any distribution for which $E\left(e^{\alpha X}\right) = \infty$ for all $0 < \alpha$. An example of such a distribution is given by the density

$$(4.2.27) \qquad p(x) = \frac{1}{4} e^{-\sqrt{|x-\theta|}}$$

which satisfies

$$(4.2.28) \qquad E\left(e^{\alpha X}\right) = \frac{1}{4} \int_{-\infty}^{\infty} e^{\alpha x - \sqrt{|x-\theta|}} dx = \infty \text{ for any } \alpha \neq 0.$$

On the other hand,

$$\sqrt{n}\left(e^{\bar{X}} - e^{\theta}\right) \xrightarrow{L} N\left(0, e^{\theta} \tau^2\right),$$

where

(4.2.29)
$$\tau^2 = \frac{1}{2}\int_0^\infty x^2 e^{-\sqrt{|x|}}dx < \infty.$$

□

This example shows that when taking the expectation of formal expansions such as (4.2.13) or (4.2.22), it is crucial to check convergence of the resulting series.

To illustrate still another approach to the approximation of means and variances that is available in some cases, let us return once more to the situation considered in Example 4.2.2.

Example 4.2.2. (continued). When the X's are i.i.d. $N(\theta,1)$, the mean $Y = \bar{X}$ is distributed as $N(\theta, 1/n)$, and hence

(4.2.30)
$$E\left(e^{a\bar{X}}\right) = \sqrt{\frac{n}{2\pi}}\int e^{ay - \frac{n}{2}(y-\theta)^2}dy,$$

which, on completing the square in the exponent, reduces to

$$E\left(e^{a\bar{X}}\right) = e^{a\theta} \cdot e^{a^2/2n} = e^{a\theta}\left[1 + \frac{a^2}{2n} + O\left(\frac{1}{n^2}\right)\right].$$

In particular therefore

(4.2.31)
$$E\left(e^{\bar{X}}\right) = e^{\theta}\left[1 + \frac{1}{2n} + O\left(\frac{1}{n^2}\right)\right]$$

and

(4.2.32)
$$E\left[\left(e^{\bar{X}}\right)^2\right] = E\left(e^{2\bar{X}}\right) = e^{2\theta}\left[1 + \frac{4}{2n} + O\left(\frac{1}{n^2}\right)\right]$$

so that finally

(4.2.33)
$$\operatorname{Var}\left(e^{\bar{X}}\right) = \frac{1}{n}e^{2\theta} + O\left(\frac{1}{n^2}\right)$$

and

(4.2.34)
$$\operatorname{Var}\left[\sqrt{n}\left(e^{\bar{X}} - e^{\theta}\right)\right] \to e^{2\theta}.$$

Formula (4.2.30) with $h(\bar{X})$ in place of $e^{a\bar{X}}$ shows that even in the normal case, the limit of the actual mean or variance need not agree with the asymptotic mean or variance. This follows from the fact that then (4.2.30) is infinite when $h(y) \to \infty$ sufficiently fast (for example, when $h(y) = e^{y^4}$ (Problem 2.16)) while $\sqrt{n}[h(\bar{X}) - h(\theta)]$ has finite asymptotic mean and variance whenever $h'(\theta) \neq 0$ and $\operatorname{Var}(X_i) < \infty$.

□

Example 4.2.4 Reciprocal. The pitfalls in the calculation of the approximate bias and variance of $h(\bar{X})$ discussed in Example 4.2.2 (continued) occur when $h(y)$ tends to ∞ too fast as $y \to \infty$. The same difficulty can arise when $h(y) \to \infty$ as y approaches some finite value y_0. To illustrate this situation, let $X_1, \ldots, X_n$ again be i.i.d. $N(\theta, 1)$ and let $h(y) = 1/y$. Then for any $\theta \neq 0$

$$(4.2.35) \qquad \sqrt{n}\left(\frac{1}{\bar{X}} - \frac{1}{\theta}\right) \to N\left(0, \frac{1}{\theta^4}\right).$$

On the other hand,

$$(4.2.36) \qquad \sqrt{\frac{2\pi}{n}} E\left(\frac{1}{\bar{X}}\right) = \int\limits_{y \neq 0} \frac{1}{|y|} e^{-\frac{n}{2}(y-\theta)^2} \, dy.$$

The integral is the sum of the integrals over $y > 0$ and $y < 0$, and for any $\epsilon > 0$, the former is

$$\geq \int\limits_0^\epsilon \frac{1}{y} e^{-\frac{n}{2}(y-\theta)^2} \, dy.$$

As $y \to 0$,

$$e^{-\frac{n}{2}(y-\theta)^2} \to e^{-\frac{n}{2}\theta^2} > \frac{1}{2} \text{ for all } \theta < \epsilon$$

when ϵ is sufficiently small, and the integral over $(0, \epsilon)$ therefore exceeds

$$\frac{1}{2} \int\limits_0^\epsilon \frac{1}{y},$$ which is infinite. Thus, despite the asymptotic normality of $1/\bar{X}$, its expectation does not exist for any n. (For a more detailed discussion of the nonexistence of $E(1/\bar{X})$, see Lehmann and Shaffer (1988).) $\square$

Summary

1. The accuracy of a (suitably normalized) estimator can be measured by its asymptotic variance or by the limit of its actual variance. These two measures will typically, but not always, agree.

2. In typical situations, both the bias and the variance of an estimator are of order $1/n$. The main contribution to the expected squared error then comes from the variance.

3. Approximate expressions for the bias and variance of an estimator $h(\bar{X})$ of $h(\theta)$ can be found by expanding $h(\bar{X})$ about $h(\theta)$. This approach and some of its difficulties are illustrated in a number of examples.

4.3   Comparing estimators

The accuracy of an estimator provides not only an indispensable measure of its reliability but also a basis for comparing different estimators of the same quantity. We shall consider here such comparisons both in terms of the limiting and asymptotic variances of the estimators.

Suppose that for two competing estimators δ_1 and δ_2 of $h(\theta)$ based on n observations, both bias and variance are of order $1/n$, say

$$(4.3.1) \qquad b_i(\theta) = E_\theta\left(\delta_i\right) - h(\theta) = \frac{a_i}{n} + O\left(\frac{1}{n^2}\right)$$

and

$$(4.3.2) \qquad \mathrm{Var}_\theta\, \delta_i = \frac{\tau_i^2}{n} + O\left(\frac{1}{n^2}\right).$$

Then up to this order of accuracy, by (1.4.6) the expected squared error of δ_i satisfies

$$(4.3.3) \qquad E_\theta\left[\delta_i - h(\theta)\right]^2 = \mathrm{Var}_\theta\left(\delta_i\right) + O\left(\frac{1}{n^2}\right).$$

The variance of δ_i can therefore be used as a measure of accuracy that is asymptotically equivalent to the expected squared error up to terms of order $1/n$.

If we then want the variance of δ_1 based on n_1 observations to be asymptotically equivalent to terms of order $1/n_1$ to the variance of δ_2 based on n_2 observations, we must have

$$\frac{\tau_1^2}{n_1} \sim \frac{\tau_2^2}{n_2}$$

and hence

$$(4.3.4) \qquad \frac{n_1}{n_2} \to \frac{\tau_1^2}{\tau_2^2}.$$

The limit of the ratio n_1/n_2 of observations required for two statistical procedures δ_1 and δ_2 to achieve the same asymptotic performance is the *asymptotic relative efficiency* (ARE) $e_{2,1}$ of δ_2 with respect to δ_1. (The motivation for this definition is given in Section 4 of Chapter 3.) If (4.3.1) and (4.3.2) hold (with n replaced by n_1 and n_2, respectively), we therefore have

$$(4.3.5) \qquad e_{2,1} = \frac{\tau_1^2}{\tau_2^2}.$$

Example 4.3.1 Estimating a normal probability. In Example 4.2.1, we considered $\delta_1 = \Phi(u - \bar{X})$ as an estimator of

$$(4.3.6) \qquad p = P(X_i \leq u) = \Phi(u - \theta)$$

when $X_1, \ldots, X_n$ are $N(\theta, 1)$, and showed that δ_1 satisfies (4.3.1) and (4.3.2) with

$$(4.3.7) \qquad \tau_1^2 = \phi^2(u - \theta).$$

An alternative estimator of p is

$$(4.3.8) \qquad \delta_2 = [\text{Number of } X_i \leq u]/n.$$

Clearly, δ_2 satisfies (4.3.1) with $a_2 = 0$ since it is unbiased, and (4.3.2) with

$$(4.3.9) \qquad \tau_2^2 = pq$$

since $\text{Var}(\delta_2) = pq/n$. It therefore follows that

$$(4.3.10) \qquad e_{2,1} = \frac{\phi^2(u - \theta)}{\Phi(u - \theta)[1 - \Phi(u - \theta)]}.$$

This can be shown to be a decreasing function of $|u - \theta|$ (see Sampford (1953)) so that we have

$$(4.3.11) \qquad e_{2,1}(\theta) \leq \frac{\phi^2(0)}{\Phi(0)[1 - \Phi(0)]} = \frac{2}{\pi} \doteq .637 \text{ for all } \theta.$$

As $\theta \to \pm\infty$, $e_{2,1}(\theta) \to 0$ (Problem 3.2). The efficiency loss resulting from the use of δ_2 instead of δ_1 is therefore quite severe. $\square$

If the accuracy of δ is measured by its asymptotic variance rather than its limiting variance $\lim(\text{Var}(\delta))$, an exactly analogous argument applies. Instead of (4.3.1) and (4.3.2), we now assume that the estimators δ_1 and δ_2 of $h(\theta)$ when based on n observations satisfy

$$(4.3.12) \qquad \sqrt{n}[\delta_i - h(\theta)] \xrightarrow{L} N\left(0, (\tau_i')^2\right).$$

Then for δ_1 and δ_2 based on n_1 and n_2 observations, respectively, to have the same limit distribution requires that $\sqrt{n_1}(\delta_1 - h(\theta))$ and $\sqrt{n_1}(\delta_2 - h(\theta)) = \sqrt{\frac{n_1}{n_2}}\sqrt{n_2}(\delta_2 - h(\theta))$ have the same limit distribution and therefore that $\tau_1'^2 = \lim \frac{n_1}{n_2}\tau_2'^2$. Thus

$$(4.3.13) \qquad e_{2,1} = \frac{\tau_1'^2}{\tau_2'^2}.$$

When the limiting and asymptotic variances agree, then $\tau_i'^2 = \tau_i^2$ and the two definitions lead to the same value.

Example 4.3.2 Estimating a normal probability (continued). Of the two estimators of $p = \Phi(u - \theta)$ considered in Example 4.3.1, δ_1 was seen in Example 4.2.1 to satisfy

$$(4.3.14) \qquad \sqrt{n}(\delta_1 - p) \to N\left(0, \phi^2(u - \theta)\right)$$

while δ_2 satisfies

$$(4.3.15) \qquad \sqrt{n}(\delta_2 - p) \to N(0, pq).$$

Application of (4.3.13) therefore shows that the ARE based on asymptotic variances leads to the same efficiency (4.3.10) which was previously obtained in terms of limiting variances. □

Example 4.3.3 Bayes estimator of binomial p. Let X be binomial $b(p, n)$ and let us compare the standard estimator $\delta_1 = X/n$ with the estimator

$$(4.3.16) \qquad \delta_2(X) = \frac{a + X}{a + b + n},$$

which is the Bayes estimator corresponding to a beta prior with parameters a and b. Then

$$\delta_2(X) = \frac{X}{n} - \frac{a + b}{a + b + n} \cdot \frac{X}{n} + \frac{a}{a + b + n}.$$

Since

$$(4.3.17) \qquad \sqrt{n}\left[-\frac{a + b}{a + b + n} \cdot \frac{X}{n} + \frac{a}{a + b + n}\right] \xrightarrow{P} 0$$

(Problem 3.3(i)), it is seen that not only $\sqrt{n}[\delta_1(X) - p]$ but also $\sqrt{n}[\delta_2(X) - p] \xrightarrow{L} N(0, pq)$ and, hence, that the ARE of δ_2 to δ_1 is equal to 1. The same result obtains if the ARE is computed in terms of the limiting rather than the asymptotic variances (Problem 3.3(ii)). □

The previous two examples were concerned with parametric situations in which the distribution is completely specified by one or more parameters. However, efficiency calculations have proved particularly useful in a number of so-called semi-parametric cases which involve not only unknown parameters but also unknown distribution functions. In many of these examples, the asymptotic variance approach is considerably easier than that based on the limiting variance. For this reason, we shall in Examples 4.3.4–4.3.8 compare estimators only in terms of their asymptotic variances.

Example 4.3.4 Center of symmetry. Suppose that $X_1, \dots, X_n$ are i.i.d. according to a distribution which is symmetric about an unknown center θ. The cdf of the X's can be written as

$$P(X_i \le x) = F(x - \theta),$$

where the distribution represented by F is symmetric about 0. We shall suppose that both θ and F are unknown and that it is desired to estimate θ. (The problem of testing θ was considered in Example 3.2.5.)

If F were known to be normal, the estimator of choice would be $\bar{X}$, which in that case has many optimum properties. However, for non-normal distributions, this estimator tends to be much less satisfactory since its asymptotic variance is unnecessarily large. Of particular interest is the case of distributions F that are more heavy-tailed than the normal. Samples from such distributions typically contain some outlying observations which strongly affect the mean. An easy way to make the estimator less sensitive to such outliers is to give them less weight.

The estimator which goes furthest in this direction is the median $\tilde{X}$, which puts all its weight on the central observation when n is odd and on the two central observations when n is even.   □

Example 4.3.5 The median. To see how successful $\tilde{X}$ is in improving $\bar{X}$ for heavy-tailed distributions, let us consider the ARE of $\tilde{X}$ to $\bar{X}$ for an arbitrary symmetric $F(x - \theta)$. We have

$$\sqrt{n}\left(\bar{X} - \theta\right) \to N\left(0, \sigma^2\right)$$

if $\sigma^2 = \mathrm{Var}\left(X_i\right) < \infty$, and by (2.4.19),

$$\sqrt{n}\left(\tilde{X} - \theta\right) \to N\left(0, 1/4 \ f^2\left(0\right)\right),$$

provided F has a density f with $f(0) > 0$. It follows that the ARE (4.3.13) of $\tilde{X}$ to $\bar{X}$ is

(4.3.18) $$e_{\tilde{X}|\bar{X}}(F) = 4\sigma^2 f^2(0).$$

It is seen that if one stretches the tail of the distribution, thereby increasing σ^2 while leaving the center of the distribution unchanged, the efficiency will increase and tend to ∞ as $\sigma^2 \to \infty$. A limiting case is the Cauchy distribution in which $\sigma^2 = \infty$, so that $e = \infty$.

Since $\bar{X}$ is optimal when F is normal, we must of course have $e < 1$ in that case. For $F = N(0, \sigma^2)$, we have in fact $f(0) = 1/\sqrt{2\pi}\sigma$ and hence $e = 2/\pi \doteq 637$. This may be unacceptably low if one believes the distribution to be near normal, and it suggests as a compromise discarding not all except the central observation, but only a proportion of the outer observations.□

Example 4.3.6 Trimmed mean. As an estimator intermediate between $\bar{X}$ and $\tilde{X}$ consider therefore the symmetrically trimmed mean, which discards the k largest and k smallest observations and uses as an estimator the average of the remaining $n - 2k$ observations. The mean and median are the two extreme cases where either no observations or all except the central one or two observations are discarded. If $\dfrac{k}{n} = \alpha < 1/2$, we shall denote the resulting *trimmed mean* by

$$(4.3.19) \qquad \bar{X}_\alpha = \frac{1}{n - 2k} \left[X_{(k+1)} + \cdots + X_{(n-k)} \right].$$

More generally, $\bar{X}_\alpha$ can be defined for any α by (4.3.19) with $k = [n\alpha]$, the largest integer $\leq n\alpha$. □

To determine the efficiency of $\bar{X}_\alpha$ for fixed α as $n \to \infty$, we require its asymptotic distribution.

Theorem 4.3.1 *Let F be symmetric about 0 and suppose there exists an interval $(-c, c)$ which may be finite or infinite such that $F(-c) = 0$, $F(c) = 1$ and that F possesses a density f which is continuous and positive on the interval $(-c, c)$. (This assumption assures that there are no gaps where the density is zero.) If $X_1, \ldots, X_n$ are i.i.d. according to $F(x - \theta)$, then for any $0 < \alpha < 1/2$,*

$$(4.3.20) \qquad \sqrt{n}\left(\bar{X}_\alpha - \theta\right) \to N\left(0, \sigma_\alpha^2\right),$$

where

$$(4.3.21) \qquad \sigma_\alpha^2 = \frac{2}{(1 - 2\alpha)^2} \left[\int\limits_0^{\xi(1-\alpha)} t^2 f(t)\,dt + \alpha \xi^2 (1 - \alpha) \right].$$

Here $\xi(\alpha)$ is the unique value for which

$$(4.3.22) \qquad F[\xi(\alpha)] = \alpha.$$

This result will not be proved here (for a proof, see Bickel (1965) and Stigler (1973)) but the following gives an idea of the argument. As will be shown in section 5.1, the joint distribution of the quantiles $X_{[\alpha n]}$ and $X_{n+1-[\alpha n]}$ is asymptotically normal; in addition the conditional distribution of $X_{[\alpha n]+1}, \ldots, X_{n-[\alpha n]}$ is that of a sample of $n - 2[\alpha n]$ observations from the conditional distribution of X_i given $X_{[\alpha n]}$ and $X_{n+1-[\alpha n]}$ and is therfore asymptotically normal by the central limit theorem. By combining these results one obtains the desired proof.

If σ^2 is the variance of the X_i, it follows from (4.3.20) and (4.3.21) that the ARE of $\bar{X}_\alpha$ to $\bar{X}$ is given by

$$(4.3.23) \qquad e_{\bar{X}_\alpha, \bar{X}}(F) = \frac{\sigma^2}{\sigma_\alpha^2}.$$

The value of this ARE for $F = $ normal, t_5, and t_3 is shown in Table 4.3.1 for various values of α. As another example, consider the ARE (4.3.23) when F is a normal mixture given by

$$(4.3.24) \qquad F(x) = (1 - \epsilon)\Phi(x) + \epsilon\Phi(x/\tau).$$

The ARE for this case is shown in Table 4.3.2 for $\tau = 3$ and a number of values of ϵ and α. These two tables suggest that against heavy-tailed distributions, the efficiency of $\bar{X}_\alpha$ holds up very well to that of $\bar{X}$ for $\alpha \leq .125$.

TABLE 4.3.1. Efficiency of $\bar{X}_\alpha$ to $\bar{X}$ for t-distributions

α	.05	.125	.25	.375	.50
Normal	.99	.94	.84	.74	.64
t_5	1.20	1.24	1.21	1.10	.96
t_3	1.70	1.91	1.97	1.85	1.62

Source: Table 4.1 of Lehmann (1983), Chapter 5.

TABLE 4.3.2. Efficiency of $\bar{X}_\alpha$ to $\bar{X}$ for normal mixtures

ϵ	α	.05	.1	.125	.25	.375	.5
.25		1.40	1.62	1.66	1.67	1.53	1.33
.05		1.20	1.21	1.19	1.09	.97	.83
.01		1.04	1.03	.98	.89	.79	.68
0		.99	.97	.94	.84	.74	.64

Source: Table 4.2 of Lehmann (1983), Chapter 5.

Since the efficiencies in Tables 4.3.1 and 4.3.2 are all fairly high, one may wonder how low the value of $e_{\bar{X}_\alpha,\bar{X}}(F)$ can get.

Theorem 4.3.2 *For all F satisfying the assumptions of Theorem 4.3.1,*

$$(4.3.25) \qquad e_{\bar{X}_\alpha,\bar{X}}(F) \geq (1 - 2\alpha)^2.$$

Proof.

$$\frac{1}{2}\sigma^2 = \int_0^\infty t^2 f(t)dt = \int_0^{\xi(1-\alpha)} t^2 f(t)dt + \int_{\xi(1-\alpha)}^\infty t^2 f(t)dt$$

$$\geq \int_0^{\xi(1-\alpha)} t^2 f(t)dt + \alpha\xi^2(1-\alpha) = \frac{1}{2}\sigma_\alpha^2(1-2\alpha)^2.$$

∎

When $\alpha = .1$, for example, the lower bound (4.3.25) is .64. As $\alpha \to 1/2$, it tends to 0 and the estimator gets close to the median.

The proof also shows for what distributions the lower bound is attained. Equality will hold provided

$$(4.3.26) \qquad \int_{\xi(1-\alpha)}^{\infty} t^2 f(t)dt = \alpha\xi^2(1-\alpha).$$

Since $t > \xi(1-\alpha)$ for all t over which the integral on the left side extends, (4.3.26) can hold only if the distribution F assigns all of its probability α in the interval $(\xi(1-\alpha), \infty)$ to the point $\xi(1-\alpha)$.

If F is restricted to be unimodal, a somewhat better lower bound can be obtained. One then has

$$(4.3.27) \qquad e_{\bar{X}_\alpha, \bar{X}}(F) \geq \frac{1}{1+4\alpha} \text{ for all unimodal } F.$$

The lower bound is attained when F is uniform. For a proof, see Bickel (1965).

The trimmed means (4.3.19) give zero weight to the outer $2k$ observations and assign equal weight to the remaining ones. Instead, one may prefer a smoother weighting which gives maximal weight to the central observation and decreases the weights as one moves away from the center. This leads to estimators of the form

$$(4.3.28) \qquad \sum_{i=1}^{n} w_i X_{(i)},$$

where $X_{(1)} < \cdots < X_{(n)}$ are the ordered X's. The linear functions (4.3.28) of the order statistics, the so-called *L-estimators*, form one of three classes that have been extensively studied in a search for more robust alternatives to $\bar{X}$ for estimating location (see, for example, the books by Huber (1981), Hampel et al. (1986), and Staudte and Sheather (1990)). We shall conclude the discussion of Example 4.3.4 with an estimator from one of the other two classes, the *R-estimators* (so called because they can be derived from rank tests).

Example 4.3.7 The Hodges-Lehmann estimator. Instead of the median $\tilde{\theta}$ of the observations X_i, consider the median of the averages

$$(4.3.29) \qquad \tilde{\tilde{\theta}} = \operatorname*{med}_{i \leq j} \left(\frac{X_i + X_j}{2} \right).$$

If the X_j are i.i.d. according to a distribution $F(x - \theta)$ where F has a density f and is symmetric about 0, then

$$(4.3.30) \qquad \sqrt{n}\left(\tilde{\tilde{\theta}} - \theta\right) \xrightarrow{L} N\left(0, \frac{1}{12\left[\int f^2(x)dx\right]^2}\right),$$

provided $\int f^2(x)dx < \infty$. This result can be obtained from the asymptotic power (3.3.44) of the one-sample Wilcoxon test in the same way that the asymptotic distribution of the median was derived (in Example 2.4.9) from that of the sign test statistic. For a proof see, for example, Hettmansperger (1984) or Lehmann (1998).

The ARE of $\tilde{\tilde{\theta}}$ to $\bar{X}$ is therefore

$$(4.3.31) \qquad 12\sigma^2 \left(\int f^2(x)dx \right)^2,$$

the same as that of the Wilcoxon to the t-test given by (3.4.14). The numerical values calculated there, and in particular the lower bound (3.4.17), thus carry over to the present situation.   □

Example 4.3.8 The two-sample shift model. As a second example of a semiparametric model, let $X_1,\dots,X_m$ and $Y_1,\dots,Y_n$ be i.i.d. according to distributions

$$(4.3.32) \qquad P(X_i \le x) = F(x) \text{ and } P(Y_j \le y) = F(y - \Delta),$$

respectively. In this model, which was already considered in (3.3.45) with θ instead of Δ, the distribution of the Y's is shifted from that of the X's by an amount Δ. We now wish to estimate the shift parameter Δ.

If F were known to be normal, the X's according to $N(\xi, \sigma^2)$ and the Y's according to $N(\eta, \sigma^2)$, we would estimate $\Delta = \eta - \xi$ by $\bar{Y} - \bar{X}$. As was the case in Example 4.3.4, this normal-theory estimator may be quite unsatisfactory when F is more heavy-tailed than the normal distribution because of its sensitivity to outlying observations. We shall consider here only one alternative, which is the two-sample analog of the estimator (4.3.29) in the one-sample case.

Note that (Problem 3.8(i))

$$(4.3.33) \qquad \bar{\Delta} = \bar{Y} - \bar{X} = \frac{1}{mn} \sum_{i=1}^{m} \sum_{j=1}^{n} (Y_j - X_j),$$

that is, $\bar{\Delta}$ is the average of the mn differences $Y_j - X_i$, $i = 1,\dots,m$; $j = 1,\dots,n$. As an alternative, consider the median of these differences

$$(4.3.34) \qquad \hat{\Delta} = \underset{i,j}{\text{med}} (Y_j - X_i).$$

Suppose the sample sizes m and n are such that

$$(4.3.35) \qquad \frac{m}{N} \to \rho, \ \frac{n}{N} \to 1 - \rho, \text{ where } N = m + n \text{ and } 0 < \rho < 1,$$

and that F has a density f. Then it can be shown that (Problem 3.8(ii))

$$(4.3.36) \qquad \sqrt{N}\left(\hat{\Delta} - \Delta\right) \xrightarrow{L} N\left(0, \tau^2\right),$$

where

$$(4.3.37) \qquad \tau^2 = \frac{1}{\rho(1-\rho)} \left(\frac{1}{\int f^2(x)dx} \right)^2.$$

On the other hand (Problem 3.8(iii))

$$(4.3.38) \qquad \sqrt{N}\left(\bar{\Delta} - \Delta\right) \to N\left(0, \frac{\sigma^2}{\rho(1-\rho)}\right),$$

and the efficiency of $\hat{\Delta}$ to $\bar{\Delta}$ is therefore

$$(4.3.39) \qquad e_{\hat{\Delta},\bar{\Delta}}(F) = 12\sigma^2 \left(\int f^2(x)dx \right)^2.$$

Formally, (4.3.39) is the same as the efficiency (3.4.14) in the one-sample problem, and the earlier results obtained for this case therefore continue to apply. However, f is now not the density of the differences $Y_j - X_i$ of which $\hat{\Delta}$ is the median but of the individual quantities $X_i - \xi$ and $Y_i - \eta$. (For a connection between these two densities, see Problem 3.9.)   □

The semiparametric models of Examples 4.3.3 and 4.3.7 represent generalizations of the normal models

$$(4.3.40) \qquad X_i : N\left(\xi, \sigma^2\right) \text{ and } X_i : N\left(\xi, \sigma^2\right), \ Y_j : N\left(\eta, \sigma^2\right)$$

in which the frequently unrealistic assumption of normality has been dropped. However, to retain the parameters of interest, the models hold on to the often still unrealistic assumptions of symmetry and shift. When these assumptions are also dropped, an entirely different approach is needed, which will be taken up in Section 6.2.

The models of Examples 4.3.3 and 4.3.7 are among the simplest semiparametric models. More general models are outlined by Oaks (1988); a survey and classification is provided by Wellner (1985). For a thorough, but more advanced, treatment, see the book by Bickel et al. (1993).

Comparisons are possible not only of different estimators of a parameter, based on the same data, but also of estimators of a common parameter occuring in different models or experiments. The following are two simple illustrations.

Example 4.3.9 Two binomial designs. In Example 2.5.1, we considered estimating p^2 by either $\delta_1 = X/n$ or $\delta_2 = (Y/n)^2$, where X and Y denote the numbers of successes in n binomial trials with success probabilities

p^2 and p, respectively. From the asymptotic distributions of $\sqrt{n}\left(\dfrac{X}{n} - p^2\right)$ and $\sqrt{n}\left(\dfrac{Y^2}{n^2} - p^2\right)$ given there, it is seen that the ARE of δ_2 to δ_1 is

$$(4.3.41) \qquad e_{2,1} = \frac{p^2\left(1 - p^2\right)}{4p^3 q} = \frac{1 + p}{4p}.$$

As noticed in the earlier example, Y^2/n^2 is more efficient than X/n when $p < 1/3$ and less efficient when $p > 1/3$. The efficiency result (4.3.41) shows that $e_{2,1}$ is a decreasing function of p and that $e_{2,1}$ tends to ∞ as $p \to 0$ and tends to $1/2$ as $p \to 1$. □

We have compared the two estimators when the parameter being estimated is p^2. Would the answer be different if we were estimating p instead? The following result shows that it would not.

Theorem 4.3.3 *Let δ_1 and δ_2 be two estimators of θ satisfying*

$$(4.3.42) \qquad \sqrt{n}\left(\delta_i - \theta\right) \xrightarrow{L} N\left(0, \tau_i^2\right),$$

and let h be a differentiable function with $h'(\theta) \neq 0$. Then the ARE of $h\left(\delta_{2n}\right)$ to $h\left(\delta_{1n}\right)$ as estimators of $h(\theta)$ has the same value τ_1^2/τ_2^2 as that of δ_{2n} to δ_{1n} as estimators of θ.

Proof. The result follows immediately from the fact that the asymptotic variance of $\sqrt{n}\left[h\left(\delta_i\right) - h\left(\theta\right)\right]^2$ is obtained by multiplying τ_i^2 by $\left[h'(\theta)\right]^2$. ■

Example 4.3.9 illustrates the comparison, not so much of two estimators, as of two designs. The following is another example of this kind; additional examples can be found in the next section.

Example 4.3.10 Sampling with and without replacement. When estimating a population mean, it is sometimes more convenient to draw the sample with rather than without replacement. Let the N population values be denoted by $v_1, \ldots, v_N$ and the n values drawn in such a sample by $X'_1, \ldots, X'_n$. Then the variables X'_i are i.i.d. with expectation $v_N.$ and variance τ_N^2 given by (2.8.47). The average $\bar{X}'_n$ has mean $v_N.$ and variance τ_N^2/n.

On the other hand, if $X_1, \ldots, X_n$ denote the n values obtained when the sample is drawn without replacement, the average $\bar{X}_n$ by (2.8.49) has variance

$$(4.3.43) \qquad \mathrm{Var}\left(\bar{X}_n\right) = \frac{N - n}{N - 1} \cdot \frac{1}{n}\tau_N^2.$$

It follows that

$$(4.3.44) \qquad e_{\bar{X}', \bar{X}} = \lim \frac{\frac{N-n}{N-1} \cdot \frac{1}{n}\tau_N^2}{\tau_N^2/n} = \lim\left(1 - \frac{n}{N}\right) \text{ as } N \to \infty,$$

if this limit exists. (Of course in this case, one could simply interpret the ratio of the variances as the finite relative efficiency since no approximations are involved.) In particular, the ARE (4.3.44) is 1 if the sampling fraction n/N tends to 0 as $N \to \infty$, but it is < 1 if n/N tends to a positive limit. Under mild assumptions (see Theorem 2.8.2), both

$$\frac{\sqrt{n}\left(\bar{X}_n - v_{N.}\right)}{\sqrt{\operatorname{Var}\left(\bar{X}_n\right)}} \text{ and } \frac{\sqrt{n}\left(\bar{X}'_n - v_{N.}\right)}{\sqrt{\operatorname{Var}\left(\bar{X}'_n\right)}} \xrightarrow{L} N(0,1).$$

Comparison of the asymptotic variances of the two estimators therefore also leads to (4.3.44) as the ARE.   □

So far, we have compared estimators for which both bias and variance are of order $1/n$ so that up to terms of that order, the bias makes no contribution to the expected squared error (4.2.9). Correspondingly, the asymptotic (normal) distributions are centered at the value being estimated and differ only in their variances. We shall now consider some situations in which the bias is of order $1/\sqrt{n}$ and the square of the bias is then of the same order as the variance. Correspondingly, the asymptotic distributions of the estimators are biased, i.e., have expectations that differs from the value being estimated. In fact, in the following examples, the asymptotic distributions of the estimators being compared have the same variance and differ only in their means. To see what happens in such situations, we must first consider the corresponding problem in the exact, non-asymptotic case.

Let δ_1 and δ_2 be two estimators of θ with common variance $\tau^2(\theta)$ and with means

(4.3.45) $\eta_i = \theta + b_i(\theta) \ (i = 1, 2),$

and suppose the accuracy of an estimator is measured by its expected squared error

(4.3.46) $R(\delta, \theta) = E(\delta - \theta)^2.$

Since then, by (4.2.9),

$$R(\delta_i, \theta) = \tau^2(\theta) + b_i^2(\theta),$$

the estimator with the smaller absolute bias also has the smaller expected squared error, and, in particular, an unbiased estimator is preferred to any biased estimator with the same variance.

An alternative measure of accuracy is provided by

(4.3.47) $P\left(|\delta - \theta| \leq a\right),$

the probability that the estimator will not differ from the estimand by more than a specified amount a. If the common distribution of $(\delta_i - \eta_i)$ is

F, then

(4.3.48) $P\left[|\delta_i - \theta| < a\right] = F\left[a - b_i\left(\theta\right)\right] - F\left[-a - b_i(\theta)\right].$

This is the area under F over an interval centered at $-b_i(\theta)$ and of length $2a$. Let us now consider the special case that

(4.3.49) F is normal with mean zero (or more generally is symmetric about 0 with a unimodal density).

Then the probability of an interval of given length decreases with the distance from its center. Thus in particular if δ_1 is unbiased and δ_2 biased, then for any $a > 0$, the probability (4.3.47) is greater for δ_1 than for δ_2 so that δ_1 is more accurate than δ_2 in this sense.

As we shall see in Example 4.3.12 below, when (4.3.49) does not hold, the conclusion reached under this assumption need no longer be valid.

These results immediately carry over to the asymptotic situation. This follows from the fact that if $k_n\left(\delta_i - \eta_i\right)$ tends in law to F, then

(4.3.50) $\lim P\left[k_n\left|\delta_i - \eta_i\right| < a\right] = F(a) - F(-a).$

It is useful to note that when $b_i(\theta) = 0$ for $i = 1$ or 2, then by (4.3.45), the limit distribution F has expectation θ.

Example 4.3.11 Let $X_1, \ldots, X_n$ be i.i.d. with mean θ and finite variance σ^2, and as estimators of θ consider $\bar{X}$ and

(4.3.51) $$\delta = \left(1 - \frac{\alpha}{\sqrt{n}}\right)\bar{X} + \frac{\alpha}{\sqrt{n}}\theta_0,$$

where θ_0 is a given value to which we think θ might be close. Then

(4.3.52) $$b_\delta(\theta) = \frac{\alpha}{\sqrt{n}}\left(\theta_0 - \theta\right), \quad b_{\bar{X}}(\theta) = 0$$

and

(4.3.53) $$\mathrm{Var}_\delta(\theta) = \left(1 - \frac{\alpha}{\sqrt{n}}\right)^2 \frac{\sigma^2}{n}, \quad \mathrm{Var}_{\bar{X}}(\theta) = \frac{\sigma^2}{n}.$$

The expected squared error of δ and $\bar{X}$ is therefore respectively

(4.3.54)
$$R\left(\delta, \theta\right) = \frac{1}{n}\left[\alpha^2\left(\theta - \theta_0\right)^2 + \sigma^2\left(1 - \frac{\alpha}{\sqrt{n}}\right)^2\right]$$
$$= \frac{1}{n}\left[\alpha^2\left(\theta - \theta_0\right)^2 + \sigma^2\right] + o\left(\frac{1}{n}\right)$$

and

(4.3.55) $$R\left(\bar{X},\theta\right) = \sigma^2/n.$$

If $\theta \neq \theta_0$, in (4.3.54) the bias term is decisive and $R\left(\bar{X},\theta\right) < R(\delta,\theta)$ to terms of order $1/n$. On the other hand, when $\theta = \theta_0$, both estimators are unbiased and the two estimators then agree to order $1/n$. When higher terms are taken into account, we find that for large n, $R\left(\delta,\theta_0\right) < R\left(\bar{X},\theta_0\right)$. This is not unexpected since δ favors θ_0.

If instead of the expected squared error of the estimators, we consider their asymptotic distributions, we see that (4.3.54) and (4.3.55) are replaced by

(4.3.56) $$\sqrt{n}\left(\delta - \theta\right) = \sqrt{n}\left(\bar{X} - \theta\right) + \alpha\left(\theta_0 - \bar{X}\right) \xrightarrow{L} N\left(\alpha\left[\theta_0 - \theta\right], \sigma^2\right)$$

and

(4.3.57) $$\sqrt{n}\left(\bar{X} - \theta\right) \xrightarrow{L} N\left(0, \sigma^2\right).$$

For $\theta \neq \theta_0$, the remark following (4.3.49) shows that $\bar{X}$ is better than δ in the sense that

$$P\left[\sqrt{n}\left|\bar{X} - \theta\right| < a\right] > P\left[\sqrt{n}\left|\delta - \theta\right| < a\right] \text{ for all } a.$$

When $\theta = \theta_0$, the asymptotic distributions of the two estimators coincide. This corresponds to the fact that $R\left(\bar{X},\theta_0\right) = R(\delta,\theta_0)$ up to terms of order $1/n$. $\qquad\square$

Example 4.3.12 Uniform. Let $X_1,\ldots,X_n$ be i.i.d. $U(0,\theta)$. Then the maximum likelihood estimator of θ is $X_{(n)}$, the largest of the X's. Since

(4.3.58) $$E\left[X_{(n)}\right] = \frac{n}{n+1}\theta,$$

$X_{(n)}$ is a biased estimator of θ. Multiplication by $\dfrac{n+1}{n}$ corrects the bias and suggests as an alternative

(4.3.59) $$\delta = \frac{n+1}{n}X_{(n)}.$$

For these estimators, we have (Problem 3.12)

(4.3.60) $$b_{X_{(n)}}(\theta) = \frac{-\theta}{n+1}, \quad \mathrm{Var}_{X_{(n)}}(\theta) = \frac{n\theta^2}{(n+1)^2(n+2)}$$

and hence

(4.3.61) $$R\left(X_{(n)},\theta\right) = \frac{(2n+2)}{(n+1)^2(n+2)}\theta^2 \sim \frac{2\theta^2}{n^2}$$

and

(4.3.62)
$$b_\delta(\theta) = 0, \; \mathrm{Var}_\delta(\theta) = \frac{\theta^2}{n(n+2)},$$

so that

(4.3.63)
$$R(\delta, \theta) = \frac{\theta^2}{n(n+2)} \sim \frac{\theta^2}{n^2}.$$

Based on the expected squared error criterion, the ARE of $X_{(n)}$ to δ is therefore

(4.3.64)
$$e_{X_{(n)}, \delta}(\theta) = \frac{1}{2} \text{ for all } \theta.$$

To make the corresponding comparison in terms of (4.3.47), note that by Example 2.3.7,

(4.3.65)
$$n\left(\theta - X_{(n)}\right) \overset{L}{\to} E(0, \theta)$$

and hence

(4.3.66)
$$n\left(X_{(n)} - \theta\right) \to -E(0, \theta),$$

i.e., the distribution of $-Y$ when Y is $E(0, \theta)$. Since the mean of this distribution is $-\theta$, the estimator $X_{(n)}$ has a negative asymptotic bias. From (4.3.66), one finds that (Problem 3.13)

(4.3.67)
$$\lim P\left\{n\left|X_{(n)} - \theta\right| < a\right\} = 1 - e^{-a/\theta}$$

and that

(4.3.68)
$$\lim P\left\{n\left|\delta - \theta\right| < a\right\} = \begin{cases} 1 - e^{-1-a/\theta} & \text{if } \theta < a \\ e^{-1+a/\theta} - e^{-1-a/\theta} & \text{if } a < \theta. \end{cases}$$

Comparison of (4.3.67) and (4.3.68) shows (Problem 3.13) that (4.3.67) is less than (4.3.68), and hence the unbiased estimator is better in this sense, when $a > c\theta$ where c is the unique positive solution of the equation

(4.3.69)
$$\frac{1}{e}e^c + \left(1 - \frac{1}{e}\right)e^{-c} = 1,$$

and is worse when $a < c\theta$. The ARE in this case therefore depends on a (Problem 3.14).   □

Summary

1. The asymptotic relative efficiency (ARE) of one estimator with respect to another is the ratio of the numbers of observations required by the two estimators to achieve the same asymptotic variance. In many situations, this equals the limiting ratio of the two actual variances. The concept is illustrated on a number of parametric and semiparametric examples. It also can be used to compare two experimental designs.

2. The definition of the ARE has to be modified when the estimators being compared are asymptotically biased.

4.4 Sampling from a finite population

Probabilistic results for simple random sampling from a finite population were obtained in Examples 2.2.4, 2.8.4, and 2.8.5. We shall here discuss some of their statistical implications and extend the results to other sampling schemes.

Example 4.4.1 Sampling from a 2-valued population. Suppose that a simple random sample of size n is drawn without replacement from a population Π of N items, of which D are defective (or have some other special characteristic) and $N - D$ are not. If X denotes the number of defectives in the sample, we saw earlier that

$$(4.4.1) \qquad E\left(\frac{X}{n}\right) = \frac{D}{N}$$

and

$$(4.4.2) \qquad \sigma_n^2 = \mathrm{Var}\left(\frac{X}{n}\right) = \frac{1}{n} \cdot \frac{N-n}{N-1} \cdot \frac{D}{N}\left(1 - \frac{D}{N}\right).$$

Since we are concerned with large-sample behavior, we consider, as in Example 2.8.4, a sequence of populations Π_N, $N = 1, 2, \ldots$. Then not only Π but also D and n require the subscript N, although we shall often suppress this subscript. It is convenient to choose the embedding sequence Π_N in such a way that

$$(4.4.3) \qquad \frac{D_N}{N} = p$$

remains constant, independent of N.

The estimator $\dfrac{X}{n}$ of p was seen in Example 2.2.4 to be consistent if

$$(4.4.4) \qquad n \text{ and } N \text{ both tend to infinity,}$$

and in Example 2.8.4 to satisfy

$$(4.4.5) \qquad \left(\frac{X}{n} - p\right) / \sigma_n \xrightarrow{L} N(0,1),$$

provided

$$(4.4.6) \qquad n \text{ and } N - n \text{ both tend to infinity.}$$

The extensive literature on refinements of the normal approximation is reviewed in Johnson, Kotz, and Kemp (1992, Chapter 6).
   Since

$$(4.4.7) \qquad E\left(\frac{X}{n}\right) = p,$$

X/n is an unbiased estimator of p and a natural measure of its accuracy is its variance σ_n^2. Let us now consider the sample size required for the variance σ_n^2 of this estimator to satisfy

$$(4.4.8) \qquad \sigma_n^2 \le v_0^2$$

for a given value of v_0^2. For fixed N and p, the variance σ_n^2 given by (4.4.2) is a decreasing function of n, and the desired value of n is therefore obtained by solving the equation

$$(4.4.9) \qquad \frac{N-n}{n} \cdot \frac{pq}{N-1} = v_0^2.$$

This has the solution

$$(4.4.10) \qquad n_0 = \frac{pq/v_0^2}{\frac{1}{N}\left[N - 1 + \frac{pq}{v_0^2}\right]}$$

and the required sample size is the smallest integer $\ge n_0$. As $N \to \infty$,

$$(4.4.11) \qquad n_0 \to \frac{pq}{v_0^2}.$$

The sample size n required to obtain a given accuracy for the estimator X/n of p therefore does not tend to infinity with the size of the population. This fact, which is often felt to be anti-intuitive, becomes plausible by noting that by (4.4.2)

$$(4.4.12) \qquad \sigma_n^2 \le \frac{pq}{n}.$$

To achieve (4.4.8), it is therefore enough to have

$$(4.4.13) \qquad \frac{pq}{n} \le v_0^2.$$

However, pq/n is the variance of X/n when sampling with replacement, which corresponds to a sequence of n binomial trials with success probability p and which therefore does not involve the population size N at all.

The value of n_0 given by (4.4.10) or (for large N) the approximate value pq/v_0^2 depend on the unknown p. An approximate sample size can be determined by replacing p by an estimate obtained by previous experience. Lacking such information, one can use the fact that $pq \leq 1/4$ for all p, to determine a conservative value of n_0 by replacing pq in (4.4.10) or (4.4.11) by $1/4$. As an example, suppose that $v_0 = .01$; then the approximate value obtained from (4.4.11) is

$$n_0 \doteq 10,000pq \leq 2,500.$$

A sample size of 2,500 therefore provides a variance $\leq .0001$ for arbitrarily large populations.

Consider next how to determine confidence intervals for p. Exact intervals can be obtained from the hypergeometric distribution of X, and for large N can be approximated from the binomial distribution (Problem 2.6 of Chapter 1). If n is also large, it is simpler to base the intervals on the normal limit of the binomial as $n \to \infty$. These are in fact the intervals given in most textbooks on sampling.

At this point, we are faced with a conceptual difficulty. We saw above that the sample size n required to achieve a given accuracy does not tend to ∞ with N but remains bounded (the binomial approximation which explained this phenomenon applies not only to the variance of the estimator but equally to confidence intervals). Now we are letting $n \to \infty$

To understand this apparent inconsistency, we must go back to the purpose of embedding a given situation in a fictitious sequence: to obtain a simple and accurate approximation. The embedding sequence is thus an artifice and has only this purpose which is concerned with a particular pair of values N and n and which need not correspond to what we would do in practice as these values change.

There is therefore no problem with replacing the (approximate) binomial confidence intervals by their normal approximation given as (4.1.23) in Example 4.1.2. Instead of using this two-stage approximation (hypergeometric $\to$ binomial $\to$ normal), we can also obtain intervals directly from the normal approximation (4.4.5) where, of course, σ_n has to be replaced by a consistent estimator. The resulting intervals are

$$(4.4.14) \qquad \frac{X}{n} - u_{\alpha/2}\hat{\sigma}_n < p < \frac{X}{n} + u_{\alpha/2}\hat{\sigma}_n,$$

where

$$(4.4.15) \qquad \hat{\sigma}_n^2 = \frac{1}{n}\frac{N-n}{N-1}\frac{X}{n}\left(1 - \frac{X}{n}\right).$$

If n is much smaller than N, one may want to choose a sequence with $n = o(N)$, and the factor $(N-n)/(N-1)$ can then be replaced by 1. This leads to the asymptotically equivalent intervals (4.1.24).   □

Example 4.4.2 Estimating a population size. The results of the preceding example provide a tool for estimating the unknown size N of an animal population by means of the following *capture-recapture method*. A first sample of size n_1 is drawn and the members of the sample are marked. The sample specimens are then released and allowed to mix with the rest of the population. Later, a second random sample of size n_2 is drawn from the population and the number X of marked members in the sample is noted. If $p = n_1/N$ denotes the (unknown) first sample fraction, we wish to estimate $N = n_1/p$.

Let X be the number of defective (i.e., marked) items in a random sample of size n_2 from a population containing a proportion p of defectives. If n_1 is large enough to represent a non-negligible fraction of the population (so that p is not too close to 0) and if n_2 is large enough for the asymptotic theory of Example 4.4.1 to be applicable, it follows that X/n_2 is a consistent estimator of p, and by Theorem 2.1.4,

$$(4.4.16) \qquad \frac{n_1 n_2}{X} \text{ is a consistent estimator of } \frac{n_1}{p} = N.$$

To obtain asymptotic confidence intervals for $N = n_1/p$, suppose that

$$(4.4.17) \qquad \frac{N - n_2}{N} \to \lambda \text{ as } n_2 \text{ and } N - n_2 \to \infty.$$

Then it follows from (4.4.2) and (4.4.5) that

$$\sqrt{n_2}\left(\frac{X}{n_2} - p\right) \to N\left(0, \lambda pq\right)$$

and hence from Theorem 2.5.2 that

$$(4.4.18) \qquad \sqrt{n_2}\left(\frac{n_2}{X} - \frac{1}{p}\right) \to N\left(0, \lambda q/p^3\right).$$

This shows that the intervals

$$(4.4.19) \qquad \frac{n_1 n_2}{X} - u_{\alpha/2}\sqrt{\frac{n_1^2 \lambda \hat{q}}{n_2 \hat{p}^3}} < N < \frac{n_1 n_2}{X} + u_{\alpha/2}\sqrt{\frac{n_1^2 \lambda \hat{q}}{n_2 \hat{p}_n^3}}$$

with $\hat{p} = \dfrac{X}{n_2}$ constitute confidence intervals for N with asymptotic level $1 - \alpha$.

An alternative approach to obtaining such intervals is to express the confidence intervals (4.4.14) for p in terms of $1/p$; the resulting intervals differ from, but are asymptotically equivalent to (4.4.19) (Problem 4.3).

For references to the extensive literature on capture-recapture methods (including applications to the estimation of human populations), see Seber (1982) and Thompson (1992). □

Example 4.4.3 Sampling from a finite population. As in Example 2.8.5, consider a simple random sample of size n from a population Π, and embed this situation in a sequence of populations Π_N ($N = 1, 2, \dots$) of elements v_{Ni}, $i = 1, \dots, N$. Then we saw in (2.8.45) and (2.8.49) that the expectation and variance of the sample mean $\bar{X}_n$ are given by

$$(4.4.20) \qquad E\left(\bar{X}_n\right) = v_N.$$

and

$$(4.4.21) \qquad \operatorname{Var}\left(\bar{X}_n\right) = \frac{1}{n}\frac{N-n}{N-1}\tau_N^2,$$

where

$$(4.4.22) \qquad \tau_N^2 = \frac{1}{N}\sum_{i=1}^{N}\left(v_{Ni} - v_N.\right)^2$$

is the *population variance* of Π_N. Also, in generalization (4.4.5), we have, by Section 2.8,

$$(4.4.23) \qquad \frac{\bar{X}_n - E\left(\bar{X}\right)}{\sqrt{\operatorname{Var}\left(\bar{X}_n\right)}} \to N(0,1),$$

provided

$$(4.4.24) \qquad n \text{ and } N - n \text{ both tend to infinity}$$

and

$$(4.4.25) \qquad \text{condition (a) or (b) of Theorem 2.8.2 holds.}$$

Note: An Edgeworth expansion for the distribution of $\bar{X}_n$ is given by Robinson (1978).

Let us now ask whether the sample average $\bar{X}_n$ is a consistent estimator of the population average $v_N.$. When considering the consistency of $\bar{X}$ for estimating the mean, based on n i.i.d. random variables $X_1, \dots, X_n$ (in Chapter 2, Section 1), we took the distribution F of the X's as fixed (i.e., not changing with n). As a result, the mean and variance of that distribution were also fixed and only the sample size n was changing. In the present situation, where we are embedding the given population Π_N

in a sequence of population $\Pi_1, \Pi_2, \ldots$, it is convenient to use an embedding sequence in which the population mean $v_{N.}$ and population variance $\tau_N^2 = \sum (v_{Ni} - v_{N.})^2 / N$ are constant, say

$$(4.4.26) \qquad v_{N.} = v \text{ and } \tau_N^2 = \tau^2 \text{ for all } N.$$

For such a sequence, $\bar{X}_n$ will be consistent for estimating v if the variance (4.4.21) tends to 0 and hence if

$$(4.4.27) \qquad \frac{1}{n} \frac{N-n}{N-1} \to 0.$$

Since $(N - n) / (N - 1) \leq 1$, this condition will hold whenever $n \to \infty$.

Note: It is seen from (4.4.21) that the same conclusion holds for any sequence Π_N with $v_{N.} = v$ for which τ_N^2 is bounded as n and $N \to \infty$.

As has been pointed out earlier, we are, of course, not dealing with a sequence of populations but with a single population Π_N for a given N. The sufficient condition for consistency that $n \to \infty$ and τ_N^2 remains bounded suggests that for the given situation $\bar{X}$ is likely to be close to $v_{N.}$ if

$$(4.4.28) \qquad \frac{1}{n} \tau_N^2 \text{ is small.}$$

Unfortunately, τ_N^2 is unknown; so how can we check (4.4.28) or the conditions for asymptotic normality which require that

$$(4.4.29) \qquad \frac{\max (v_{Ni} - v_{N.})^2}{\sum (v_{Nj} - v_{N.})^2} \text{ be small?}$$

Such a check may be possible if, as is usually the case, some information about the v's is available, either from past experience or from the physical situation. Suppose, in particular, that we know some bounds within which the v's must lie, say

$$|v_{Ni}| \leq M \text{ for all } i.$$

Then it follows that $\tau_N^2 \leq 4M^2$, which provides a bound for (4.4.28). Under the same conditions, the numerator of (4.4.29) is $\leq 4M^2$. Since the denominator is a sum of non-negative terms, it cannot be very small unless the v's are highly concentrated. (It is, for example, $\geq R^2/4$, where R is the range of the v's). In this way, one may be able to obtain a check also on (4.4.29).

If the limit result (4.4.23) holds, we can use (4.4.21) to obtain approximate confidence intervals for $v_{N.}$. As before, consider a sequence of populations satisfying (4.4.26). Then (4.4.23) states that

$$(4.4.30) \qquad \frac{\sqrt{n} \, (\bar{X}_n - v)}{\tau} \sqrt{\frac{N-1}{N-n}} \xrightarrow{L} N(0, 1),$$

so that

(4.4.31) $$P\left[\sqrt{n}\,|\bar{X}_n - v| < u_{\alpha/2}\tau\sqrt{\frac{1}{n}\frac{N-n}{N-1}}\right] \to 1 - \alpha.$$

This limit relation provides confidence intervals for $v_{N\cdot} = v$ when a consistent estimator $\hat{\tau}_n$ is substituted for τ.

Now

(4.4.32) $$\tau^2 = \frac{1}{N}\sum_{i=1}^{N} v_{Ni}^2 - v^2,$$

and a natural estimator of τ^2 is therefore

(4.4.33) $$\hat{\tau}_n^2 = \frac{1}{n}\sum_{i=1}^{n} X_i^2 - \bar{X}_n^2 = \frac{1}{n}\sum_{i=1}^{n}\left(X_i - \bar{X}_n\right)^2.$$

A sufficient condition for $\bar{X}_n$ to be a consistent estimator of v (and hence $\bar{X}_n^2$ of v^2) was earlier seen to be that $n \to \infty$. The same argument shows that in a sequence of populations in which not only $v_{N\cdot}$ and τ_N^2 are independent of N but also the corresponding quantities when the v_{Ni} are replaced by v_{Ni}^2, the condition $n \to \infty$ will be sufficient for $\hat{\tau}_n^2$ to be a consistent estimator of τ^2. $\square$

Example 4.4.4 Stratified sampling. In the situation of Example 4.4.3, more accurate estimators can typically be obtained by dividing the population into more homogenous subpopulations, called *strata*, and drawing a random sample from each. Let there be s strata of sizes $N_1, \ldots, N_s$ $\left(\sum N_i = N\right)$, and let independent samples of sizes $n_1, \ldots, n_s$ be drawn from these strata.

We shall denote the values attached to the N_i elements of the ith stratum by v_{ij} $(j = 1, \ldots, N_i; i = 1, \ldots, s)$. The notation differs from that in Example 4.4.3, where one of the subscripts referred to the fact that we are not dealing with a single population, but with a sequence of populations of increasing size. Such a sequence is needed of course also in the present case, where we are now concerned with a sequence of populations $\Pi^{(k)}$ $(k = 1, 2, \ldots)$ subdivided into strata $\Pi_i^{(k)}$ $(i = 1, \ldots, s)$ of sizes $N_i^{(k)}$ consisting of elements $v_{ij}^{(k)}$ $(j = 1, \ldots, N_i; i = 1, \ldots, s; k = 1, 2, \ldots)$. For the sake of convenience, we shall in the following usually suppress the superscript k. Let us denote the variance of the ith stratum by

(4.4.34) $$\tau_i^2 = \frac{1}{N_i}\sum_{j=1}^{N_i}\left(v_{ij} - v_{i\cdot}\right)^2,$$

where

$$(4.4.35) \qquad v_{i\cdot} = \frac{1}{N_i} \sum_{j=1}^{N_i} v_{ij}.$$

When the v's in the ith stratum take on only two values, τ_i^2 reduces to

$$(4.4.36) \qquad \tau_i^2 = \frac{D_i}{N_i} \left(1 - \frac{D_i}{N_i} \right),$$

where D_i denotes the number of defectives or other special items in the ith stratum.

If the v-values in the sample from the ith stratum are denoted by X_{ij} $(j = 1, \dots, n_i)$, the estimator of $v_{i\cdot}$ corresponding to the estimator $\bar{X}_n$ considered in Example 4.4.3 is $X_{i\cdot} = \sum_{j=1}^{N_i} X_{ij}/n_i$ and the natural estimator of the population average $v_{\cdot\cdot} = \sum N_i v_{i\cdot}/N$ is

$$(4.4.37) \qquad X^* = \sum \frac{N_i}{N} X_{i\cdot\cdot}$$

Since the $X_{i\cdot}$ are independent, it follows that

$$(4.4.38) \qquad \mathrm{Var}\,(X^*) = \sum \left(\frac{N_i}{N} \right)^2 V_i^2,$$

where by (4.4.21)

$$(4.4.39) \qquad V_i^2 = \mathrm{Var}\,(X_{i\cdot}) = \frac{1}{n_i} \frac{N_i - n_i}{N_i - 1} \tau_i^2.$$

We shall now determine conditions for consistency and asymptotic normality of X^* as $k \to \infty$ under two different assumptions:

(a) The number s of strata is fixed and the sample sizes n_i tends to infinity for all i for which N_i/N does not tend to 0.

(b) The number $s = s^{(k)}$, $k = 1, 2, \dots$ of strata tends to ∞ as $k \to \infty$. $\square$

Example 4.4.5 Stratified sampling with a fixed number of strata.
Since $E\,(X^*) = v_{\cdot\cdot}$, a sufficient condition for X^* to be a consistent estimator of $v_{\cdot\cdot}$ is that $\mathrm{Var}\,(X^*) \to 0$ and hence that

$$(4.4.40) \qquad \sum_{i=1}^{s} \left(\frac{N_i}{N} \right)^2 V_i^2 \to 0.$$

Let us suppose that $[\tau_i^{(k)}]^2$ is bounded as $k \to \infty$ for each i. Then $V_i^2 = O(1/n_i)$ by (4.4.39), and a sufficient condition for (4.4.40) and hence for consistency of X^* is that

(4.4.41)
$$\frac{1}{n_i} \left(\frac{N_i}{N} \right)^2 \to 0 \text{ for each } i.$$

This condition will hold in particular if

(4.4.42)
$$n_i \to \infty \text{ for all } i.$$

Note however that (4.4.42) is not required for any value of i for which

(4.4.43)
$$N_i/N \to 0.$$

Since τ_i^2 is bounded, such a stratum makes a negligible contribution to $\text{Var}(X^*)$ regardless of the value of n_i.

In order to find conditions for the asymptotic normality of X^*, let us write

(4.4.44)
$$\frac{X^* - v_{..}}{\sqrt{\text{Var}X^*}} = \sum w_i \frac{X_i - v_i.}{\sqrt{\text{Var}X_i.}},$$

where

(4.4.45)
$$w_i = \frac{N_i}{N} \sqrt{\frac{\text{Var}X_i.}{\text{Var}X^*}} = \frac{N_i}{N} V_i \Bigg/ \sqrt{\sum_{j=1}^{s} \left(\frac{N_j}{N} \right)^2 V_j^2}$$

so that

(4.4.46)
$$\sum_{i=1}^{s} w_i^2 = 1.$$

Asymptotic normality of (4.4.44) will now follow from the following lemma, the proof of which will be considered in Problem 1.11 of Chapter 5. □

Lemma 4.4.1 *For each $k = 1, 2, \ldots$, let $w_i^{(k)}$, $i = 1, \ldots, s$, be constants satisfying*

(4.4.47)
$$\sum_{i=1}^{s} w_i^{(k)^2} = 1$$

and let $U_i^{(k)}$ ($i = 1, \ldots, s$) be s independent random variables satisfying

(4.4.48)
$$U_i^{(k)} \overset{L}{\to} N(0, 1) \text{ as } k \to \infty$$

for each i for which $w_i^{(k)}$ does not tend to 0 as $k \to \infty$. Then

(4.4.49)
$$\sum_{i=1}^{s} w_i^{(k)} U_i^{(k)} \overset{L}{\to} N(0, 1) \text{ as } k \to \infty.$$

Theorem 4.4.1 *The estimator X^* given by (4.4.37), for fixed s satisfies*

(4.4.50)
$$\frac{X^* - v_{..}}{\sqrt{\operatorname{Var} X^*}} \xrightarrow{L} N(0,1) \ as \ k \to \infty,$$

provided $n_i^{(k)}$, $N_i^{(k)}$, and the values $v_{ij}^{(k)}$ $(j = 1, \ldots, N_i)$ satisfy (4.4.24) and (4.4.25) for each $i = 1, \ldots, s$ for which N_i/N does not tend to 0, with n, N, and v_{Ni} replaced by $n_i^{(k)}$, $N_i^{(k)}$, and $v_{ij}^{(k)}$, respectively.

Proof. The result follows immediately from Lemma 4.4.1 and (4.4.23). ∎

Example 4.4.6 Stratified sampling with a large number of strata.
When s remains fixed as $k \to \infty$, the estimator X^* is the sum of a fixed number of independent terms and asymptotic normality can be expected only if these terms individually tend to normality. To model the situation with a large number of strata, we shall let $s \to \infty$. We are then dealing with the sum of an increasing number of independent terms so that, under suitable restrictions, the central limit theorem (CLT) applies. Since the N_i, n_i, and τ_i^2 all depend on i, the terms are not identically distributed. Conditions for asymptotic normality can therefore be obtained from Liapounov's theorem (Theorem 2.7.2) and somewhat better conditions from the Lindeberg CLT of Section A.1 of the Appendix. The latter leads to the following result which is proved in Bickel and Freedman (1984).

Theorem 4.4.2 *In the notation of Theorem 4.4.1, the normal limit (4.4.50) holds as $s \to \infty$, provided*

(4.4.51)
$$\max_{i=1,\ldots,s} \frac{\left(\dfrac{N_i}{N}\right)^2 V_i^2/\rho_i}{\sum \left(\dfrac{N_j}{N}\right) V_j^2} \to 0,$$

where $\rho_i = n_i (N_i - 1) / (N_i - n_i)$, and if there exists a constant M (independent of k) such that

(4.4.52)
$$\max_{i=1,\ldots,s} \left[\frac{1}{N_i^{(k)}} \sum_{j=1}^{N_i} \left| v_{ij}^{(k)} - v_{i\cdot}^{(k)} \right|^3 \right] \leq M \ for \ all \ k.$$

Since $\dfrac{1}{\rho_i} \leq \dfrac{1}{n_i}$, condition (4.4.51) will hold, in particular, when $\max(1/n_i^{(k)}) \to 0$ and hence when

(4.4.53)
$$\min_{i=1,\ldots,s(k)} \left(n_i^{(k)} \right) \to \infty \ as \ k \to \infty.$$

In analogy with (4.4.42), condition (4.4.53) is not required for values of i for which $w_i^{(k)}$ (given by (4.4.45)) tends to 0 (Problem 4.5).

Condition (4.4.52) holds whenever the $v_{ij}^{(k)}$ are uniformly bounded, i.e., there exists M' such that

$$\left| v_{ij}^{(k)} \right| \leq M' \text{ for all } i, j, \text{ and } k.$$

It is thus satisfied in particular when the populations are 2-valued.   □

Example 4.4.7 Ratio estimators. Example 4.4.3 was concerned with the asymptotic behavior of the sample mean $\bar{X}_n$ as an estimator of the population mean $v_N.$ in simple random sampling. In many sampling situations, each unit has attached to it not only its v-value but also a second auxiliary value w; for example, the v-value at some previous time. The w's provide additional information which, it is hoped, will lead to more accurate estimation of $\bar{v}$.

Consider therefore a population Π consisting of N units to each of which are attached two numbers $(v_1, w_1), \ldots, (v_N, w_N)$. The w's are assumed to be positive. A random sample of n units is drawn from Π and the pairs of values of the sampled units are denoted by $(X_1, Y_1), \ldots, (X_n, Y_n)$. A natural estimator of $\bar{v}/\bar{w}$ is $\bar{X}/\bar{Y}$. If $\bar{w}$ is known, this suggests

$$(4.4.54) \qquad R = \bar{w}\frac{\bar{X}}{\bar{Y}}$$

as an estimator of $\bar{v}$. Both R and $\bar{X}/\bar{Y}$ are called ratio estimators.

In order to discuss the asymptotic behavior of these estimators, let us—as in Example 4.4.3—embed the given population in a sequence of populations $\Pi_1, \Pi_2, \ldots$. Suppose that Π_N consists of N units with values

$$(v_{N1}, w_{N1}), \ldots, (v_{NN}, w_{NN}).$$

It will be convenient to keep the population means constant,

$$v_N. = \bar{v}, \ w_N. = \bar{w}$$

say, and we shall denote the population variances by[†]

$$(4.4.55) \qquad S_v^2 = \frac{1}{N}\sum (v_{Ni} - \bar{v})^2 \text{ and } S_w^2 = \frac{1}{N}\sum (w_{Ni} - \bar{w})^2.$$

To prove asymptotic normality of $R_n = R$, write

$$(4.4.56) \qquad R_n - \bar{v} = \frac{\bar{w}}{\bar{Y}_n}\left[\bar{X}_n - \frac{\bar{v}}{\bar{w}}\bar{Y}_n\right] = \frac{\bar{w}}{\bar{Y}_n}\bar{Z}_n,$$

[†]In many books on sampling, S^2 is defined with the denominator for N replaced by $N - 1$. The asymptotic theory is not affected by this change.

where

(4.4.57)
$$Z_i = X_i - \frac{\bar{v}}{\bar{w}} Y_i, \ i = 1, \dots, n.$$

Thus, if

(4.4.58)
$$r_{Ni} = v_{Ni} - \frac{\bar{v}}{\bar{w}} w_{Ni},$$

the Z's are a random sample from the population $\{r_{N1}, \dots, r_{NN}\}$. We shall denote the associated population variance by

(4.4.59)
$$S_r^2 = \frac{1}{N} \sum (r_{Ni} - r_{N.})^2.$$

Theorem 4.4.3 *The ratio estimator R_n of $\bar{v}$ satisfies*

(4.4.60)
$$\frac{\sqrt{n}(R_n - \bar{v})}{\sqrt{\frac{N-n}{N-1} \cdot \frac{1}{n} S_r^2}} \to N(0,1),$$

provided n and $N - n$ tend to ∞, and conditions (a) or (b) of Theorem 2.8.2 are satisfied with v_{Ni} replaced by r_{Ni}.

Proof. Under the stated assumptions, it follows from Theorem 2.8.2 that

(4.4.61)
$$\frac{\bar{Z} - E(\bar{Z})}{\sqrt{\operatorname{Var} \bar{Z}}} \xrightarrow{L} N(0,1).$$

Now

$$E(\bar{Z}) = E(Z_i) = \bar{v} - \frac{\bar{v}}{\bar{w}} \bar{w} = 0$$

and

$$\operatorname{Var}(\bar{Z}) = \frac{N-n}{N-1} \cdot \frac{1}{n} S_r^2;$$

thus (4.4.60) follows from (4.4.56), (4.4.61), and the fact that the factor $\bar{w}/\bar{Y}_n$ in (4.4.56) tends to 1 in probability. ∎

Corollary 4.4.1 *Under the assumptions of Theorem 4.4.3,*

(4.4.62)
$$\frac{\sqrt{n} \left[\frac{\bar{X}_n}{\bar{Y}_n} - \frac{\bar{v}}{\bar{w}} \right]}{\sqrt{\frac{N-n}{N-1} \cdot \frac{1}{n} \frac{S_r}{\bar{w}}}} \xrightarrow{L} N(0,1).$$

Proof. This follows immediately from Theorem 4.4.3 since

$$\frac{\bar{X}_n}{\bar{Y}_n} - \frac{\bar{v}}{\bar{w}} = \frac{R_n - \bar{v}}{\bar{w}}.$$

■

For Theorem 4.4.3 to be applicable in practice, it is necessary to replace the unknown S_r in the denominator of (4.4.60) by a consistent estimator. Since

$$(4.4.63) \qquad S_r^2 = \frac{1}{N} \sum v_{Ni}^2 - \frac{2}{N} \frac{\bar{v}}{\bar{w}} \sum v_{Ni} w_{Ni} + \frac{\bar{v}^2}{N\bar{w}^2} \sum w_{Ni}^2,$$

this only requires consistent estimators of

$$(4.4.64) \qquad \frac{1}{N} \sum v_{Ni}^2, \ \frac{1}{N} \sum v_{Ni} w_{Ni}, \text{ and } \frac{1}{N} \sum w_{Ni}^2,$$

and of $\bar{v}/\bar{w}$. Conditions for

$$(4.4.65) \qquad \frac{1}{n} \sum X_i^2, \ \frac{1}{n} \sum X_i Y_i, \text{ and } \frac{1}{n} \sum Y_i^2$$

to be consistent estimators of (4.4.64) are given by (4.4.26), (4.4.27), and the discussion following (4.4.33) with the obvious substitutions of v_{Ni}^2, $v_{Ni} w_{Ni}$, and w_{Ni}^2 for v_{Ni}. Consistency of $\bar{X}/\bar{Y}$ for estimating $\bar{v}/\bar{w}$ follows from Corollary 4.4.1. □

Example 4.4.8 Comparison of ratio estimator and sample mean. When an auxiliary variable is available, the ratio estimator R_n and the sample mean $\bar{X}_n$ provide two alternative estimators of $\bar{v}$. To determine the conditions under which each is preferred, consider the asymptotic relative efficiency of $\bar{X}_n$ to R_n. In Section 4.3, we discussed two different ways of handling such a comparison: in terms of the actual and the asymptotic variances. In the present case, the asymptotic variance approach is more convenient. However, it requires a slight extension of the result (4.3.13) for the ARE, namely the fact that (4.3.13) remains valid when assumption (4.3.12) is replaced by

$$(4.4.66) \qquad \frac{\sqrt{n}}{k_{in}} [\delta_i - h(\theta)] \xrightarrow{L} N\left(0, \tau_i'^2\right),$$

provided

$$(4.4.67) \qquad \frac{k_{1n}}{k_{2n}} \to 1 \text{ as } n \to \infty \text{ (Problem 4.6)},$$

If $\delta_1 = R_n$ and $\delta_2 = \bar{X}_n$, (4.4.66) holds with $h(\theta) = \bar{v}$,

$$(4.4.68) \qquad \tau_1'^2 = S_r^2, \ \tau_2'^2 = \tau^2,$$

and

$$k_{1n} = k_{2n} = \sqrt{\frac{N-n}{N-1}}.$$

Condition (4.4.67) is therefore satisfied.

Assuming (4.4.66), it follows from (4.3.13) and (4.4.68) that

(4.4.69) $$e_{\bar{X}_n|R_n} = \lim \frac{\sum \left(v_{Ni} - \frac{\bar{v}}{\bar{w}}w_{Ni}\right)^2}{\sum (v_{Ni} - \bar{v})^2}.$$

The numerator can be rewritten as

$$\sum \left[(v_{Ni} - \bar{v}) + \bar{v}\left(1 - \frac{w_{Ni}}{\bar{w}}\right)\right]^2$$
$$= \sum (v_{Ni} - \bar{v})^2 + \frac{\bar{v}^2}{\bar{w}^2}\sum (w_{Ni} - \bar{w})^2 - 2\frac{\bar{v}}{\bar{w}}\sum (v_{Ni} - \bar{v})(w_{Ni} - \bar{w})$$

and hence

(4.4.70) $$e_{\bar{X}_n|R_n} = 1 + \frac{\bar{v}^2}{\bar{w}^2}\frac{S_w^2}{S_v^2} - 2\frac{\bar{v}}{\bar{w}}\frac{S_w}{S_v}\rho,$$

where

(4.4.71) $$\rho = \frac{\sum (v_{Ni} - \bar{v})(w_{Ni} - \bar{w})}{S_v S_w}$$

is the population correlation coefficient of v and w.

Thus the ratio estimator R is more efficient than $\bar{X}_n$ if and only if

$$\frac{\bar{v}^2}{\bar{w}^2}\frac{S_w}{S_v} < 2\rho\frac{\bar{v}}{\bar{w}}$$

or equivalently if

(4.4.72) $$\frac{1}{2}\frac{\bar{v}}{\bar{w}}\bigg/ \frac{S_v}{S_w} < \rho \text{ when } \frac{\bar{v}}{\bar{w}} > 0,$$
$$\rho < \frac{1}{2}\frac{\bar{v}}{\bar{w}}\bigg/ \frac{S_v}{S_w} \text{ when } \frac{\bar{v}}{\bar{w}} < 0;$$

that is, if ρ is large positive when $\frac{\bar{v}}{\bar{w}} > 0$ and large negative when $\frac{\bar{v}}{\bar{w}} < 0$.□

Example 4.4.9 Cluster sampling. When interviewing a member of a household, it may require relatively little additional effort to interview all members of the household. More generally, suppose that a population is divided into groups, called *clusters*, that a simple random sample is drawn not of n individual members but of n clusters, and that the v-values of all members of each sampled cluster are obtained. On the basis of this information, it is desired to estimate the total or average v-value of the population.

Let there be N' clusters and let the ith cluster contain w_i' members with v-values v_{ij}, $j = 1, \ldots, w_i'$, $i = 1, \ldots, N'$. The population size is

$$(4.4.73) \qquad N = \sum_{i=1}^{N'} w_i',$$

and the total v-value of the ith cluster will be denoted by

$$(4.4.74) \qquad v_i' = \sum_{j=1}^{w_i'} v_{ij}.$$

We shall assume that N'/N tends to a finite positive limit. If X_i' denotes the v'-value of the ith sampled cluster, then $X_1', \ldots, X_n'$ is a simple random sample from $\{v_1', \ldots, v'_{N'}\}$ and $\bar{X}_n'$ is a consistent, asymptotically normal estimator of the average

$$(4.4.75) \qquad \bar{v}' = \frac{1}{N'} \sum_{i=1}^{N'} v_i' = \frac{1}{N'} \sum_{i=1}^{N'} \sum_{j=1}^{w_i'} v_{ij}$$

of the cluster totals under conditions given in Example 4.4.3. However, the quantity we wish to estimate is not (4.4.75), but the population average

$$(4.4.76) \qquad \bar{v} = \frac{1}{N} \sum_{i=1}^{N'} \sum_{j=1}^{w_i'} v_{ij} = \frac{N'}{N} \bar{v}'.$$

It follows from the properties of $\bar{X}_n'$ that under the same conditions, the estimator

$$(4.4.77) \qquad \delta_1 = \frac{N'}{N} \bar{X}_n'$$

of $\bar{v}$ is also consistent and asymptotically normal (Problem 4.7).

An alternative to $\bar{X}_n$ for estimating $\bar{v}'$ is the ratio estimator which utilizes the information provided by the cluster sizes w_i'. If Y_i' denotes the value of w' associated with the sample value X_i', this ratio estimator is $R' = \bar{w}' \bar{X}_n' / \bar{Y}_n'$. The corresponding estimator of $\bar{v}$ is by (4.4.76)

$$(4.4.78) \qquad \delta_2 = \bar{w}' \frac{N'}{N} \frac{\bar{X}_n'}{\bar{Y}_n'}.$$

Now

$$\bar{w}' = \frac{1}{N'} \sum_{i=1}^{N'} w_i' = \frac{N}{N'},$$

so that (4.4.78) reduces to

$$(4.4.79) \qquad\qquad \delta_2 = \frac{\bar{X}_n'}{\bar{Y}_n'}.$$

Conditions for the asymptotic normality of δ_2 are given in Theorem 4.4.3; the ARE of δ_2 to δ_1 is obtained as in (4.4.71).

It is interesting to note that if N' is known (as is often the case) the estimator δ_2 of $\bar{v}$ requires no knowledge of the population size while this is needed for δ_1. The situation is just the opposite for the estimators $\delta_1^* = N'\bar{X}_n'$ and $\delta_2^* = N\frac{\bar{X}_n'}{\bar{Y}_n'}$ of the population total $N\bar{v}$.   □

Summary

Consistency and asymptotic normality are discussed for the standard estimators of the total and average of a finite population Π, based on various sampling schemes. To obtain an asymptotic theory, in each case it is necessary to embed Π in a sequence of populations $\Pi^{(k)}$ of increasing size tending to infinity.

(i) *Simple random sampling from a population of two kinds of elements.* In this case, the embedding sequence Π_k is chosen so that the proportion D/N remains constant. In addition to considering the estimation of the total and average of Π, it is also shown how to estimate the size of Π using the capture-recapture method.

(ii) *Simple random sampling from a population Π of items $v_1, \ldots, v_N$.* The sequence $\Pi^{(k)}$ is now chosen so that the population mean and variance remain constant. The implications of the asymptotic results for the given finite population are discussed.

(iii) *Stratified sampling.* Asymptotic normality of the estimator is established for two cases: (a) A fixed (possibly small) number of large strata and (b) a large number of strata of arbitrary size.

(iv) *Ratio estimators under simple random sampling when an auxiliary variable is available.* Conditions for the asymptotic normality of the resulting ratio estimator are given and it is shown that this estimator is more efficient than the sample mean, provided the auxiliary variable is sufficiently highly correlated with the primary variable of interest.

(v) *Cluster sampling.* When clusters are sampled instead of individuals, two estimators are proposed for estimating the population average or total. One is proportional to the sample mean of the clusters, the other to a ratio estimator which uses cluster size as an auxilliary variable. The two estimators differ not only in their efficiency but also in whether they require knowledge of the population size (when the number of clusters is known).

4.5   Problems

Section 1

1.1 Show that (4.1.8), (4.1.11), and (4.1.12) each differs from the others only by terms of order $O(1/n)$.

1.2 (i) The left endpoints of the intervals (4.1.8) are always positive.

(ii) Both the left and right end points of the intervals (4.1.11) are to the left of the corresponding end points of (4.1.8).

1.3 Prove (4.1.16).

1.4 Show that the first term of (4.1.17) tends to $\alpha/2$.

1.5 Determine lower confidence bounds for λ with asymptotic confidence level $1-\alpha$ corresponding to the intervals (4.1.8), (4.1.11), and (4.1.12).

1.6 In the preceding problem, determine whether these lower bounds

(i) can take on negative values,

(ii) constitute strong confidence bounds.

1.7 (i) Solve (4.1.21) to obtain approximate confidence intervals for p.

(ii) Show that neglecting higher order terms in the intervals of part (i) leads to the intervals (4.1.23).

1.8 Prove that the infimum of the coverage probability of the binomial intervals (4.1.23) is zero.

1.9 Show that the infimum of the coverage probability of the preceding problem tends to $1-\alpha$ when it is taken over $a_1 < p < a_2 (0 < a_1 < a_2 < 1)$ rather than over the full interval $0 < p < 1$.

1.10 (i) Determine whether the lower and upper end points of the intervals (4.1.23) can take on values less than 0 and greater than 1, respectively.

(ii) Answer the same question for the intervals obtained by solving (4.1.21) for p_0, and replacing p_0 by p.

1.11 Use (3.6.1) to obtain asymptotic confidence intervals for p in the negative binomial situation in which binomial trials are continued until m successes are obtained.

1.12 Let $X_1, \ldots, X_n$ be i.i.d. according to the exponential distribution $E(\xi, a)$. Determine approximate confidence intervals for a based on $\sum [X_i - X_{(1)}]$.

[**Hint**: Problem 1.8(i) of Chapter 3.]

1.13 In the preceding problem, show that the intervals

$$X_{(1)} - \frac{c_1}{n} \sum [X_i - X_{(1)}] < \xi < X_{(1)} - \frac{c_2}{n} \sum [X_i - X_{(1)}],$$
$$0 < c_2 < c_1$$

are approximate confidence intervals for ξ with asymptotic confidence coefficient $\gamma = e^{-c_1} - e^{-c_2}$.

[**Hint**: Problem 1.8(ii) of Chapter 3.]

1.14 Let $X_1, \ldots, X_m$ and $Y_1, \ldots, Y_n$ be i.i.d. according to $E(\xi, a)$ and $E(\eta, a)$, respectively. Determine approximate confidence intervals for $\Delta = \eta - \xi$.

1.15 Let $X_1, \ldots, X_n$ be i.i.d. according to the uniform distribution $U(0, \theta)$. Use (3.1.17) to obtain asymptotic confidence intervals for θ.

1.16 Let $X_1, \ldots, X_n$ be i.i.d. according to the uniform distribution $U\left(\xi - \frac{a}{2}, \xi + \frac{a}{2}\right)$, with both ξ and a unknown. Determine asymptotic confidence intervals for ξ.

[**Hint**: A test of $H : \xi = \xi_0$ can be based on the statistic $\frac{1}{2}\left(X_{(1)} + X_{(n)}\right) / \left(X_{(n)} - X_{(1)}\right)$.]

1.17 Show that the intervals (4.1.56) with k given by (4.1.55) are strong confidence intervals for θ.

1.18 Let $X_1, \ldots, X_m$ and $Y_1, \ldots, Y_n$ be i.i.d. according to distributions

$$F(x) = P(X_i \leq x) \text{ and } G(y) = P(Y_j \leq y) = F(y - \theta),$$

respectively, and let $D_{(1)} < \cdots < D_{(mn)}$ denote the ordered set of mn differences $Y_j - X_i$ ($i = 1, \ldots, m; j = 1, \ldots, n$). Determine an approximate value for k so that the intervals

$$D_{(k)} \leq \theta \leq D_{(mn+1-k)}$$

constitute confidence intervals for θ at asymptotic level $1 - \alpha$.

[**Hint**: Follow the method of Example 4.1.7 and use the asymptotic normality result (3.2.9).]

1.19 Prove (4.1.42).

Section 2

2.1 If ϕ denotes the standard normal density, then

(i) the kth derivative of ϕ is of the form $P(x)\phi(x)$, where P is a polynomial of degree k, for $k = 1, 2, \ldots$;

(ii) all the derivatives $\phi^{(k)}(x)$ are bounded functions of x.

2.2 Prove

(i) (4.2.16),

(ii) (4.2.18).

2.3 Prove Theorem 4.2.1.

[**Hint**: Use the fact that $E\left(\bar{X} - \theta\right)^3 = E\left(X_1 - \theta\right)^3 / n^2$.]

2.4 Let X have the binomial distribution $b(p, n)$. Use Theorem 4.2.1 to determine up to order $1/n$ the bias and variance of

(i) $\delta = \left(\dfrac{X}{n}\right)^m$ as an estimator of p^m (m fixed; $n \to \infty$);

(ii) $h\left(\dfrac{X}{n}\right)$ as an estimator of $h(p) = \dbinom{m}{k} p^k q^{m-k}$ at all points $0 < p < 1$ at which $\dfrac{k}{p} \neq \dfrac{m-k}{q}$.

[**Hint**: $\dfrac{X}{n} = \bar{X}$ where $X_i = 1$ or 0 as the ith trial is success or failure.]

2.5 Solve part (ii) of the preceding problem when $h'(p) = 0$ by carrying the expansion of $h(\bar{X})$ about $h(p)$ one step further.

2.6 Let $X_1, \ldots, X_n$ be i.i.d. according to the Poisson distribution $P(\lambda)$. Determine up to order $1/n$ the bias and variance of

(i) $\delta = e^{-\bar{X}}$,

(ii) $\dfrac{\bar{X}^k}{k!} e^{-\bar{X}}$

as estimators of $e^{-\lambda}$ and $\dfrac{\lambda^k}{k!} e^{-\lambda}$, respectively.

2.7 Let Y have the negative binomial distribution given in Table 1.6.2 and Problem 2.4.12.

(i) Show that $\delta = \dfrac{1}{1 + Y/m}$ is a consistent estimator of p as $m \to \infty$.

(ii) Determine the bias and variance of δ up to terms of order $1/m$.

2.8 Under the assumptions of Theorem 4.2.1, consider the estimator $\delta' = h\left(c_n \bar{X}\right)$ of $h(\theta)$, where $c_n = 1 + \dfrac{c}{n} + O\left(\dfrac{1}{n^2}\right)$. Then the bias of δ' satisfies

$$E\left[h\left(c_n \bar{X}\right)\right] - h(\theta) = \frac{1}{n}\left[c\theta h'(\theta) + \frac{1}{2}h''(\theta)\sigma^2\right] + O\left(\frac{1}{n^2}\right)$$

while the variance of δ' is given by (4.2.20).

2.9 Determine the bias of δ up to terms of order $1/n^2$ for the estimators of

(i) Problem 2.4 (i);

(ii) Problem 2.6 (i).

2.10 Prove (4.2.24).

2.11 Show that the limiting variance (4.2.24) agrees with the asymptotic variance of $e^{\bar{X}}$.

2.12 Let $X_1, \ldots, X_n$ be i.i.d. according to the Poisson distribution $P(\lambda)$. Determine up to order $1/n$ the bias and variance of $\delta = e^{\bar{X}}$ as an estimator of e^λ by the method of Example 4.2.2 (continued).

[**Hint:** Use the fact that $\bar{X} = Y/n$, where Y is distributed as $P(n\lambda)$ and that $e^{y/n} = \left(e^{1/n}\right)^y$.]

2.13 Obtain a variant of Theorem 4.2.1 which requires existence and boundedness of only $h^{(iii)}$ instead of $h^{(iv)}$, but which asserts of the remainder term only that it is of order $O\left(n^{-3/2}\right)$.

[**Hint:** Use the fact that $\bar{X} = Y/n$, where Y is distributed as $P(n\lambda)$ and that $e^{y/n} = \left(e^{1/n}\right)^y$.]

2.14 Let $X_1, \ldots, X_n$ be i.i.d. according to the uniform distribution $U(0, \theta)$ and let h be a function satisfying the conditions of Theorem 4.2.1. Then

(i) $E\left[h\left(X_{(n)}\right)\right] - h(\theta) = -\dfrac{\theta}{n}h'(\theta) + \dfrac{1}{n^2}\left[\theta h'(\theta) + \theta^2 h''(\theta)\right] + O\left(\dfrac{1}{n^3}\right)$

and

(ii) $\mathrm{Var}\left[X_{(n)}\right] = \dfrac{\theta^2}{n^2}\left[h'(\theta)\right]^2 + O\left(\dfrac{1}{n^3}\right)$.

2.15 Compare the variance in part (ii) of the preceding problem with the asymptotic variance, i.e., the variance of the limiting distribution of $n\left[h\left(X_{(n)}\right) - h(\theta)\right]$.

2.16 Show that the integral in (4.2.30) becomes infinite when e^{ay} is replaced by e^{y^4}.

2.17 Under the assumptions of Problem 2.14,

(i) obtain the limiting distribution of $n \left[\dfrac{1}{X_{(n)}} - \dfrac{1}{\theta} \right]$;

(ii) show that $E\left[1/X_{(n)}\right] = \infty$.

Section 3

3.1 If n_1 and $n_2(n_1)$ are two sequences of sample sizes for which

$$n_1/n_2 \to c > 0 \text{ as } n_1 \to \infty,$$

then $O\left(\dfrac{1}{n_1^k}\right)$ implies $O\left(\dfrac{1}{n_2^k}\right)$ and vice versa.

3.2 The efficiency $e_{2,1} = e_{2.1}(\theta)$ given by (4.3.10) tends to 0 as $\theta \to \pm\infty$.

3.3 (i) Prove (4.3.17).

(ii) In Example 4.3.3, show that $\mathrm{Var}\left(\delta_2\right)/\mathrm{Var}\left(\delta_1\right) \to 1$ as $n \to \infty$.

3.4 (i) Determine the ARE (4.3.18) of $\tilde{X}$ to $\bar{X}$ for the normal mixture (4.3.24).

(ii) Use (i) to check the last column of Table 4.3.2.

(iii) Determine the range of values the ARE of (i) can take on for varying values of ϵ and τ.

3.5 Let $X_1, \ldots, X_n$ be i.i.d. Poisson $P(\lambda)$. For estimating

$$p = P\left(X_i = 0\right) = e^{-\lambda},$$

obtain the ARE of $\delta = e^{-\bar{X}}$ with respect to

$$\hat{p} = (\text{Number of } X\text{'s} = 0)/n,$$

and discuss the behavior of the ARE as λ varies from 0 to ∞.

3.6 Solve the preceding problem when $X_1, \ldots, X_m$ are i.i.d. according to the negative binomial distribution discussed in Problem 2.4.12; the quantity being estimated is

$$h(p) = P\left(X_1 = 0\right) = p^m q$$

and the estimators are

$$\delta_2 = \hat{p}^m \hat{q} \text{ with } \hat{p} = \frac{m}{m + \sum X_i}$$

and

$$\delta_1 = \text{Number of } X\text{'s equal to zero}/m.$$

3.7 Let $X_1, \dots, X_n$ be i.i.d. as $N\left(0, \sigma^2\right)$.

(i) Show that $\delta_1 = k \sum |X_i| / n$ is a consistent estimator of σ if and only if $k = \sqrt{\pi/2}$.

(ii) Determine the ARE of δ_1 with respect to $\delta_2 = \sqrt{\sum X_i^2 / n}$.

3.8 (i) Verify the identity (4.3.33).

(ii) Prove (4.3.36).

(iii) Prove (4.3.38).

3.9 If X and Y are independent, each with density f, the density of $Z = Y - X$ is $f^*(z) = \int\limits_{-\infty}^{\infty} f(z+x) f(x) dx$ and hence $f^*(0) = \int\limits_{-\infty}^{\infty} f^2(x) dx$.

[**Hint:** Make the transformation from (x, y) to $x = x$, $z = y - x$.]

3.10 Determine the efficiency $e_{2,1}$ of Example 4.3.9 when X and Y are distributed as Poisson $P(\lambda^2)$ and $P(\lambda)$, respectively, and it is desired to estimate λ^2.

3.11 In Example 4.3.11

(i) determine the ARE of δ to $\bar{X}$ based on expected squared error;

(ii) solve the preceding problem when expected squared error is replaced by the expected fourth power of the error, and compare the two AREs.

3.12 Verify (4.3.60)–(4.3.63).

3.13 (i) Verify (4.3.67) and (4.3.68).

(ii) Show that (4.3.67) < (4.3.68) provided $a/\theta > c$, where $c = \log(e - 1)$ is the unique positive solution of (4.3.69).

3.14 Make a table showing the ARE of $X_{(n)}$ to δ in Example 4.3.12 for varying values of a/θ and compare it with (4.3.64).

3.15 Under the assumptions of Example 4.3.12

(i) find the bias of $X_{(n-1)}$ as an estimator of θ and determine a constant c_n for which $\delta = c_n X_{(n-1)}$ is unbiased;

(ii) determine the ARE of $X_{(n-1)}$ to (a) δ and (b) $X_{(n)}$ based on expected squared error;

(iii) discuss the AREs of part (ii) based on (4.3.50) for suitable k_n.

3.16 Let $X_1, \ldots, X_n$ be i.i.d. according to the exponential distribution $E(\theta, 1)$.

(i) Find the bias of $X_{(1)}$ as an estimator of θ and determine c_n so that $\delta = X_{(1)} - c_n$ is unbiased.

(ii) Determine the ARE of $X_{(1)}$ with respect to δ (a) based on squared error and (b) based on (4.3.50) for suitable k_n.

Section 4

4.1 Under the assumptions of Example 4.4.1, suppose that instead of the confidence intervals (4.4.14) for p, we are interested in intervals of the form

$$\left| \frac{X}{n} - p \right| \leq d.$$

For given d and α, find the (approximate) sample size n required for these intervals to hold with probability $1 - \alpha$.

4.2 Make a table showing the approximate sample sizes required in the preceding problem when $\alpha = .05$, $N = 100, 1,000, 10,000$ and for $N = \infty$ (given by (4.1.25)) as a function of p.

4.3 (i) In Example 4.4.2, obtain confidence intervals for N by expressing the confidence intervals (4.4.14) in terms of $1/p$.

(ii) Show that the intervals of (i) are asymptotically equivalent to those given by (4.4.19).

4.4 (i) In Example 4.4.3, show that $\dfrac{n}{n-1} \hat{\tau}_n^2$ is an unbiased estimator of $\dfrac{N}{N-1} \tau^2$.

(ii) Determine the bias of $\hat{\tau}_N^2$ as an estimator of τ^2.

4.5 Show that the requirement (4.4.53) can be relaxed under conditions analogous to (4.4.43).

4.6 Prove that (4.3.13) remains valid when (4.3.12) is replaced by (4.4.66) provided (4.4.67) holds.

4.7 Prove the asymptotic normality and consistency of the estimators (4.4.77) and (4.4.79) under the conditions stated there.

4.8 (i) In Example 4.4.2, if $\dfrac{n_1}{n} = p$ and $\dfrac{n_2}{n} \to 1 - \lambda$, show that the length of the intervals (4.4.19) with p instead of $\hat{p}$ is asymptotically equivalent to

$$2\sqrt{N} \sqrt{\left(1 - \frac{n_1}{N}\right)\left(1 - \frac{n_2}{N}\right)} \Big/ \frac{n_1}{N} \cdot \frac{n_2}{N},$$

and hence is of order $\sqrt{N}$ if $0 < p$ and $\lambda < 1$.

(ii) Show that this approximate length is minimized, for a fixed total sample size $n = n_1 + n_2$ when $n_1 = n_2$.

Bibliographic Notes

Examples of large-sample confidence intervals (without that terminology and without a clear interpretation) can be found in Laplace's *Théorie Analytique des Probabilités* (3rd ed., 1820). The difficulty encountered in Examples 4.1.1 and 4.1.2 was pointed out for the binomial case by Blyth and Still (1983) and is discussed further in Lehmann and Loh (1990).

The large-sample theory of point estimation was given a general formulation by Fisher (1922a, 1925b), with Edgeworth (1908/9) an important forerunner.[‡] An early rigorous innovative exposition is provided by Cramér (1946).

The asymptotic theory of estimation in finite population was initiated by Madow (1948), Erdös and Renyi (1959), and Hajek (1960). Sampling is a large topic involving many issues (and including many designs) not covered here. They are discussed in books on the subject such as Kish (1965), Cochran (1977), Särndal et al. (1992), and Thompson (1992).

[‡] For a discussion of this history, see Pratt (1976).

5
Multivariate Extensions

Preview

The basic concepts and results of Chapters 1 and 2 are extended from the univariate to the multivariate case, i.e., from random variables to random vectors. This includes in particular: convergence in probability and in law, the normal distribution, the central limit theorem, and the delta method. The theory is applied to the multivariate one- and two-sample problems and regression, to goodness-of-fit tests, and to testing independence in a 2×2 table.

5.1 Convergence of multivariate distributions

Multivariate problems arise not only when the observations are vector-valued and their distributions therefore multivariate but also in many univariate problems involving the joint distribution of more than one statistic. Large-sample theory for either case requires generalization of the basic results of Chapters 1 and 2 to the multivariate situation. We begin with some definitions and results concerning convergence in probability and in law for general k-variate distributions.

We shall be concerned with functions of k-tuples $\underline{x} = (x_1, \dots, x_k)$ which may be viewed as points or vectors in k-dimensional space R^k. The distance

$\| \underline{x} - \underline{y} \|$ between two points $\underline{x}$ and $\underline{y}$ is defined by

$$(5.1.1) \qquad \| \underline{x} - \underline{y} \| = \sqrt{\sum (y_i - x_i)^2},$$

and a neighborhood of a point $\underline{x}^{(0)}$ is the set N of all $\underline{x}$ satisfying

$$(5.1.2) \qquad \| \underline{x} - \underline{x}^{(0)} \| < \delta$$

for some $\delta > 0$.

Definition 5.1.1 A sequence $\underline{x}^{(n)}$, $n = 1, 2, \dots$, is said to converge to $\underline{x}^{(0)}$,

$$(5.1.3) \qquad \underline{x}^{(n)} \to \underline{x}^{(0)},$$

if $\| \underline{x}^{(n)} - \underline{x}^{(0)} \| \to 0$, and $\underline{x}^{(0)}$ is then called the limit of the sequence $\underline{x}^{(n)}$, $n = 1, 2, \dots$. This is equivalent to the convergence of each of the k coordinates (Problem 1.1(i)).

The probability distribution of a random vector $(X_1, \dots, X_k)$ can be characterized by its (multivariate) *cumulative distribution function* (cdf)

$$(5.1.4) \qquad F(x_1, \dots, x_k) = P(X_1 \le x_1, \dots, X_k \le x_k).$$

For properties of cumulative distribution functions corresponding to those in the univariate case given in Theorem 1.6.1 of Chapter 1, see, for example, Billingsley (1986, Section 20). If there exists a function f such that

$$(5.1.5) \qquad F(x_1, \dots, x_k) = \int_{-\infty}^{x_1} \cdots \int_{-\infty}^{x_k} f(u_1, \dots, u_k)\, du_1, \dots, du_k,$$

then any f satisfying (5.1.5) for all $(x_1, \dots, x_k)$ is a probability density of F. As in the univariate case, a density is not unique defined by (5.1.5), but it is determined up to sets of probability zero by

$$(5.1.6) \qquad \left. \frac{\partial^k F(x_1, \dots, x_k)}{\partial x_1, \dots, \partial x_k} \right|_{u_1, \dots, u_k} = f(u_1, \dots, u_k).$$

Definition 5.1.2 A sequence of random vectors $\underline{X}^{(n)}$ is said to converge to $\underline{c}$ in probability if

$$(5.1.7) \qquad P\left(\underline{X}^{(n)} \in N \right) \to 1$$

for every neighborhood N of $\underline{c}$. This is equivalent to the convergence in probability of each component $X_i^{(n)}$ to c_i (Problem 1.1.(ii)).

Definition 5.1.3 A real-valued function f of $\underline{x}$ is said to be continuous at $\underline{x}^{(0)}$ if for every $\epsilon > 0$, there exists a neighborhood N of $\underline{x}^{(0)}$ such that

$$(5.1.8) \qquad \left| f(\underline{x}) - f\left(\underline{x}^{(0)}\right) \right| < \epsilon \text{ for all } \underline{x} \in N.$$

Note: Continuity of f at $\underline{x}^{(0)}$ implies that f is continuous at $\underline{x}^{(0)}$ in each coordinate when the other coordinates are held fixed, but the converse is not true (Problem 1.2).

In generalization of Theorem 2.1.4 (and its proof in Problem 1.3 of Chapter 2) we have

Theorem 5.1.1 *If* $\underline{X}^{(n)} \xrightarrow{P} \underline{c}$ *and if* f *is continuous at* $\underline{c}$ *then* $f\left(\underline{X}^{(n)}\right) \xrightarrow{P} f(\underline{c})$.

Note: Definition 5.1.3 and Theorem 5.1.1 can be generalized further to the case where not only $\underline{x}$ but also f is vector-valued. For this purpose, it is only necessary to change $\left| f(\underline{x}) - f\left(\underline{x}^{(0)}\right) \right|$ to $\| f(\underline{x}) - f\left(\underline{x}^{(0)}\right) \|$.

A point $\underline{x}$ at which a cdf H is continuous is called a *continuity point* of H. An interesting characterization of such points can be given in terms of the following definition.

Definition 5.1.4 A point $\underline{a}$ is a *boundary point* of a set S in R^k if $\underline{a}$ is either a point of S or a limit point of a sequence of points of S, and also a point or a limit of a sequence of points of the complement of S. The set of all boundary points of S is the *boundary* of S, denoted by ∂S.

Theorem 5.1.2 *A point* $\underline{a}$ *is a continuity point of a cdf* H *if and only if the boundary* B *of the set*

$$\{(x_1, \ldots, x_k): \quad x_1 \leq a_1, \ldots, x_k \leq a_k\}$$

(shown in Figure 5.1.1 for the cases $k = 2$ *and* $k = 1$*) satisfies*

$$(5.1.9) \qquad\qquad P_H(B) = 0.$$

Proof. That (5.1.9) is necessary follows directly from the definition of continuity since B is a subset of the set

$$(5.1.10)$$
$$\{(x_1, \ldots, x_k): \ a_1 - \delta < x_1 < a_1 + \delta, \ldots, a_k - \delta < x_k < a_k + \delta\}$$

for all $\delta > 0$ and since the probability of (5.1.10) under H tends to 0 as $\delta \to 0$ if H is continuous at $\underline{a}$. That (5.1.9) is also sufficient is shown for example in Billingsley (1986, Section 20). ∎

In the univariate case ($k = 1$), the boundary of the set

$$\{x: \ x \leq a\}$$

is the point a and condition (5.1.9) reduces to $P(X = a) = 0$. However, Theorem 5.1.2 shows that when $k > 1$, $P_H(\underline{X} = \underline{a}) = 0$ is no longer sufficient for $\underline{a}$ to be a continuity point of H. For example, if H assigns proba-

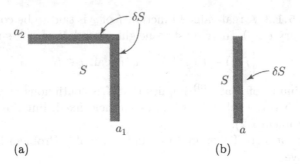

FIGURE 5.1.1. Boundary of (a) $S = \{x : x_1 \leq a_1, x_2 \leq a_2\}$, (b) $S = \{x : x \leq a\}$

bility $\dfrac{1}{2}$ each to the points $(0, -1)$ and $(-1, 0)$, then $P(X_1 = X_2 = 0)$ is 0, but $(0, 0)$ is not a continuity point of H (Problem 1.5).

Convergence in law is now defined as in the univariate case (Definition 2.3.2).

Definition 5.1.5 A sequence of distributions with cdf's H_n is said to converge to a distribution with cdf H if

(5.1.11) $H_n(\underline{a}) \to H(\underline{a})$ at all continuity points $\underline{a}$ of H.

The terminology and notation regarding convergence in law is used exactly as in the univariate case.

The following result is an extension of Example 2.3.5 to the multivariate case. For a proof and a generalization, see Problems 1.3 and 1.4.

Theorem 5.1.3 If the random vectors $\underline{Y}^{(n)} = \left(Y_1^{(n)}, \ldots, Y_k^{(n)}\right)$ converge in probability to a constant vector $\underline{c}$, then

$$\underline{Y}^{(n)} \overset{L}{\to} \underline{Y},$$

where $\underline{Y}$ is equal to $\underline{c}$ with probability 1.

Let $\underline{X}^{(n)}$ be a sequence of random vectors with cdf H_n converging in law to $\underline{X}$ with cdf H. One then often needs to know whether for some set S in R^k,

(5.1.12) $P\left(\underline{X}^{(n)} \in S\right) \to P(\underline{X} \in S).$

That (5.1.12) need not be true for all S is seen from the case $k = 1$, $S = \{x : x \leq a\}$. Then (5.1.12) can only be guaranteed when a is a continuity point of H.

Theorem 5.1.4 A sufficient condition for (5.1.12) to hold is that

(5.1.13) $P(\underline{X} \in \partial S) = 0.$

For a proof, see, for example, Billingsley (1986, Section 29).

Corollary 5.1.1 *If*

$$\left(X_1^{(n)}, \ldots, X_k^{(n)}\right) \xrightarrow{L} (X_1, \ldots, X_k),$$

then

$$X_i^{(n)} \xrightarrow{L} X_i \ for \ every \ i.$$

The proof is left to Problem 1.9.

In applications of Theorem 5.1.4, the set S frequently takes the form $f\left(\underline{X}^{(n)}\right) \le c$. Then the following result provides a convenient way of checking (5.1.12).

Theorem 5.1.5 *Let $\underline{X}^{(n)} = \left(X_1^{(n)}, \ldots, X_k^{(n)}\right)$ be a sequence of random vectors that converges in law to a random vector $\underline{X} = (X_1, \ldots, X_k)$. Then for any continuous real- or vector-valued function f of k arguments, we have*

(5.1.14) $$f\left(\underline{X}^{(n)}\right) \xrightarrow{L} f(\underline{X}).$$

For a proof, see, for example, Serfling (1980, p. 24–25).

If f is real-valued and $P[f(\underline{X}) = c] = 0$, so that c is a continuity point of the distribution of $f(\underline{X})$ and if $S = \{x : f(\underline{X}) \le c\}$, then (5.1.14) implies that $P\left[\underline{X}^{(n)} \in S\right] \to P[\underline{X} \in S]$.

Example 5.1.1 Difference of means. Let $X_1, \ldots, X_m$ and $Y_1, \ldots, Y_n$ be independently distributed according to distributions F and G, with means ξ and η and finite variances σ^2 and τ^2, respectively. Then

$$\sqrt{m} \left(\bar{X} - \xi\right) \xrightarrow{L} N\left(0, \sigma^2\right) \text{ and } \sqrt{n} \left(\bar{Y} - \eta\right) \to N\left(0, \tau^2\right)$$

If $\dfrac{m}{m+n} \to \lambda (0 < \lambda < 1)$, it follows that

$$\sqrt{m+n} \left(\bar{X} - \xi\right) = \sqrt{\frac{m+n}{m}} \sqrt{m} \left(\bar{X} - \xi\right) \xrightarrow{L} N\left(0, \frac{\sigma^2}{\lambda}\right)$$

and $\sqrt{m+n} \left(\bar{Y} - \eta\right) \xrightarrow{L} N\left(0, \dfrac{\tau^2}{1-\lambda}\right)$ and hence that

(5.1.15) $$\left(\sqrt{m+n} \left(\bar{X} - \xi\right), \sqrt{m+n} \left(\bar{Y} - \eta\right)\right) \xrightarrow{L} (X, Y),$$

where X and Y are independent random variables with distributions $N\left(0, \frac{\sigma^2}{\lambda}\right)$ and $N\left(0, \frac{\tau^2}{1-\lambda}\right)$, respectively. Since $Y - X$ is a continuous function of (X, Y), it follows that

$$\sqrt{m+n}\left[(\bar{Y} - \bar{X}) - (\eta - \xi)\right] \xrightarrow{L} N\left(0, \frac{\sigma^2}{\lambda} + \frac{\tau^2}{1-\lambda}\right)$$

or, equivalently, that

(5.1.16)
$$\frac{(\bar{Y} - \bar{X}) - (\eta - \xi)}{\sqrt{\dfrac{\sigma^2}{m} + \dfrac{\tau^2}{n}}} \xrightarrow{L} N(0, 1).$$

The problem of testing $H : \eta = \xi$ against $\eta > \xi$ was discussed in Example 3.1.6 and more generally in the comments following Lemma 3.1.1. We now see that the conclusions stated there follow from the present Theorems 5.1.4 and 5.1.5. More specifically, consider the probability

$$P\left\{\sqrt{m+n}\left[(\bar{Y} - \bar{X}) - (\eta - \xi)\right] \leq z\right\}.$$

By Theorem 5.1.4, this tends to

$$P((Y - X) \leq z) = \Phi\left(\frac{z}{\sqrt{\dfrac{\sigma^2}{\lambda} + \dfrac{\tau^2}{1-\lambda}}}\right)$$

since $P(Y - X = z) = 0$. $\square$

Example 5.1.2 Uniform distribution. Let $X_1, \ldots, X_n$ be i.i.d. according to the uniform distribution $U(\xi, \eta)$. From Example 2.3.7 it follows (Problem 1.12(i)) that the marginal distributions of $n\left[\eta - X_{(n)}\right]$ and $n\left[X_{(1)} - \xi\right]$ both tend to the exponential distribution $E(0, \eta - \xi)$. Consider now the joint distribution of these two variables. For any $\xi < \eta$ and any positive y, z, one has $y + z < n(\eta - \xi)$ for n sufficiently large, and then

$$P\left[n\left(X_{(1)} - \xi\right) > y \text{ and } n\left(\eta - X_{(n)}\right) > z\right]$$
$$= P\left[\frac{y}{n} + \xi < X_{(1)} \text{ and } X_{(n)} < \eta - \frac{z}{n}\right]$$
$$= \frac{\left[(\eta - \frac{z}{n}) - (\frac{y}{n} + \xi)\right]^n}{(\eta - \xi)^n} = \left[1 - \frac{y+z}{n(\eta - \xi)}\right]^n \to e^{\frac{-y}{\eta - \xi}} \cdot e^{\frac{-z}{\eta - \xi}}.$$

It follows that the variables $n\left[X_{(1)} - \xi\right]$ and $n\left[\eta - X_{(n)}\right]$—although dependent for each finite n—are independent in the limit, each with distribution $E(0, \eta - \xi)$.

If $\eta - \xi = 1$ and $R = X_{(n)} - X_{(1)}$ denotes the range of the X's, the limiting distribution of $n(1 - R)$ has probability density xe^{-x}, $x > 0$ (Problem 1.12(ii)).

The following result generalizes Theorem 2.3.3.

Theorem 5.1.6 (Multivariate Slutsky theorem). *If*

$$(5.1.17) \qquad \left(X_1^{(n)}, \ldots, X_k^{(n)}\right) \overset{L}{\to} (X_1, \ldots, X_k)$$

and if $A_i^{(n)}$ and $B_i^{(n)}$ $(i = 1, \ldots, k)$ are random variables which tend in probability to constants a_i and b_i respectively, then

$$(5.1.18)$$
$$\left(A_1^{(n)} + B_1^{(n)} X_1^{(n)}, \ldots, A_k^{(n)} + B_k^{(n)} X_k^{(n)}\right) \overset{L}{\to} (a_1 + b_1 X_1, \ldots, a_k + b_k X_k).$$

For a proof, see Problem 1.14.

Closely related is the following result.

Theorem 5.1.7 *If*

$$(5.1.19) \qquad \sum a_i^{(n)} X_i^{(n)} \overset{L}{\to} T$$

and if

(i) $A_i^{(n)} - a_i^{(n)} \overset{P}{\to} 0$ *for each* $i = 1, \ldots, k$ *as* $n \to \infty$

and

(ii) *the variables $X_i^{(n)}$ are bounded in probability as $n \to \infty$, then*

$$(5.1.20) \qquad \sum A_i^{(n)} X_i^{(n)} \overset{L}{\to} T.$$

Proof. It follows from Lemma 2.3.1 that

$$\left(A_i^{(n)} - a_i^{(n)}\right) X_i^{(n)} \overset{P}{\to} 0 \text{ for each } i$$

and hence that

$$\sum A_i^{(n)} X_i^{(n)} - \sum a_i^{(n)} X_i^{(n)} \overset{P}{\to} 0.$$

Therefore $\sum A_i^{(n)} X_i^{(n)}$ has the same limit distribution as $\sum a_i^{(n)} X_i^{(n)}$. ∎

Note: Condition (ii) holds in particular if the variables $X_i^{(n)}$ converge in law to some limiting variable X_i. On the other hand, Theorem 5.1.7 is not valid without assumption (ii) (Problem 1.19).

A convenient tool for establishing convergence in law of a sequence $\underline{X}^{(n)}$ of random vectors is the following result due to Cramér and Wold, which reduces the problem to the one-dimensional case.

Theorem 5.1.8 *A necessary and sufficient condition for*

$$(5.1.21) \qquad \left(X_1^{(n)}, \ldots, X_k^{(n)}\right) \xrightarrow{L} (X_1, \ldots, X_k)$$

is that

$$(5.1.22) \qquad \sum_{i=1}^{k} c_i X_i^{(n)} \xrightarrow{L} \sum_{i=1}^{k} c_i X_i \text{ for all constants } (c_1, \ldots, c_k),$$

i.e., that every linear combination of the $X_i^{(n)}$ converges in law to the corresponding combination of the X_i.

For a proof, see, for example, Serfling (1980) or Billingsley (1986).

Example 5.1.3 Orthogonal linear combinations. Let $Y_1, Y_2, \ldots$ be i.i.d. with mean $E(Y_i) = 0$ and variance $\text{Var}(Y_i) = \sigma^2$, and consider the joint distribution of the linear combinations

$$X_1^{(n)} = \sum_{j=1}^{n} a_{nj} Y_j \text{ and } X_2^{(n)} = \sum_{j=1}^{n} b_{nj} Y_j$$

satisfying the orthogonality conditions (see Section 5.3)

$$(5.1.23) \qquad \sum_{j=1}^{n} a_{nj}^2 = \sum_{j=1}^{n} b_{nj}^2 = 1 \text{ and } \sum_{j=1}^{n} a_{nj} b_{nj} = 0.$$

Then we shall show that, under the additional assumption (5.1.28), the relation (5.1.21) holds with $k = 2$ and with (X_1, X_2) independently distributed, each according to the normal distribution $N(0, \sigma^2)$.

To prove this result, it is by Theorem 5.1.8 enough to show that

$$(5.1.24) \qquad c_1 X_1^{(n)} + c_2 X_2^{(n)} = \sum (c_1 a_{nj} + c_2 b_{nj}) Y_j \xrightarrow{L} c_1 X_1 + c_2 X_2,$$

where the distribution of $c_1 X_1 + c_2 X_2$ is $N\left(0, [c_1^2 + c_2^2] \sigma^2\right)$. The sum on the left side of (5.1.24) is of the form $\sum d_{nj} Y_j$ with

$$(5.1.25) \qquad d_{nj} = c_1 a_{nj} + c_2 b_{nj}.$$

The asymptotic normality of such sums was stated in Theorems 2.7.3 and 2.7.4 under the condition that

$$(5.1.26) \qquad \max_j d_{nj}^2 / \sum_{j=1}^{n} d_{nj}^2 \to 0.$$

It follows from (5.1.23) that

(5.1.27) $$\sum_{j=1}^{n} d_{nj}^2 = c_1^2 + c_2^2;$$

furthermore

$$\max d_{nj}^2 \leq 2 \max \left[c_1^2 a_{nj}^2 + c_2^2 b_{nj}^2 \right] \leq 2 \left[c_1^2 \max a_{nj}^2 + c_2^2 \max b_{nj}^2 \right].$$

Condition (5.1.26) therefore holds, provided

(5.1.28) $$\max_j a_{jn}^2 \to 0 \text{ and } \max_j b_{jn}^2 \to 0 \text{ as } n \to \infty.$$

Under this condition, it thus follows that

$$\sum d_{nj} Y_j / \sqrt{\sum d_{jn}^2} \xrightarrow{L} N\left(0, \sigma^2\right)$$

and hence by (5.1.27) that

$$\sum d_{nj} Y_j \xrightarrow{L} N\left(0, \left(c_1^2 + c_2^2\right)\sigma^2\right),$$

as was to be proved.   □

Example 5.1.4 Joint confidence sets for regression coefficients. Let

(5.1.29) $$X_i = \alpha + \beta v_i + E_i \ (i = 1, \dots, n),$$

where the E_i are i.i.d. (not necessarily normal) with zero mean and common variance σ^2, and let

(5.1.30) $$\hat{\alpha} = \sum a_{ni} X_i, \ \hat{\beta} = \sum b_{ni} X_i$$

with

(5.1.31) $$a_{ni} = \frac{1}{n} - \frac{\bar{v}(v_i - \bar{v})}{\sum (v_j - \bar{v})^2} \text{ and } b_{ni} = \frac{v_i - \bar{v}}{\sum (v_j - \bar{v})^2}$$

be the estimators of α and β considered in Example 2.7.4 and 2.7.6. Then (Problem 1.17)

(5.1.32) $$\hat{\alpha} - \alpha = \sum a_{ni} Y_i, \ \hat{\beta} - \beta = \sum b_{ni} Y_i,$$

where the Y's are i.i.d. with $E(Y_i) = 0$, $\mathrm{Var}(Y_i) = \sigma^2$, and with a distribution which is independent of α and β.

It was seen in Example 2.7.6 that $\hat{\beta} - \beta$, suitably normalized, is asymptotically normal if

(5.1.33)
$$\frac{\max_i (v_i - \bar{v})^2}{\sum_{j=1}^{n} (v_j - \bar{v})^2} \to 0 \text{ as } n \to \infty$$

and the corresponding result for $\hat{\alpha} - \alpha$ was indicated in Problem 2.7.11(ii). These results provide asymptotic tests and confidence intervals separately for each of these parameters. We shall now consider joint confidence sets for α and β, and the problem of testing $H : \alpha = \alpha_0,\ \beta = \beta_0$.

The result of the preceding example is not directly applicable since (5.1.23) will typically not hold. We shall therefore replace $\hat{\alpha} - \alpha$ and $\hat{\beta} - \beta$ by

(5.1.34)
$$Y_{1n} = \sqrt{n}\left[\hat{\alpha} - \alpha + \bar{v}\left(\hat{\beta} - \beta\right)\right] = \sum a'_{ni} Y_i$$

and

(5.1.35)
$$Y_{2n} = \left(\hat{\beta} - \beta\right)\sqrt{\sum (v_j - \bar{v})^2} = \sum b'_{ni} Y_i,$$

where the a'_{ni} and b'_{ni} are given by

(5.1.36)
$$a'_{ni} = \frac{1}{\sqrt{n}},\ b'_{ni} = \frac{v_i - \bar{v}}{\sqrt{\sum (v_j - \bar{v})^2}},$$

which do satisfy (5.1.23). (For a method for deriving the linear functions (5.1.34) and (5.1.35), see Problem 1.15.) The assumptions of Example 5.1.3 therefore hold with a'_{ni} and b'_{ni} in place of a_{ni} and b_{ni}, respectively, and it follows that

(5.1.37)
$$(Y_{1n}, Y_{2n}) \xrightarrow{L} (X_1, X_2),$$

where X_1, X_2 are independent $N(0, \sigma^2)$, provided (5.1.28) holds. This will be the case if the v's satisfy (5.1.33), which is therefore a sufficient condition for the validity of (5.1.37). Under (5.1.33), it then follows from Theorem 5.1.5 that the distribution of $\left(Y_{1n}^2 + Y_{2n}^2\right)/\sigma^2$ tends to χ^2 with 2 degrees of freedom and hence that

(5.1.38)
$$n\left[(\hat{\alpha} - \alpha) + \left(\hat{\beta} - \beta\right)\bar{v}\right]^2 + \left(\hat{\beta} - \beta\right)^2 \sum (v_i - \bar{v})^2 \xrightarrow{L} \sigma^2 \chi_2^2.$$

The ellipses

(5.1.39)
$$n(\alpha - \hat{\alpha})^2 + 2n\bar{v}(\alpha - \hat{\alpha})\left(\beta - \hat{\beta}\right) + \left(\beta - \hat{\beta}\right)^2 \sum v_i^2 \le C_\gamma \sigma^2,$$

where C_γ is the lower γ critical value of χ_2^2 and the left side of (5.1.39) is a slight simplification of the left side of (5.1.38), therefore contain the unknown parameter vector (α, β) with a probability that tends to γ as $n \to \infty$. It follows from Example 3.5.4 that

$$(5.1.40) \qquad \hat{\sigma}^2 = \frac{1}{n} \sum \left(X_i - \hat{\alpha} - \hat{\beta} v_i \right) \overset{P}{\to} \sigma^2,$$

and hence that the sets (5.1.39) with σ^2 replaced by $\hat{\sigma}^2$ constitute joint asymptotic level γ confidence sets for (α, β). Finally, the set obtained from (5.1.39) by replacing α and β by α_0 and β_0, respectively, then provides an acceptance region for testing $H: \alpha = \alpha_0, \beta = \beta_0$ at asymptotic level $1 - \gamma$.
□

Summary

1. The definitions and properties of convergence in probability and in law are extended from random variables to random vectors.

2. Convergence in law in the multivariate case can be reduced to the univariate case by a result of Cramér and Wold which states that $\left(X_1^{(n)}, \ldots, X_k^{(n)} \right)$ converges in law to $(X_1, \ldots, X_k)$ if and only if

$$\sum_{i=1}^{k} c_i X_i^{(n)} \overset{L}{\to} \sum_{i=1}^{k} c_i X_i \text{ for all } (c_1, \ldots, c_k).$$

3. In generalization of Theorem 2.7.3, conditions are given for the convergence of $\left(\sum a_{nj} Y_j, \sum b_{nj} Y_j \right)$, where the Y's are i.i.d. As an application, joint confidence sets are obtained for the regression coefficients in simple linear regression.

5.2   The bivariate normal distribution

The two most important multivariate distributions are the multivariate normal and the multinomial distributions. We shall in the present section consider the special case of the bivariate normal which has the advantage that its density can be stated explicitly, namely as

$$(5.2.1)$$

$$p(x, y) \quad = \quad \frac{1}{2\pi\sigma\tau\sqrt{1-\rho^2}} \times$$

$$\exp\left(-\frac{1}{2(1-\rho^2)} \left[\frac{(x-\xi)^2}{\sigma^2} - 2\rho\frac{(x-\xi)(y-\eta)}{\sigma\tau} + \frac{(y-\eta)^2}{\tau^2} \right] \right).$$

From (5.2.1), the following properties can be obtained by direct calculation instead of requiring more abstract algebraic arguments.

(i) If (X, Y) has the joint density (5.2.1), then the marginal distributions of X and Y are $N\left(\xi, \sigma^2\right)$ and $N\left(\eta, \tau^2\right)$, respectively (Problem 2.1(i)).

(ii) From (i), it follows that

(5.2.2) $\xi = E(X), \ \sigma^2 = \mathrm{Var}(X), \ \eta = E(Y), \ \tau^2 = \mathrm{Var}(Y).$

It can further be shown (Problem 2.1(ii)) that ρ is the coefficient of correlation between X and Y, i.e.,

(5.2.3) $\rho = \mathrm{Cov}\,(X, Y)\,/\sigma\tau.$

(iii) Under (5.2.1), lack of correlation between X and Y is equivalent to independence between X and Y. That independence implies $\rho = 0$ is true without the assumption of normality (i.e., for any distribution with finite second moments). The converse is not true in general; in the normal case, it follows from (5.2.1).

Formula (5.2.1) breaks down when $\left(1 - \rho^2\right)\sigma^2\tau^2 = 0$ since then the denominator of the constant factor is zero. This occurs when either $\rho = \pm 1$ or when σ or τ or both are zero. A correlation of $+1$ or -1 implies a linear relationship $aX + bY = c$ so that the distribution assigns probability 1 to the line $ax + by = c$ and hence probability 0 to any set that does not intersect this line. For this distribution there cannot exist a density f satisfying (5.1.5). Roughly, the argument is that $f(x, y)$ would have to be zero for all points off the line, so that the right side of (5.1.5) for $x_1 = x_2 = \infty$ would reduce to

(5.2.4) $$\int\!\!\int\limits_{ax+by=c} f\,(x, y)\,dx dy$$

but that the set $\{(x, y) : ax + by = c\}$ is "too thin" for the integral (5.2.4) to be positive. A bivariate distribution which assigns probability 1 to a line or curve or point in the plane is said to be *degenerate* or *singular*.

The case $\sigma^2 = \mathrm{Var}(X) = 0$ corresponds to the situation in which X is constant, i.e., the distribution assigns probability 1 to a line $x = c$. Finally, if $\sigma = \tau = 0$, all probability is concentrated on a single point $x = c, y = d$. Such distributions are not often encountered as distributions of data, but they cannot be ignored in asymptotic work since the limit of a sequence of non-degenerate distributions can be degenerate.

Example 5.2.1 Degenerate limit distribution. Suppose that

$$X_n = X, \ Y_n = X + \frac{1}{n}Z,$$

where X and Z are independent normal $N(0,1)$. Then the joint distribution of (X_n, Y_n) is the non-degenerate bivariate normal distribution with means $\xi_n = \eta_n = 0$ and (Problem 2.3(i))

$$\sigma_n^2 = 1, \quad \tau_n^2 = 1 + \frac{1}{n^2}, \quad \rho_n = \frac{1}{\sqrt{1 + \dfrac{1}{n^2}}}.$$

On the other hand, the joint distribution H_n of (X_n, Y_n) satisfies (Problem 2.3(ii))

$$H_n(x, y) \to H(x, y) \text{ for all } (x, y),$$

where H is the distribution of (X, X). The limit distribution H is degenerate since it assigns probability 1 to the line $y = x$; it distributes this probability over the line according to a normal density.   □

Two important properties of bivariate normal distributions are stated without proof in the following lemma; the corresponding results for general multivariate normal distributions will be discussed in Section 5.4.

Lemma 5.2.1

(i) *If* (X, Y) *has the bivariate normal distribution with mean* (ξ, η), *variances* σ^2 *and* τ^2, *and correlation coefficient* ρ, *then the joint distribution of*

$$(X^*, Y^*) = (a_1 X + b_1 Y, a_2 X + b_2 Y)$$

is again bivariate normal, with mean

$$(\xi^*, \eta^*) = (a_1 \xi + b_1 \eta, a_2 \xi + b_2 \eta),$$

variances

(5.2.5) $\mathrm{Var}\,(a_i X + b_i Y) = a_i^2 \sigma^2 + 2 a_i b_i \rho \sigma \tau + b_i^2 \tau^2 \quad (i = 1, 2),$

and covariance

(5.2.6)
$\mathrm{Cov}\,(a_1 X + b_1 Y,\ a_2 X + b_2 Y) = a_1 a_2 \sigma^2 + (a_1 b_2 + a_2 b_1)\,\rho \sigma \tau + b_1 b_2 \tau^2.$

(ii) *If the distribution* (X, Y) *is non-singular, then so is the distribution of* (X^*, Y^*), *provided*

(5.2.7) $a_1 b_2 - a_2 b_1 \neq 0.$

□

We shall refer to $(\sigma^2, \tau^2, \rho\sigma\tau)$ as the *covariance structure* of a bivariate distribution and shall frequently denote it by $(\sigma_{11}, \sigma_{22}, \sigma_{12})$. The covariance structure will be called *non-singular* if

$$(5.2.8) \qquad \sigma^2\tau^2 \left(1 - \rho^2\right) = \sigma_{11}\sigma_{22} - \sigma_{12}^2 > 0.$$

Example 5.2.2 Simultaneous confidence intervals for regression coefficients. In Example 5.1.4, we obtained a joint (elliptical) confidence set for the coefficients (α, β) in the simple linear regression model (5.1.29). We shall now use the results of that example and Lemma 5.2.1 to derive simultaneous separate intervals

$$(5.2.9) \qquad \underline{L}_1 \leq \alpha \leq \bar{L}_1, \quad \underline{L}_2 \leq \beta \leq \bar{L}_2.$$

With Y_{1n} and Y_{2n} defined by (5.1.34) and (5.1.35), respectively, one finds

$$\sqrt{n}\left(\hat{\alpha} - \alpha\right) = Y_{1n} - \frac{\sqrt{n}\bar{v}}{\sqrt{\sum (v_j - \bar{v})^2}} Y_{2n},$$

$$\sqrt{n}\left(\hat{\beta} - \beta\right) = \frac{\sqrt{n}Y_{2n}}{\sqrt{\sum (v_j - \bar{v})^2}}.$$

Assuming (5.1.33), it therefore follows from (5.1.37) and Lemma 5.2.1 that for large n, the joint distribution of $(\sqrt{n}\left(\hat{\alpha} - \alpha\right), \sqrt{n}(\hat{\beta} - \beta))$ is approximately the bivariate normal distribution (5.2.1) with $\xi = \eta = 0$,

$$(5.2.10) \qquad \begin{aligned} \text{Var}\left[\sqrt{n}\left(\hat{\alpha} - \alpha\right)\right] &= n\sigma^2 \left[\frac{1}{n} + \frac{\bar{v}^2}{\sum (v_j - \bar{v})^2}\right], \\ \text{Var}\left[\sqrt{n}\left(\hat{\beta} - \beta\right)\right] &= \frac{n\sigma^2}{\sum (v_j - \bar{v})^2} \end{aligned}$$

and correlation coefficient

$$(5.2.11) \qquad \rho = -\sqrt{n}\bar{v}/\sqrt{\sum v_j^2}.$$

Limits $\underline{L}_1, \bar{L}_1, \underline{L}_2,$ and $\bar{L}_2$ for which the probability of (5.2.9) is approximately equal to the desired confidence level γ can now be determined from tables or computer programs for the bivariate normal distribution with parameters (5.2.10) and (5.2.11). Here σ^2 is replaced by $\hat{\sigma}^2$ given by (5.1.40) and one may choose the L's, for example, so that the probabilities of the two intervals (5.2.9) are the same.   □

Let us next state the form of the bivariate central limit theorem corresponding to the univariate Theorem 2.4.1.

Theorem 5.2.1 Bivariate central limit theorem. *Let (X_i, Y_i), $i = 1, 2, \ldots$, be i.i.d. with $E(X_i) = \xi$, $E(Y_i) = \eta$, $\mathrm{Var}(X_i) = \sigma^2$, $\mathrm{Var}(Y_i) = \tau^2$ and correlation coefficient ρ satisfying (5.2.8). Then the joint distribution of*

$$\left(\sqrt{n}\,(\bar{X} - \xi),\ \sqrt{n}\,(\bar{Y} - \eta)\right)$$

tends in law to the bivariate normal distribution with density (5.2.1) in which the means ξ and η are zero.

For a proof see Theorem 5.4.4.

Example 5.2.3 Joint limit distribution of uncorrelated, dependent variables. Let $X_1, \ldots, X_n$ be i.i.d. $N(0, \sigma^2)$ and consider the joint distribution of $\sqrt{n}\bar{X}$ and $\sqrt{n}\left(\sum X_i^2/n\right)$. Since the correlation coefficient of X_i and X_i^2 is 0, it follows from Theorem 5.2.1 that, in the limit, $\sqrt{n}\bar{X}$ and $\sqrt{n}\left(\sum X_i^2/n\right)$ are independent despite the fact that X_i^2 is completely determined by X_i.

The following argument gives an idea of how this asymptotic independence comes about. Consider the conditional distribution of $\bar{X}$ given not only $\sum x_i^2$ but the values of $X_1^2, \ldots, X_n^2$. Given $X_i^2 = x_i^2$, the random variable X_i takes on the values $|x_i|$ and $-|x_i|$ with probability $1/2$ each, independently for $i = 1, \ldots, n$. If $Y_i = \pm 1$ with probability $1/2$ each, $\sum X_i$ can be written as

$$\sum X_i = \sum_{i=1}^{n} |x_i| Y_i$$

and Theorem 2.7.3 applies with $d_{ni} = |x_i|$ to show that $\sum X_i / \sqrt{\sum x_i^2} \xrightarrow{L} N(0, 1)$, provided (2.7.17) holds. When the X_i are independent $N(0, \sigma^2)$, it can be shown that with probability tending to 1, both (2.7.17) holds and $\sum X_i^2/n$ is close to σ^2 and hence the conditional distribution of $\sqrt{n}\bar{X}$ is close to $N(0, 1)$. $\qquad\square$

Example 5.2.4 Simultaneous inference for mean and variance. The following slight modification of the preceding example has some statistical interest. Let $X_1, \ldots, X_n$ be i.i.d. $N(\xi, \sigma^2)$ and consider the problem of simulaneously estimating or testing ξ and σ^2. We shall then be interested in the joint distribution of

(5.2.12)
$$\frac{\sqrt{n}\,(\bar{X} - \xi)}{\sigma} \quad \text{and} \quad \sqrt{n}\left(\frac{\sum (X_i - \bar{X})^2}{n\sigma^2} - 1\right).$$

Since this distribution is independent of ξ and σ, suppose that $\xi = 0$. As pointed out in the preceding example,

(5.2.13) $$\sqrt{n}\bar{X}/\sigma \text{ and } \sqrt{n}\left[\left(\sum X_i^2/n\sigma^2\right) - 1\right]$$

are then asymptotically independently distributed by Theorem 5.2.1 according to $N(0,1)$ and $N(0,2)$, respectively, the latter since in the normal case $\text{Var}\left(X_i^2/\sigma^2\right) = 2$. This is then also the joint limit distribution of (5.2.12), which provides the limiting probability of the simultaneous confidence sets (Problem 2.4)

(5.2.14)
$$\bar{X} - \frac{a\hat{\sigma}}{\sqrt{n}} \le \xi \le \bar{X} + \frac{a\hat{\sigma}}{\sqrt{n}},$$
$$\frac{\sum\left(X_i - \bar{X}\right)^2}{n} - \frac{\sqrt{2}b\hat{\sigma}^2}{\sqrt{n}} \le \sigma^2 \le \frac{\sum\left(X_i - \bar{X}\right)^2}{n} + \frac{\sqrt{2}b\hat{\sigma}^2}{\sqrt{n}}$$

where in the lower and upper limits we have replaced σ^2 by $\hat{\sigma}^2$.   □

Example 5.2.5 Joint and simultaneous confidence sets for two means. Let $(X_1, Y_1), \ldots, (X_n, Y_n)$ be a sample from a bivariate (not necessarily normal) distribution with means (ξ, η) and covariance structure $(\sigma^2, \tau^2, \rho\sigma\tau)$ and consider the problem of simultaneous confidence intervals for ξ and η. By the central limit theorem (CLT), the limit of

(5.2.15) $$P\left[\frac{\sqrt{n}\left|\bar{X} - \xi\right|}{\sigma} \le a, \frac{\sqrt{n}\left|\bar{Y} - \eta\right|}{\tau} \le b\right]$$

as $n \to \infty$ is equal to

(5.2.16) $$P_\rho\left[|V| \le a, |W| \le b\right],$$

where (V, W) have the bivariate normal distribution with 0 means and covariance structure $(1, 1, \rho)$. For each value of ρ, we can therefore determine constants $a(\rho)$, $b(\rho)$ for which the probability (5.2.16) is equal to the preassigned confidence coefficient γ. To be specific, suppose we take $b(\rho) = a(\rho)$. Then $a(\rho)$ is a continuous function of ρ and it follows from Theorem 5.1.2 and the fact that the sample correlation coefficient R tends to ρ in probability (Problem 4.4) that if σ and τ are replaced by consistent estimators $\hat{\sigma}$ and $\hat{\tau}$ and if ρ is replaced by R, then

(5.2.17) $$P\left[\left|\sqrt{n}\left(\bar{X} - \xi\right)\right| \le \hat{\sigma}a(R), \left|\sqrt{n}\left(\bar{Y} - \eta\right)\right| \le \hat{\tau}a(R)\right] \to \gamma.$$

If we prefer intervals of the same length for ξ and η, we can proceed slightly differently by considering the simultaneous intervals

(5.2.18) $$\sqrt{n}\left|\bar{X} - \xi\right| \le a, \sqrt{n}\left|\bar{Y} - \eta\right| \le a.$$

The probability of (5.2.18) is a continuous function of σ, τ, and ρ, and for each (σ, τ, ρ), we can determine $a(\sigma, \tau, \rho)$ such that the limiting probability of (5.2.18) is equal to γ. It then follows as before that the probability of the simultaneous intervals

$$\sqrt{n}\left|\bar{X} - \xi\right| \leq a(\hat{\sigma}, \hat{\tau}, R), \ \sqrt{n}\left|\bar{Y} - \eta\right| \leq a(\hat{\sigma}, \hat{\tau}, R)$$

tends to γ.

For joint confidence sets for (ξ, η), see Problem 2.5.   □

Example 5.2.6 Testing for independence and lack of correlation.
In the bivariate normal model (5.2.1), the standard test of the hypothesis $H : \rho = 0$, or, equivalently, $H : X, Y$ are independent, rejects H in favor of the alternatives $\rho \neq 0$ when the sample correlation coefficient

$$(5.2.19) \qquad R = \frac{\sum (X_i - \bar{X}) (Y_i - \bar{Y}) / n}{S_X S_Y}$$

is sufficently large in absolute value, where

$$S_X^2 = \frac{1}{n} \sum (X_i - \bar{X})^2, \ S_Y^2 = \frac{1}{n} \sum (Y_i - \bar{Y})^2.$$

Under H, we have

$$(5.2.20) \qquad \sqrt{n}R = \frac{\sqrt{n}\sum X_i Y_i / n}{S_X S_Y} - \frac{\sqrt{n}\bar{X}\bar{Y}}{S_X S_Y} \xrightarrow{L} N(0, 1).$$

To see this limit result, note that the distribution of R is independent of $\xi = E(X)$ and $\eta = E(Y)$, so that we can assume $\xi = \eta = 0$. Since, under H, the variables X and Y are independent, it follows that

$$(5.2.21) \qquad \text{Var}(XY) = E\left(X^2 Y^2\right) = E\left(X^2\right) E\left(Y^2\right) = \text{Var}(X)\text{Var}(Y)$$

and hence from the CLT that the first term on the right side of (5.2.20) tends in law to $N(0, 1)$. On the other hand, the second term tends in probability to zero, and the result follows. The rejection region

$$(5.2.22) \qquad \left|\sqrt{n}R\right| \geq u_{\alpha/2}$$

therefore has asymptotic level α for testing H.

Let us now consider what happens to the level of (5.2.22) when the assumption of normality is not justified. Suppose that $(X_1, Y_1), \ldots, (X_n, Y_n)$ is a sample from some bivariate distribution F with finite second moments and let its correlation coefficient be ρ. In the normal case, the hypothesis

$$(5.2.23) \qquad H_1 : \rho = 0$$

is equivalent to the hypothesis

(5.2.24) $H_2 : X$ and Y are independent.

This is not true in general and it then becomes necessary to distinguish between H_1 and H_2.

Under H_2 nothing changes; both (5.2.20) and (5.2.21) remain valid, and the asymptotic level of (5.2.22) therefore continues to be α. However, under the weaker hypothesis H_1, equation (5.2.21) will in general not hold any more. In fact, even when $\rho = 0$,

$$(5.2.25) \qquad \gamma^2 = \frac{\mathrm{Var}(XY)}{\mathrm{Var}(X)\mathrm{Var}(Y)}$$

can take on any value between 0 and ∞. This can be seen, for example, by putting (i) $Y = X$ and (ii) $Y = 1/X$. If X is symmetric about 0, then in case (i),

$$\gamma^2 = \frac{E\left(X^4\right) - \left[E\left(X^2\right)\right]^2}{\left[E\left(X^2\right)\right]^2},$$

which can be made arbitrarily large (including ∞) by putting enough weight in the tail of the distribution of X. On the other hand, in case (ii),

$$\gamma^2 = \frac{1}{\mathrm{Var}(X)\mathrm{Var}\left(\dfrac{1}{X}\right)}.$$

Here the denominator can be made arbitrarily large by putting enough weight near the origin, thereby making $\mathrm{Var}(1/X)$ large without changing $\mathrm{Var}(X)$ much (Problem 2.7).

Since it follows from the argument leading to (5.2.20) under H that under H_1

(5.2.26) $\sqrt{n}R \to N\left(0, \gamma^2\right),$

we see that the asymptotic level of (5.2.22) can now take on any value $\alpha(\gamma)$ between 0 and 1. Thus the level of the normal theory test is asymptotically robust against non-normality under H_2, but not under H_1.

Another important aspect of (5.2.22) as a test of independence is the behavior of its power against a fixed alternative $\rho \neq 0$ as $n \to \infty$. This power can be written as

$$(5.2.27) \qquad P\left[\left|\sqrt{n}\frac{\left[\dfrac{1}{n}\sum X_i Y_i - \rho\sigma\tau\right]}{S_X S_Y} + \frac{\sqrt{n}\rho\sigma\tau}{S_X S_Y}\right| > u_{\alpha/2}\right].$$

Since the first term inside the brackets tends in law to $N\left(0, \gamma^2\right)$, the probability (5.2.27) tends to 1 as $n \to \infty$. The test (5.2.23) is therefore consistent for testing H_2 against any alternative $\rho \neq 0$. On the other hand, when $\rho = 0$ but X and Y are not necessarily independent, the limiting probability of (5.2.27) is just $\alpha(\gamma)$ and the test is therefore not consistent against such alternatives. A non-parametric test of independence which is consistent against all continuous alternatives to independence is given in Hoeffding (1948b).   □

Let us next extend the delta method (Theorem 2.5.2) to the bivariate case. To this end, we require first an extension of Taylor's theorem (Theorem 2.5.1). Recall that the partial derivative $\dfrac{\partial}{\partial x_i} f\left(x_1, \ldots, x_k\right)$ at the point $\underline{a} = (a_1, \ldots, a_k)$ is defined as

$$\left.\frac{\partial}{\partial x_i} f\left(x_1, \ldots, x_k\right)\right|_{(a_1, \ldots, a_k)} = \left.\frac{d}{d x_i} f\left(a_1, \ldots, a_{i-1}, x_i, a_{i+1}, \ldots, a_k\right)\right|_{x_i = a_i},$$

that is, as the ordinary derivative with respect to x_i when the remaining variables are being held constant.

Theorem 5.2.2 *Let f be a real-valued function of k variables for which the k first partial derivatives exist in a neighborhood of a point $\underline{a}$. Then*

$$(5.2.28) \qquad \begin{aligned} & f\left(a_1 + \Delta_1, \ldots, a_k + \Delta_k\right) \\ & = f\left(a_1, \ldots, a_k\right) + \sum \Delta_i \left.\frac{\partial f}{\partial x_i}\right|_{\underline{x} = \underline{a}} + o\left(\sqrt{\sum \Delta_i^2}\right). \end{aligned}$$

We have given here only the statement corresponding to the case $r = 1$ of Theorem 2.5.1(i). Expansions corresponding to higher values of r and to part (ii) of the theorem are, of course, also available if the needed higher derivatives exist. See, for example, Serfling (1980) or Courant (1927, 1988).

The following result extends Theorem 2.5.2 to the bivariate case. In the statement and proof, all derivatives $\partial f / \partial u$, $\partial f / \partial v$, $\partial g / \partial u$ and $\partial g / \partial v$ are understood to be evaluated at the point $u = \xi$, $v = \eta$.

Theorem 5.2.3 *Suppose that*

$$(5.2.29) \qquad \sqrt{n}\left(U_n - \xi\right), \ \sqrt{n}\left(V_n - \eta\right) \overset{L}{\to} N(0, \Sigma),$$

where $N(0, \Sigma)$ denotes the bivariate normal distribution with mean (0,0) and covariance structure $\sigma_{11} = \sigma^2$, $\sigma_{12} = \rho \sigma \tau$, and $\sigma_{22} = \tau^2$. Let f and g be two real-valued functions of two variables for which the expansion (5.2.28) with $k = 2$ is valid at the point (ξ, η). Then the joint distribution of

$$(5.2.30) \qquad \sqrt{n}\left[f\left(U_n, V_n\right) - f(\xi, \eta)\right], \ \sqrt{n}\left[g\left(U_n, V_n\right) - g(\xi, \eta)\right]$$

tends in law to the bivariate normal distribution with mean (0,0) and covariance structure $(\tau_{11}, \tau_{22}, \tau_{12})$ given by

$$\tau_{11} = \left(\frac{\partial f}{\partial u}\right)^2 \sigma_{11} + 2\frac{\partial f}{\partial u}\frac{\partial f}{\partial v}\sigma_{12} + \left(\frac{\partial f}{\partial v}\right)^2 \sigma_{22},$$

$$(5.2.31) \qquad \tau_{12} = \frac{\partial f}{\partial u}\frac{\partial g}{\partial u}\sigma_{11} + \left(\frac{\partial f}{\partial u}\frac{\partial g}{\partial v} + \frac{\partial f}{\partial v}\frac{\partial g}{\partial u}\right)\sigma_{12} + \frac{\partial f}{\partial v}\frac{\partial g}{\partial v}\sigma_{22},$$

$$\tau_{22} = \left(\frac{\partial g}{\partial u}\right)^2 \sigma_{11} + 2\frac{\partial g}{\partial u}\frac{\partial g}{\partial v}\sigma_{12} + \left(\frac{\partial g}{\partial v}\right)^2 \sigma_{22},$$

provided

$$(5.2.32) \qquad \frac{\partial f}{\partial u}\frac{\partial g}{\partial v} \neq \frac{\partial f}{\partial v}\frac{\partial g}{\partial u}.$$

Proof. By Theorem 5.2.2,

$$f(u_n, v_n) - f(\xi, \eta) = (u_n - \xi)\frac{\partial f}{\partial u} + (v_n - \eta)\frac{\partial f}{\partial v} + R_n,$$

$$g(u_n, v_n) - g(\xi, \eta) = (u_n - \xi)\frac{\partial g}{\partial u} + (v_n - \eta)\frac{\partial g}{\partial v} + R_n',$$

where R_n and R_n' are both $o\left(\sqrt{(u_n - \xi)^2 + (v_n - \eta)^2}\right)$. By (5.2.29) and Lemma 5.2.1, the joint limit distribution of

$$\sqrt{n}\left[(U_n - \xi)\frac{\partial f}{\partial u} + (V_n - \eta)\frac{\partial f}{\partial v}\right] \text{ and } \sqrt{n}\left[(U_n - \xi)\frac{\partial g}{\partial u} + (V_n - \eta)\frac{\partial g}{\partial v}\right]$$

is the bivariate normal distribution with mean (0,0) and covariance matrix (5.2.31). To complete the proof, it is only necessary to show that $\sqrt{n}R_n$ and $\sqrt{n}R_n'$ tend to 0 in probability. Now

$$(5.2.33) \qquad \sqrt{n}R_n = o\left(\sqrt{n\left[(u_n - \xi)^2 + (v_n - \eta)^2\right]}\right).$$

It follows from (5.2.29) that

$$n\left[(U_n - \xi)^2 + (V_n - \eta)^2\right]$$

is bounded in probability and hence that $\sqrt{n}R_n \xrightarrow{P} 0$, and the same argument applies to $\sqrt{n}R_n'$. ∎

Corollary 5.2.1 *Under the assumptions of Theorem 5.2.3, the distribution of*

$$(5.2.34) \qquad \sqrt{n}\left[f(U_n, V_n) - f(\xi, \eta)\right] \xrightarrow{L} N(0, \tau_{11}),$$

where τ_{11} is given by (5.2.31).

Proof. (5.2.29) and Corollary 5.1.1.   ∎

Example 5.2.7 The sample moments. If $X_1, \ldots, X_n$ are i.i.d. with mean ξ and finite central k^{th} moment

$$(5.2.35) \qquad \mu_k = E(X_i - \xi)^k,$$

it was shown in Example 2.1.3 that

$$(5.2.36) \qquad M_k = \frac{1}{n} \sum (X_i - \bar{X})^k$$

is consistent for estimating μ_k. We shall now consider the asymptotic distribution of

$$(5.2.37) \qquad \sqrt{n}(M_k - \mu_k),$$

under the assumption that the $2k^{\text{th}}$ moment of X_i exists.

For the quantities

$$(5.2.38) \qquad M'_k = \frac{1}{n} \sum (X_i - \xi)^k,$$

it follows from the central limit theorem that

$$(5.2.39) \qquad \sqrt{n}(M'_k - \mu_k) \xrightarrow{L} N\left(0, \mu_{2k} - \mu_k^2\right).$$

In the special case $k = 2$, it was seen in Example 2.4.4 that

$$(5.2.40) \qquad \sqrt{n}(M_2 - M'_2) \xrightarrow{P} 0,$$

and that, therefore, (5.2.36) has the same limit distribution as (5.2.38). As we shall see, the corresponding result no longer holds when $k > 2$. To obtain the asymptotic distribution of (5.2.36) for general k, note that

$$\sum (X_i - \bar{X})^k = \sum (X_i - \xi)^k - k(\bar{X} - \xi)\sum (X_i - \xi)^{k-1} + \cdots + (-1)^k (\bar{X} - \xi)^k$$

so that

$$(5.2.41) \qquad \begin{aligned} \sqrt{n}(M_k - \mu_k) &= \sqrt{n}(M'_k - \mu_k) \\ &- k\sqrt{n}(\bar{X} - \xi)\frac{\sum (X_i - \xi)^{k-1}}{n} + \cdots + (-1)^k \frac{\sqrt{n}(\bar{X} - \xi)^k}{n}. \end{aligned}$$

Since

$$\sqrt{n}(\bar{X} - \xi)^i = \frac{\left[\sqrt{n}(\bar{X} - \xi)\right]^i}{(\sqrt{n})^{i-1}} \xrightarrow{P} 0 \text{ for } i > 1,$$

all terms on the right side of (5.2.41) except the first two tend to 0 in probability and $\sqrt{n}\,(M_k - \mu_k)$ therefore has the same limit distribution as

$$(5.2.42) \qquad \sqrt{n}\,(M_k' - \mu_k) - k\sqrt{n}\,(\bar{X} - \xi)\,M_{k-1}'.$$

For $k = 2$, $M_{k-1}' = \bar{X} - \xi \xrightarrow{P} 0$ and the second term becomes negligible in the limit. This is no longer true when $k > 2$. To determine the limit distribution of (5.2.42) in that case, consider the joint distribution of

$$(5.2.43) \qquad \sqrt{n}\,(M_k' - \mu_k) \text{ and } \sqrt{n}\,(\bar{X} - \xi).$$

By the bivariate central limit theorem, (5.2.43) tends in law to the bivariate normal distribution with mean (0,0) and covariance structure

$$(5.2.44) \qquad \sigma_{11} = \mu_{2k} - \mu_k^2, \quad \sigma_{22} = \sigma^2 = \mu_2, \quad \sigma_{12} = \mu_{k+1}.$$

Since $M_{k-1}' \xrightarrow{P} \mu_{k-1}$, the joint limit distribution of

$$\sqrt{n}\,(M_k' - \mu_k) \text{ and } k\sqrt{n}\,(\bar{X} - \xi)\,M_{k-1}'$$

is therefore bivariate with mean (0,0) and covariance structure

$$(5.2.45) \qquad \sigma_{11} = \mu_{2k} - \mu_k^2, \quad \sigma_{22} = k^2\mu_{k-1}^2\mu_2, \quad \sigma_{12} = k\mu_{k-1}\mu_{k+1}.$$

Thus, finally, via (5.2.41) it is seen that

$$(5.2.46) \qquad \sqrt{n}\,(M_k - \mu_k) \to N\left(0, \tau^2\right),$$

where

$$(5.2.47) \qquad \tau^2 = \mu_{2k} - \mu_k^2 + k^2\mu_{k-1}^2\mu_2 - 2k\mu_{k-1}\mu_{k+1}.$$

$\square$

Example 5.2.8 Effect size. In a paired comparison experiment, let V_i and W_i denote the control and treatment responses within the ith pair. We assume that the differences $X_i = W_i - V_i$ are i.i.d. with mean ξ and variance σ^2 and we are interested in the *effect size* $\theta = \xi/\sigma$, defined by Cohen (1969). A natural estimator of θ is $\bar{X}/S$, where $S^2 = \sum (X_i - \bar{X})^2 / (n-1)$. To set approximate confidence limits for θ requires the asymptotic distribution of

$$(5.2.48) \qquad \sqrt{n}\left(\frac{\bar{X}}{S} - \frac{\xi}{\sigma}\right).$$

As starting point for obtaining the limit distribution of (5.2.48), consider the joint distribution of

$$\sqrt{n}\,(\bar{X} - \xi) \text{ and } \sqrt{n}\,(S^2 - \sigma^2),$$

which has the same limit distribution as

(5.2.49) $$\sqrt{n}\,(\bar{X} - \xi) \text{ and } \sqrt{n}\left[\frac{\sum (X_i - \xi)^2}{n} - \sigma^2\right].$$

By Theorem 5.2.1, the joint distribution of (5.2.49) tends to the bivariate normal distribution with mean $(0,0)$ and covariance structure

$$\sigma_{11} = \text{Var}\,(X_i) = \sigma^2, \quad \sigma_{22} = \text{Var}\,(X_i - \xi)^2 = \mu_4 - \sigma^4$$

and

$$\sigma_{12} = \text{Cov}\left(X_i, (X_i - \xi)^2\right) = \text{Cov}\left(X_i - \xi, (X_i - \xi)^2\right) = E\,(X_i - \xi)^3 = \mu_3.$$

We can now apply Corollary 5.2.1 with $U_n = \bar{X}$, $V_n = S^2$, and $f(u,v) = u/\sqrt{v}$. Then $\partial f/\partial u = 1/\sqrt{v}$ and $\partial f/\partial v = \dfrac{-u}{2v\sqrt{v}}$, and hence (5.2.48) has a normal limit distribution with mean 0 and with variance

(5.2.50)
$$\frac{1}{v}\sigma^2 - \frac{u}{v^2}\mu_3 + \frac{u^2}{4v^3}\,\text{Var}\,(X_i - \xi)^2\bigg|_{u=\xi,v=\sigma^2}$$
$$= 1 - \frac{\xi}{\sigma^4}\mu_3 + \frac{\xi^2}{4\sigma^6}\text{Var}\,(X_i - \xi)^2.$$

Consistent estimators of (5.2.50) which are needed to obtain confidence intervals for $\theta = \xi/\sigma$ can be obtained in the usual way .   □

Note: Since $\sigma_{12} = \mu_3 = 0$ when the distribution of the X's is symmetric about ξ, it follows that the two variables (5.2.49) are then asymptotically independent. For the special case of normal variables, this was already seen in Example 5.2.4.

Summary

1. The (non-degenerate) bivariate normal distribution is defined by its density. An example shows that a sequence of such distributions can have a degenerate limit which assigns probability 1 to a line, curve, or point in a plane.

2. The univariate CLT is generalized to the bivariate case and is applied to obtain some joint and simultaneous confidence sets for two parameters.

3. Taylor's theorem is generalized to the multivariate case and used to obtain a bivariate delta method. The latter result is used to obtain limit distributions for sample moments and for the effect size in a paired comparison experiment.

5.3   Some linear algebra

The present section summarizes some results from linear algebra needed for the rest of the chapter. We shall assume familiarity with the concepts of matrix and determinant and with the addition and multiplication of matrices.

Definition 5.3.1

(i) A set of k vectors $\underline{a}_i = (a_{i1}, \dots, a_{in})$ in R^n ($i = 1, \dots, k; k \leq n$) is *linearly independent* if

$$c_1 \underline{a}_1 + \cdots + c_k \underline{a}_k = \underline{0} \text{ implies } c_1 = \cdots = c_k = 0.$$

(ii) A $k \times n$ matrix $A = (a_{ij})$, $k \leq n$, is of maximal rank if the k vectors $\underline{a}_i = (a_{i1}, \dots, a_{in})$ formed by its rows are linearly independent; a square matrix of maximal rank is said to be *non-singular*.

Lemma 5.3.1

(i) A square matrix A is non-singular if and only if its determinant $|A| \neq 0$.

(ii) Let A be $k \times k$ and consider the linear transformation

$$(5.3.1) \qquad \underline{y} = A\underline{x} \text{ with } \underline{x} = \begin{pmatrix} x_1 \\ \vdots \\ x_k \end{pmatrix}, \ \underline{y} = \begin{pmatrix} y_1 \\ \vdots \\ y_k \end{pmatrix}$$

as a transformation from k-dimensional Euclidean space R^k to itself. Then A is non-singular if and only if the transformation (5.3.1) is $1:1$.

(For a proof of this and other results stated in this section without proof, see any book on linear algebra. A recent account particularly oriented toward statistical applications is Harville (1997).)

As an illustration of (i), note that (5.2.7) states that the 2×2 matrix

$$(5.3.2) \qquad A = \begin{pmatrix} a_1 & b_1 \\ a_2 & b_2 \end{pmatrix}$$

is non-singular.

Lemma 5.3.2 *If the $k \times k$ matrix A is non-singular, there exists a unique $k \times k$ matrix A^{-1}, the inverse of A, satisfying*

$$(5.3.3) \qquad AA^{-1} = I,$$

where I is the $k \times k$ identity matrix

(5.3.4)
$$
\begin{pmatrix}
1 & 0 & \cdot & \cdot & \cdot & 0 \\
0 & 1 & 0 & \cdot & \cdot & \cdot \\
 & & \cdot & \cdot & & \\
 & & \cdot & \cdot & & \cdot \\
 & & \cdot & & \cdot & 0 \\
0 & 0 & \cdot & \cdot & \cdot & 1
\end{pmatrix}.
$$

The matrix A^{-1} also satisfies

(5.3.5) $\qquad\qquad\qquad A^{-1}A = I.$

Example 5.3.1 Inverse of a 2×2 matrix. Let $\Delta = a_1 b_2 - a_2 b_1 \neq 0$ be the determinant of the matrix A given by (5.3.2). Then

(5.3.6) $\qquad\qquad A^{-1} = \begin{pmatrix} b_2/\Delta & -b_1/\Delta \\ -a_2/\Delta & a_1/\Delta \end{pmatrix},$

as is seen by forming the product AA^{-1}. $\square$

Example 5.3.2 Linear equations. The system of k linear equations in k unknowns

(5.3.7) $\qquad\qquad\qquad B\underline{x} = \underline{c},$

where B is a non-singular $k \times k$ matrix and $\underline{x}$ and $\underline{c}$ are $k \times 1$ column matrices, has the unique solution

(5.3.8) $\qquad\qquad\qquad \underline{x} = B^{-1}\underline{c}.$

If $\underline{c} = \underline{0}$, the unique solution of (5.3.7) is $\underline{x} = \underline{0}$. $\square$

Definition 5.3.2

(i) The *transpose* A' of a $k \times l$ matrix A with elements a_{ij} is the $l \times k$ matrix the (i,j)-th element of which is a_{ji}.

(ii) A square matrix A is said to be *symmetric* if $A' = A$, i.e., if $a_{ij} = a_{ji}$ for all i and j.

Elementary properties of transposed matrices are (Problem 3.1)

(5.3.9) $\qquad\qquad\qquad (AB)' = B'A'$

and

(5.3.10) $\qquad\qquad\qquad (A^{-1})' = (A')^{-1}.$

A sequence of matrices $A_n = \left(a_{ij}^{(n)}\right)$ is said to converge to a matrix $A = (a_{ij})$; in symbols, $A_n \to A$, if $a_{ij}^{(n)} \to a_{ij}$ for all i,j.

Lemma 5.3.3

(i) *If a sequence of non-singular matrices A_n converges to a non-singular matrix A, then $A_n^{-1} \to A^{-1}$.*

(ii) *If a sequence of random non-singular matrices A_n converges in probability to a constant non-singular matrix A (i.e., if $a_{ij}^{(n)} \overset{P}{\to} a_{ij}$ for all i, j), then $A_n^{-1} \overset{P}{\to} A^{-1}$.*

Proof. Part (i) follows from the fact, not proved here, that each element of A^{-1} is the ratio of two determinants which are sums and differences of products of the elements a_{ij}. Part (ii) is an immediate consequence of (i).∎

Definition 5.3.3 A $k \times k$ matrix $A = (a_{ij})$ is *positive definite* if the quadratic form

$$(5.3.11) \qquad \sum_{i=1}^{k} \sum_{j=1}^{k} a_{ij} u_i u_j > 0 \text{ for all } (u_1, \dots, u_k) \neq (0, \dots, 0),$$

and is *positive semi-definite* if (5.3.11) holds with $>$ replaced by $\geq$. The corresponding terms are also applied to the quadratic form itself.

Example 5.3.3 Covariance matrix. Let $(X_1, \dots, X_k)$ be a random vector with covariance matrix $\Sigma = (\sigma_{ij})$. Then for any constant vector $\underline{u}' = (u_1, \dots, u_k)$,

$$(5.3.12) \qquad \underline{u}' \Sigma \underline{u} = \sum \sum \sigma_{ij} u_i u_j = \mathrm{Var}\left(\sum u_i X_i\right) \geq 0,$$

so that Σ is positive semi-definite. Furthermore, $\mathrm{Var}\left(\sum u_i X_i\right) > 0$ unless

$$(5.3.13) \qquad \sum u_i X_i = \text{constant}$$

with probability 1. Thus, Σ is positive definite unless the joint distribution of $(X_1, \dots, X_k)$ assigns probability 1 to some hyperplane (5.3.13). □

Note: When considering a quadratic form,

$$\sum \sum a_{ij} u_i u_j,$$

we can assume without loss of generality that the matrix A is symmetric. (Otherwise put $b_{ij} = b_{ji} = (a_{ij} + a_{ji})/2$. Then B is symmetric and

$$\sum \sum b_{ij} u_i u_j = \sum \sum a_{ij} u_i u_j \text{ for all } (u_1, \dots, u_k).)$$

This symmetry assumption for quadratic forms will therefore be made from now on.

We shall next consider the behavior of a quadratic form

$$(5.3.14) \qquad Q = \sum\sum a_{ij}x_i x_j = \underline{x}'A\underline{x}; \quad \underline{x}' = (x_1, \dots, x_k)$$

under a linear transformation

$$(5.3.15) \qquad \underline{y} = B\underline{x},$$

where B is a non-singular $k \times k$ matrix. Since

$$(5.3.16) \qquad \underline{x}'A\underline{x} = \underline{y}'\left(B^{-1}\right)' AB^{-1}\underline{y},$$

it follows that in terms of the $\underline{y}$'s, Q is a quadratic form with matrix

$$(5.3.17) \qquad \left(B^{-1}\right)' AB^{-1}.$$

Theorem 5.3.1 *For any positive definite symmetric matrix A, there exists a non-singular matrix B such that*

$$(5.3.18) \qquad \left(B^{-1}\right)' AB^{-1} = I$$

and hence, if $\underline{y} = B\underline{x}$, such that

$$(5.3.19) \qquad \sum_{i=1}^{k}\sum_{j=1}^{k} a_{ij}x_i x_j = \sum_{i=1}^{k} y_i^2.$$

Example 5.3.4 Expectations and covariance matrices under non-singular linear transformations. Consider the linear transformation

$$(5.3.20) \qquad \underline{Y} = B\underline{X},$$

where $\underline{X}$ and $\underline{Y}$ are random column matrices with elements $X_1, \dots, X_k$ and $Y_1, \dots, Y_k$, and B is a non-singular $k \times k$ matrix of constants.

Theorem 5.3.2

(i) *If $\xi_i = E(X_i)$ and $\eta_i = E(Y_i)$, the expectation vectors $\underline{\xi}$ and $\underline{\eta}$ are related by*

$$(5.3.21) \qquad \underline{\eta} = B\underline{\xi}.$$

(ii) *The covariance matrices of $\underline{X}$ and $\underline{Y}$*

$$(5.3.22) \qquad \Sigma = E(X - \xi)(X - \xi)' \text{ and } T = E(Y - \eta)(Y - \eta)'$$

are related by

$$(5.3.23) \qquad T = B\Sigma B'.$$

Proof.

(i) By definition, $y_i = \sum b_{ij} x_j$ and hence

$$E(Y_i) = \sum b_{ij} \xi_j.$$

(ii) The proof is left to Problem 3.5.   □

Definition 5.3.4

(i) The length of a vector $\underline{x} = (x_1, \dots, x_k)$ is $\sqrt{\sum x_i^2} = \sqrt{\underline{x}'\underline{x}}$, and the inner product of two vectors $\underline{x}$ and $\underline{y}$ is $\sum x_i y_i = \underline{x}'\underline{y}$, where $\underline{x}$ and $\underline{y}$ are the column matrices with elements $x_1, \dots, x_k$ and $y_1, \dots, y_k$, respectively.

(ii) Two vectors $\underline{x}$ and $\underline{y}$ in R^k are said to be *orthogonal* if

(5.3.24) $$\underline{x}'\underline{y} = \sum x_i y_i = 0.$$

Note: The length of $\underline{x}$ defined by (i) is equal to the distance (defined by (5.1.1)) of $\underline{x}$ from the origin.

Definition 5.3.5 A $k \times k$ matrix A is said to be *orthogonal* if (a) its row vectors $\underline{a}_j$ are all of length 1 and (b) all pairs of row vectors $\underline{a}_i$, $\underline{a}_j$ $(i \neq j)$ are orthogonal.

Note: Some authors call such a matrix orthonormal and use the term orthogonal more broadly for any matrix satisfying (b).

It follows immediately from Definition 5.3.5 that A is orthogonal if and only if

(5.3.25) $$AA' = I \text{ or } A' = A^{-1}.$$

This shows further, by (5.3.5), that if A is orthogonal, so is A'. From (5.3.25), it follows that the determinant $|A|$ of an orthogonal matrix A satisfies $|A|^2 = 1$ and hence that

(5.3.26) $$|A| = \pm 1.$$

Lemma 5.3.4 *An orthogonal transformation* $\underline{y} = Q\underline{x}$ *of the vectors in* R^k *leaves the lengths of vectors* $\underline{x}$ *and the inner products of pairs* $\underline{x}, \underline{y}$ *unchanged.*

Proof. The inner product of $Q\underline{x}$ and $Q\underline{y}$ is

$$(Q\underline{x}, Q\underline{y}) = \underline{x}'Q'Q\underline{y} = \underline{x}'\underline{y}$$

and the result for length follows by putting $\underline{x} = \underline{y}$. ∎

It follows, in particular, that if $\underline{x}$ and $\underline{y}$ are orthogonal, so are $Q\underline{x}$ and $Q\underline{y}$.

Theorem 5.3.3 *For any symmetric matrix A there exists an orthogonal matrix Q such that*

$$(5.3.27) \qquad (Q^{-1})'AQ^{-1} = QAQ^{-1} \text{ is diagonal,}$$

that is, has all of its off-diagonal elements equal to 0.

Corollary 5.3.1 *If the diagonal matrix (5.3.27) is denoted by*

$$(5.3.28) \qquad \Lambda = \begin{pmatrix} \lambda_1 & & 0 \\ & \ddots & \\ 0 & & \lambda_k \end{pmatrix}$$

and if $\underline{y} = Q\underline{x}$, then

$$(5.3.29) \qquad \sum_{i=1}^{k}\sum_{j=1}^{k} a_{ij}x_ix_j = \sum_{i=1}^{k} \lambda_i y_i^2.$$

Note: The set of λ's in (5.3.28) and (5.3.29) is unique except for their order, and the λ's are in fact the k roots of the equation

$$(5.3.30) \qquad\qquad |A - \lambda I| = 0,$$

the *eigenvalues* of A. Here the symmetry of A implies that the roots $\lambda_1, \ldots, \lambda_k$ of the equation (5.3.30) are all real.

It is interesting to note that Theorem 5.3.1 is an easy consequence of Theorem 5.3.3. To see this, note first that if A is positive definite, it follows from (5.3.29) that all the λ's are positive since otherwise the quadratic form would be able to take on the value 0 or negative values. If we now let

$$D = \begin{pmatrix} 1/\sqrt{\lambda_1} & & 0 \\ & \ddots & \\ 0 & & 1/\sqrt{\lambda_k} \end{pmatrix},$$

we have

$$DQAQ'D' = I$$

and DQ will therefore serve for the matrix $(B^{-1})'$ of Theorem 5.3.1.

Theorem 5.3.4 Let $(X_1, \dots, X_k)$ be a random vector with covariance matrix Σ.

(i) There exists an orthogonal transformation $\underline{Y} = Q\underline{X}$ such that the components $Y_1, \dots, Y_k$ of Y are uncorrelated.

(ii) If Σ is non-singular, there exists a non-singular transformation $\underline{Y} = B\underline{X}$ such that the random variables $Y_1, \dots, Y_k$ are uncorrelated and all have variance 1.

Proof. These results follow immediately from (5.3.23) and Theorems 5.3.1 and 5.3.3. ∎

The covariance matrix of a random vector is the k-dimensional generalization of the variance of a random variable. It is sometimes useful to have a corresponding generalization of the standard deviation. For this purpose, we now define the square root $A^{1/2}$ of a positive definite symmetric matrix A as the unique symmetric, positive definite matrix C satisfying

(5.3.31) $$C \cdot C = A.$$

To see that such a matrix exists, consider first the case that A is the positive definite diagonal matrix with diagonal elements $\lambda_1, \dots, \lambda_k$, all of which are positive. Then if D is the diagonal matrix with diagonal elements $\sqrt{\lambda_1}, \dots, \sqrt{\lambda_k}$, clearly $D \cdot D = \Lambda$. If A is any positive definite, symmetric matrix, there exists by Theorem 5.3.3 an orthogonal matrix Q such that QAQ^{-1} is diagonal, say $QAQ^{-1} = \Lambda$, and hence

$$A = Q^{-1}\Lambda Q = Q'\Lambda Q.$$

Substituting $D \cdot D$ for Λ, we therefore have

$$A = Q'DDQ = Q'DQ \cdot Q'DQ.$$

Since $C = Q'DQ$ is symmetric and positive definite, it satisfies the requirement for $C = A^{1/2}$.

Let us finally show that there can only be one symmetric positive definite matrix C satisfying (5.3.31). If C_1 and C_2 are two such matrices, they satisfy $C_1 C_1' = C_2 C_2'$ and hence

$$C_1^{-1} C_2 C_2' (C_1')^{-1} = I.$$

It follows that $C_1^{-1} C_2$ is positive definite, symmetric, and orthogonal. This implies (Problem 3.9(ii)) that $C_1^{-1} C_2 = I$ and hence $C_1 = C_2$.

An important application of matrices occurs when changing variables in multiple integrals and probability densities. Let

$$(5.3.32) \qquad \underline{g}(\underline{x}) = (g_1(\underline{x}), \dots, g_n(\underline{x})),$$

where the g_i are real-valued continuous functions defined over a region S in R^n. Suppose $\underline{g}$ defines a 1:1 mapping of S onto $\underline{g}(S)$, and that the g_i have continuous partial derivatives $\partial g_j/\partial x_i$ in S. For any point $\underline{y}$ in $\underline{g}(S)$, let the unique point $\underline{x}$ for which $\underline{g}(\underline{x}) = \underline{y}$ be given by

$$(5.3.33) \qquad x_i = h_i(y_1, \dots, y_n).$$

Finally, let

$$(5.3.34) \qquad \frac{\partial(y_1, \dots, y_n)}{\partial(x_1, \dots, x_n)} = \left(\frac{\partial y_j}{\partial x_i} \right)$$

be the $n \times n$ matrix whose (i,j)-th element is $\partial y_j/\partial x_i$. The determinant J of the matrix (5.3.34)

$$(5.3.35) \qquad J = \text{Determinant of } \left(\frac{\partial g_j}{\partial x_i} \right)$$

is called the *Jacobian* of the transformation (5.3.32).

Example 5.3.5 Linear transformations. Let

$$(5.3.36) \qquad y_j = \sum_{i=1}^{n} a_{ij} x_i \quad (j = 1, \dots, n).$$

Then $\partial y_j/\partial x_i = a_{ij}$, and the Jacobian of the transformation (5.3.36) is therefore the determinant of the matrix (a_{ij}). In particular, if the transformation (5.3.36) is orthogonal, it follows from (5.3.26) that $J = \pm 1$. $\qquad \square$

Suppose next that $\underline{X} = (X_1, \dots, X_n)$ is a random vector with probability density $p_{\underline{X}}(x_1, \dots, x_n)$ defined over a region S in R^n, and that $\underline{g}$ satisfies the assumptions made for (5.3.32). Then the probability density of $\underline{Y} = \underline{g}(\underline{X})$ is given by

$$(5.3.37) \qquad \begin{aligned} p_{\underline{Y}}(y_1, \dots, y_n) &= p_{\underline{X}}(x_1, \dots, x_n) \cdot |J|^{-1} \\ &= p_{\underline{X}}[h_1(\underline{y}), \dots, h_n(\underline{y})] \cdot |J|^{-1}, \end{aligned}$$

where $|J|$ denotes the absolute value of J. (See, for example, Cramér (1946, Section 22.2).)

Example 5.3.6 Joint distribution of sample mean and variance.
Let $X_1, \ldots, X_n$ be i.i.d. $N\left(0, \sigma^2\right)$ so that

(5.3.38) $$p_X\left(x_1, \ldots, x_n\right) = \frac{1}{\left(\sqrt{2\pi}\sigma\right)^n} e^{-\frac{1}{2\sigma^2}\sum_{i=1}^{n} x_i^2}.$$

To obtain the distribution of the sample mean $\bar{X}$, make an orthogonal transformation

(5.3.39) $$y_j = \sum_{i=1}^{n} a_{ij} x_i$$

such that $y_1 = \sqrt{n}\bar{x}$. The existence of such a transformation can be shown by the construction used in Problems 1.15 and 1.16, which is known as the *Gram-Schmidt orthogonalization process*. It is interesting to note that in the following derivation we do not require explicit knowledge of this orthogonal transformation. It follows from (5.3.37) and the fact that $|J| = 1$ that

$$p_Y\left(y_1, \ldots, y_n\right) = \frac{1}{\left(\sqrt{2\pi}\sigma\right)^n} e^{-\frac{1}{2\sigma^2}\sum_{i=1}^{n} y_i^2}$$

since $\sum x_i^2 = \sum y_i^2$. Thus the Y's are again independent normal $N\left(0, \sigma^2\right)$, and, in particular, $Y_1 = \sqrt{n}\bar{X}$ is distributed as $N\left(0, \sigma^2\right)$. Furthermore,

$$\sum\left(X_i - \bar{X}\right)^2 = \sum X_i^2 - n\bar{X}^2 = \sum_{i=1}^{n} Y_i^2 - Y_1^2,$$

so that $\sum\left(X_i - \bar{X}\right)^2$ is distributed as $\sigma^2 \chi_{n-1}^2$ and is independent of Y_1 and hence of $\bar{X}$.   □

Summary

1. A square matrix is non-singular if its rows are linearly independent. A non-singular matrix has a unique inverse A^{-1} satisfying $AA^{-1} = I$.

2. A $k \times k$ matrix $A = (a_{ij})$ is positive definite if the quadratic form $\sum\sum a_{ij} u_i u_j$ is positive for all $(u_1, \ldots, u_k) = (0, \ldots, 0)$. The covariance matrix of a random vector $(X_1, \ldots, X_k)$ is positive definite unless $(X_1, \ldots, X_k)$ lies in some hyperplane with probability 1. The square root $A^{1/2}$ of a positive definite symmetric matrix A is the unique positive definite symmetric matrix C satisfying $C \cdot C = A$.

3. Any positive definite quadratic form in $(x_1, \ldots, x_k)$ can be reduced to $\sum_{i=1}^{k} y_i^2$ by a non-singular linear transformation $\underline{y} = B\underline{x}$.

4. A $k \times k$ matrix $A = (a_{ij})$ is orthogonal if each of its row vectors is of length 1 and if each pair of row vectors satisfies the orthogonally condition $\sum_{\nu=1}^{k} a_{i\nu} a_{j\nu} = 0$, i.e., if it satisfies $AA' = 1$. Any quadratic form $\sum a_{ij} x_i x_j$ can be reduced to $\sum \lambda_i y_i^2$ by an orthogonal transformation $y = Qx$, where the λ's are the roots of the equation $|A - \lambda I| = 0$.

5. The Jacobian of a transformation $y = g(x)$ from R^n to R^n is the determinant of the matrix of partial derivatives $(\partial y_i / \partial x_j)$. The Jacobian of a linear transformation is a constant; that of an orthogonal transformation is ± 1.

6. The probability density of $Y = g(X)$ is the probability density of X (expressed in terms of Y) multiplied by the absolute value of the reciprocal of the Jacobian of the transformation.

5.4   The multivariate normal distribution

In Section 5.2, attention was restricted to the bivariate normal distribution and the associated bivariate central limit theorem. Let us now consider the general k-variate normal distribution, the density of which is of the form

$$(5.4.1) \qquad Ce^{-\frac{1}{2} \sum_{i=1}^{k} \sum_{j=1}^{k} a_{ij}(x_i - \xi_i)(x_j - \xi_j)},$$

where without loss of generality the matrix $A = (a_{ij})$ will be taken to be symmetric. In addition, we shall assume A to be positive definite. To see why this restriction is needed, consider the orthogonal transformation

$$(5.4.2) \qquad y = Qx, \quad \eta = Q\xi,$$

which according to (5.3.27)–(5.3.29) (with a change of notation) results in

$$\sum \sum a_{ij} (x_i - \xi_i)(x_j - \xi_j) = \sum_{i=1}^{k} \lambda_i (y_i - \eta_i)^2.$$

By (5.3.37), the density of $Y = QX$ is then

$$(5.4.3) \qquad Ce^{-\frac{1}{2} \sum_{i=1}^{k} \lambda_i (y_i - \eta_i)^2}$$

since $|J| = 1$. Now the integral over R^k of (5.4.3), and therefore of (5.4.1), will be finite only if all the λ's are positive and hence if A is positive definite. In any other case, (5.4.1) therefore cannot be a probability density.

The reduction to (5.4.3) also enables us to evaluate the constant C. Since

$$\sqrt{\frac{\lambda}{2\pi}} \int_{-\infty}^{\infty} e^{-\frac{1}{2}\lambda(y-\eta)^2} \, dy = 1,$$

it follows from (5.4.3) that

$$C = \frac{1}{\left(\sqrt{2\pi}\right)^k} \sqrt{\Pi \lambda_i}.$$

By taking determinants on both sides of the equation

(5.4.4) $QAQ' = \Lambda,$

where Λ is the diagonal matrix with diagonal elements $\lambda_1, \ldots, \lambda_k$, we see that $|A| = \Pi \lambda_i$ and hence that

(5.4.5) $C = \dfrac{\sqrt{|A|}}{\left(\sqrt{2\pi}\right)^k}.$

Since $E(Y_i) = \eta_i$, it follows from (5.4.2) that

(5.4.6) $E(X_i) = \xi_i.$

From (5.4.3), it is seen that the covariance matrix of $\underline{Y}$ is Λ^{-1} and then from (5.3.23) that the covariance matrix Σ of $\underline{X} = Q^{-1}\underline{Y}$ is $Q^{-1}\Lambda^{-1}Q$, which by (5.4.4) implies that the covariance matrix of $(X_1, \ldots, X_k)$ is

(5.4.7) $\Sigma = A^{-1}.$

We shall denote the multivariate normal distribution given by (5.4.1) and (5.4.5) with $A = \Sigma^{-1}$ by $N(\xi, \Sigma)$.

Theorem 5.4.1 *If $\underline{X} = (X_1, \ldots, X_k)$ is distributed according to $N(\xi, \Sigma)$ and if $\underline{Y} = B\underline{X}$, where B is an $r \times k$ matrix $(r \le k)$ of maximal rank, then $\underline{Y}$ has the multivariate normal distribution $N(\underline{\eta}, B\Sigma B')$ with $\underline{\eta} = B\underline{\xi}$.*

Proof. We shall here prove only the two most important special cases: (i) $r = k$ and (ii) $r = 1$. For a proof of the general result see Problem 4.2(ii). Different proofs can be found in Anderson (1984) and Tong (1990).

(i) $\underline{r = k}$.

By (5.3.17), the exponent in (5.4.1) satisfies

$$(\underline{x} - \underline{\xi})' A (\underline{x} - \underline{\xi}) = (\underline{y} - \underline{\eta})' (B^{-1})' AB^{-1} (\underline{y} - \underline{\eta})$$

and by Example 5.3.5, the Jacobian of the transformation $y = B\underline{x}$ is a constant. It follows that $\underline{Y}$ is multivariate normal with mean η and covariance matrix

$$T = \left[\left(B^{-1} \right)' A B^{-1} \right]^{-1} = B A^{-1} B' = B \Sigma B',$$

as was to be proved.

(ii) $\underline{r = 1}$

This part of the theorem states that any linear combination

$$Y = c_1 X_1 + \cdots + c_k X_k \quad \text{(at least one } c \neq 0\text{)}$$

is normal with

$$E(Y) = \sum c_i \xi_i \text{ and } \text{Var}(Y) = \sum \sum c_i c_j \sigma_{ij}.$$

The expectation and variance of Y are obvious. We shall prove normality with the help of Theorem 5.3.1. Let B be a matrix satisfying condition (5.3.18) of that theorem. Then QB also satisfies (5.3.18) for any orthogonal matrix Q. Now there exists a rotation in the plane containing the vectors $(b_{11}, \ldots, b_{1k})$ and $(c_1, \ldots, c_k)$ which takes the vector $(b_{11}, \ldots, b_{1k})$ into a vector proportional to $(c_1, \ldots, c_k)$. This rotation is represented by an orthogonal matrix Q satisfying

$$Q(b_{11} \cdots b_{1k})' = (dc_1, \ldots, dc_k) \text{ for some } d \neq 0.$$

Let $\underline{Z} = QB\underline{X}$ and $\underline{\zeta} = QB\underline{\xi}$. Then

$$\left(\underline{x} - \underline{\xi} \right)' A \left(\underline{x} - \underline{\xi} \right) = \left(\underline{z} - \underline{\zeta} \right)' \left[(QB)^{-1} \right]' A (QB)^{-1} \left(\underline{z} - \underline{\zeta} \right)$$

and, by (5.3.18),

$$\left[(QB)^{-1} \right]' A (QB)^{-1} = I,$$

so that

$$\left(\underline{x} - \underline{\xi} \right)' A \left(\underline{x} - \underline{\xi} \right) = \sum_{i=1}^{k} (z_i - \zeta_i)^2 .$$

This shows that

$$Z_1 = \sum_{i=1}^{k} dc_i X_i$$

is distributed as $N(0,1)$ and hence proves the normality of $Y = \sum_{i=1}^{k} c_i X_i.\blacksquare$

Corollary 5.4.1 *If* $(X_1, \ldots, X_k)$ *is distributed as* $N\left(\underline{\xi}, \Sigma\right)$, *then*

(i) X_i *is distributed as* $N\left(\xi_i, \sigma_{ii}\right)$

and, more generally,

(ii) *any subset* $(X_{i_1}, \ldots, X_{i_r})$ *has an r-variate normal distribution with mean* $(\xi_{i_1}, \ldots, \xi_{i_r})$ *and with covariances* $\mathrm{Cov}\,(X_i, X_j) = \sigma_{ij}$.

Proof. (i) follows from the case $r = 1$ by letting $c_1 = 1$, $c_2 = \cdots = c_k = 0$ and the general case (ii) follows analogously. ∎

The following theorem establishes two properties of the multivariate normal distribution that are of great importance for statistical applications.

Theorem 5.4.2 *Let* $(X_1, \ldots, X_k)$ *be distributed according to (5.4.1).*

(i) *The linear function*

(5.4.8) $$\underline{Y} = A^{1/2}\left(\underline{X} - \underline{\xi}\right)$$

has the k-variate normal distribution $N(0, I)$, *that is, the joint distribution of k independent* $N(0, 1)$ *variables.*

(ii) *The quadratic form*

(5.4.9) $$\sum\sum a_{ij}\left(X_i - \xi_i\right)\left(X_j - \xi_j\right) = \left(\underline{X} - \underline{\xi}\right)' A \left(\underline{X} - \underline{\xi}\right)$$

is distributed as χ^2 *with k degrees of freedom.*

Proof.

(i) By Theorem 5.4.1, the linear function (5.4.8) is distributed according to the k-variable normal distribution with mean 0 and covariance matrix

$$A^{1/2}\Sigma A^{1/2} = A^{1/2}A^{-1}A^{1/2} = A^{1/2}\left(A^{1/2}\right)^{-1}\left(A^{1/2}\right)^{-1}A^{1/2} = I$$

where the second equality follows from Problem 3.13.

(ii) The quadratic form (5.4.9) is equal to

$$\left[A^{1/2}\left(\underline{X} - \underline{\xi}\right)\right]'\left[A^{1/2}\left(\underline{X} - \underline{\xi}\right)\right] = \underline{Y}'\underline{Y} = \sum_{i=1}^{k} Y_i^2.$$

Since the Y's are independent $N(0, 1)$ by (i), the result follows. (For an alternative proof not depending on (i), see Problem 4.2.) ∎

Theorem 5.4.2 provides the necessary information for calculating the confidence coefficient of confidence sets and the null-distribution of test statistics to be considered in the next section. Calculation of the power of the tests requires the following extension.

Theorem 5.4.3 *Let $(X_1, \ldots, X_k)$ be distributed according to (5.4.1).*

(i) *The linear function*

$$A^{1/2}\left(\underline{X} - \eta\right)$$

has the k-variable normal distribution $N(A^{1/2}[\eta - \underline{\xi}], \underline{\tau})$, that is, the joint distribution of k independent normal variables with means $(A^{1/2}(\eta_1 - \xi_1), \ldots, A^{1/2}(\eta_k - \xi_k))$ and each with unit variance.

(ii) *The quadratic form*

$$\sum\sum a_{ij}\left(X_i - \eta_i\right)\left(X_j - \eta_j\right) = \left(\underline{X} - \eta\right)' A \left(\underline{X} - \eta\right)$$

is distributed as $\sum_{i=1}^{k}(Y_i + c_i)^2$ where $Y_1, \ldots, Y_k$ are independent $N(0,1)$ and $\sum c_i^2 = (\eta - \underline{\xi})' A (\eta - \underline{\xi})$. This distribution is the non-central χ^2-distribution with non-centrality parameter $\sum c_i^2$.

The proof is exactly analogous to that of Theorem 5.4.2 (Problem 4.3).

Theorem 5.4.4 Multivariate central limit theorem. *Let $\underline{X}^{(j)} = \left(X_1^{(j)}, \ldots, X_k^{(j)}\right), j = 1, \ldots, n$, be n i.i.d. k-variable row vectors with mean $\underline{\xi} = (\xi_1, \ldots, \xi_k)$ and positive definite covariance matrix Σ. If*

(5.4.10) $$\bar{X}_i = \frac{1}{n}\left[X_i^{(1)} + \cdots + X_i^{(n)}\right],$$

then

(5.4.11) $$\left(\sqrt{n}\left(\bar{X}_1 - \xi_1\right), \ldots, \sqrt{n}\left(\bar{X}_k - \xi_k\right)\right) \overset{L}{\to} N(0, \Sigma).$$

Proof. Let $\underline{Y}$ be a k-variate random vector with distribution $N(0, \Sigma)$. In order to prove that the left side of (5.4.11) tends in law to $\underline{Y}$, it is by Theorem 5.1.8 enough to show that

(5.4.12) $$\sqrt{n}\sum c_i\left(\bar{X}_i - \xi_i\right) \overset{L}{\to} N(0, \underline{c}'\Sigma\underline{c})$$

for every constant vector $\underline{c} = (c_1, \ldots, c_k)$. The left side of (5.4.12) is equal to $\sqrt{n}\sum_{j=1}^{n}\sum_{i=1}^{k} c_i\left(X_i^{(j)} - \xi_i\right)/n$ and the result now follows from the univariate central limit theorem.   ∎

Example 5.4.1 Multinomial. Consider n multinomial trials, that is, independent trials, each with possible outcomes $O_1, \ldots, O_{k+1}$ having probabilities $p_1, \ldots, p_{k+1}$ $\left(\sum p_i = 1 \right)$, the same for each trial. Let $X_i^{(\nu)} = 1$ if the νth trial results in outcome O_i and $= 0$ otherwise, so that $Y_i = \sum_{\nu=1}^{n} X_i^{(\nu)}$ is the total number of trials resulting in outcome O_i. Since $Y_1 + \cdots + Y_{k+1} = n$, we shall restrict attention to the variables $Y_1, \ldots, Y_k$. In the notation of Theorem 5.4.4, $\bar{X}_i = Y_i/n$ and $\xi_i = E\left(X_i^{(\nu)} \right) = p_i$ so that

$$(5.4.13) \qquad \left(\sqrt{n} \left(\frac{Y_1}{n} - p_1 \right), \ldots, \sqrt{n} \left(\frac{Y_k}{n} - p_k \right) \right) \xrightarrow{L} N(0, \Sigma),$$

where $\Sigma = (\sigma_{ij})$ is the covariance matrix of $\left(X_1^{(1)}, \ldots, X_k^{(1)} \right)$ given by (Problem 4.6(i))

$$(5.4.14) \qquad \sigma_{ij} = \begin{cases} p_i(1 - p_i) & \text{if } j = i \\ -p_i p_j & \text{if } j \neq i. \end{cases}$$

The limiting density of (5.4.13) is therefore given by (5.4.1) with $\xi_1 = \cdots = \xi_k = 0$ and $A = \Sigma^{-1}$. It is easily checked (Problem 4.6(ii)) that $A = (a_{ij})$ is given by

$$(5.4.15) \qquad a_{ij} = \begin{cases} \dfrac{1}{p_i} + \dfrac{1}{p_{k+1}} & \text{if } j = i \\[2mm] \dfrac{1}{p_{k+1}} & \text{if } j \neq i. \end{cases}$$

$\square$

Example 5.4.2 Quantiles. The asymptotic distribution of the median, which was established in Example 2.4.9 and was extended to any quantile in Problem 4.8 of Chapter 2, generalizes to the joint distribution of several quantiles.

Theorem 5.4.5 *Let $X_1, \ldots, X_n$ be i.i.d. according to a distribution with cdf F having probability density f. For $0 < \lambda_1 < \cdots < \lambda_r < 1$, let $F(\xi_i) = \lambda_i$, and suppose that f is continuous and positive at the r points $\xi_1, \ldots, \xi_r$. Then if $n_1 < \cdots < n_r < n$ are such that*

$$(5.4.16) \qquad \frac{n_i}{n} = \lambda_i + o\left(\frac{1}{\sqrt{n_i}} \right), \quad i = 1, \ldots, r,$$

the joint distribution of

$$\sqrt{n} \left(X_{(n_1)} - \xi_1 \right), \ldots, \sqrt{n} \left(X_{(n_r)} - \xi_r \right)$$

is asymptotically normal with mean zero and covariance matrix $\Sigma = (\sigma_{ij})$
given by

(5.4.17)
$$\sigma_{ij} = \frac{\lambda_i \left(1 - \lambda_j\right)}{f\left(\xi_i\right) f\left(\xi_j\right)} \ \text{for all } i \le j.$$

This result can be proved, although we shall not do so here, by the method of Example 2.4.8 and a multivariate Berry-Essen theorem (for such a theorem, see, for example, Götze (1991)). An alternative proof can be found in David (1981).   □

We turn next to the k-variate version of the delta method, which was stated for the case $k = 2$ in Theorem 5.2.3 and for $k = 1$ in Theorem 2.5.2.

Theorem 5.4.6 *Suppose that*

(5.4.18)
$$\left(\sqrt{n}\left(Y_1^{(n)} - \eta_1\right), \dots, \sqrt{n}\left(Y_k^{(n)} - \eta_k\right)\right) \xrightarrow{L} N\left(0, \Sigma\right).$$

Let $f_i \ (i = 1, \dots, k)$ *be* k *real-valued functions of* k *variables for each of which the expansion (5.2.28) is valid at the point* $(\eta_1, \dots, \eta_k)$. *Then the joint distribution of the* k *variables*

(5.4.19)
$$\sqrt{n}\left[f_i\left(Y_1^{(n)}, \dots, Y_k^{(n)}\right) - f_i\left(\eta_1, \dots, \eta_k\right)\right], \quad i = 1, \dots, k,$$

tends in law to the k-*variate normal distribution with mean 0 and with covariance matrix* $T = (\tau_{ij})$ *given by*

(5.4.20)
$$\tau_{ij} = \sum_{s=1}^{k} \sum_{t=1}^{k} \sigma_{st} \left. \frac{\partial f_i}{\partial y_s} \cdot \frac{\partial f_j}{\partial y_t} \right|_{y = \eta}$$

provided the Jacobian matrix with (i, s)-*th element equal to*

(5.4.21)
$$\left. \frac{\partial f_i}{\partial y_s} \right|_{y = \eta}$$

is non-singular.

The proof is exactly analogous to that of the bivariate result.

Corollary 5.4.2 *Under the assumptions of Theorem 5.4.6, the distribution of*

(5.4.22)
$$\sqrt{n}\left[f_1\left(Y_1^{(n)}, \dots, Y_k^{(n)}\right) - f_1\left(\eta_1, \dots, \eta_k\right)\right]$$

tends in law to $N\left(0, \tau_{11}\right)$, *where* τ_{11} *is given by (5.4.20).*

Example 5.4.3 Correlation coefficient. Let $(X_1, Y_1), \ldots, (X_n, Y_n)$ be i.i.d. according to some bivariate distribution with means $E(X) = \xi$, $E(Y) = \eta$, variances $\sigma^2 = \text{Var} X$, $\tau^2 = \text{Var} Y$, and correlation coefficient ρ. Let

$$(5.4.23) \qquad S_X^2 = \frac{1}{n} \sum (X_i - \bar{X})^2, \quad S_Y^2 = \frac{1}{n} \sum (Y_i - \bar{Y})^2$$

and let

$$(5.4.24) \qquad R_n = \frac{\frac{1}{n} \sum (X_i - \bar{X})(Y_i - \bar{Y})}{S_X S_Y}$$

denote the sample correlation coefficient. We shall assume that the moments

$$(5.4.25) \qquad \mu_{ij} = E\left[(X - \xi)^i (Y - \mu)^j\right]$$

are finite for all (i, j) with $i + j \leq 4$, and shall use Corollary 5.4.2 to determine the limit distribution of

$$(5.4.26) \qquad \sqrt{n}(R_n - \rho).$$

(For $\rho = 0$, this limit was obtained in Example 5.2.6.)

We begin by recalling from (5.2.20) that

$$\sqrt{n} R_n = \frac{\frac{\sqrt{n}}{n} \sum (X_i - \xi)(Y_i - \eta)}{S_X S_Y} - \frac{\sqrt{n}(\bar{X} - \xi)(\bar{Y} - \eta)}{S_X S_Y}$$

and that the second term tends in probability to 0 since

$$\sqrt{n}(\bar{X} - \xi)(\bar{Y} - \eta) = \frac{1}{\sqrt{n}}\left[\sqrt{n}(\bar{X} - \xi)\sqrt{n}(\bar{Y} - \eta)\right]$$

and the term in square brackets is bounded in probability. The limit distribution of (5.4.26) is therefore the same as that of

$$(5.4.27) \qquad \sqrt{n}(R'_n - \rho),$$

where

$$(5.4.28) \qquad R'_n = \frac{1}{n} \sum (X_i - \xi)(Y_i - \eta) / S_X S_Y.$$

To determine the limit distribution of (5.4.27), we shall apply Corollary 5.4.2 with

$$(5.4.29) \qquad Y_1^{(n)} = \frac{1}{n} \sum (X_i - \xi)(Y_i - \eta), \quad Y_2^{(n)} = S_X^2, \quad Y_3^{(n)} = S_Y^2,$$

and

(5.4.30) $$f_1(u, v, w) = \frac{u}{\sqrt{vw}}.$$

From the argument leading to (5.4.27), the joint limit distribution of

(5.4.31) $$\sqrt{n}\left[Y_1^{(n)} - E(XY)\right], \quad \sqrt{n}\left[Y_2^{(n)} - \sigma^2\right], \quad \sqrt{n}\left[Y_3^{(n)} - \tau^2\right]$$

is seen to be unchanged if (5.4.29) is replaced by

(5.4.32)
$$Y_1^{(n)} = \frac{1}{n}\sum(X_i - \xi)(Y_i - \eta),$$
$$Y_2^{(n)} = \frac{1}{n}\sum(X_i - \xi)^2, \quad Y_3^{(n)} = \frac{1}{n}\sum(Y_i - \eta)^2$$

and hence, by Theorem 5.4.4, is $N(0, \Sigma)$, where the elements of σ_{ij} of Σ are

(5.4.33)
$$\sigma_{11} = \mathrm{Var}\left[(X - \xi)(Y - \eta)\right] = E\left[(X - \xi)^2(Y - \eta)^2 - \rho^2\sigma^2\tau^2\right],$$

$$\sigma_{12} = \sigma_{21} = \mathrm{Cov}\left[(X - \xi)(Y - \eta), (X - \xi)^2\right] = E\left[(X - \xi)^3(Y - \eta) - \rho\sigma^3\tau\right],$$

$$\sigma_{22} = \mathrm{Var}\,(X - \xi)^2 = E\,(X - \xi)^4 - \sigma^4,$$

$$\sigma_{13} = \sigma_{31} = \mathrm{Cov}\left[(X - \xi)(Y - \eta), (Y - \eta)^2\right] = E\left[(X - \xi)(Y - \eta)^3\right] - \rho\sigma\tau^3,$$

$$\sigma_{23} = \sigma_{32} = \mathrm{Cov}\left[(X - \xi)^2, (Y - \eta)^2\right] = E\left[(X - \xi)^2(Y - \eta)^2\right] - \sigma^2\tau^2,$$

$$\sigma_{33} = \mathrm{Var}\,(Y - \eta)^2 = E\,(Y - \eta)^4 - \tau^4.$$

It now follows from Corollary 5.4.2 that

(5.4.34) $$\sqrt{n}\left(\frac{\frac{1}{n}\sum(X_i - \xi_i)(Y_i - \eta_i)}{S_X S_Y} - \rho\right) \xrightarrow{L} N(0, \gamma^2)$$

with

(5.4.35)
$$\begin{aligned}
\gamma^2 =\ & \left(\frac{\partial f}{\partial u}\right)^2\sigma_{11} + \frac{\partial f}{\partial u}\frac{\partial f}{\partial v}\sigma_{12} + \frac{\partial f}{\partial u}\frac{\partial f}{\partial w}\sigma_{13} \\
& + \frac{\partial f}{\partial u}\frac{\partial f}{\partial v}\sigma_{12} + \left(\frac{\partial f}{\partial v}\right)^2\sigma_{22} + \frac{\partial f}{\partial v}\frac{\partial f}{\partial w}\sigma_{23} \\
& + \frac{\partial f}{\partial u}\frac{\partial f}{\partial w}\sigma_{13} + \frac{\partial f}{\partial v}\frac{\partial f}{\partial w}\sigma_{23} + \left(\frac{\partial f}{\partial w}\right)^2\sigma_{33},
\end{aligned}$$

which in matrix notation can be written as

(5.4.36) $$\left(\frac{\partial f}{\partial u}\ \frac{\partial f}{\partial v}\ \frac{\partial f}{\partial w}\right)\Sigma\left(\begin{matrix}\partial f/\partial u \\ \partial f/\partial v \\ \partial f/\partial w\end{matrix}\right).$$

Here

$$(5.4.37) \qquad \frac{\partial f}{\partial u} = \frac{1}{\sqrt{vw}}, \quad \frac{\partial f}{\partial v} = -\frac{1}{2}\frac{u}{\sqrt{v^3 w}}, \quad \frac{\partial f}{\partial w} = -\frac{1}{2}\frac{u}{\sqrt{vw^3}}$$

are evaluated at

$$(5.4.38) \qquad u = \rho\sigma\tau, \ v = \sigma^2, \ w = \tau^2.$$

$\square$

Example 5.4.4 The normal correlation coefficient. Of particular interest is the distribution of the correlation coefficient in the case that the bivariate distribution is normal. Evaluation of γ^2 in this situation requires knowledge of the moments (5.4.33). Since the distribution of R_n is independent of ξ, η, σ, and τ, we can assume without loss of generality that $\xi = \eta = 0$ and $\sigma = \tau = 1$. The moments are then functions of ρ only and are given by (see Problem 4.9 or, for example, Lehmann and Casella (1998, p. 68))

$$(5.4.39) \qquad \sigma_{11} = 1 + \rho^2, \ \sigma_{12} = \sigma_{13} = 2\rho, \ \sigma_{22} = \sigma_{33} = 2, \ \sigma_{23} = 2\rho^2$$

and (5.4.35) reduces to (Problem 4.10)

$$(5.4.40) \qquad \gamma^2 = \left(1 - \rho^2\right)^2.$$

Under the assumption of normality we can obtain a variance-stabilizing transformation for R_n by the method of Chapter 2, Section 5. By (5.4.40), we set

$$f'(\rho) = \frac{1}{1 - \rho^2} = \frac{1}{2}\left[\frac{1}{1 - \rho} + \frac{1}{1 + \rho}\right].$$

to find

$$(5.4.41) \qquad f(\rho) = \frac{1}{2}\log\frac{1 + \rho}{1 - \rho}.$$

It follows that

$$(5.4.42) \qquad \sqrt{n}\left[f\left(R_n\right) - f(\rho)\right] \xrightarrow{L} N\left(0, 1\right).$$

The transformation (5.4.41) is called Fisher's z-transformation.[*]

The limit result (5.4.42) can be used to obtain approximate confidence intervals and tests for ρ (Problem 4.7). Since, in the normal case, $\rho = 0$

---

[*]For a more detailed discussion of this transformation with references, see Mudholkar (1983) and Stuart and Ord, Vol. 1 (1987). Much additional material on the distribution of R_n is given in Chapter 32 of Johnson, Kotz, and Balakrishnan (1995).

is equivalent to independence of X and Y, the test of $H : \rho = 0$ also serves as a test of independence. If normality cannot be assumed, one can obtain asymptotic inferences for ρ from (5.4.34) where it is then necessary to replace the covariances σ_{ij} given by (5.4.33) by consistent estimators. Such an approach for the general k-variate case is discussed, for example, by Steiger and Hakstian (1982). $\qquad\square$

Summary

1. The non-singular multivariate distribution $N\left(\underline{\xi},\Sigma\right)$ is defined by its density (5.4.1); its expectation vector and covariance matrix are shown to be $\underline{\xi}$ and Σ, respectively.

2. If $\underline{X} = (X_1,\dots,X_k)$ is distributed as $N\left(\underline{\xi},\Sigma\right)$ and B is an $r \times k\,(r \leq k)$ matrix of constants of maximal rank r, then $\underline{Y} = B\underline{X}$ is distributed as $N\left(B\underline{\xi},B\Sigma B'\right)$. In particular, X_i is distributed as $N\left(\xi_i,\sigma_{ii}\right)$, and, more generally, any subset of the X's has a normal distribution.

3. If $\underline{X} = (X_1,\dots,X_k)$ is distributed as $N\left(\underline{\xi},\Sigma\right)$, then the quadratic form $\left(\underline{X}-\underline{\xi}\right)' A \left(\underline{X}-\underline{\xi}\right)$ with $A = \Sigma^{-1}$ is distributed as χ^2 with k degrees of freedom and $\left(\underline{X}-\underline{\eta}\right)' A \left(\underline{X}-\underline{\eta}\right)$ with $\underline{\eta} \neq \underline{\xi}$ as non-central χ^2.

4. If $\underline{X}^{(j)}$, $j = 1,\dots,n$, are i.i.d. k-variate random vectors with common mean $\underline{\xi}$ and covariance matrix Σ and if $\underline{\bar{X}}$ denotes the average $\sum \underline{X}^{(j)}/n$, then the multivariate central limit theorem states that $\sqrt{n}\left(\underline{\bar{X}}-\underline{\xi}\right) \xrightarrow{L} N(0,\Sigma)$. Two important applications are the joint limiting behavior of (i) the outcomes of n multinomial trials and (ii) r quantiles of n i.i.d. random variables.

5. The delta method (Theorem 2.5.2) is extended to the multivariate case. The result is used to obtain the asymptotic distribution of the sample correlation coefficient with and without the assumption of normality.

5.5   Some applications

As applications of the multivariate techniques of the preceding sections, let us first consider the multivariate one- and two-sample problems.

Example 5.5.1 Confidence sets and test for a multivariate mean. Let $\left(X_1^{(j)},\dots,X_k^{(j)}\right), j = 1,\dots n$, be a sample from a k-variate distribution with mean $(\xi_1,\dots,\xi_k)$ and non-singular covariance matrix Σ, and consider the problem of estimating or testing the mean. By the multivariate central limit theorem,

$$(5.5.1) \qquad \sqrt{n}\left(\bar{X}_1 - \xi_1\right), \ldots, \sqrt{n}\left(\bar{X}_k - \xi_k\right) \xrightarrow{L} N\left(0, \Sigma\right) \text{ as } n \to \infty,$$

where

$$(5.5.2) \qquad \bar{X}_i = \frac{1}{n} \sum_{j=1}^{n} X_i^{(j)}.$$

From this it follows by Theorem 5.4.2 and Theorem 5.1.5 that

$$(5.5.3) \qquad \sqrt{n}\left(\underline{\bar{X}} - \underline{\xi}\right)' A \sqrt{n}\left(\underline{\bar{X}} - \underline{\xi}\right) \xrightarrow{L} \chi_k^2,$$

where $\underline{\bar{X}} = \left(\bar{X}_1, \ldots, \bar{X}_k\right)$, $\underline{\xi} = \left(\xi_1, \ldots, \xi_k\right)$ and $A = \Sigma^{-1}$. If Σ and, hence, A are known, this leads to the confidence sets

$$(5.5.4) \qquad n\left(\underline{\xi} - \underline{\bar{X}}\right)' A\left(\underline{\xi} - \underline{\bar{X}}\right) \leq C_k$$

for $\underline{\xi}$ with asymptotic confidence coefficient γ, where

$$(5.5.5) \qquad \int_0^{C_k} \chi_k^2 = \gamma.$$

For $k = 2$ and in the notation of (5.2.1), the sets (5.5.4) reduce to

$$(5.5.6)$$
$$n\left[\frac{1}{\sigma^2}\left(\xi - \bar{X}\right)^2 - \frac{2\rho}{\sigma\tau}\left(\xi - \bar{X}\right)\left(\eta - \bar{Y}\right) + \frac{1}{\tau^2}\left(\eta - \bar{Y}\right)^2\right] \leq C_2\left(1 - \rho^2\right),$$

which are ellipses centered on $(\bar{X}, \bar{Y})$. In the general case, the sets (5.5.4) are k-dimensional ellipsoids centered on the point $(\bar{X}_1, \ldots, \bar{X}_k)$.

In applications, the covariance matrix Σ is typically not known. However, it follows from Problem 4.4 that the sample covariance matrix $\hat{\Sigma}$ is a consistent estimator of Σ, and from Lemma 5.3.3 and Problem 4.5 that, therefore, $\hat{A} = \hat{\Sigma}^{-1}$ is a consistent estimator of A. This shows that the sets

$$(5.5.7) \qquad n\left(\underline{\xi} - \underline{\bar{X}}\right)' \hat{A}\left(\underline{\xi} - \underline{\bar{X}}\right) \leq C_k$$

constitute asymptotically valid confidence sets for $(\xi_1, \ldots, \xi_k)$ at level γ for any non-singular covariance matrix Σ and any fixed shape of the distribution F of $(X_1, \ldots, X_k)$. However, the difficulty pointed out in (4.1.32) for the case $k = 1$ of course persists in the present more general situation.

Note: If F is known to be multivariate normal, the exact distribution of the left side of (5.5.7) is proportional to Hotelling's T^2. Our asymptotic result therefore shows that the confidence intervals or tests based on Hotelling's T^2 are asymptotically robust against non-normality in the same sense in which this was shown for Student's t (in the case $k = 1$) in Example 4.1.3.

Let us next consider the power of the test of $H : \underline{\xi} = \underline{\xi}^{(0)}$ with rejection region

$$(5.5.8) \qquad n \left(\underline{X} - \underline{\xi}^0 \right)' \hat{A} \left(\bar{X} - \underline{\xi}^{(0)} \right) \geq C_k.$$

The situation is quite analogous to that considered in Example 3.3.3. As in the univariate case, the test can be shown to be consistent against any fixed alternative $\underline{\xi} \neq \underline{\xi}^{(0)}$. For this reason, we shall consider the power not against a fixed alternative but against a sequence

$$(5.5.9) \qquad \underline{\xi}^{(k)} = \underline{\xi}_0 + \frac{\underline{\Delta}}{\sqrt{n}}$$

with $\underline{\Delta} = (\Delta_1, \dots, \Delta_k) \neq (0, \dots, 0)$.

The univariate Berry-Esseen theorem and Theorem 5.1.6 show that for any fixed Σ and any sequence of vectors $\underline{\xi}^{(n)} \to \underline{\xi}^{(0)}$,

$$(5.5.10) \qquad \sqrt{n} \hat{A}^{1/2} \left(\bar{X} - \underline{\xi}^{(n)} \right) \xrightarrow{L} N(0, I).$$

Here application of the Berry-Esseen theorem requires finiteness of the third moment of $\sum c_i X_i$ for all $(c_1, \dots, c_k)$, which is guaranteed by that of the third moment of $\sum X_i$. With $\underline{\xi}^{(n)}$ given by (5.5.9), it follows that

$$(5.5.11) \qquad \sqrt{n} \hat{A}^{1/2} \left(\bar{X} - \underline{\xi}^{(n)} \right) \xrightarrow{L} N \left(A^{1/2} \underline{\Delta}, I \right)$$

and therefore that

$$(5.5.12) \qquad n \left(\bar{X} - \underline{\xi}^{(0)} \right)' \hat{A} \left(\bar{X} - \underline{\xi}^{(0)} \right) \xrightarrow{L} \underline{Y}'\underline{Y} = \sum_{i=1}^{k} Y_i^2,$$

where Y_i is distributed as $N(\eta_i, 1)$ and $\sum \eta_i^2 = \underline{\Delta}'A\underline{\Delta}$. By Theorem 5.4.3, the distribution of $\sum_{i=1}^{k} Y_i^2$ is the non-central χ^2-distribution with k degrees of freedom and non-centrality parameter $\sum \eta_i^2$. If we denote a random variable with this distribution by $\chi^2 \left(\sum \eta_i^2 \right)$, the power of the test (5.5.8) against the alternatives (5.5.9) tends to $P \left[\chi^2 \left(\sum \eta_i^2 \right) \geq C_k \right]$.   □

Example 5.5.2 Inference for the difference of two mean vectors. In generalization of Example 4.1.4, consider samples $\left(X_1^{(r)}, \dots, X_k^{(r)} \right), r =$

$1, \ldots, m$, and $\left(Y_1^{(s)}, \ldots, Y_k^{(s)}\right)$, $s = 1, \ldots, n$, from k-variate distributions F and G with means $\underline{\xi} = (\xi_1, \ldots, \xi_k)$ and $\underline{\eta} = (\eta_1, \ldots, \eta_k)$ and covariance matrices Σ and $\mathcal{T}$, respectively. We shall be concerned with testing $H : \underline{\eta} = \underline{\xi}$ or, more generally, with obtaining approximate confidence sets for $\underline{\eta} - \underline{\xi}$.

Let $\bar{X}_i$ be defined by (5.5.2) and $\bar{Y}_i$ correspondingly, let $N = m + n$, and suppose that

(5.5.13) $\dfrac{m}{N} \to \rho, \ \dfrac{n}{N} \to 1 - \rho$ as m and $n \to \infty$ with $0 < \rho < 1$.

Then by the central limit theorem,

$$\sqrt{N} \left(\bar{\underline{X}} - \underline{\xi}\right) \xrightarrow{L} N\left(0, \frac{1}{\rho}\Sigma\right)$$

and

$$\sqrt{N} \left(\bar{\underline{Y}} - \underline{\eta}\right) \xrightarrow{L} N\left(0, \frac{1}{1-\rho}\mathcal{T}\right)$$

and hence

(5.5.14) $$\sqrt{N} \left[\left(\bar{\underline{Y}} - \bar{\underline{X}}\right) - \left(\underline{\eta} - \underline{\xi}\right)\right] \xrightarrow{L} N\left(0, \frac{1}{\rho}\Sigma + \frac{1}{1-\rho}\mathcal{T}\right).$$

It follows in analogy with (5.5.3) that

(5.5.15)

$$\left[\left(\underline{\eta} - \underline{\xi}\right) - \left(\bar{\underline{Y}} - \bar{\underline{X}}\right)\right]' \left(\frac{1}{m}\Sigma + \frac{1}{n}\mathcal{T}\right)^{-1} \left[\left(\underline{\eta} - \underline{\xi}\right) - \left(\bar{\underline{Y}} - \bar{\underline{X}}\right)\right] \xrightarrow{L} \chi_k^2.$$

If Σ and $\mathcal{T}$ are known, (5.5.15) provides confidence sets analogous to those given by (5.5.4) in the one-sample case.

When Σ and $\mathcal{T}$ are unknown, we simply replace them by $\hat{\Sigma}$ and $\hat{\mathcal{T}}$ and obtain the confidence sets

(5.5.16)

$$\left[\left(\underline{\eta} - \underline{\xi}\right) - \left(\bar{\underline{Y}} - \bar{\underline{X}}\right)\right]' \left(\frac{1}{m}\hat{\Sigma} + \frac{1}{n}\hat{\mathcal{T}}\right)^{-1} \left[\left(\underline{\eta} - \underline{\xi}\right) - \left(\bar{\underline{Y}} - \bar{\underline{X}}\right)\right] \leq C_k$$

with C_k given by (5.5.5). As usual, (5.5.16) permits a more explicit representation when $k = 2$ (Problem 5.1).

An assumption that is frequently made in this two-sample situation, but that should not be made lightly, is that

(5.5.17) $$\Sigma = \mathcal{T}.$$

The ij-th element of the common covariance matrix is then estimated by

(5.5.18)
$$S_{ij} = \left[\sum_{r=1}^{m} \left(X_i^{(r)} - \bar{X}_i \right) \left(X_j^{(r)} - \bar{X}_j \right) \right.$$
$$\left. + \sum_{s=1}^{n} \left(Y_i^{(s)} - \bar{Y}_i \right) \left(Y_j^{(s)} - \bar{Y}_j \right) \right] / (m+n)$$

and if $S = (S_{ij})$ is the resulting estimated covariance matrix, the left side of (5.5.16) is replaced by

(5.5.19)
$$\left[(\underline{\eta} - \underline{\xi}) - (\underline{\bar{Y}} - \underline{\bar{X}}) \right]' \frac{1}{m+n} S^{-1} \left[(\underline{\eta} - \underline{\xi}) - (\underline{\bar{Y}} - \underline{\bar{X}}) \right].$$

If F and G are known to be normal, then, except for a constant factor, (5.5.19) is distributed as Hotelling's T^2. It follows that under assumption (5.5.17), the confidence intervals and tests based on Hotelling's T^2 are asymptotically robust against non-normality. On the other hand, as in the univariate case (see Problem 5.15 of Chapter 3), they are not asymptotically robust against inequality of the covariance matrices Σ and T unless $\Sigma = T$ or $\dfrac{m}{n} \to 1$ (Problem 5.2). □

Example 5.5.3 Simple linear regression. In generalization of Example 2.7.4, suppose that

(5.5.20) $X_{i\nu} = \alpha_i + \beta_i v_{i\nu} + E_{i\nu} \quad (i = 1, \ldots, k; \; \nu = 1, \ldots, n),$

so that for each of the k components of the response vector $(X_{1\nu}, \ldots, X_{k\nu})$, we assume the simple linear regression structure of the univariate model. In particular, the v's are known constants while the errors $(E_{1\nu}, \ldots, E_{k\nu})$ are assumed to be n i.i.d. k-vectors distributed according to some k-variate distribution F with mean $(0, \ldots, 0)$ and covariance matrix Σ.

The standard estimators of α_i and β_i, in generalization of (2.7.11), are

(5.5.21)
$$\hat{\beta}_i = \frac{\sum\limits_{\nu=1}^{n} \left(X_{i\nu} - \bar{X}_i \right) \left(v_{i\nu} - \bar{v}_i \right)}{\sum\limits_{\nu=1}^{n} \left(v_{i\nu} - \bar{v}_i \right)^2}, \quad \hat{\alpha}_i = \bar{X}_i - \hat{\beta}_i \bar{v}_i,$$

where

(5.5.22)
$$\bar{X}_i = \sum_{\nu=1}^{n} X_{i\nu}/n, \quad \bar{v}_i = \sum_{\nu=1}^{n} v_{i\nu}/n.$$

To obtain the joint asymptotic distribution of $\left(\hat{\beta}_1 - \beta_1, \ldots, \hat{\beta}_k - \beta_k \right)$ in generalization of the univariate result given in Example 2.7.6, we require the following multivariate version of Theorem 2.7.3.

Theorem 5.5.1 *Let* $(Y_{1\nu}, \dots, Y_{k\nu})$, $\nu = 1, \dots, n$, *be* n *i.i.d.* k-*vectors with* $E(Y_{i\nu}) = 0$ *for all* i *and covariance matrix* Σ, *and for each* i, *let* $\left\{ d_{n\nu}^{(i)} \right\}$, $\nu = 1, \dots, n$, *be a double array of constants satisfying* $\sum_{\nu} d_{n\nu}^{(i)^2} = 1$ *and*

$$(5.5.23) \qquad \max_{i,\nu} d_{n\nu}^{(i)^2} \to 0 \text{ as } n \to \infty.$$

Then

$$(5.5.24) \qquad \left(\sum_{\nu=1}^{n} d_{n\nu}^{(1)} Y_{1\nu}, \dots, \sum_{\nu=1}^{n} d_{n\nu}^{(k)} Y_{k\nu} \right) \xrightarrow{L} N(0, \Sigma).$$

The proof, which will not be given here, utilizes Theorem 5.1.8 and then applies an argument, analogous to that used to prove Theorem 2.7.3, to linear combinations $\sum_{i=1}^{k} c_i \sum_{\nu=1}^{n} d_{n\nu}^{(i)} Y_{i\nu}$. For details, see Arnold (1981, Theorem 19.16). ∎

From Theorem 5.5.1 with

$$(5.5.25) \qquad d_{n\nu}^{(i)} = \frac{v_{i\nu} - \bar{v}_i}{\sqrt{\sum_{\nu} (v_{k\nu} - \bar{v}_k)^2}},$$

it is seen that

$$(5.5.26) \qquad \left(\frac{\sqrt{n} \left(\hat{\beta}_1 - \beta_1 \right)}{\sqrt{\sum_{\nu} (v_{1\nu} - \bar{v}_1)^2}}, \dots, \frac{\sqrt{n} \left(\hat{\beta}_k - \beta_k \right)}{\sqrt{\sum_{\nu} (v_{k\nu} - \bar{v}_k)^2}} \right) \to N(0, \Sigma),$$

provided (5.5.23) holds.

Theorem 5.4.2 shows that

$$(5.5.27) \qquad n \sum \sum a_{ij} \frac{\left(\hat{\beta}_i - \beta_i \right) \left(\hat{\beta}_j - \beta_j \right)}{\sqrt{\sum (v_{i\nu} - \bar{v}_i)^2 \sum (v_{j\nu} - \bar{v}_j)^2}} \xrightarrow{L} \chi_k^2,$$

where $A = (a_{ij}) = \Sigma^{-1}$. If Σ is known, this provides confidence sets and tests for β. If Σ is unknown, we can use the fact that the limit result (5.5.27) remains valid when $A = \Sigma^{-1}$ is replaced by the consistent estimator Σ^{-1} where the ij-th element of Σ is given by (Problem 5.4)

$$(5.5.28) \qquad \hat{\sigma}_{ij} = \sum \left(X_{i\nu} - \hat{\alpha}_i - \hat{\beta}_i v_{i\nu} \right) \left(X_{j\nu} - \hat{\alpha}_j - \hat{\beta}_j v_{j\nu} \right) / n.$$

Note: As was the case in Example 5.5.1 and with (5.5.19), when the distribution of the E's is k-variate normal, then, except for a constant factor, the quantity (5.5.27) is distributed as Hotelling's T^2. This shows that the level of the T^2-test for regression coefficients is asymptotically robust against non-normality.   □

Example 5.5.4 The multinomial one-sample problem. Consider a sequence of n multinomial trials with $k+1$ possible outcomes and, in the notation of Example 5.4.1, let $Y_1, \ldots, Y_{k+1}$ be the numbers of trials resulting in these outcomes. Then the joint distribution of $(Y_1, \ldots, Y_{k+1})$ is the multinomial distribution $M(p_1, \ldots, p_{k+1}; n)$ given by

$$(5.5.29) \qquad P(Y_1 = y_1, \ldots, Y_{k+1} = y_{k+1}) = \frac{n!}{y_1! \cdots y_{k+1}!} p_1^{y_1}, \ldots, p_{k+1}^{y_{k+1}}.$$

Here the p's denote the probabilities of the outcomes and we have

$$(5.5.30) \qquad \sum p_i = 1, \quad \sum y_i = n.$$

Let us now consider testing the hypothesis

$$(5.5.31) \qquad H: \ p_i = p_i^{(0)}, \quad i = 1, \ldots, k+1$$

against the alternatives that $p_i \neq p_i^{(0)}$ for at least some i. The standard test for this problem is *Pearson's* χ^2-*test*, which rejects H when

$$(5.5.32) \qquad Q = n \sum_{i=1}^{k+1} \left(\frac{Y_i}{n} - p_i^{(0)} \right)^2 / p_i^{(0)} \geq C_k.$$

The distribution of Q under a fixed alternative to H, suitably normalized, can be shown to be asymptotically normal (Chapter 6, Problem 3.13). The following theorem provides the asymptotic behavior of Q under H.

Theorem 5.5.2 *The distribution of Q under H tends to the χ^2-distribution with k degrees of freedom as $n \to \infty$.*

Proof. It follows from (5.4.14) and Theorem 5.4.2(ii) that

$$(5.5.33) \qquad n \sum_{i=1}^{k} \sum_{j=1}^{k} a_{ij} \left(\frac{Y_i}{n} - p_i^{(0)} \right) \left(\frac{Y_j}{n} - p_j^{(0)} \right) \xrightarrow{L} \chi_k^2,$$

where a_{ij} is given by (5.4.15). The left side of (5.5.33) is equal to

$$n \sum_{i=1}^{k} \frac{1}{p_i^{(0)}} \left(\frac{Y_i}{n} - p_i^{(0)} \right)^2 + \frac{n}{p_{k+1}^0} \sum_{i=1}^{k} \sum_{j=1}^{k} \left(\frac{Y_i}{n} - p_i^{(0)} \right) \left(\frac{Y_j}{n} - p_j^0 \right).$$

The last term is equal to

$$n \left[\sum_{i=1}^{k} \left(\frac{Y_i}{n} - p_i^0 \right) \right]^2 / p_{k+1}^0 = n \left(\frac{Y_{k+1}}{n} - p_{k+1}^0 \right)^2 / p_{k+1}^0$$

and the result follows.   ∎

Thus (5.5.32) defines a test of H with asymptotic level α if we determine C_k so that

(5.5.34)
$$\int_{C_k}^{\infty} \chi_k^2 = \alpha.$$

For a book-length discussion of this test with many references, see Greenwood and Nikulin (1996).

Some idea of the accuracy of the χ^2-approximation can be obtained from Table 5.5.1, which compares the exact value of $P(Q \geq C)$ with the approximate value $P(\chi_k^2 \geq C)$ for $k = 3$, $p_1^{(0)} = .2$, $p_2^{(0)} = .3$, $p_3^{(0)} = .5$ and $n = 10, 25, 50$ for a few values of C that give values of the probability between .01 and .05.

TABLE 5.5.1. Exact and approximate rejection probabilities for the χ^2-test

	$n = 10$		$n = 25$			$n = 50$		
Exact	.050	.029	.047	.019	.015	.051	.030	.017
Approx.	.049	.035	.043	.021	.015	.047	.032	.017

Source: Radlow and Alf (1975).

Corollary 5.5.1 *The test (5.5.32) with C_k given by (5.5.34) is consistent against any fixed alternative $p \neq p^{(0)}$.*

Proof. Let $(p_1, \ldots, p_{k+1})$ be any alternative with $p_i \neq p_i^0$ for at least some i, and suppose, in particular, that $p_i \neq p_i^0$. Write

(5.5.35)
$$\frac{\sqrt{n}\left(\dfrac{Y_i}{n} - p_i^0\right)}{\sqrt{p_i^0}} = \frac{\sqrt{n}\left(\dfrac{Y_i}{n} - p_i\right)}{\sqrt{p_i^0}} + \sqrt{n}\frac{p_i - p_i^0}{\sqrt{p_i^0}}.$$

Since Y_i has the binomial distribution $b(p_i, n)$, the first term on the right side converges in law to $N(0, p_i q_i / p_i^0)$, and (5.5.35) therefore converges in probability to $+\infty$ or $-\infty$, depending on the sign of $p_i - p_i^0$. In either case, $Q \xrightarrow{P} \infty$, and the probability of (5.5.32) therefore tends to 1.   ∎

To obtain an approximation for the power of the test (5.5.32), let us proceed as in Section 2 of Chapter 3 and consider the power not against a fixed alternative but against a sequence of alternatives $p_i^{(n)}$ tending to p_i^0 at the rate $1/\sqrt{n}$, so that

(5.5.36)
$$\sqrt{n}\left(p_i^{(n)} - p_i^{(0)}\right) \to \Delta_i,$$

where $\sum_{i=1}^{k+1} \Delta_i = 0$ since $\sum \left(p_i^{(n)} - p_i^{(0)}\right) = 0$.

Theorem 5.5.3 *The limit distribution of Q under the alternatives (5.5.36) is non-central χ^2 with k degrees of freedom and non-centrality parameter*

$$(5.5.37) \qquad \lambda = \sum_{i=1}^{k+1} \Delta_i^2 / p_i^0.$$

Proof. In analogy with (5.5.35), write

$$(5.5.38) \qquad \frac{\sqrt{n}\left(\dfrac{Y_i}{n} - p_i^0\right)}{\sqrt{p_i^0}} = \frac{\sqrt{n}\left[\dfrac{Y_i}{n} - p_i^{(n)}\right]}{\sqrt{p_i^0}} + \frac{\Delta_i}{\sqrt{p_i^0}} + R_n,$$

where $R_n \xrightarrow{P} 0$ as $n \to \infty$. Then the joint limit distribution of the variables on the left side of (5.5.38) is the same as that of

$$(5.5.39) \qquad Z_i = \frac{\sqrt{n}\left[\dfrac{Y_i}{n} - p_i^{(n)}\right] + \Delta_i}{\sqrt{p_i^0}} \qquad (i = 1, \ldots, k).$$

It is intuitively plausible from (5.4.13) that the joint distribution of the variables

$$(5.5.40) \qquad \sqrt{n}\left(\frac{Y_i}{n} - p_i^{(n)}\right)$$

will tend in law to $N(0, \Sigma)$, where Σ is given by (5.4.14), and hence that $T = (T_1, \ldots, T_k)$ with

$$(5.5.41) \qquad T_i = \sqrt{n}\left(\frac{Y_i}{n} - p_i^{(n)}\right) + \Delta_i \quad (i = 1, \ldots, k)$$

will tend in law to $N(\Delta, \Sigma)$, where $\Delta = (\Delta_1, \ldots, \Delta_k)$. Once this is proved, it follows from Theorem 5.4.3 that

$$(5.5.42) \qquad \sqrt{n} A^{1/2} T \xrightarrow{L} N(\Delta, I)$$

and

$$(5.5.43) \qquad \left(\sqrt{n} A^{1/2} T\right)' \left(\sqrt{n} A^{1/2} T\right) \xrightarrow{L} \chi_k^2 \left(\sum_{i=1}^{k} \sum_{j=1}^{k} a_{ij} \Delta_i \Delta_j\right),$$

as claimed.

To prove the above statements concerning the variables (5.5.40) and (5.5.41), we proceed as in Example 2.4.8. For this purpose, we can either utilize a multivariate Berry-Esseen theorem (given, for example, by Götze

(1991)) or reduce the problem to the univariate case by means of Theorem 5.1.8. We shall follow the second route here.

To this end, define the variables $X_i^{(j)}$, $i = 1, \ldots, k+1$, $j = 1, \ldots, n$, as in Example 5.4.1 so that $Y_i/n = \bar{X}_i$, and consider the variables

$$(5.5.44) \qquad \sqrt{n} \sum_{i=1}^{k} c_i \left(\bar{X}_i - p_i^{(n)} \right) \Big/ \sqrt{\operatorname{Var} \sum c_i \left(\bar{X}_i - p_i^{(n)} \right)}.$$

According to Corollary 2.4.1, the distribution of (5.5.44) will tend to the standard normal distribution provided

$$(5.5.45) \qquad \frac{E_n \sum_{i=1}^{k} c_i \left| X_i^{(1)} - p_i^{(n)} \right|^3}{\left\{ \operatorname{Var}_n \left[\sum_{i=1}^{k} c_i \left(X_i^{(1)} - p_i^{(n)} \right) \right] \right\}^{3/2}} \quad \text{is bounded.}$$

Since for any fixed $c_1, \ldots, c_k$ the numerator is obviously bounded, it is only necessary to check that the denominator is bounded away from 0 as $n \to \infty$. The denominator tends to

$$(5.5.46) \qquad \left\{ \operatorname{Var}_0 \left[\sum_{i=1}^{k} c_i \left(X_i^{(1)} - p_i^0 \right) \right] \right\}^{3/2}$$

and (5.5.45) will be proved if we can show that (5.5.46) is > 0 for all $(c_1, \ldots, c_k)$. Now the variance in (5.5.46) is equal to $\sum_{i=1}^{k} \sum_{j=1}^{k} c_i c_j \sigma_{ij}$ so that we need to show that Σ or, equivalently, $A = \Sigma^{-1}$ is positive definite. By the identity following (5.5.33),

$$(5.5.47) \qquad \sum \sum a_{ij} u_i u_j = \sum_{i=1}^{k} \frac{u_i^2}{p_i} + \frac{1}{p_{k+1}} \left(\sum_{i=1}^{k} u_i \right)^2,$$

which is positive unless all u's are 0. This completes the proof of (5.5.45). It follows that

$$\sqrt{n} \sum_{i=1}^{k} c_i \left(\bar{X}_i - p_i^{(n)} \right) \xrightarrow{L} N\left(0, \underline{c}' \Sigma \underline{c}\right)$$

and application of Theorem 5.1.8 now proves that (5.5.40) and (5.5.41) have the claimed limit distributions. ∎

For a discussion of the accuracy of the non-central χ^2 approximation to the power of Pearson's χ^2-test, see Slakter (1968). A different approximation will be considered in Chapter 6, Problem 3.13. □

Example 5.5.5 The multinomial two-sample problem. Consider two sequences of m and n multinomial trials each with $k+1$ outcomes, and let the numbers of trials resulting in outcomes $1, \ldots, k+1$ be $(X_1, \ldots, X_{k+1})$ and $(Y_1, \ldots, Y_{k+1})$ and their distributions $M(p_1, \ldots, p_{k+1}; m)$ and $M(q_1, \ldots, q_{k+1}; n)$, respectively. To test the hypothesis

$$(5.5.48) \qquad H: \; p_i = q_i \text{ for all } i = 1, \ldots, k+1,$$

consider the joint null distribution of the differences

$$(5.5.49) \qquad \frac{Y_i}{n} - \frac{X_i}{m}, \; i = 1, \ldots, k.$$

It is easily checked (Problem (5.5(i)) that under H, their covariance matrix is $\left(\dfrac{1}{m} + \dfrac{1}{n}\right) \Sigma$, where Σ is the multinomial covariance matrix given by (5.4.14). It follows from the proof of (5.4.13) that if $m/n \to \rho \; (0 < \rho < \infty)$ and $N = m + n$, then (Problem 5.5(ii))

$$(5.5.50) \qquad \sqrt{N\frac{m}{N}\frac{n}{N}} \left(\frac{Y_1}{n} - \frac{X_1}{m}, \ldots, \frac{Y_k}{n} - \frac{X_k}{m}\right) \xrightarrow{L} N(0, \Sigma)$$

and hence from the proof of Theorem 5.5.2 that under H,

$$(5.5.51) \qquad Q = N\frac{m}{N}\frac{n}{N} \sum_{i=1}^{k+1} \left(\frac{Y_i}{n} - \frac{X_i}{m}\right)^2 / p_i \xrightarrow{L} \chi_k^2.$$

Let C_k be defined by (5.5.34), and let $\hat{Q}$ be obtained from Q by replacing p_i by a consistent estimator $\hat{p}_i$. Then the rejection region

$$(5.5.52) \qquad \hat{Q} \geq C_k$$

provides a test of H with asymptotic level α. $\qquad \square$

Summary

Multivariate limit theory is applied to the following problems:

1. Developing asymptotic inference for

    (i) the mean vector of a multivariate distribution;

    (ii) the difference of the mean vectors of two distributions, both when the covariance matrices are unequal and when they are equal;

    (iii) a set of regression coefficients.

2. (i) Obtaining the limit distribution of Pearson's χ^2-statistic for testing a multinomial distribution, and the asymptotic power of the resulting test.

    (ii) Treating the corresponding problem for testing the equality of two multinomial distributions.

5.6   Estimation and testing in 2×2 tables

Consider N independent trials, the outcome of each classified according to two criteria, as A or $\bar{A}$, and as B or $\bar{B}$; for example, a series of operations classified according to the gender of the patient and the success or failure of the treatment. The results can be displayed in a 2×2 table, as shown in Table 5.6.1, where N_{AB} is the number of cases having both attributes

TABLE 5.6.1. 2×2 table

	B	B	
A	N_{AB}	$N_{A\bar{B}}$	N_A
$\bar{A}$	$N_{\bar{A}B}$	$N_{\bar{A}\bar{B}}$	$N_{\bar{A}}$
	N_B	$N_{\bar{B}}$	N

A and B, and so on. The joint distribution of the four cell entries is then multinomial, corresponding to N trials and four possible outcomes with probabilities, say p_{AB}, $p_{\bar{A}B}$, $p_{A\bar{B}}$, and $p_{\bar{A}\bar{B}}$.

A standard measure of the degree of association of the attributes A and B is the *cross-product ratio* (also called *odds ratio*)

$$(5.6.1) \qquad \rho = \frac{p_{AB}p_{\bar{A}\bar{B}}}{p_{\bar{A}B}p_{A\bar{B}}}.$$

An alternative form for ρ is obtained by using the fact that

$$p_{AB} = p_A p_{B|A},$$

where p_A and $p_{B|A}$ denote the probability of A and the conditional probability of B given A, respectively, and analogous equations for the other cell probabilities. Substituting these expressions in (5.6.1) leads to

$$(5.6.2) \qquad \rho = \frac{p_{B|A}p_{\bar{B}|\bar{A}}}{p_{B|\bar{A}}p_{\bar{B}|A}}.$$

The attributes A and B are said to be positively associated if

$$(5.6.3) \qquad p_{B|A} > p_{B|\bar{A}} \text{ and } p_{\bar{B}|\bar{A}} > p_{\bar{B}|A},$$

and these conditions imply that

$$(5.6.4) \qquad \rho > 1.$$

It can be shown that also conversely $\rho > 1$ implies (5.6.3) (Problem 6.1). In the case of negative dependence, the inequalities (5.6.3) and (5.6.4) are reversed. Independence of A and B is characterized by equality instead of inequality and hence by $\rho = 1$.

The odds ratio ρ is estimated by replacing the cell probabilities $p_{AB}, \ldots$ by the corresponding frequencies $N_{AB}/N, \ldots$, and this leads to the estimator

$$(5.6.5) \qquad \hat{\rho} = \frac{N_{AB}N_{\bar{A}\bar{B}}}{N_{A\bar{B}}N_{\bar{A}B}}.$$

The following result gives the asymptotic distribution of the *log odds ratio* $\log \hat{\rho}$ and of $\hat{\rho}$.

Theorem 5.6.1 *Under the multinomial distribution*

$$M\left(p_{AB}, p_{A\bar{B}}, p_{\bar{A}B}, p_{\bar{A}\bar{B}}; N\right)$$

assumed for the entries of Table 5.6.1, we have

(5.6.6) $$\sqrt{N}\left(\log \hat{\rho} - \log \rho\right) \overset{L}{\to} N\left(0, \tau^2\right)$$

with

(5.6.7) $$\tau^2 = \frac{1}{p_{AB}} + \frac{1}{p_{A\bar{B}}} + \frac{1}{p_{\bar{A}B}} + \frac{1}{p_{\bar{A}\bar{B}}},$$

and

(5.6.8) $$\sqrt{N}\left(\hat{\rho} - \rho\right) \overset{L}{\to} N\left(0, \rho^2\tau^2\right).$$

Proof. To simplify the notation, put

$$Y_1 = \frac{N_{AB}}{N}, \ Y_2 = \frac{N_{A\bar{B}}}{N}, \ Y_3 = \frac{N_{\bar{A}B}}{N},$$

and

$$Y_4 = \frac{N_{\bar{A}\bar{B}}}{N} = 1 - Y_1 - Y_2 - Y_3$$

so that

$$\log \hat{\rho} = \log Y_1 + \log\left(1 - Y_1 - Y_2 - Y_3\right) - \log Y_2 - \log Y_3.$$

Similarly, put

$$\pi_1 = p_{AB}, \ \pi_2 = p_{A\bar{B}}, \ \pi_3 = p_{\bar{A}B},$$

and

$$\pi_4 = p_{\bar{A}\bar{B}} = 1 - \pi_1 - \pi_2 - \pi_3.$$

We shall prove (5.6.6) by applying Corollary 5.4.2 to the function

(5.6.9)
$$f\left(y_1, \ y_2, \ y_3\right) = \log y_1 + \log\left(1 - y_1 - y_2 - y_3\right) - \log y_2 - \log y_3.$$

The partial derivatives of f evaluated at $y_i = \pi_i \ (i = 1, 2, 3)$ are then

(5.6.10)
$$\left.\frac{\partial f}{\partial y_1}\right|_{y=\pi} = \frac{1}{\pi_1} - \frac{1}{\pi_4}, \ \left.\frac{\partial f}{\partial y_2}\right|_{y=\pi} = -\frac{1}{\pi_2} - \frac{1}{\pi_4},$$
$$\left.\frac{\partial f}{\partial y_3}\right|_{y=\pi} = -\frac{1}{\pi_3} - \frac{1}{\pi_4}.$$

It follows from Example 5.4.1 that the joint asymptotic distribution of

$$\sqrt{N}\,(Y_1 - \pi_1),\ \sqrt{N}\,(Y_2 - \pi_2),\ \sqrt{N}\,(Y_3 - \pi_3)$$

is normal with covariance matrix given by (5.4.14) with $k = 3$ and π in place of p. On substituting (5.6.10) and (5.4.14) into (5.4.20) with $i = j = 1$, we find for the asymptotic variance τ^2 of $\sqrt{N}\,(\log\hat{\rho} - \log\rho)$,

(5.6.11)

$$(\pi_1 - \pi_1^2)\left(\frac{1}{\pi_1} - \frac{1}{\pi_4}\right)^2 + (\pi_2 - \pi_2^2)\left(\frac{1}{\pi_2} + \frac{1}{\pi_4}\right)^2$$

$$+ (\pi_3 - \pi_3^2)\left(\frac{1}{\pi_3} + \frac{1}{\pi_4}\right)^2 - \sum\sum_{i\neq j}\pi_i\pi_j\,\frac{\partial f}{\partial y_i}\frac{\partial f}{\partial y_i}\bigg|_{\underline{y}=\underline{\pi}}$$

and hence

(5.6.12)
$$\tau^2 = A - B,$$

where

(5.6.13) $$A = \pi_1\left(\frac{1}{\pi_1} - \frac{1}{\pi_4}\right)^2 + \pi_2\left(\frac{1}{\pi_2} + \frac{1}{\pi_4}\right)^2 + \pi_3\left(\frac{1}{\pi_3} + \frac{1}{\pi_4}\right)^2$$

and

(5.6.14) $$B = \sum_{i=1}^{3}\sum_{j=1}^{3}\pi_i\pi_j\,\frac{\partial f}{\partial y_i}\frac{\partial f}{\partial y_j}\bigg|_{\underline{y}=\underline{\pi}} = \left(\sum_{i=1}^{3}\pi_i\,\frac{\partial f}{\partial y_i}\bigg|_{\underline{y}=\underline{\pi}}\right)^2.$$

An easy calculation (Problem 6.2) now shows that

(5.6.15) $$A = \frac{1}{\pi_1} + \frac{1}{\pi_2} + \frac{1}{\pi_3} + \frac{1}{\pi_4} + \frac{1}{\pi_4^2}\ \text{and}\ B = \frac{1}{\pi_4^2},$$

which proves (5.6.7).

The limit relation (5.6.8) follows from (5.6.6) by applying the delta method (Theorem 2.5.2) to the function

(5.6.16) $$\rho = f(\log\rho) = e^{\log\rho}.$$

∎

The results (5.6.6) and (5.6.8) can be used to obtain confidence intervals for $\log\rho$ and ρ. For this purpose, it is only necessary to replace the unknown asymptotic variances τ^2 and $\rho^2\tau^2$ by the consistent estimators $\hat{\tau}^2$ and $\hat{\rho}^2\hat{\tau}^2$, respectively, where $\hat{\rho}$ is given by (5.6.5) and $\hat{\tau}^2$ by

(5.6.17) $$\hat{\tau}^2 = N\left(\frac{1}{N_{AB}} + \frac{1}{N_{A\bar{B}}} + \frac{1}{N_{\bar{A}B}} + \frac{1}{N_{\bar{A}\bar{B}}}\right).$$

The intervals are then respectively

(5.6.18) $$|\log \hat{\rho} - \log \rho| \leq u_{\alpha/2}\hat{\tau}/\sqrt{N}$$

and

(5.6.19) $$|\hat{\rho} - \rho| \leq u_{\alpha/2}\hat{\rho}\hat{\tau}/\sqrt{N}.$$

Besides the measures ρ and $\log \rho$, another popular measure of association is Yule's Q, defined as

(5.6.20) $$Q = \frac{p_{AB}p_{\bar{A}\bar{B}} - p_{A\bar{B}}p_{\bar{A}B}}{p_{AB}p_{\bar{A}\bar{B}} + p_{A\bar{B}}p_{\bar{A}B}}.$$

Q can be expressed in terms of ρ through the equation (Problem 6.3(i))

(5.6.21) $$Q = \frac{\rho - 1}{\rho + 1}.$$

A consistent estimator is

(5.6.22) $$\hat{Q} = \frac{\hat{\rho} - 1}{\hat{\rho} + 1},$$

which has the asymptotic distribution (Problem 6.3(ii))

(5.6.23) $$\sqrt{N}\left(\hat{Q} - Q\right) \xrightarrow{L} N\left(0, \frac{4\rho^2}{(1+\rho)^4}\tau^2\right).$$

Confidence intervals for Q can be obtained as in (5.6.17) and (5.6.18).

For additional measures of association and the properties of such measures, see Agresti (1990).

A problem that is frequently of interest in a 2 × 2 table is testing the hypothesis of independence

(5.6.24) $$H: \rho = 1.$$

A natural test of H against the alternatives $\rho > 1$ of positive dependence rejects H when $\hat{\rho} - 1$ is sufficiently large; more specifically, by (5.6.8), at asymptotic level α when

(5.6.25) $$\frac{\sqrt{N}\,(\hat{\rho} - 1)}{\rho\tau} \geq u_\alpha$$

or equivalently when

(5.6.26) $$\frac{\sqrt{N}\,(N_{AB}N_{\bar{A}\bar{B}} - N_{A\bar{B}}N_{\bar{A}B})}{N_{A\bar{B}}N_{\bar{A}B}\rho\tau} \geq u_\alpha.$$

For (5.6.26) to become a usable test, it is still necessary to replace $\rho\tau$ by a consistent estimator. One possibility is to estimate it by $\hat{\rho}\hat{\tau}$, with $\hat{\rho}$ and $\hat{\tau}$ given by (5.6.5) and (5.6.17), respectively. Alternatively, since the probability of (5.6.25) needs to tend to α only under H, we can replace ρ by its hypothetical value 1 and note that τ^2 under H reduces to (Problem 5.6.5)

$$(5.6.27) \qquad \tau^2 = \frac{1}{p_A p_B} + \frac{1}{p_A p_{\bar{B}}} + \frac{1}{p_{\bar{A}} p_B} + \frac{1}{p_{\bar{A}\bar{B}}} = \frac{1}{p_A p_{\bar{A}} p_B p_{\bar{B}}}.$$

A consistent estimator is obtained by replacing the probabilities $p_A, \dots$ by the corresponding frequencies.

The standard test of H is in fact neither of these two tests but is given by the rejection region

$$(5.6.28) \qquad \frac{\sqrt{N}\,(N_{AB}N_{\bar{A}\bar{B}} - N_{A\bar{B}}N_{\bar{A}B})}{\sqrt{N_A N_{\bar{A}} N_B N_{\bar{B}}}} \geq u_\alpha.$$

This test also has asymptotic level α and, under H, is asymptotically equivalent to (5.6.26). When the test is carried out against the two-sided alternatives $\rho \neq 1$, H is rejected when the absolute value of (5.6.28) is $\geq u_{\alpha/2}$ or, equivalently, when

$$(5.6.29) \qquad \frac{N\,(N_{AB}N_{\bar{A}\bar{B}} - N_{A\bar{B}}N_{\bar{A}B})^2}{N_A N_{\bar{A}} N_B N_{\bar{B}}} \geq C,$$

where C is the upper α critical value of χ^2 with one degree of freedom.

It is interesting to compare the foregoing results regarding the multinomial model for the 2×2 table with the corresponding results for the two-binomial model of Example 3.1.5(ii). The present situation differs from the earlier one (after suitable adjustment of notation) in that the marginal totals N_A and $N_{\bar{A}}$, which are random in the multinomial case, are fixed, and are equal to m and n, in the case of two binomials. In the latter, the odds ratio is

$$(5.6.30) \qquad \frac{p_{B|A} p_{\bar{B}|\bar{A}}}{p_{B|\bar{A}} p_{\bar{B}|A}},$$

which by (5.6.2) agrees with its multinomial definition (5.6.1).

The test with rejection region

$$(5.6.31) \qquad \frac{\dfrac{X}{m} - \dfrac{Y}{n}}{\sqrt{\dfrac{X}{m}\left(1 - \dfrac{X}{m}\right) + \dfrac{Y}{n}\left(1 - \dfrac{Y}{n}\right)}} \geq u_\alpha$$

proposed in (3.1.35) does not agree with the corresponding multinomial test (5.6.28). An alternative test which does agree with (5.6.28) is obtained by replacing the denominator of (5.6.31) by the estimator of the standard deviation of the numerator under the hypothesis that $p_{B|A}$ and $p_{B|\bar{A}}$ are equal (Problem 6.7).

Two-by-two tables are the simplest examples of *contingency tables*, which comprise not only larger two-way but also higher-dimensional tables. Their description, modeling, and analysis are treated in books such as Bishop, Fienberg, and Holland (1975), Reynolds (1977), and Agresti (1990). The results of the present section generalize to larger tables, but the methods used here become cumbersome and the problems are treated more easily by a more general approach, which is discussed in Section 7.8.

Summary

1. Three measures of association in a 2×2 table, the odds ratio, the log odds ratio, and Yule's Q are shown to be asymptotically normal and their asymptotic variances are derived.

2. A test of independence in a 2×2 table is obtained under the assumption of a multinomial model and is compared with the corresponding test for the two-binomial model in which the row totals are fixed.

5.7 Testing goodness of fit

A basic problem concerning a sample $X_1, \ldots, X_n$ of i.i.d. observations is to test whether the X's have been drawn from a specified distribution or family of distributions. Of the many procedures for testing the hypothesis that the common distribution F of the X's satisifies

$$(5.7.1) \qquad\qquad H : F = F_0,$$

let us consider first Pearson's χ^2-test which has the advantage of simplicity and flexibility. It forms the principal subject of the present section. Some other tests will be discussed more briefly later in the section.

Suppose first that the X's take on only a finite number of values $a_1, \ldots, a_{k+1}$ with probabilities $p_1, \ldots, p_{k+1}$ $\left(\sum p_i = 1 \right)$ and let Y_i be the number of X's equal to a_i. Then attention can be restricted to the Y's,[†] whose joint distribution is the multinomial distribution $M(p_1, \ldots, p_{k+1}; n)$ given by (5.5.29). In terms of the Y's, the problem reduces to that of testing

$$(5.7.2) \qquad\quad H : p_i = p_i^{(0)} \text{ for all } i = 1, \ldots, k+1,$$

which was treated in Example 5.5.4 by means of Pearson's χ^2-test given by (5.5.32).

[†]They constitute sufficient statistics for the X's.

Application of Pearson's test is not restricted to the case in which the X's take on only a finite number of values, but can be reduced to that case by grouping. To test, for example, that n independent variables $X_1, \ldots, X_n$ come from a specified Poisson distribution $P(\lambda_0)$, one might group together all observations exceeding some value k_0. If $Y_0, Y_1, \ldots, Y_{k_0}, Y_{k_0+1}$ then denote the numbers of X's equal to $0, 1, \ldots, k_0$, or exceeding k_0, the test (5.5.32) becomes applicable to the hypothesis $P(\lambda_0)$. Analogously, if the hypothesis to be tested is that $X_1, \ldots, X_n$ are i.i.d. according to the standard normal distribution or any other specified continous distribution F_0, one can divide the real axis into $k+1$ intervals

$$(-\infty, a_1), \ (a_1, a_2), \ldots, (a_{k-1}, a_k), \ (a_k, \infty)$$

and let $a_0 = -\infty$, $a_{k+1} = \infty$. If Y_i is the number of observations falling into the interval $J_i = (a_{i-1}, a_i)$, $i = 1, \ldots, k+1$, then $(Y_1, \ldots, Y_{k+1})$ has the multinomial distribution $M(p_1, \ldots, p_{k+1}; n)$, where

(5.7.3) $$p_i = F(a_i) - F(a_{i-1}).$$

With $p_i^{(0)}$ denoting the probabilities (5.7.3) computed under H, the test (5.5.32) is therefore a test of H_0 with asymptotic level α.

It is interesting to note that this test is obviously not consistent against all alternatives $F \neq F_0$ since there are many distributions other than F_0 for which $p_i = p_i^0$ for all $i = 1, \ldots, k+1$, and for which therefore the rejection probability tends to α rather than to 1 as $n \to \infty$. A test of H_0 that is consistent against all $F \neq F_0$ will be given in Theorem 5.7.2.

So far, it has been assumed that the hypothesis H given by (5.7.1) completely specifies the distribution F_0 and hence the probabilities $p_i^{(0)}$, but in applications, it more typically specifies a parametric family F_θ and, accordingly, probabilities, say

(5.7.4) $$H: \ p_i = p_i^{(0)}(\theta_1, \ldots, \theta_r).$$

If the θ's were known, the test would be based on

(5.7.5) $$\mathbf{X}^2 = \sum_{i=1}^{k+1} \frac{\left[Y_i - np_i^{(0)}(\theta_1, \ldots, \theta_r)\right]^2}{np_i^{(0)}(\theta_1, \ldots, \theta_r)}$$

with $k > r$ and asymptotic distribution χ_k^2 under H. Since they are unknown, it is natural to replace them by consistent estimators $\hat{\theta}_1, \ldots, \hat{\theta}_r$. Such a replacement will change the distribution of $\mathbf{X}^2$, with the change depending on the particular estimators used.

Perhaps the most natural method from the present point of view is to estimate the θ_i by the values $\hat{\theta}_i$ that minimize (5.7.5), the so-called minimum χ^2-estimators. Under suitable regularity conditions, the distribution of the resulting test statistic

$$(5.7.6) \quad \hat{\mathbf{X}}^2 = \sum \frac{\left[Y_i - np_i^{(0)}\left(\hat{\theta}_1, \ldots, \hat{\theta}_r\right)\right]^2}{np_i^{(0)}\left(\hat{\theta}_1, \ldots, \hat{\theta}_r\right)} = n\sum \left(\frac{Y_i}{n} - \hat{p}_i^{(0)}\right)^2 / \hat{p}_i^{(0)}$$

tends to the χ^2-distribution with $k - r$ degrees of freedom under H. A rigorous treatment can be found in Cramér (1946, Section 30.3). For more general discussions, see Rao (1973), Bishop, Fienberg, and Holland (1975), Read and Cressie (1988), Agresti (1990), and Greenwood and Nikulin (1996).

The χ^2-approximation for the statistic Q of (5.5.32) or the statistic (5.7.6) typically will not work well if the expectations $E(Y_i)$ of any of the cell frequencies are very small. In this case, one may therefore wish to combine cells with small expectations into larger ones not suffering from this defect. If the expectations np_i depend on unknown parameters, such a combination rule has to be based on estimated probabilities $\hat{p}_i$, and the resulting cell boundaries are therefore random instead of constant, as has been assumed so far. The same difficulty arises with a slightly different type of combination rule frequently used in practice in which the classes are determined by combining adjoining cells containing too few observations.

Unfortunately, with a data-based combination rule, the multinomial distribution of the Y's given by (5.5.29), and the theory based on it, no longer apply. To see what modifications are needed, consider a particularly simple rule which might be used to determine the classes when the hypothesis specifies that the distribution of the X's is a given continuous distribution F_0 (so that, in particular, the X's will be distinct with probability 1).

Suppose for the sake of simplicity that the sample size is a multiple of $k + 1$, $n = m(k + 1)$ say, and divide the real axis into $k + 1$ intervals, $K_i = (A_{i-1}, A_i)$, $i = 1, \ldots, k + 1$, such that each interval contains exactly m observations. Here we let $A_0 = -\infty$, $A_{k+1} = \infty$ and can let

$$A_i = X_{(im)},$$

where $X_{(1)} < \cdots < X_{(n)}$ denote the ordered observations. Then the statistic (5.5.32) becomes

$$(5.7.7) \quad \sum_{i=1}^{k+1} \frac{\left(m - nP_i^{(0)}\right)^2}{nP_i^{(0)}} = n\sum_{i=1}^{k+1} \frac{\left(\frac{1}{k+1} - P_i^{(0)}\right)^2}{P_i^{(0)}},$$

where

$$P_i^{(0)} = F_0(A_i) - F_0(A_{i-1})$$

is the probability of an observation falling into the interval J_i and is now a random variable.

More generally, let the numbers of observations m_i in K_i be fixed but not necessarily equal. (Although these numbers depend on n, we shall denote them by m_i rather than the more accurate but cumbersome $m_i^{(n)}$.) If $A_i = X_{(m_1+\cdots+m_i)}$, the natural analog to (5.5.32) becomes

$$(5.7.8) \qquad Q' = n \sum_{i=1}^{k+1} \left[\frac{m_i}{n} - P_i^{(0)} \right]^2 / P_i^{(0)}.$$

It differs from (5.5.32) in that the frequencies Y_i/n which were random earlier have now been replaced by the constant frequencies m_i/n, while the fixed probabilities $p_i^{(0)}$ have been replaced by the random $P_i^{(0)}$.

To determine the limit distribution of Q', let us compare the situation of the constant intervals $J_i = (a_{i-1}, a_i)$ and the random intervals $K_i = (A_{i-1}, A_i)$, which are shown in Figures 5.7.1(a) and 5.7.1(b). In this comparison, we shall suppress the superscript zero in the p's and P's.

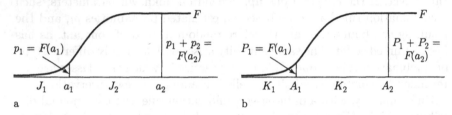

a   b

FIGURE 5.7.1. Correspondence between fixed and random intervals

To make the two situations comparable, we must relate the A's to the a's, which we do by requiring that the fixed numbers m_i falling into the random intervals K_i are equal to the expected numbers $E(Y_i) = np_i$ falling into the fixed intervals J_i. Because of the integer nature of the m's, we shall require this condition to hold only up to terms of order $o(\sqrt{n})$, so that the condition relating to two situations becomes

$$(5.7.9) \qquad \frac{m_i}{n} = p_i + o\left(\frac{1}{\sqrt{n}}\right).$$

It follows from (5.7.9) that $A_i \xrightarrow{P} a_i$, and hence that $P_i = F(A_i) - F(A_{i-1}) \xrightarrow{P} p_i = F(a_i) - F(a_{i-1})$.

Under assumption (5.7.9), we shall now prove the following equivalence result.

Theorem 5.7.1 Let $X_1, \ldots, X_n$ be i.i.d. according to a continuous distribution F_0. Let

$$a_0 = -\infty < a_1, < \cdots < a_k < a_{k+1} = \infty$$

be arbitrary constants with

$$(5.7.10) \qquad p_i = F(a_i) - F(a_{i-1}),$$

and let Y_i be the number of X's falling into the interval $J_i = (a_{i-1}, a_i)$. Also, let $m_1, \ldots, m_k$ be integers with $\sum m_i = n$ and let $A_i = (X_{m_1 + \cdots + m_i})$ and

(5.7.11) $P_i = F(A_i) - F(A_{i-1})$.

Then if (5.7.9) holds, the joint distribution of

(5.7.12) $\sqrt{n}\left(\dfrac{Y_1}{n} - p_1\right), \ldots, \sqrt{n}\left(\dfrac{Y_k}{n} - p_k\right)$

is the same as that of

(5.7.13) $\sqrt{n}\left(\dfrac{m_1}{n} - P_1\right), \ldots, \sqrt{n}\left(\dfrac{m_k}{n} - P_k\right)$.

Proof. The joint limit distribution of (5.7.12) is, by Example 5.4.1, the multivariate normal distribution $N(0, \Sigma)$ with the elements σ_{ij} of Σ given by (5.4.14).

That (5.7.13) has the same limit distribution follows from Theorem 5.4.5. Unfortunately for the present purpose, that theorem is stated not in terms of the P_i and m_i but in terms of their sums

(5.7.14) $P_1 + \cdots + P_i = F(A_i) = F\left(X_{(n_i)}\right)$ and $m_1 + \cdots + m_i = n_i$.

Let us put

(5.7.15) $p_1 + \cdots + p_i = \lambda_i$,

so that

(5.7.16) $\lambda_i = \dfrac{n_i}{n} + o\left(\dfrac{1}{\sqrt{n}}\right)$,

and $\xi_i = a_i$ so that

(5.7.17) $F(\xi_i) = \lambda_i$.

Since the variables $F(X_1), \ldots, F(X_n)$ constitute a sample from the uniform distribution $U(0, 1)$, it follows from Theorem 5.4.5 that the variables

(5.7.18) $\sqrt{n}\left[F\left(X_{(n_1)}\right) - \lambda_1\right], \ldots, \sqrt{n}\left[F\left(X_{(n_k)}\right) - \lambda_k\right]$

have a joint normal limit $N(0, T)$ with

(5.7.19) $\tau_{ij} = \lambda_i(1 - \lambda_j)$ for all $i \leq j$,

and by (5.7.16), the same is therefore true for the variables

(5.7.20)

$(Z_1, \ldots, Z_k) = \sqrt{n}\left[F\left(X_{(n_1)}\right) - \dfrac{n_1}{n}\right], \ldots, \sqrt{n}\left[F\left(X_{(n_k)}\right) - \dfrac{n_k}{n}\right]$.

Let the variables (5.7.12) be denoted by

$$(5.7.21) \qquad Y_i' = \sqrt{n}\left(\frac{Y_i}{n} - p_i\right)$$

and let

(5.7.22)

$$Z_i' = \sqrt{n}\left(P_i - \frac{m_i}{n}\right) = \sqrt{n}\left[F\left(X_{(m_i)}\right) - F\left(X_{(m_i-1)}\right) - \frac{n_i - n_{i-1}}{m}\right]$$

and

$$(5.7.23) \qquad Z_i = \sqrt{n}\left[F\left(X_{(n_i)}\right) - \frac{n_i}{n}\right].$$

Then

$$(5.7.24) \qquad Z_1' = Z_1,\; Z_2' = Z_2 - Z_1,\ldots,Z_k' = Z_k - Z_{k-1}$$

and it remains to determine the joint limit distribution of the Z_i''s from that of the Z_i's.

Now a linear transformation takes an asymptotically normal vector with distribution $N(0, T)$ into a vector with asymptotic distribution $N(0, T')$, where T' is obtained by applying the transformation (5.7.24) to variables with covariance matrix T. In the present case, for example,

$$\tau_{12}' = \tau_{12} - \tau_{11} = \lambda_1(1 - \lambda_2) - \lambda_1(1 - \lambda_1) = \lambda_1(\lambda_1 - \lambda_2) = -p_1 p_2$$

and

$$\begin{aligned}\tau_{23}' &= \tau_{23} - \tau_{22} - \tau_{13} + \tau_{12}\\ &= \lambda_2(1 - \lambda_3) - \lambda_2(1 - \lambda_2) - \lambda_1(1 - \lambda_3) + \lambda_1(1 - \lambda_2)\\ &= \lambda_2(\lambda_2 - \lambda_3) - \lambda_1(\lambda_2 - \lambda_3) = -p_2 p_3.\end{aligned}$$

In the same way, it is easily checked (Problem 7.2) that quite generally

$$(5.7.25) \qquad \tau_{ij}' = \sigma_{ij},$$

where σ_{ij} is given by (5.4.15). This completes the proof of the theorem. ∎
An easy consequence is

Corollary 5.7.1 *Under the assumptions of Theorem 5.7.1, the goodness-of-fit measure Q' defined by (5.7.8) has a limiting χ^2-distribution with k degrees of freedom.*

Proof. Since $P_i \overset{P}{\to} p_i$ and since the variables $\sqrt{n}\left(\frac{m_i}{n} - P_i\right)$ are bounded in probability, Q' has the same limit distribution as

$$Q'' = n\sum\left(P_i - \frac{m_i}{n}\right)^2 / p_i.$$

Now Q'' is the same quadratic form in the variables (5.7.13), as Q is in the variables (5.7.12). It thus follows from Theorem 5.7.1 that Q'', and hence Q', has the same limit distribution as Q, which is χ^2 with k degrees of freedom. ∎

The asymptotic equivalence of Q and Q' generalizes to the case where parameters have to be estimated and to other ways of determining the random cell boundaries. Results of this kind are discussed for example by Bofinger (1973) and by Moore and Spruill (1975).

The asymptotic theory for the χ^2-test (5.5.32) based on a division of the real axis into $k + 1$ intervals assumed k to be fixed as $n \to \infty$. However, the argument showing lack of consistency suggests that at least for some purposes it might be better to let k increase indefinitely with n so that the subdivision becomes finer as n increases. The hypothesis $H : p_i = p_i^{(0)}$, $i = 1, \ldots , k+1$, then tests a number of parameters that increases and tends to infinity as $n \to \infty$. The resulting limit process is quite different from those considered so far in this book in which the number of parameters being tested or estimated was considered fixed. If $k = k(n) \to \infty$ as $n \to \infty$, it can be shown that under appropriate conditions, the distribution of Q in (5.5.32), suitably normalized, tends to a normal rather than a χ^2-distribution (see, for example, Morris (1975)). This is plausible but does not follow from the fact that $\sqrt{k}\left(\dfrac{\chi_k^2}{k} - 1\right) \overset{L}{\to} N\,(0,2)$ (Example 2.4.2 and Problem 7.3).

The problem of the best number of class intervals in the χ^2-test has been extensively treated in the literature (for references, see Stuart and Ord (1991, Chapter 30)), but unfortunately it is very complex and has no simple answer. One complicating issue is the fact that as k changes, the dividing points $a_1, \ldots , a_k$ also change so that the problem really is that of determining the best choice of both k and the points $a_1, \ldots , a_k$. To avoid this difficulty, much of the literature has dealt with the case in which the a's are determined so that $p_1^{(0)} = \cdots = p_{k+1}^{(0)} = 1/(k+1)$, this despite some evidence in favor of unequal p's (for a discussion and references, see, for example, Rayner and Best (1989, p. 25)). Even with this simplifying assumption, the answer depends strongly on the type of alternatives against which good power is desired.

An alternative test of H that has more satisfactory power properties but which is less flexible is based on a comparison of the hypothetical cdf F_0 with the empirical (or sample) cdf defined by

(5.7.26) $$\hat{F}_n(x) = \frac{\text{Number of } X_i \leq x}{n}.$$

The resulting Kolmogorov test rejects $H : F = F_0$ when

(5.7.27) $$D_n = \sup_x \left| \hat{F}_n(x) - F_0(x) \right| > C_n.$$

An important property of D_n which we shall not prove is given in the following lemma. (For a proof, see, for example, Lehmann (1998).)

Lemma 5.7.1 *The null distribution of the Kolmogorov statistic D_n, i.e., the distribution of D_n when the X's are i.i.d. according to F_0, is the same for all continuous distributions F_0.*

The null distribution of D_n therefore depends only on n. A table for $n \leq 100$ is provided, for example, by Owen (1962). A limit distribution for large n requires magnification of D_n by a factor $\sqrt{n}$, and it is not normal but given by (Kolmogorov, 1933)

$$(5.7.28) \qquad P\left(\sqrt{n}D_n \leq z\right) \rightarrow L(z) = 1 - 2\sum_{j=1}^{\infty}(-1)^{j-1}e^{-2j^2z^2}.$$

Formulas for the density and moments of the distribution can be found in Johnson and Kotz, Vol. 2 (1970). Convergence to the limit distribution is slow. For $n = 64$, for example, $P\left(\sqrt{n}D_n \leq \frac{1}{2}\right) = .0495$, but from a table of the limit distribution (5.7.28) in Owen (1962), one obtains $\lim P\left(\sqrt{n}D_n \leq \frac{1}{2}\right) = .036$. The limit result (5.7.28) can be used to establish the following two properties of the sample cdf $\hat{F}_n$ as an estimator of a continuous distribution F.

1. Since $\sqrt{n}D_n$ has a limit distribution, it follows from Theorem 2.3.4 that $D_n \xrightarrow{P} 0$. This shows that the sample cdf is a consistent estimator of the true continuous cdf in the sense that the maximum difference between the two curves tends to zero in probability.

2. Let $c = L^{-1}(\gamma)$ be the value for which $L(c) = \gamma$. Then the probability is approximately γ that

$$(5.7.29) \qquad \sup_x \left|\hat{F}_n(x) - F(x)\right| < \frac{c}{\sqrt{n}},$$

i.e., that

$$(5.7.30) \qquad \hat{F}_n(x) - \frac{c}{\sqrt{n}} < F(x) < \hat{F}_n(x) + \frac{c}{\sqrt{n}} \text{ for all } x.$$

The inequalities (5.7.29) thus provide approximate confidence bands for the unknown continuous cdf F.

Theorem 5.7.2 *For testing (5.7.1), the Kolmogorov test—unlike Pearson's χ^2-test—is consistent in the sense that its power against any fixed alternative $F_1 \neq F_0$ tends to 1 as $n \rightarrow \infty$.*

Proof. If α is the level of the test (5.7.27),

$$(5.7.31) \qquad \sqrt{n} C_n \to L^{-1} (1 - \alpha).$$

Let a be any value for which $F_1(a) \neq F_0(a)$. Since $\left| F_0(a) - \hat{F}_n(a) \right| \leq D_n$, it follows that the power $\beta_n (F_1)$ against F_1 satisfies

$$(5.7.32) \quad \beta_n (F_1) = P_{F_1} (D_n > C_n) \geq P \left[\sqrt{n} \left| \hat{F}_n(a) - F_0(a) \right| \geq \sqrt{n} C_n \right].$$

Now $n\hat{F}_n(a)$ has the binomial distribution $b(p, n)$ with $p = F_1(a)$, and hence $\sqrt{n} \left[\hat{F}_n(a) - F_1(a) \right] \to N(0, pq)$. Therefore

$$\sqrt{n} \left[\hat{F}_n(a) - F_0(a) \right] = \sqrt{n} \left[\hat{F}_n - F_1(a) \right] + \sqrt{n} [F_1(a) - F_0(a)]$$

tends in probability to $+\infty$ or $-\infty$, as $F_1(a) > F_0(a)$ or $F_1(a) < F_0(a)$. In either case, $\sqrt{n} \left| \hat{F}_n(a) - F_0(a) \right|$ tends to ∞ in probability, and the result follows from (5.7.31) and (5.7.32). ∎

The Kolmogorov test of $H : F = F_0$ is one of a large class of tests that are based on some measure of distance of the sample cdf from the hypothetical F_0. Two other examples are the Cramér-von Mises test which rejects when

$$(5.7.33) \qquad W_n = n \int \left[\hat{F}_n(x) - F_0(x) \right]^2 f_0(x) dx \geq C_n,$$

where f_0 is the density of F_0, and the Anderson-Darling test with rejection region

$$(5.7.34) \qquad A_n = n \int \frac{\left[\hat{F}_n(x) - F_0(x) \right]^2}{F_0(x) [1 - F_0(x)]} f_0(x) dx \geq C_n.$$

These tests share the property of the Kolmogorov test stated in Lemma 5.7.1 that their null distribution is independent of F_0. Both tend to be more powerful than the Kolmogorov test. The A_n-test, in particular, is more sensitive to departures from F_0 in the tails of the distribution than either D_n or W_n since the factor $1/F_0(x) [1 - F_0(x)] \to \infty$ as $x \to \pm\infty$.

Tests based on statistics measuring the discrepancy between the empirical and the hypothesized distribution functions (the so-called EDF (empirical distribution function) statistics) such as the Kolmogorov, Cramér-von Mises, and Anderson-Darling statistics tend to be more powerful than Pearson's χ^2-test; in turn, A_n and W_n tend to perform better than the Kolmogorov statistic D_n. The statistics A_n and W_n are examples of V-statistics which will be defined in Example 6.4.5 of Chapter 6. Their asymptotic distribution is normal under alternatives to H while degeneracy occurs under the hypothesis and the resulting asymptotic null distribution is then a mixture of χ^2-distributions. (See, for example, Durbin (1973), Hajek and Sidak (1967), or Serfling (1980, Section 5.5.2).)

Complications arise when the hypothesis does not completely specify the distribution F_0 but instead only a parametric family F_0^θ. Then not only the exact distribution of the EDF statistics but also their asymptotic distribution depend on the shape of F_0. A general discussion of the effect of estimating parameters on the asymptotic distribution of a statistic is given by De Wet and Randles (1987), who also provide references to the literature on this problem.

To conclude this section, we shall briefly discuss some tests of the hypothesis that F_0 is a member of the family of normal distributions. Let $X_1, \ldots, X_n$ be i.i.d. according to a distribution F and consider the hypothesis

(5.7.35) $H : F$ is a member of the family of normal distributions
$\{N(\xi, \sigma^2), \ -\infty < \xi < \infty, \ 0 < \sigma\}$.

We shall discuss only some of the many tests that have been proposed for this problem. For a fuller treatment of the literature, see, for example, D'Agostino (1982, 1986) and Shapiro (1990).

(i) *Moment tests.*

The first tests of normality go back to Karl Pearson and are based on the third and fourth sample moments. Since the normal distribution is symmetric about its mean, its third central moment is zero. The third standardized moment

$$(5.7.36) \qquad \sqrt{\beta_1} = \frac{E(X_i - \xi)^3}{\sigma^{3/2}}$$

is a measure of the skewness of the distribution F. It is therefore natural to reject the hypothesis of normality if the standardized third sample moment

$$(5.7.37) \qquad \sqrt{b_1} = M_3 / M_2^{3/2}$$

is too large in absolute value, where M_k is defined in (5.2.36).

The asymptotic distribution of $\sqrt{b_1}$ can be obtained by the method used in Example 5.2.8 and is given by (Problem 7.4)

$$(5.7.38) \qquad \sqrt{n}\left(\sqrt{b_1} - \sqrt{\beta_1}\right) \xrightarrow{L} N(0, \tau^2),$$

provided F has finite moments up to sixth order. Here

(5.7.39)

$$\tau^2 = \text{Var}(M_3)\left(\frac{\partial\sqrt{b_1}}{\partial M_3}\right)^2 + 2\text{Cov}(M_2, M_3)\frac{\partial\sqrt{b_1}}{\partial M_2}\frac{\partial\sqrt{b_1}}{\partial M_3}$$
$$+\text{Var}(M_2)\left(\frac{\partial\sqrt{b_1}}{\partial M_2}\right)^2,$$

where the derivatives are evaluated at the population moments. In the normal case, the central moments

(5.7.40)
$$\mu_k = E(X_i - \xi)^k$$

are equal to

(5.7.41)
$$\mu_1 = \mu_3 = \cdots = 0 \text{ and } \mu_{2k} = 1\cdot 3\cdot 5\cdots(2k-1)\mu_2^k$$

and it can be seen that $\tau^2 = 6$. D'Agostino (1986) suggests that the normal approximation can be used when $n \geq 150$ and provides tables for smaller sizes. For further work, see Ramsey and Ramscy (1990).

A completely analogous development is possible for testing normality against departures of tailheaviness, based on the standardized fourth moment

(5.7.42)
$$b_2 = M_4/M_2^2,$$

and shows that under the hypothesis of normality (Problem 7.5(ii)),

(5.7.43)
$$\sqrt{n}(b_2 - 3)/\sqrt{24}$$

is asymptotically distributed as $N(0,1)$. Unfortunately, the convergence to normality is very slow. For small and moderate n, the distribution of (5.7.43) is highly skewed and the normal approximation is not recommended for $n < 1000$. To obtain faster convergence, Anscombe and Glynn (1983) propose a function

(5.7.44)
$$\Psi\left(\frac{b_2 - 3}{\sqrt{24}}\right),$$

which is adequate for $n \geq 20$ (D'Agostino (1986, p. 390); see also Ramsey and Ramsey (1993)).

It can also be shown that jointly $\left(\sqrt{n}(\sqrt{b_1} - \sqrt{\beta_1}), \sqrt{n}(b_2 - E(b_2))\right)$ have a bivariate normal limit distribution (Problem 7.6). If the distribution of the X's is symmetric, and hence in particular under the hypothesis of normality, $\sqrt{b_1}$ and b_2 are uncorrelated and asymptotically independent. This fact can be used to obtain tests of normality that are effective against both asymmetry and non-normal tail behavior. In particular, under the assumption of normality,

$$(5.7.45) \qquad n\left(\frac{b_1}{6} + \frac{(b_2 - 3)^2}{24}\right)$$

has a limiting χ^2-distribution with 2 degrees of freedom. The same is true of

$$(5.7.46) \qquad n\left[\frac{b_1}{6} + \Psi^2\left(\frac{b_2 - 3}{\sqrt{24}}\right)\right],$$

which converges to its χ^2 limit much faster.

The tests of normality based on (5.7.45) or (5.7.46) are not consistent against all alternatives but will give good power against the alternatives that are often of greatest interest in which the alternative exhibits skewness and/or non-normal tail behavior. For a further discussion of this and related tests with references to the literature, see Bowman and Shenton (1986) and D'Agostino (1986).

(ii) χ^2-*Test.*

As a second possibility, consider the classical χ^2-approach with a fixed number $k + 1$ of cells defined by k fixed division points $a_1, \dots, a_k$. If the two nuisance parameters ξ and σ^2 are estimated so as to minimize (5.7.5), the resulting $\hat{X}^2$ given by (5.7.6) has a limiting χ^2_{k-2} distribution. In the present context, it seems more natural to use the standard estimator

$$(5.7.47) \qquad \hat{\xi} = \bar{X} \text{ and } \hat{\sigma}^2 = \sum(X_i = \bar{X})^2/n$$

which are known to be asymptotically efficient (see Chapter 7, Section 4) rather than the minimum χ^2-estimators. However, the limit distribution of the resulting statistic is then no longer χ^2. For a discussion of this difficulty, see, for example, Stuart and Ord (1991, 5th Ed., Sections 30.17–30.19). We shall not consider χ^2-type tests any further here since better tests are available for testing normality.

(iii) *Tests based on EDF statistics.*

The Kolmogorov, Cramér-von Mises, and Anderson-Darling statistics defined by (5.7.27), (5.7.33), and (5.7.34), respectively, can be used for testing normality by applying them to the variables $(X_i - \bar{X})/S$. The asymptotic theory of the resulting tests is beyond the scope of this book; references to the literature can be found in Stephens (1986), which also provides rules and tables for calculating critical values.

(iv) *Shapiro-Wilk-type tests.*

This last class which provides powerful tests of normality against omnibus alternatives is based on expected normal order statistics. We begin by stating a few properties of expected order statistics.

Let $X_1, \ldots, X_n$ be a sample from a distribution F_0 and let $X_{(1)} \leq \cdots \leq X_{(n)}$ denote the ordered sample. Then the expected order statistics

$$(5.7.48) \qquad a_{in} = E\left[X_{(i)}\right], \quad i = 1, \ldots, n,$$

depend on both i and n. However, for the sake of simplicity we shall sometimes suppress the second subscript.

(a) If $E\,|X_1| < \infty$, then the expectations (5.7.48) exist for all $i = 1, \ldots, n$. (For a proof see David (1981).)

(b) $\displaystyle\sum_{i=1}^{n} a_{in} = nE\,(X_i).$

This is immediate from the fact that $\sum X_{(i)} = \sum X_i.$

(c) *Hoeffding's theorem* (Hoeffding (1953)) Let $G_n(x)$ denote the cdf of the n constants $a_{1n}, \ldots, a_{nn}$, i.e.,

$$(5.7.49) \qquad G_n(x) = (\text{Number of } a's \leq x)\,/n.$$

Then

$$(5.7.50) \qquad G_n(x) \to F_0(x) \text{ at all continuity points of } F_0.$$

Let us now return to the case that $X_1, \ldots, X_n$ are i.i.d. and that we wish to test the hypothesis of normality given by (5.7.35). The expected order statistics b_{in} from $N\left(\xi, \sigma^2\right)$ satisfy

$$(5.7.51) \qquad b_{in} = E\left[X_{(i)}\right] = \xi + \sigma a_{in},$$

where the a's are the expected order statistics from $N(0,1)$. [The inadequacy of our notation is seen from the fact that the sample size n occurs on the left and right side of (5.7.51) but not the middle term. The sample should be denoted by $X_1^{(n)}, \ldots, X_n^{(n)}$ and the order statistics by $X_{(1)}^{(n)}, \ldots, X_{(n)}^{(n)}$]. The a's have been extensively tabled (for example, in Harter (1961).

It is intuitively plausible that under the hypothesis, the $X_{(i)}$'s should be close to the b_{in}'s and this idea receives support from Hoeffding's theorem.

This suggests testing H by means of the correlation coefficient of the b_{in}'s and the $X_{(i)}$'s

$$(5.7.52) \qquad W' = \frac{\sum (b_{in} - \bar{b}_n)(X_{(i)} - \bar{X})}{\sqrt{\sum (b_{in} - \bar{b}_n)^2}\sqrt{\sum (X_{(i)} - \bar{X})^2}}.$$

Since a correlation coefficient is unchanged under linear transformations of the variables, we can in (5.7.52) replace the b's by the a's in which case by (i) we have $\bar{a}_n = 0$. Note also that $\sum (X_{(i)} - \bar{X})^2 = \sum (X_i - \bar{X})^2$ and that $\sum a_{in}\bar{X} = 0$, so that W' can be written as

$$(5.7.53) \qquad W' = \frac{\sum a_{in} X_{(i)}}{\sqrt{a_{in}^2}\sqrt{\sum (X_i - \bar{X})^2}}.$$

Since $|W'| \leq 1$ and under the hypothesis we would expect W' to be close to 1, H is rejected for small values of W'. This is the Shapiro-Francia test.

A natural alternative to W' proposed by De Wet and Venter (1972) replaces the expected order statistics by

$$(5.7.54) \qquad a'_{in} = \Phi^{-1}\left(\frac{i}{n+1}\right).$$

(It is easy to see that the a'_{in}'s satisfy (5.7.50); Problem 7.7.)

Still another alternative for the constants a_{in} is

$$(5.7.55) \qquad \begin{pmatrix} a''_{1n} \\ \vdots \\ a''_{1n} \end{pmatrix} = \begin{pmatrix} a_{1n} \\ \vdots \\ a_{1n} \end{pmatrix} V^{-1},$$

where V is the matrix of covariances of the $X_{(i)}$ given by

$$(5.7.56) \qquad v_{ij} = E\left(X_{(i)} - a_{in}\right)\left(X_{(j)} - a_{jn}\right).$$

The resulting test is the Shapiro-Wilk (1965) test, the first and still one of the most widely used of these correlation tests. For tables of critical values for these tests, see Stephens (1986) and D'Agostino (1986). The asymptotic theory of the test statistics was obtained for W in Leslie, Stephens, and Fotopoulos (1986) and for W' in Verrill and Johnson (1987). The limit distributions of $n(W - E(W))$ and $n(1 - W')$ belong to the family of distributions of the variables

$$(5.7.57) \qquad \sum_{i=1}^{\infty} \gamma_i\left(Y_i^2 - 1\right),$$

where the γ's are real numbers and the Y's are independent $N(0,1)$. For general discussions of distributions of this kind, see, for example, DeWet and Ventner (1973) and Gregory (1977). Under a fixed alternative to H, the appropriate normalizing factor is not n but $\sqrt{n}$ and the limit distribution is normal (Sarkadi (1985)). As a result, one would expect the tests to be consistent against general fixed alternatives. Consistency was proved against all non-normal alternatives with finite variance for W' by Sarkadi (1975) and for W by Leslie, Stephens, and Fotopoulos.

Power comparisons of many of the tests considered in this section have been made in a number of studies, among them Shapiro, Wilk, and Chen (1968) and Pearson, D'Agostino, and Bowman (1977). A summary of results is provided by D'Agostino (1986), who dismisses χ^2-tests and the Kolmogorov test as not sufficiently powerful and recommends a moment test based on the statistics b_1 and b_2, the Anderson-Darling test, and the Shapiro-Wilk test, with the choice among them depending on the specific alternatives against which power is sought.

Summary

1. Pearson's χ^2-test of goodness of fit is modified to deal with the situation in which fixed intervals containing random numbers of observations are replaced by random intervals determined so that the number of observations in each has a predetermined fixed value. It is shown that the test statistic has the same limiting χ^2 distribution in both cases.

2. As alternatives to Pearson's χ^2-test, several goodness-of-fit tests are proposed which are based on a comparison of the empirical distribution function (EDF) with the hypothetical cdf. One such test, due to Kolmogorov, is shown to be consistent against all alternatives and is used as a basis for confidence bounds for an unknown continuous cdf.

3. Four classes of tests of normality are briefly discussed. They are based respectively on (i) third and fourth sample moments, (ii) Pearson's χ^2, (iii) EDF statistics, and (iv) correlation coefficients between the n order statistics and n constants which correspond to the positions at which one might expect the order statistics if the hypothesis were true.

5.8   Problems

Section 1

1.1 (i) A sequence of vectors $\underline{x}^{(n)} = \left(x_1^{(n)}, \ldots, x_k^{(n)} \right)$ tends to $\underline{x}^{(0)} = \left(x_1^{(0)}, \ldots, x_k^{(0)} \right)$ if and only if $x_i^{(n)} \to x_i^{(0)}$ for each $i = 1, \ldots, k$.

(ii) A sequence of random vectors $\underline{X}^{(n)}$ converges in probability to a constant vector $\underline{c}$ if and only if $X_i^{(n)} \to c_i$ for each i.

1.2 (i) If a real-valued function f of $\underline{x}$ is continuous at $\underline{a}$, then $f(a_1, \ldots, a_{i-1}, x_i, a_{i+1}, \ldots, a_k)$ is a continuous function of the variable x_i at a_i for each $i = 1, \ldots, k$.

(ii) That the converse of part (i) is not true can be seen by considering the behavior of the function

$$
\begin{aligned}
f(x, y) &= \frac{xy}{x^2 + y^2} \quad \text{when } (x, y) \neq (0, 0) \\
&= 0 \quad\quad\;\; \text{when } (x = y = 0)
\end{aligned}
$$

at the point $\underline{a} = (0, 0)$.

1.3 Prove Theorem 5.1.3.

1.4 Prove the following generalization of Theorem 5.1.3

Theorem. Let $\underline{X}^{(n)} = (X_1^{(n)}, \ldots, X_r^{(n)}) \overset{L}{\to} \underline{X}$ and $\underline{Y}^{(n)} = (Y_1^{(n)}, \ldots, Y_s^{(n)}) \overset{P}{\to} \underline{c}$, then $(\underline{X}^{(n)}, \underline{Y}^{(n)}) \overset{L}{\to} (\underline{X}, \underline{Y})$ where $\underline{Y}$ is equal to $\underline{c}$ with probability 1.

1.5 (i) If $P(X_1 = -1, X_2 = 0) = P(X_1 = 0, X_2 = -1) = 1/2$, show that $(0, 0)$ is not a continuity point of the distribution H of (X_1, X_2).

(ii) Construct a bivariate distribution H of (X_1, X_2) such that

(a) X_1 and X_2 are independent,

(b) $P(X_1 = X_2 = 0) = 0$,

(c) the point $(0, 0)$ is not a continuity point of H.

1.6 Let X be $N(0, 1)$, let H be the bivariate distribution of (X, X), and let $S = \{(x, y) : y = x\}$. Then

(i) $\partial S = S$ and every point of ∂S is a continuity point of H,

(ii) $P_H[(X, X) \in \partial S] = 1$.

[**Hint for (ii):** For any point (a, a) in ∂S, evaluate $P(X \leq a - \epsilon, X \leq a - \epsilon)$ and $P(X \leq a, X \leq a)$.]

1.7 Let $X_n = U + V_n$, $Y_n = U - V_n$, where U and V_n are independent normal with means 0 and variances $\text{Var}(U) = 1$, $\text{Var}(V_n) = \frac{1}{n}$, respectively.

(i) Show that (X_n, Y_n) converges in law to a distribution H; determine H and find its points of continuity and discontinuity.

(ii) Find sets S and S' such that the probabilities $P\left[(X_n, Y_n) \in S\right]$ and $P\left[(X_n, Y_n) \in S'\right]$, respectively, do and do not converge to the corresponding probabilities under the limit distribution.

1.8 (i) Generalize the definition of "bounded in probability" (Definition 2.3.1) to the bivariate case.

(ii) Prove the analog of Theorem 2.3.2 for the bivariate case.

1.9 Prove Corollary 5.1.1

[**Hint**: Let a be a continuity point of the distribution of X_i, and apply Theorem 5.1.4 to the set $S = \{\underline{x} : x_i \leq a\}$.]

1.10 Under the assumptions of Corollary 5.1.1, show that

$$\left(X_i^{(n)}, X_j^{(n)}\right) \xrightarrow{L} (X_i, X_j) \text{ for all } i, j.$$

1.11 (i) Under the assumptions of Example 5.1.1, consider the rejection region

$$\frac{m\left(\bar{X} - \xi_0\right)^2}{\sigma^2} + \frac{n\left(\bar{Y} - \eta_0\right)^2}{\tau^2} \geq C_{m,n}$$

for testing $H : \xi = \xi_0, \eta = \eta_0$ at asymptotic level α. Use Theorem 5.1.5 to show that $C_{m,n} \to v_\alpha$, where v_α is determined by

$$P\left(\chi_2^2 \geq v_\alpha\right) = \alpha.$$

(ii) For testing the hypothesis H of part (i) at asymptotic level α, consider the alternative test which accepts H when

$$\max\left\{\frac{\sqrt{m}\,|\bar{X} - \xi_0|}{\sigma}, \frac{\sqrt{n}\,|\bar{Y} - \eta_0|}{\tau}\right\} \leq C'_{m,n}.$$

Determine the limit of $C'_{m,n}$ as $m, n \to \infty$.

1.12 Under the assumptions of Example 5.1.2, show that

(i) the marginal distributions of $n\left[\eta - X_{(n)}\right]$ and $n\left[X_{(1)} - \xi\right]$ both tend to the exponential distribution $E(0, \eta - \xi)$,

(ii) if $\eta - \xi = 1$ and $R = X_{(n)} - X_{(1)}$, the limiting distribution of $n(1 - R)$ has probability density xe^{-x}, $x > 0$.

1.13 Under the assumptions of Example 5.1.2, determine the joint limit distribution of $\left(n\left[X_{(2)} - \xi\right], n\left[\eta - X_{(n-1)}\right]\right)$.

1.14 Prove Theorem 5.1.6.

[**Hint:** In the theorem of Problem 5.1.4, let

$$Y_i^{(n)} = \begin{cases} A_i^{(n)} & \text{for } i = 1, \ldots, k \\ B_i^{(n)} & \text{for } i = k+1, \ldots, 2k \end{cases}$$

and $f(\underline{x}, \underline{y}) = (y_1 + x_1 y_{k+1}, \ldots, y_k + x_k y_{2k})$.]

1.15 Consider the variables $X_1^{(n)}$ and $X_2^{(n)}$ of Example 5.1.3 but without assuming (5.1.23). There exist constants a_n, β_n, and γ_n such that $a'_{nj} = \alpha_n a_{nj}$ and $b'_{nj} = \beta_n a_{nj} + \gamma_n b_{nj}$ satisfy (5.1.23), provided at least one of the a's and at least one of the b's is $\neq 0$, and the vectors $(a_{n1}, \ldots, a_{nk})$, $(b_{n1}, \ldots, b_{nk})$ are not proportional. The conclusion of Example 5.1.3 then holds for the variables $\left(\sum a'_{nj} X_j, \sum b'_{nj} X_j\right)$ provided (5.1.28) holds with a_{nj} and b_{nj} replaced by a'_{nj} and b'_{nj}, respectively.

1.16 (i) Extend the result of Example 5.1.3 to three variables

$$X_1^{(n)} = \sum a_{nj} Y_j, \quad X_2^{(n)} = \sum b_{nj} Y_j, \quad X_3^{(n)} = \sum c_{nj} Y_j.$$

(ii) Extend the result of Problem 1.15 to the situation of part (i) under the assumption that the a's, b's, and c's satisfy no relation of the form

$$k_{1n} \sum a_{nj} + k_{2n} \sum b_{nj} + k_{3n} \sum c_{nj} = 0.$$

1.17 Verify equation (5.1.32).

1.18 (i) For each n, let $U_i^{(n)}$, $i = 1, \ldots, s$, be s independent random variables converging in law to U_i as $n \to \infty$, and let $w_i^{(n)}$ be constants converging to w_i. Then

$$\sum_{i=1}^{s} w_i^{(n)} U_i^{(n)} \xrightarrow{L} \sum_{i=1}^{s} w_i U_i.$$

(ii) If, in (i), the U_i are normal $N(0, 1)$ and if

$$\sum_{i=1}^{s} [w_i^{(n)}]^2 = 1 \text{ for all } n,$$

then $\sum w_i^{(n)} U_i^{(n)} \xrightarrow{L} N(0, 1).$

[**Hint for (i):** Write $w_i^{(n)}U_i$ as $w_iV_i^{(n)}$ with $V_i^{(n)} = \dfrac{w_i^{(n)}}{w_i}U_1^{(n)}$ and apply Theorem 5.1.6.]

Note: The conclusion (ii) remains valid without the assumption $w_i^{(n)} \to w_i$. This fact, which we shall not prove here, provides a proof of Lemma 4.4.1.

1.19 Construct a counterexample to Theorem 5.1.7 if the variables $X_i^{(n)}$ are not required to be bounded in probability.

[**Hint:** Let $k = 2$, $X_1^{(n)} = Y_1 - k_n$, $X_{2n} = Y_1 + k_n$, and $a_{1n} = a_{2n} = 1$.]

Section 2

2.1 By completing the square in the expression

$$\frac{(y - \eta)^2}{\tau^2} - 2\rho\frac{(x - \xi)(y - \eta)}{\sigma\tau}$$

in the exponent of (5.2.1), show that

(i) the variables

$$Z = \frac{Y - \eta}{\tau} - \rho\frac{(X - \xi)}{\sigma} \text{ and } \frac{X - \xi}{\sigma}$$

are independently normally distributed;

(ii) $E(X - \xi)(Y - \eta)/\sigma\tau = \rho$;

(iii) the quadratic form

$$\frac{1}{1 - \rho^2}\left[\frac{(X - \xi)^2}{\sigma^2} - 2\rho\frac{(X - \xi)(Y - \eta)}{\sigma\tau} + \frac{(Y - \eta)^2}{\tau^2}\right]$$

has a χ^2-distribution with 2 degrees of freedom.

2.2 (i) If X and Y are independent with expectations ξ and η and variances σ^2, τ^2, then

$$\rho = E(X - \xi)(Y - \eta)/\sigma\tau = 0.$$

(ii) Give an example in which $\rho = 0$ but X and Y are not independent.

[**Hint for (ii):** Let $Y = g(X)$ for a suitable function g.]

2.3 In Example 5.2.1

(i) verify the correlation coefficient ρ_n;

(ii) show that $H_n(x, y) \to H(x, y)$ for all x, y.

2.4 Determine the limiting probability of the simultaneous confidence intervals (5.2.14).

2.5 In Example 5.2.5, determine joint asymptotic confidence sets for (ξ, η) along the lines of Example 5.1.4.

[**Hint**: Assume first that σ, τ, and ρ are known and find a constant c such that $\bar{X}$ and $\bar{Y} - c\bar{X}$ are uncorrelated.]

2.6 Show that $\sum (X_i - \bar{X}) (Y_i - \bar{Y}) = \sum X_i Y_i - n\bar{X}\bar{Y}$.

2.7 Determine a random variable X for which $\mathrm{Var}(X)\mathrm{Var}(1/X)$ is arbitrarily large.

[**Hint**: Let $X = \pm\epsilon$ with probability $p/2$ each and $= \pm 1$ with probability $q/2$ each.]

2.8 Let $\{(X_n, Y_n), n = 1, 2, \dots\}$ be a sequence of pairs of random variables such that $Y_n > 0$ and

$$\sqrt{n}\,(X_n - \xi), \quad \sqrt{n}\,(Y_n - \eta)$$

tends in law to a bivariate normal distribution with zero means and covariance structure $(\sigma_{11}, \sigma_{12}, \sigma_{22})$. Obtain the limit distribution of

$$\sqrt{n}\left(\frac{X_n}{Y_n} - \frac{\xi}{\eta}\right).$$

2.9 (i) If $(X_1, Y_1), \dots, (X_n, Y_n)$ is a sample from (5.2.1), determine the joint limit distribution of

$$\sqrt{n}\,(\bar{X} - \xi) \quad \text{and} \quad \sqrt{n}\left[\frac{1}{n}\sum (Y_i - \eta)^2 - \sigma^2\right].$$

(ii) Suppose that the joint density of X and Y is $p(x, y)$ instead of being given by (5.2.1). Find symmetry conditions on $p(x, y)$ under which the limit distribution of (i) continues to hold.

[**Hint for (i)**: Show that $\bar{X}$ and $\sum (Y_i - \eta)^2$ are uncorrelated.]

2.10 **Bivariate exponential distribution.**

The following distribution due to Marshall and Olkin (1967) arises when two components of a system are subject to certain independent events A, B, and C that terminate the life of the first component, the second component, and both components, respectively. The joint distribution of the lifetimes of the two components is given by

(5.8.1)
$$\bar{F}(x, y) = P(X > x, Y > y) = e^{-\lambda x - \mu y - \nu \max(x, y)}, \quad x, y > 0,$$

where $\lambda, \mu, \nu \geq 0$ and $\lambda + \nu, \mu + \nu > 0$.

(i) Show that at all points $x \neq y$, the distribution has a density

(5.8.2) $\qquad -\dfrac{\partial^2 \bar{F}(x, y)}{\partial x, \partial y} = \begin{cases} \lambda(\mu + \nu)e^{-\lambda x - (\mu + \nu)y} & \text{if } x < y \\ (\lambda + \nu)\mu e^{-(\lambda + \nu)x - \mu y} & \text{if } x > y. \end{cases}$

(ii) $P(X < Y) = \dfrac{\lambda}{\lambda + \mu + \nu}$, $P(Y < X) = \dfrac{\mu}{\lambda + \mu + \nu}$, and hence $P(X = Y) = \dfrac{\nu}{\lambda + \mu + \nu} > 0$ if $\nu > 0$, so that the distribution has a singular component.

2.11 For a sample $X_1, \ldots, X_n$ of i.i.d. variables distributed as $U(0, \theta)$, the limit distribution of $U_n = n(\theta - X_{(n)})$ and $V_n = n(\theta - X_{(n-1)})$ were considered in Example 2.3.7 and in Problem 3.12 of Chapter 2, respectively.

(i) Show that (U_n, V_n) has a joint limit distribution given by

(5.8.3) $\qquad \begin{aligned} \bar{F}(a, b) &= P[U \geq a, V \geq b] \\ &= \begin{cases} e^{-b/\theta}\left[1 + \dfrac{b - a}{\theta}\right] & \text{when } a < b \\ e^{-a/\theta} & \text{when } b < a. \end{cases} \end{aligned}$

(ii) The probability density of variables (U, V) with distribution given by (5.8.3) is

(5.8.4) $\qquad p(u, v) = \begin{cases} \dfrac{1}{\theta^2} e^{-v/\theta} & \text{if } u < v \\ 0 & \text{otherwise.} \end{cases}$

(iii) The joint limit distribution of $Y_n = n(\theta - X_n)$ and $Z_n = n[X_{(n)} - X_{(n-1)}]$ has the density

$$p_{Y,Z}(y, z) = \dfrac{1}{\theta^2} e^{-(y+z)/\theta},$$

so that in the limits Y_n and Z_n are independent.

Section 3

3.1 (i) Prove (5.3.9) and (5.3.10).

(ii) If A is symmetric, then so is A^{-1}.

3.2 If A is symmetric and positive definite, then so is A^{-1}.

[**Hint:** If the diagonal elements of (5.3.27) are $\lambda_1, \ldots, \lambda_k$, then those of A^{-1} are $1/\lambda_1, \ldots, 1/\lambda_k$. Taking the inverse of $QAQ^{-1} = \Lambda$ shows that $QA^{-1}Q^{-1} = \Lambda^{-1}$.]

3.3 If A is positive definite, so is BAB' for any non-singular B.

[**Hint:** Use (5.3.14)–(5.3.16).]

3.4 (i) Show that the matrix $\begin{pmatrix} 2 & 1 \\ 1 & 2 \end{pmatrix}$ is positive definite;

(ii) Let $A = (a_{ij})$ be positive semidefinite and let $B = (b_{ij})$ be such that $b_{ij} = a_{ij}$ when $i \neq j$ and $b_{ii} > a_{ii}$ for all i. Then B is positive definite.

3.5 Verify (5.3.23).

3.6 (i) Show that the matrix $\begin{pmatrix} 1/\sqrt{2} & 1/\sqrt{2} \\ 1/\sqrt{2} & -1/\sqrt{2} \end{pmatrix}$ is orthogonal.

(ii) Use the scheme $\begin{pmatrix} + & + & + & + \\ + & + & - & - \\ + & - & + & - \\ + & - & - & + \end{pmatrix}$ to construct a 4×4 orthogonal matrix.

3.7 Find a, b, c, and d so that the matrix $\begin{pmatrix} a & a & a \\ a & a & b \\ a & c & d \end{pmatrix}$ is orthogonal.

3.8 Let

(5.8.5) $\qquad\qquad y_j = x_{i_j} \; (j = 1, \dots, n),$

where $(i_1, \dots, i_n)$ is a permutation of $(1, \dots, n)$. Show that (5.8.5) is an orthogonal transformation.

[**Hint:** Determine the matrix A if (5.8.5) is written as $y = Ax$.]

3.9 (i) A diagonal matrix is orthogonal if and only if all the diagonal elements are ± 1.

(ii) If A is positive definite, symmetric, and orthogonal, then it is the identity matrix.

[**Hint:** (ii) Use Theorem 5.3.3 and part (i).]

3.10 If Q_1 and Q_2 are orthogonal, so is $Q_1 Q_2$.

3.11 (i) Use Theorem 5.3.3 to prove the following generalization of Theorem 5.3.1. For any symmetric matrix A (not necessarily positive definite or even non-singular), there exists a non-singular matrix B such that $y = Bx$ implies

(5.8.6) $\qquad \sum \sum a_{ij} y_i y_j \equiv x_1^2 + \cdots + x_p^2 - x_{p+1}^2 - \cdots - x_{p+q}^2,$

where p, q, and $r = k - p - q$ are respectively the number of λ's in (5.3.28) that are positive, negative, and zero.

(ii) The matrix A of part (i) is positive semi-definite if and only if $q = 0$, and non-singular if and only if $r = 0$.

3.12 Prove the following converse of the result of Example 5.3.3. Given any symmetric, positive definite $k \times k$ matrix A, there exist random variables $Y_1, \ldots, Y_k$ which have A as their covariance matrix.

3.13 If A is symmetric and positive definite, then so is A^{-1} by Problem 3.2. Show that

$$(5.8.7) \qquad \left(A^{-1}\right)^{1/2} = \left(A^{1/2}\right)^{-1}.$$

3.14 (i) If $A = \begin{pmatrix} \overset{k}{A_{11}} & \overset{m}{0} \\ 0 & A_{22} \end{pmatrix} \begin{matrix} k \\ m \end{matrix}$ is a non-singular $(k+m) \times (k+m)$

matrix, then A_{11} and A_{22} are non-singular and $A^{-1} = \begin{pmatrix} A_{11}^{-1} & 0 \\ 0 & A_{22}^{-1} \end{pmatrix}$.

(ii) If A is a non-singular $(m+k) \times (m+k)$ matrix of the form $A = \begin{pmatrix} A_{11} & A_{12} \\ 0 & A_{22} \end{pmatrix}$, then A^{-1} is of the same form.

3.15 (i) Condition (5.3.26) is not sufficient for A to be orthogonal.

(ii) The two properties of Q asserted in Lemma 5.3.4 are not only necessary but also sufficient for Q to be orthogonal.

[**Hint for (ii)**: Evaluate QQ'.]

3.16 (i) Let X and Y be independent $N(0,1)$ and let

$$X = r\cos\theta, \ Y = r\sin\theta \ (0 < r, \ 0 < \theta < 2\pi)$$

be (X, Y) expressed in polar coordinates. Use (5.3.37) to find the joint distribution of r and θ, and hence show that $r^2 = X^2 + Y^2$ is distributed as χ^2 with 2 degrees of freedom.

(ii) Let X, Y, and Z be independent $N(0,1)$ and let

$$X = r\cos\theta, \ Y = r\cos\phi\sin\theta, \ Z = r\sin\phi\sin\theta$$
$$(0 < r, \ -\pi/2 < \theta < \pi/2, \ 0 < \phi < 2\pi)$$

be (X, Y, Z) expressed in polar coordinates. Use the method of (i) to show that $r^2 = X^2 + Y^2 + Z^2$ is distributed as χ^2 with 3 degrees of freedom.

Section 4

4.1 If $X = (X_1, \dots, X_k)$ and $Y = (Y_1, \dots, Y_k)$ are two independent k-vectors with covariance matrices Σ and T, respectively, then the covariance matrix of $X + Y$ is $\Sigma + T$.

4.2 (i) Prove Theorem 5.4.2(ii) directly without reference to (i) by making an orthogonal transformation $Y = QX$ which reduces (5.4.9) to
$$\sum \lambda_i (Y_i - \eta_i)^2.$$

(ii) Prove Theorem 5.4.1 for $1 < r < k$.

[Hint for (ii): The case $r = 2$ asserts a joint normal distribution for $Y_1 = \sum c_i X_i$ and $Y_2 = \sum d_i X_i$. To prove this, choose the transformation QB of the case $r = 1$ in such a way that the plane spanned by the first two row vectors of B coincide with the plane spanned by the vectors $(c_1, \dots, c_k)$ and $(d_1, \dots, d_k)$ and apply part (i) of Theorem 5.4.1 with $k = 2$.]

4.3 Prove Theorem 5.4.3.

4.4 In a sample from a bivariate distribution with finite second moments, the sample covariance and sample correlation coefficient are consistent estimators of their population analogs.

4.5 (i) In a sample from a bivariate distribution with non-singular covariance matrix, the probability that the sample covariance is non-singular tends to 1 as the sample size tends to infinity.

(ii) The result of (i) remains true if bivariate is replaced by k-variate for any $k \geq 2$.

[Hint: (i) By Lemma 5.3.1 and (5.2.8), the population covariance matrix is non-singular if and only if $\sigma > 0$, $\tau > 0$, and $-1 < \rho < 1$. The result now follows from the preceding problem.

(ii) The covariance matrix Σ is non-singular if and only if its determinant is $\neq 0$ and hence > 0. The result now follows from the fact that this determinant is a continuous function of the covariances $(\sigma_{11}, \sigma_{12}, \dots, \sigma_{k-1k}, \sigma_{kk})$.]

Note: If the underlying bivariate or k-variate distribution has a continuous density, it can be shown that the sample covariance matrix is non-singular with probability 1 for every finite sample size n.

4.6 (i) Verify (5.4.14);

(ii) Verify (5.4.15).

4.7 Let $(X_1, Y_1), \dots, (X_n, Y_n)$ be a sample from the bivariate normal distribution (5.2.1). Use (5.4.34) and (5.4.40) to obtain approximate confidence intervals for ρ.

4.8 Let $(X_1, Y_1), \ldots, (X_n, Y_n)$ be a sample from the normal distribution (5.2.1).

(i) Obtain the joint limit distribution of

$$\sqrt{n}\left(\frac{\bar{X}}{S_X} - \frac{\xi}{\sigma}\right), \quad \sqrt{n}\left(\frac{\bar{Y}}{S_Y} - \frac{\eta}{\tau}\right),$$

where S_X^2 and S_Y^2 are given by (5.4.23).

(ii) Use the result of (i) to obtain joint and simultaneous confidence sets for the effect sizes ξ/σ and η/τ.

[**Hint:** For (i), see Problem 2.8.]

4.9 (i) If (X, Y) are bivariate normal with $\xi = \eta = 0$ and $\sigma = \tau = 1$, then $U = Y - X$ and $V = X + Y$ are independent normal $N(0, 2(1 - \rho))$ and $N(0, 2(1 + \rho))$, respectively.

(ii) Use (i) to prove (5.4.39).

4.10 Verify (5.4.40).

4.11 Consider samples $(X_1, Y_1), \ldots, (X_m, Y_m)$ and $(X_1', Y_1'), \ldots, (X_n', Y_n')$ from two bivariate normal distributions with parameters $(\xi, \eta, \sigma, \tau, \rho)$ and $(\xi', \eta', \sigma', \tau', \rho')$, respectively. Use (5.4.42) to obtain a test of the hypothesis $H : \rho = \rho'$.

4.12 If in the k-variate normal density (5.4.1) all covariances σ_{ij} with $i \leq r$, $j > r$ $(1 \leq r < k)$ are zero, then $(X_1, \ldots, X_r)$ and $(X_{r+1}, \ldots, X_n)$ are independent.

[**Hint:** It follows from Problem 3.14(ii) that $a_{ij} = 0$ for all $i \leq r, j > r$ and hence that the density (5.4.1) is the product of the densities of $(X_1, \ldots, X_r)$ and $(X_{r+1}, \ldots, X_k)$.]

Section 5

5.1 For the case $k = 2$, write out the matrices

(i) $\frac{1}{m}\hat{\Sigma} + \frac{1}{n}\hat{T}$ in (5.5.16);

(ii) S^{-1} in (5.5.19).

5.2 Show that the level of the test based on (5.5.19) is asymptotically robust against inequality of the covariance matrices Σ and T if and only if $m/n \to 1$.

5.3 (i) Use the confidence sets (5.5.16) to obtain a test of $H : \eta = \xi$.

(ii) Determine the asymptotic power of the test of part (i) by proving a result analogous to (5.5.12).

5.4 Verify (5.5.28).

5.5 (i) Determine the covariance matrix of the variables (5.5.49).

(ii) Verify (5.5.50) and (5.5.51).

5.6 With the assumptions and notation of Example 5.2.7, determine the limit of the joint distribution of

$$\sqrt{n}\,(M_k - \mu_k) \text{ and } \sqrt{n}\,(M_l - \mu_l), \quad k \neq l, \ k, l \geq 2.$$

5.7 Let $(X_1, Y_1), \dots, (X_n, Y_n)$ be i.i.d. with covariance structure $(\sigma_{11}, \sigma_{22}, \sigma_{12})$. Determine the limit distribution of

$$\sqrt{n}\left[\frac{1}{n}\sum (X_i - \bar{X})(Y_i - \bar{Y}) - \sigma_{12}\right].$$

5.8 Extend Example 5.2.5 and Problem 2.5 to samples (X_i, Y_i, Z_i) from a trivariate distribution and to the problem of estimating the mean vector (ξ, η, ζ).

5.9 In the notation of Example 5.2.5, show that the vector of sample means $(\bar{X}, \bar{Y})$ is asymptotically independent of the sample variances $\hat{\sigma}^2$ and $\hat{\tau}^2$ and the sample covariance $\sum (X_i - \bar{X})(Y_i - \bar{Y})/n$, provided $(X_i - \xi, Y_i - \eta)$ has the same distribution as $(\xi - X_i, \eta - Y_i)$.

5.10 Let $\left(X_1^{(\nu)}, \dots, X_p^{(\nu)}\right)$, $\nu = 1, \dots, n$, be a sample from a p-variate distribution with mean $(\xi_1, \dots, \xi_p)$, covariance matrix $\Sigma = (\sigma_{ij})$, and finite fourth moments

$$\tau_{ijkl} = E\,(X_i - \xi_i)(X_j - \xi_j)(X_k - \xi_k)(X_l - \xi_l), \quad i \leq j \leq k \leq l.$$

If $S_{ij} = \displaystyle\sum_{\nu=1}^{n} \left(X_i^{(\nu)} - \bar{X}_i\right)\left(X_j^{(\nu)} - \bar{X}_j\right)/n$ denotes the (i,j)-th sample covariance, show that the set of variables

$$\sqrt{n}\,(S_{ij} - \sigma_{ij}), \quad i \leq j,$$

has a normal limit distribution with 0 mean and express the covariance matrix of the limit distribution in terms of the τ_{ijkl}. (For asymptotic inferences concerning Σ, see, for example, Seber (1984, Section 3.5.8)).

Note: When the distribution of the $\left(X_1^{(\nu)}, \dots, X_p^{(\nu)}\right)$ is normal, the τ's are determined by the σ's and are given by

$$\tau_{ijkl} = \sigma_{ik}\sigma_{jl} + \sigma_{il}\sigma_{jk}.$$

For a proof, see, for example, Anderson (1984, pp. 81–82).

5.11 Determine the asymptotic power of the test (5.5.52) by obtaining a result analogous to Theorem 5.5.3.

Section 6

6.1 Show that (5.6.4) implies (5.6.3).

[**Hint:** Use the fact that $P_{\bar{B}|A} = 1 - P_{B|A}$ and $P_{\bar{B}|\bar{A}} = 1 - P_{B|\bar{A}}$.]

6.2 Verify (5.6.15).

6.3 Verify (i) (5.6.21); (ii) (5.6.23).

6.4 If ρ is defined by (5.6.1), show that $\rho = 1$ implies

$$(5.8.8) \qquad P_{AB} = P_A P_B, \quad P_{A\bar{B}} = P_A P_{\bar{B}}, \text{ etc.}$$

[**Hint:** Write $P_B = P_A P_{B|A} + P_{\bar{A}} P_{B|\bar{A}}$ and use the fact that by Problem 6.1, $\rho = 1$ implies $P_{B|\bar{A}} = P_{B|A}$.]

6.5 Prove (5.6.27).

[**Hint:** Use the result of the preceding problem.]

6.6 Find asymptotic confidence intervals for $P_{B|\bar{A}} - P_{B|A}$

(i) in the 2-binomial model

(ii) in the multinomial model assumed in Table 5.6.1.

6.7 Under the hypothesis (5.6.24), show that $\text{Var}\left(\dfrac{X}{m} - \dfrac{Y}{n}\right) = \dfrac{Npq}{mn}$, which can be estimated by $D^2 = N\dfrac{X+Y}{mn}\left(1 - \dfrac{X+Y}{mn}\right)$. The test obtained from (5.6.31) by replacing its denominator by D agrees with the test (5.6.28).

Section 7

7.1 Let Y_{ij}, $i = 1, \ldots, r+1$, $j = 1, \ldots, s$, have the multinomial distribution $M(p_{11}, \ldots, p_{r+1,1}; \ldots; p_{1s}, \ldots, p_{r+1,s}; n)$ and consider the hypothesis H that p_{ij} is independent of j for all i. Under H, the probabilities $p_{ij} = p_i$ are then functions of the $r+1$ unknown parameters $p_1, \ldots, p_{r+1}$. Determine the statistic (5.7.6).

7.2 Check equation (5.7.25).

7.3 (i) Explain why the fact that

$$\sqrt{k}\left[\frac{\chi^2}{k} - 1\right] \xrightarrow{L} N(0, 2)$$

does not prove that the distribution of the quadratic form Q defined by (5.5.32), suitably normalized, tends to a normal distribution as $k \to \infty$.

(ii) Construct a specific counterexample to the incorrect assertion of (i).

7.4 (i) Prove (5.7.38) with τ^2 given by (5.7.39).

(ii) Check that in the normal case $\tau^2 = 6$.

7.5 (i) Obtain the limit distribution of $\sqrt{n} \left(b_2 - E(b_2) \right)$ when $X_1, \ldots, X_n$ is a sample from a distribution possessing moments of sufficiently high order.

(ii) Verify that (5.7.43) tends in law to $N(0, 1)$ when F is normal.

7.6 (i) Show that $\sqrt{n} \left(\sqrt{b_1} - \sqrt{\beta_1} \right)$ and $\sqrt{n} \left(b_2 - E(b_2) \right)$ have a joint limit distribution that is bivariate normal.

(ii) If the distribution of the X's is symmetric, show that the correlation of the limit distribution of (i) is zero so that the two statistics are asymptotically independent.

7.7 Show that the constants a'_{in} defined by (5.7.54) satisfy (5.7.50).

Bibliographic Notes

The history of the multivariate normal distribution, which goes back to the work of Gauss and Laplace, is discussed in Stigler (1986) and Hald (1998). Modern treatments of the distribution are given in books on multivariate analysis such as Anderson (1984) and in Tong (1990). Convergence in probability and in law and the central limit in the multivariate case are treated, for example, in Feller (Vol. 2) (1966), Billingsley (1986), and Ferguson (1996). The asymptotic theory of the multivariate one- and two-sample problem and of regression is treated in Arnold (1981).

Discussions of the asymptotics of 2×2 and more general contingency tables can be found in Agresti (1990) and in Bishop, Fienberg, and Holland (1975). Chi-squared tests for goodness of fit and for testing independence in contingency tables were intitiated by Karl Pearson (1900), who obtained limiting χ^2-distributions for his test statistic. His result was correct for the case of a simple hypothesis but failed to allow for the decrease in degrees of freedom required when parameters have to be estimated. The necessary correction was provided by Fisher (1922b). EDF-based alternatives to χ^2 were introduced by Cramér (1928) and von Mises (1931), followed by Kolmogorov's test of 1933. A rigorous treatment of EDF theory is provided by Hajek and Sidak (1967). A survey of goodness-of-fit testing from a methodological point of view can be found in D'Agostino and Stephens (1986).

6

Nonparametric Estimation

Preview

Chapters 3–5 were concerned with asymptotic inference in parametric models, i.e., models characterized by a small number of parameters which it was desired to estimate or test. The present chapter extends such inferences to nonparametric families $\mathcal{F}$ over which the unknown distribution F is allowed to roam freely subject only to mild restrictions such as smoothness or existence of moments. The parameters to be estimated are functionals $h(F)$, that is, real-valued functions $h(F)$ defined over $\mathcal{F}$. The estimator of $h(F)$ will often be the plug-in estimator $h\left(\hat{F}_n\right)$, the functional h evaluated at the sample cdf.

Section 1 deals with the special case of functionals $h(F) = E\phi(X_1, \ldots, X_a)$ where the X's are i.i.d. according to F, and with their unbiased estimators $U = U(X_1, \ldots, X_n)$ based on a sample of size $n > a$. These so-called U-statistics differ slightly from the corresponding $V = h\left(\hat{F}_n\right)$, but this difference becomes negligible as $n \to \infty$.

The general theory of statistical functionals $h\left(\hat{F}_n\right)$ is discussed in Sections 2 and 3. The principal tool is the extension of the delta method (Sections 2.5 and 5.4) to the distribution of $h\left(\hat{F}_n\right) - h(F)$. This is achieved by means of a Taylor expansion of this difference, the first term of which turns out to be a sum of i.i.d. random variables. If the remainder R_n satisfies $\sqrt{n}R_n \xrightarrow{P} 0$, it follows that $\sqrt{n}\left[h\left(\hat{F}_n\right) - h(F)\right]$ is asymptotically normal. The asymptotic variance is found to be the integral of the square of the influence function which plays a central role in robustness theory.

Section 4 deals with density estimation, i.e., with estimating the probability density $h(F) = F' = f$. Since $h\left(\hat{F}_n\right)$ is not defined in this case, certain kernel estimators are proposed as approximations. Due to this additional approximation, the rate of convergence of these estimators to $h(F)$ is slower than the standard rate $1/\sqrt{n}$.

When the functional to be estimated is a measure of the performance of $h\left(\hat{F}_n\right)$ such as its bias, variance, or cdf, it depends not only on F but also on the sample size n. The estimators $\lambda_n\left(\hat{F}_n\right)$ of such functionals $\lambda_n(F)$ are called bootstrap estimators and are treated in Section 5. Conditions are given for consistency, and examples are provided in which the bootstrap estimator is not consistent. The estimator $\lambda_n\left(\hat{F}_n\right)$ is often hard to calculate, but it can be approximated arbitrarily closely by drawing a sufficiently large sample from the known distribution $\hat{F}_n$. The bootstrap is of wide applicability even to problems which are otherwise quite intractable. It has the additional advantage that when $\lambda_n(F)$ tends to a limit λ that does not depend on F, then $\lambda_n\left(\hat{F}_n\right)$ typically provides a better approximation to $\lambda_n(F)$ than does the limit λ.

6.1 U-Statistics

Suppose that $X_1, \ldots, X_n$ are i.i.d. with cdf F. We shall assume that F, instead of being restricted to a parametric family, is completely unknown, subject only to some very general conditions such as continuity or existence of moments. The "parameter" $\theta = \theta(F)$ to be estimated is a real-valued function defined over this nonparametric class $\mathcal{F}$, for example, the expectation, variance, standard deviation or coefficient of variation, or the median, the probability $F(a) = P(X_1 \leq a)$, and so on. Other possibilities for θ may require more than one X for their definition, for example $E|X_2 - X_1|$, the probability $p = P[(X_1, X_2) \in S]$ that the pair (X_1, X_2) falls into some given set S, or some function of p such as p^2, $\sqrt{p}$, or $1/p$. A real-valued function θ defined over a set of distributions is called a *statistical functional*.

In the present section, we shall restrict attention to a particularly simple class of such functionals or parameters θ. To define this class, note that for several of the examples given in the preceding paragraph, $\theta(F)$ can be written as an expectation. This is obviously the case for $E(X_1)$ and $E|X_2 - X_1|$, but applies also to $P(X_1 \leq a)$, which can be written as $E\phi(X_1)$ where $\phi(x)$ is 1 when $x \leq a$ and is 0 otherwise, and analogously for $p = P[(X_1, X_2) \in S]$. On the other hand, it can be shown that neither the median nor the standard deviation or coefficient of variation has this property, and the same is true for any function of p that is not a polynomial, for example, $1/p$ or $\sqrt{p}$. For proof and further discussion, see Bickel and Lehmann (1969).

The θ's to be considered in this section are those for which there exists an integer a and a function ϕ of a arguments such that

(6.1.1)
$$\theta = E\left[\phi\left(X_1, \ldots, X_a\right)\right].$$

We shall call such θ's *expectation functionals*. The case of general functionals θ not restricted to this class will be taken up in the next section.

Before proceeding, let us note that without loss of generality we can assume ϕ to be symmetric in its a arguments. For if it is not, $\phi\left(X_{i_1}, \ldots, X_{i_a}\right)$ also satisfies (6.1.1) for any permutation $(i_1, \ldots, i_a)$ of $(1, \ldots, a)$ and so therefore does the symmetric function

$$\phi^*\left(X_1, \ldots, X_a\right) = \frac{1}{a!} \sum_{(i_1, \ldots, i_a)} \cdots \sum \phi\left(X_{i_1}, \ldots, X_{i_a}\right),$$

where the sum extends over all $a!$ such permutations.

Let us now turn to the estimation of θ by means of n observations $X_1, \ldots, X_n$ from F, where we shall assume that $a \leq n$. Clearly, $\phi(X_1, \ldots, X_a)$ is an unbiased estimator of θ and so is $\phi\left(X_{i_1}, \ldots, X_{i_a}\right)$ for any a-tuple

(6.1.2)
$$1 \leq i_1 < \cdots < i_a \leq n.$$

This shows that also

(6.1.3)
$$U = \frac{1}{\binom{n}{a}} \sum_{(i_1, \ldots, i_a)} \cdots \sum \phi\left(X_{i_1}, \ldots, X_{i_a}\right)$$

is an unbiased estimator of θ, where this time the sum extends over all a-tuples satisfying (6.1.2). The unbiased estimator (6.1.3) is characterized by the fact that it is symmetric in the n variables $(X_1, \ldots, X_n)$ (Problem 1.4). It is the only symmetric estimator which is unbiased for all F for which $\theta(F)$ exists, and it can be shown to have smaller variance than any other such unbiased estimator. We shall in the next section find that U is closely related (and asymptotically equivalent) to the plug-in estimator $\theta\left(\hat{F}_n\right)$.

Our main concern in this section is the asymptotic behavior of U, but let us first consider some examples of θ and U.

$\underline{a = 1}$. In this case, $\theta = E\phi\left(X_1\right)$ and

(6.1.4)
$$U = \frac{1}{n} \sum_{i=1}^{n} \phi\left(X_i\right).$$

Since this is the average of n i.i.d. random variables, asymptotic normality follows from the classical central limit theorem (Theorem 2.4.1) provided $0 < \mathrm{Var}\phi(X_1) < \infty$. Examples are $\theta = E(X_1)$ and $\theta = P(X_1 \leq a)$.

$\underline{a = 2.}$

(a) If $\phi(x_1, x_2) = |x_2 - x_1|$, the statistic

$$(6.1.5) \qquad U = \frac{1}{\binom{n}{2}} \sum_{i<j} \sum |X_j - X_i|$$

is known as *Gini's mean difference*.

(b) If

$$(6.1.6) \qquad \phi(x_1, x_2) = \begin{cases} 1 & \text{when } x_1 + x_2 > 0 \\ 0 & \text{otherwise,} \end{cases}$$

then

$$(6.1.7) \qquad W = \binom{n}{2} U$$

is the number of pairs $i < j$ for which $X_i + X_j > 0$ and

$$(6.1.8) \qquad \theta = P(X_1 + X_2 > 0).$$

The statistic W is closely related to the one-sample Wilcoxon statistic (3.2.31).

(c) When the distribution F is discrete and $X_i + X_j$ takes on the value 0 with positive probability, one may wish to replace (6.1.6) with

$$(6.1.9) \qquad \phi(x_1, x_2) = \begin{cases} 1 & \text{if } x_1 + x_2 > 0 \\ 1/2 & \text{if } x_1 + x_2 = 0 \\ 0 & \text{if } x_1 + x_2 < 0. \end{cases}$$

In this case,

$$(6.1.10) \qquad \theta = P(X_1 + X_2 > 0) + \frac{1}{2} P(X_1 + X_2 = 0).$$

Before stating any asymptotic results, let us speculate about what to expect. For a sum of n i.i.d. variables with variance σ^2, the variance of the sum is $n\sigma^2$, and the variance of their average is therefore σ^2/n. In the case $a = 1$, the variance of (6.1.4) therefore tends to 0 at rate $1/n$. Consider next the case $a = 2$, so that

$$(6.1.11) \qquad U = \frac{1}{\binom{n}{2}} \sum_{i<j} \sum \phi(X_i, X_j).$$

Since the right side is an average of

$$\binom{n}{2} \sim \frac{1}{2}n^2$$

terms, one might be tempted to conjecture that the variance of U tends to 0 at the rate $1/n^2$. There is, however, an important difference between the sums (6.1.4) and (6.1.11): In the latter case, the terms, although identically distributed, are no longer independent.

To see the effect of this dependence, note that the variance of $\binom{n}{2}U$ is

(6.1.12) $$\sum_{i<j}\sum\sum_{k<l}\sum \text{Cov}\left[\phi\left(X_i, X_j\right),\ \phi\left(X_k, X_l\right)\right].$$

This sum contains three kinds of terms. When all four subscripts i, j, k, l are distinct, the pair (X_i, X_j) is independent of the pair (X_k, X_l) and the terms of (6.1.12) corresponding to these pairs are therefore 0. A second possibility is that only three of the four subscripts are distinct, that, for example, $i = k$ with j and l being different. Without counting these terms, it is intuitively plausible that their number is of order n^3. This leaves the terms with $i = k$ and $j = l$ in which only two of the subscripts are distinct and whose number is of order n^2. Thus (6.1.12) is the sum of two terms of order n^3 and n^2 respectively and Var(U) is obtained by dividing this sum by $\binom{n}{2}^2$. The result is of order $1/n$, as it was in the case $a = 1$. This is actually not so surprising since the information in U, regardless of the value of a, is provided by just n i.i.d. variables, namely $X_1, \dots, X_n$. (A discussion of the amount of information contained in a data set will be given in Chapter 7.)

The argument suggesting that Var(U) is of order $1/n$ can be made precise by counting the various covariance terms and for general a then leads to Theorem 6.1.1. Before stating this result, we require the following lemma.

Lemma 6.1.1

(i) If

(6.1.13) $$\phi_i\left(x_1, \dots, x_i\right) = E\phi\left(x_1, \dots, x_i,\ X_{i+1}, \dots, X_a\right),$$

then

(6.1.14) $$E\phi_i\left(X_1, \dots, X_i\right) = \theta \ \text{for all } 1 \le i < a.$$

(ii) If

(6.1.15) $$\text{Var } \phi_i\left(X_1, \dots, X_i\right) = \sigma_i^2,$$

then

(6.1.16)
$$\mathrm{Cov}\left[\phi\left(X_1,\dots,X_i,\ X_{i+1},\dots,X_a\right),\phi\left(X_1,\dots,X_i,\ X'_{i+1},\dots,X'_a\right)\right]$$
$$= \sigma_i^2,$$

where $X_1,\dots,X_i,\ X_{i+1},\dots,X_a,\ X'_{i+1},\dots,X'_a$ are i.i.d. according to F.

Proof. For the sake of simplicity we shall give the proof for the case $a = 2$, $i = 1$ and under the assumption that F has a density f.

(i) Since

$$\phi_1\left(x_1\right) = \int \phi\left(x_1,\ x_2\right) f\left(x_2\right) dx_2,$$

we have

$$E\phi_1\left(X_1\right) = \int \left[\int \phi\left(x_1,\ x_2\right) f\left(x_2\right) dx_2\right] f\left(x_1\right) dx_1$$
$$= \int \int \phi\left(x_1,\ x_2\right) f\left(x_1\right) f\left(x_2\right) dx_1 dx_2 = \theta.$$

(ii) Analogously,

$$E\phi_1^2\left(X_1\right)$$
$$= \int \left[\int \phi\left(x_1,\ x_2\right) f\left(x_2\right) dx_2\right] \left[\int \phi\left(x_1,\ x_3\right) f\left(x_3\right) dx_3\right] f\left(x_1\right) dx_1$$
$$= \int \int \int \phi\left(x_1,\ x_2\right) \phi\left(x_1,\ x_3\right) f\left(x_1\right) f\left(x_2\right) f\left(x_3\right) dx_1 dx_2 dx_3,$$

and (6.1.16) follows by subtracting θ^2 from both sides and using part (i). ∎

Theorem 6.1.1
(i) The variance of the U-statistic (6.1.3) is equal to

(6.1.17)
$$\mathrm{Var}\left(U\right) = \sum_{i=1}^a \binom{a}{i} \binom{n-a}{a-i} \sigma_i^2 \Big/ \binom{n}{a}$$

with σ_i^2 given by (6.1.16). Here $X_1,\dots,X_i,\ X_{i+1},\dots,X_a,\ X'_{i+1},\dots,X'_a$ are independently distributed according to the common distribution F.
(ii) If $\sigma_1^2 > 0$ and $\sigma_i^2 < \infty$ for all $i = 1,\dots,a$, then

(6.1.18)
$$\mathrm{Var}\left(\sqrt{n}U\right) \to a^2 \sigma_1^2$$

Proof.

(i) For the case $a = 2$, the proof is outlined above and given in more detail in Problem 1.5. For the general case, see Lee (1990, Section 1.3).

(ii) Since

$$(6.1.19) \qquad \binom{n-a}{k} = \frac{1}{k!}(n-a)(n-a-1)\cdots(n-a-k+1) \sim \frac{n^k}{k!},$$

the terms in the sum (6.1.17) corresponding to $i > 1$ are all of smaller order than the term corresponding to $i = 1$. This latter term is asymptotically equivalent to

$$\sigma_1^2 \frac{an^{a-1}}{(a-1)!} \bigg/ \frac{n^a}{a!} = \frac{a^2\sigma_1^2}{n},$$

and this completes the proof. ∎

Theorem 6.1.2 *If*

$$(6.1.20) \qquad 0 < \sigma_i^2 < \infty \quad \text{for all } i = 1,\ldots,a,$$

then as $n \to \infty$

$$(6.1.21) \qquad \sqrt{n}\,(U - \theta) \xrightarrow{L} N\left(0, a^2\sigma_1^2\right);$$

and

$$(6.1.22) \qquad \frac{U - \theta}{\sqrt{\operatorname{Var} U}} \xrightarrow{L} N(0,1).$$

We shall postpone discussion of the proof to the end of the section and Problem 1.17.

Note 1: Condition (6.1.21) can be replaced by the seemingly weaker condition $\sigma_a^2 < \infty$. This follows from the fact that $\sigma_i^2 \leq \sigma_a^2$ since the square of a covariance is less than or equal to the product of the variances.

Note 2: Since the asymptotic variance in (6.1.18) involves only σ_1^2, it may seem surprising that it is not enough in Theorem 6.1.2(ii) to require that $\sigma_1^2 < \infty$. The apparent paradox is explained by the possible difference between the asymptotic variance $a^2\sigma_1^2$ and the limit of the actual variance. It is seen from (6.1.17) that the actual variance of U involves not only σ_1^2 but also the variances $\sigma_2^2,\ldots,\sigma_a^2$. If any of these latter variances is infinite but $\sigma_1^2 < \infty$, we have a case where the actual variance of $\sqrt{n}U$ and therefore its limit is infinite, while the asymptotic variance is finite. We shall see another example of this phenomenon in Example 6.2.5 of the next section.

Example 6.1.1 One-sample Wilcoxon statistics. Consider the U-statistics with $a = 2$ and with ϕ and θ given by (6.1.6) and (6.1.8), respectively. Then

$$
\begin{aligned}
(6.1.23) \quad \sigma_1^2 &= \mathrm{Cov}\left[\phi\left(X_1, X_2\right), \phi\left(X_1, X_2'\right)\right] \\
&= P\left[X_1 + X_2 > 0 \text{ and } X_1 + X_2' > 0\right] - \left[P\left(X_1 + X_2 > 0\right)\right]^2.
\end{aligned}
$$

For any given F, the probabilities on the right side can be evaluated numerically and, in a few cases, analytically (Problem 1.6).

The calculations greatly simplify when F is symmetric about 0. Then $-X_1$ has the same distribution as X_1 and hence

$$P\left(X_1 + X_2 > 0\right) = P\left(X_2 > -X_1\right) = P\left(X_2 > X_1\right).$$

If F is continuous (so that $P\left(X_1 = X_2\right) = 0$), it therefore follows from symmetry that

$$(6.1.24) \qquad\qquad P\left(X_1 + X_2 > 0\right) = 1/2.$$

Analogously,

$$P\left(X_1 + X_2 > 0 \text{ and } X_1 + X_2' > 0\right) = P\left(X_1 > X_2 \text{ and } X_1 > X_2'\right)$$

is the probability that X_1 is the largest of the three i.i.d. variables X_1, X_2, and X_2'. For continuous F, we thus have

$$(6.1.25) \qquad P\left(X_1 + X_2 > 0 \text{ and } X_1 + X_2' > 0\right) = 1/3$$

and hence

$$(6.1.26) \qquad\qquad \sigma_1^2 = 1/12.$$

The opposite of distributions symmetric about 0 in a certain sense are distributions for which

$$(6.1.27) \qquad\qquad P\left(X > 0\right) \text{ is 0 or 1.}$$

For such distributions, both terms on the right side of (6.1.23) are 0 or 1 and hence $\sigma_1^2 = 0$. In fact, the variable U then degenerates to the constant 0 or 1.

As stated after (6.1.7), the statistic $W = \binom{n}{2} U$ is the number of pairs $i < j$ for which $X_i + X_j > 0$. It is interesting to compare W with the Wilcoxon statistic V_s of Example 3.2.5, which by (3.2.31) is the number of pairs $i \le j$ for which $X_i + X_j > 0$. If S denotes the number of positive X's, we therefore have

$$(6.1.28) \qquad\qquad V_S = W + S.$$

From this relation it can be seen that the standardized variables

$$\frac{V_S - E\left(V_S\right)}{\sqrt{\text{Var}\left(V_S\right)}} \quad \text{and} \quad \frac{W - E(W)}{\sqrt{\text{Var}(W)}}$$

have the same limit distribution (Problem 1.7).   □

Expectation functionals and *U*-statistics were defined in (6.1.1) and (6.1.3) for the case that $X_1, \ldots, X_n$ are n i.i.d. random variables. However, they apply equally when the X's are n i.i.d. random vectors. Then $\theta(F)$ may, for example, be a covariance, or if $X = (Y_1, \ldots, Y_k)$, it may be the probability $P\left(Y_1 \le a_1, \ldots, Y_k \le a_k\right)$.

Another extension, which does require some changes, is from a single sample to the s-sample case. Let

(6.1.29) $\phi\left(x_{11}, \ldots, x_{1a_1}; \ldots; x_{s1}, \ldots, x_{sa_s}\right)$

be a function of $a_1 + a_2 + \cdots + a_s$ arguments which is symmetric in each of the s groups of arguments, and let

(6.1.30) $\theta = E\phi\left(X_{11}, \ldots, X_{1a_1}; \ldots; X_{s1}, \ldots, X_{sa_s}\right),$

where $X_{i1}, \ldots, X_{ia_i}$ are i.i.d., according to distributions F_i, $i = 1, \ldots, s$, and the s groups are independent of each other.

As an example, suppose that $s = 2$, $a_1 = a_2 = 1$, and

$$\theta = P\left(X_{11} < X_{21}\right),$$

where X_{11} and X_{21} are independent with distributions F_1 and F_2, respectively. Then θ is given by (6.1.30) with

(6.1.31) $\phi\left(x_{11}, x_{21}\right) = \begin{cases} 1 & \text{when } x_{11} < x_{21} \\ 0 & \text{otherwise.} \end{cases}$

Another example with $s = 2$ and $a_1 = a_2 = 1$ is

$$\theta = E\left(X_{21} - X_{11}\right),$$

which is given by (6.1.30) with

$$\phi\left(x_{11}, x_{21}\right) = x_{21} - x_{11}.$$

Let us now consider estimating θ on the basis of independent observations $X_{11}, \ldots, X_{n_1}$, from F_1; $X_{21}, \ldots, X_{2n_2}$ from F_2; etc. Suppose that

$$a_i \le n_i \text{ for all } i = 1, \ldots, s$$

and that the distributions $F_1, \ldots, F_s$ are unknown, subject only to some very general conditions. By (6.1.30),

$$\phi\left(X_{11}, \ldots, X_{1a_1}; \ldots; X_{s1}, \ldots, X_{sa_s}\right)$$

is an unbiased estimator of θ, and so therefore is

$$\phi\left(X_{1i_1}, \ldots, X_{1i_{a_1}}; \ldots; X_{sr_1}, \ldots, X_{sr_{a_s}}\right)$$

for any of the $\prod_{i=1}^{s} \binom{n_i}{a_i}$ sets of combinations

(6.1.32) $$1 \le i_1 < \cdots < i_{a_1} \le n_1; \ldots; 1 \le r_1 < \cdots < r_{a_s} \le n_s.$$

It follows that also

(6.1.33) $$U = \frac{1}{\binom{n_1}{a_1} \cdots \binom{n_s}{a_s}} \sum \phi\left(X_{1i}, \ldots, X_{1i_{a_1}}; \cdots; X_{sr_1}, \ldots, x_{sr_{a_s}}\right)$$

is an unbiased estimator of θ, where the summation extends over all $\binom{n_1}{a_1} \cdots \binom{n_s}{a_s}$ combinations (6.1.32), so that

(6.1.34) $$E(U) = \theta.$$

The estimator U is symmetric in each of its s groups of variables and can be shown to have the smallest variance among all estimators that are unbiased for all $F = (F_1, \ldots, F_s)$ for which $\theta(F)$ exists.

It will be convenient in the following to denote the sum on the right side of (6.1.33) by

(6.1.35) $$W = \sum \phi\left(X_{1i_1}, \ldots, X_{1i_{a_1}}; \cdots; X_{sr_1}, \ldots, X_{sr_{a_1}}\right).$$

As an example, let $s = 2$, $a_1 = a_2 = 2$, and

(6.1.36) $$\phi\left(x_{11}, x_{12}; x_{21}, x_{22}\right) = \begin{cases} 1 & \text{if } |x_{12} - x_{11}| < |x_{22} - x_{21}| \\ 0 & \text{otherwise.} \end{cases}$$

Then W is the number of quadruples $i < j$; $k < l$ for which

$$|X_{1j} - X_{1i}| < |X_{2l} - X_{2k}|,$$

a statistic that has been proposed for testing that F_1 and F_2 differ only by a shift against the alternatives that F_2 is more spread out than F_1.

As a second example, let $a_1 = a_2 = \cdots = a_s = 1$ and let

(6.1.37)
$$\phi\left(x_{11}; \ldots; x_{s1}\right) = \text{the number of pairs } 1 \le \alpha < \beta \le s \text{ for which } x_{\alpha 1} < x_{\beta 1}.$$

Then W is the so-called Jonckheere-Terpstra statistic which is used to test $H: F_1 = \cdots = F_s$ against the alternatives that the responses expected under F_{i+1} are larger than those expected under F_i for all $i = 1, \ldots, s - 1$.

Theorem 6.1.1 concerning the variance of U generalizes in a natural way to several samples. Since the notation gets complicated, we shall state the extension here only for the case $s = 2$. It is then convenient to make a change of notation, replacing a_1, a_2 by $a, b; n_1, n_2$ by m, n; the variables X_{1i}, X_{2j} by X_i, Y_j and their distributions F_1, F_2 by F, G. The U-statistic (6.1.33) now becomes

$$(6.1.38) \qquad U = \frac{1}{\binom{m}{a}\binom{n}{b}} \sum \phi(X_{i_1}, \dots, X_{i_a}; Y_{j_1}, \dots, Y_{j_b})$$

with the sum extending over all

$$(6.1.39) \qquad 1 \le i_1 < \cdots < i_a \le m; \ 1 \le j_1 < \cdots < j_b \le n.$$

Theorem 6.1.3

(i) The variance of the U-statistic (6.1.38) is

$$(6.1.40) \qquad \mathrm{Var}(U) = \sum_{i=1}^{a}\sum_{j=1}^{b} \frac{\binom{a}{i}\binom{m-a}{a-i}}{\binom{m}{a}} \frac{\binom{b}{j}\binom{n-b}{b-j}}{\binom{n}{b}} \sigma_{ij}^2,$$

where

(6.1.41)
$$\sigma_{ij}^2 = \mathrm{Cov}\left[\phi(X_1, \dots, X_i, X_{i+1}, \dots, X_a; Y_1, \dots, Y_j, Y_{j+1}, \dots, Y_b),\right.$$
$$\left.\phi(X_1, \dots, X_i, X'_{i+1}, \dots, X'_a; Y_1, \dots, Y_j, Y'_{j+1}, \dots, Y'_b)\right].$$

Here the X's and Y's are independently distributed according to F and G, respectively, and the covariances (6.1.41) are ≥ 0 by an extension of Lemma 6.1.1 (Problem 1.9).
(ii) If

$$\sigma_{10}^2 = \mathrm{Cov}[\phi(X_1, X_2, \dots, X_a; Y_1, \dots, Y_b),$$
$$\phi(X_1, X'_2, \dots, X'_a; Y'_1, \dots, Y'_b)] > 0,$$
$$\sigma_{01}^2 = \mathrm{Cov}[\phi(X_1, \dots, X_a; Y_1, Y_2, \dots, Y_b),$$
$$\phi(X'_1, \dots, X'_a; Y_1, Y'_2, \dots, Y'_b)] > 0,$$

and $\sigma_{aa}^2 > 0$, and if $N = m + n$ and

$$(6.1.42) \qquad \frac{m}{N} \to \rho, \ \frac{n}{N} \to 1 - \rho \ \text{with } 0 \le \rho \le 1,$$

then

$$(6.1.43) \qquad \mathrm{Var}\left(\sqrt{N}U\right) \to \sigma^2 = \frac{a^2}{\rho}\sigma_{10}^2 + \frac{b^2}{1-\rho}\sigma_{01}^2.$$

Note: If $0 < \rho < 1$, it is enough in (ii) to require that at least one of σ_{10}^2 and σ_{01}^2 is positive.

Example 6.1.2 Wilcoxon two-sample statistic. Consider the U-statistic (6.1.38) with $a = b = 1$ and ϕ given by (6.1.31), so that

$$(6.1.44) \qquad U = \frac{1}{mn} \sum_{i=1}^{m} \sum_{j=1}^{n} \phi(X_i, Y_j)$$

with

$$(6.1.45) \qquad \phi(x,y) = 1 \text{ if } x < y \text{ and } = 0 \text{ otherwise}$$

and with

$$(6.1.46) \qquad \theta = P(X < Y).$$

Then

$$(6.1.47) \qquad \begin{aligned} \sigma_{10}^2 &= \text{Cov}\left[\phi(X_1, Y_1),\ \phi(X_1, Y_1')\right] \\ &= P(X_1 < Y_1 \text{ and } X_1 < Y_1') - [P(X_1 < Y_1)]^2 \end{aligned}$$

and, analogously,

$$(6.1.48) \qquad \sigma_{01}^2 = P(X_1 < Y_1,\ X_1' < Y_1) - [P(X_1 < Y_1)]^2.$$

For any given distributions F of the X's and G of the Y's, these variances can be evaluated numerically and, in a few cases, analytically (Problems 1.10 and 1.11).

The calculations greatly simplify when $F = G$. Then, by symmetry, if F is continuous,

$$P(X_1 < Y_1) = 1/2$$

and

$$P(X_1 < Y_1 \text{ and } X_1 < Y_1') = P(X_1 < Y_1 \text{ and } X_1' < Y_1) = 1/3$$

so that

$$(6.1.49) \qquad \sigma_{01}^2 = \sigma_{10}^2 = 1/12$$

and hence

$$(6.1.50) \qquad \text{Var}\left(\sqrt{N}U\right) \rightarrow \frac{1}{12}\left(\frac{1}{\rho} + \frac{1}{1-\rho}\right) = \frac{1}{12\rho(1-\rho)}.$$

Another special case occurs when

$$(6.1.51) \qquad P(X < Y) \text{ is 0 or 1.}$$

Then not only are $\sigma_{01}^2 = \sigma_{10}^2 = 0$, but the variable U degenerates to the constant 0 or 1 and hence $\text{Var}(U) = 0$. $\qquad\square$

Let us now return to the general U-statistics (6.1.33) and for this purpose define

$$\sigma_{i_1,\ldots,i_s} \quad (0 \le i_1 \le a_1, \ldots, 0 \le i_s \le a_s)$$

in obvious generalization of (6.1.41). Then we have the basic limit theorem for s-sample U-statistics.

Theorem 6.1.4 *Let the sample sizes $n_1, \ldots, n_s$ all tend to infinity in such a way that*

$$(6.1.52) \qquad \qquad \frac{n_i}{N} \to \rho_i, \quad 0 \le \rho_i \le 1,$$

where $N = n_1 + \cdots + n_s$, and let $0 < \sigma^2_{i_1,\ldots,i_s} < \infty$ for all

$$0 \le i_j \le a_j, \quad j = 1, \ldots, s.$$

Then

$$(6.1.53) \qquad \qquad U'_N = \sqrt{N}\,(U - \theta) \xrightarrow{L} N\left(0, \sigma^2\right)$$

with

$$(6.1.54) \qquad \qquad \sigma^2 = \frac{a_1^2}{\rho_1}\sigma^2_{10\ldots0} + \cdots + \frac{a_s^2}{\rho_s}\sigma^2_{00\ldots1}.$$

and also

$$(6.1.55) \qquad \qquad \mathrm{Var}\left(\sqrt{N}U\right) \to \sigma^2$$

and $(U - \theta)/\sqrt{\mathrm{Var}U} \to N(0,1)$.

The proof will be postponed to the end of the section.

Another useful extension deals with the joint distribution of two U-statistics. For the sake of simplicity, we shall suppose that they are both one-sample U-statistics, say

$$U^{(1)} = \frac{1}{\binom{n}{a}} \sum \cdots \sum \phi^{(1)}\left(X_{i_1}, \ldots, X_{i_a}\right),$$

$$(6.1.56)$$

$$U^{(2)} = \frac{1}{\binom{n}{b}} \sum \cdots \sum \phi^{(2)}\left(X_{i_1}, \ldots, X_{i_b}\right)$$

with the summations defined as in (6.1.3) for an i.i.d. sample $X_1, \ldots, X_n$ (real- or vector-valued) and with $a, b \le n$. Let

$$\theta_i = EU^{(i)}, \quad i = 1, 2,$$

and

$$\sigma_{ij} = \text{Cov}[\phi^{(1)}(X_1,\dots,X_i, X_{i+1},\dots,X_a),$$
$$\phi^{(2)}(X_1,\dots,X_j, X'_{j+1},\dots,X'_b)],$$

where $i \le a$, $j \le b$. Then we have in generalization of Theorem 6.1.1:

Theorem 6.1.5 *If $a \le b$, the covariance of $U^{(1)}$ and $U^{(2)}$ is*

$$(6.1.57) \qquad \text{Cov}\left(U^{(1)}, U^{(2)}\right) = \sum_{i=1}^{a} \binom{b}{i}\binom{n-b}{a-i}\sigma_{ii}\Big/\binom{n}{a}$$

and

$$(6.1.58) \qquad n\,\text{Cov}\left(U^{(1)}, U^{(2)}\right) \to ab\,\sigma_{11}.$$

For a proof, see Hoeffding (1948a) or Lee (1990). Also, in generalization of Theorem 6.1.2, we have

Theorem 6.1.6 *If*

$$\text{Var}\,\phi^{(1)}(X_1,\dots,X_a) \text{ and } \text{Var}\,\phi^{(2)}(X_1,\dots,X_b) \text{ are both } < \infty,$$

then

$$(6.1.59) \qquad \left(\sqrt{n}\left(U^{(1)} - \theta_1\right),\ \sqrt{n}\left(U^{(2)} - \theta_2\right)\right) \to N(0, \Sigma).$$

Here Σ is the limiting covariance matrix of $(\sqrt{n}(U^{(1)} - \theta_1)$, $\sqrt{n}(U^{(2)} - \theta_2))$, the entries of which are given by (6.1.18) and (6.1.58), provided Σ is positive definite.

Proof. If $a = b$, the result follows immediately from Theorem 5.1.8 and the limit theorem 6.1.2 for a single U-statistic. The case $a \ne b$ can be handled by noting that an a-dimensional U-statistic can also be represented as a U-statistic of dimension b for any $a \le b$. (See hint for Problem 1.2.) Thus $U^{(1)}$ and $U^{(2)}$ can be represented as U-statistics of common dimension b.∎

Example 6.1.3 Wilcoxon and t. Suppose that $X_1,\dots,X_n$ are i.i.d. according to a distribution F that is symmetric about ξ. Two widely used tests of $H : \xi = 0$ are the one-sample t-test and Wilcoxon tests discussed in Sections 3.1 and 3.2. It is tempting (and although frowned upon, not unheard of) to compute both test statistics but report only the result of the more significant of the two. If both tests are carried out at level α, the true level α' resulting from such practice will, of course, exceed α. Let us see by how much.

For this purpose, we require the joint asymptotic null distribution of $\bar{X}/\hat{\sigma}$ and of the one-sample Wilcoxon statistic discussed in Chapter 3 and in Example 6.1.1 of the present chapter. These two statistics are asymptotically equivalent to respectively $U_1 = \bar{X}/\sigma$ and the U-statistic U_2 given by (6.1.11) with

$$\phi(x_1, x_2) = 1 \text{ if } x_1 + x_2 > 0, \text{ and } = 0 \text{ otherwise.}$$

Lemma 6.1.2 *Under H, the joint limit distribution of $\sqrt{n}U_1$ and $\sqrt{n}(U_2 - \theta)$ with θ given by (6.1.8) is bivariate normal with mean zero and with covariance matrix $\Sigma = \begin{pmatrix} 1 & \gamma \\ \gamma & 1/3 \end{pmatrix}$, where*

$$(6.1.60) \qquad \gamma = E\left[\max(X_1, X_2)\right].$$

Proof. From Theorem 6.1.5 and Example 6.1.1, it follows that the desired limit distribution is bivariate normal with

$$\sigma^2 = \operatorname{Var}\left(\sqrt{n}U_1\right) = 1 \text{ and } \tau^2 = \lim \operatorname{Var}\left(\sqrt{n}U_2\right) = 1/3.$$

It remains to evaluate the covariance of $(\sqrt{n}U_1, \sqrt{n}U_2)$ which, by (6.1.58), is equal to

$$\lim[n \operatorname{Cov}(U_1, U_2)] = 2 \operatorname{Cov}\left[\frac{X_1}{\sigma}, \phi(X_1, X_2)\right] = 2E\left[\frac{X_1}{\sigma}\phi(X_1, X_2)\right].$$

Since $\phi(x_1, x_2) = \phi(x_2, x_1)$ and since $X_1, -X_1, X_2$, and $-X_2$ all have the same distribution, we have

$$(6.1.61) \qquad E[X_1\phi(X_1, X_2)] = \frac{1}{2}E[X_1\phi(X_1, -X_2) + X_2\phi(-X_1, X_2)].$$

Now

$$X_1\phi(X_1, -X_2) = \begin{cases} X_1 & \text{if } X_1 > X_2 \\ 0 & \text{otherwise;} \end{cases}$$

$$X_2\phi(-X_1, X_2) = \begin{cases} X_2 & \text{if } X_2 > X_1 \\ 0 & \text{otherwise;} \end{cases}$$

and hence by (6.1.61)

$$E(X_1\phi(X_1, X_2)) = \frac{1}{2}E\left[\max(X_1, X_2)\right].$$

∎

The distributions of the two test statistics $\sqrt{n}U_1$ and $\sqrt{3n}U_2$ both have a standard normal limit under H and the procedure under consideration therefore rejects H when at least one of them exceeds the standard normal critical value u_α. The rejection probability under H is therefore

$$(6.1.62) \qquad \alpha' = P\left[\max\left(\sqrt{n}U_1, \sqrt{3n}U_2\right) > u_\alpha\right].$$

The asymptotic value of α' can be obtained from the bivariate normal distribution given in Lemma 6.1.2. When $\alpha = .05$, some typical values of α' are

$$\begin{array}{ccc} \text{Normal} & \text{Uniform} & \text{Double Exponential} \\ .059 & .050 & .066 \end{array}$$

Since it follows from (6.1.62) that $\alpha' \geq \alpha$, the value .05 for the uniform distribution is the smallest α' can be. On the other hand, it can be shown (Jiang (1997)) that an upper bound for α' is $1 - (1 - \alpha)^2$, the value of α' when the two test statistics in question are independent (Problem 1.15), so that for $\alpha = .05$, we have $\alpha' \leq .0975$. (This upper bound is sharp.) $\quad\square$

To conclude the section, we shall now consider the proof of Theorem 6.1.4. The proof in the general case is somewhat tedious because of the complexity of the notation. We shall therefore carry out the proof only for the case $s = 2$, $a_1 = a_2 = 1$. The case $s = 1$, $a_1 = 2$ is treated in Problem 1.17. A general proof can be found in Lee (1990), Lehmann (1998), and Serfling (1980).

Proof of Theorem 6.1.4 when $s = 2$, $a_1 = a_2 = 1$. Let $X_1, \ldots, X_m$ and $Y_1, \ldots, Y_n$ be independently distributed with distributions F and G, respectively, and with m and n satisfying (6.1.38). For

$$(6.1.63) \qquad U = \frac{1}{mn} \sum_{i=1}^{m} \sum_{j=1}^{n} \phi(X_i, Y_j)$$

and

$$(6.1.64) \qquad \theta = E\phi(X, Y) = E(U),$$

we wish to prove (6.1.53).

The proof utilizes the following lemma.

Lemma 6.1.3 *Let T_N^*, $N = 1, 2, \ldots$, be a sequence of random variables, the distributions of which tend to a limit distribution L, and let T_N be another sequence satisfying*

$$(6.1.65) \qquad E(T_N^* - T_N)^2 \to 0.$$

Then the distribution of T_n also tends to L.

Proof. Let $R_N = T_N - T_N^*$. Then it follows from Theorem 2.1.1 that $R_N \xrightarrow{P} 0$ and hence from Theorem 2.3.3 that T_N has the same limit distribution as T_N^*.

The theorem will be proved by applying the lemma to

$$(6.1.66) \qquad T_N = \sqrt{N}(U - \theta)$$

and a sum T_N^* of independent terms so that its limit follows from the central limit theorem, and satisfying

$$(6.1.67) \qquad E(T_N^*) = 0.$$

We then have

$$E\left(T_N - T_N^*\right)^2 = \text{Var}\left(T_N\right) + \text{Var}\left(T_N^*\right) - 2\,\text{Cov}\left(T_N, T_N^*\right).$$

It follows from (6.1.43) with $a = b = 1$ that

(6.1.68) $$\text{Var}\left(T_N\right) \to \sigma^2 = \frac{\sigma_{10}^2}{\rho} + \frac{\sigma_{01}^2}{1-\rho}$$

and the proof will be completed by showing that also

(6.1.69) $$\text{Var}\left(T_N^*\right) \to \sigma^2 \text{ and } \text{Cov}\left(T_N, T_N^*\right) \to \sigma^2.$$

To carry out this program, i.e., to prove asymptotic normality of T_N^* and the relations (6.1.67) and (6.1.69), we must next define T_N^*. For this purpose, we introduce the functions

(6.1.70) $$\phi_{10}(x) = E\phi(x, Y) \text{ and } \phi_{01}(y) = E\phi(X, y).$$

For example, if ϕ is given by (6.1.45), we have

(6.1.71) $$\begin{aligned} \phi_{10}(x) &= P(x < Y) = 1 - G(x), \\ \phi_{10}(y) &= P(X < y) = F(y). \end{aligned}$$

By Problem 1.9, these functions satisfy

(i) $E\phi_{10}(X) = E\phi_{01}(Y) = \theta$

and

(ii) $\text{Var } \phi_{10}(X) = \sigma_{10}^2$ and $\text{Var } \phi_{01}(Y) = \sigma_{01}^2$, where σ_{ij}^2 is defined by (6.1.41), so that

(6.1.72) $$\sigma_{10}^2 = \text{Cov}\left[\phi\left(X_1, Y_1\right), \phi\left(X_1, Y_1'\right)\right]$$

and

(6.1.73) $$\sigma_{01}^2 = \text{Cov}\left[\phi\left(X_1, Y_1\right), \phi\left(X_1', Y_1\right)\right].$$

To apply Lemma 6.1.3, let us now define*

(6.1.74) $$T_N^* = \sqrt{N}\left\{\frac{1}{m}\left[\sum \phi_{10}\left(X_i\right) - \theta\right] + \frac{1}{n}\sum\left[\phi_{01}\left(Y_j\right) - \theta\right]\right\}.$$

Then (6.1.67) follows from (i), and hence by the central limit theorem

(6.1.75) $$\sqrt{m}\left[\frac{1}{m}\sum\phi_{10}\left(X_i\right) - \theta\right] \xrightarrow{L} N\left(0, \sigma_{10}^2\right)$$

---

*An explanation for this choice of T_N^* will be given at the end of Section 6.3.

and

$$(6.1.76) \qquad \sqrt{n} \left[\frac{1}{n} \sum \phi_{01}(Y_j) - \theta \right] \xrightarrow{L} N\left(0, \sigma_{01}^2\right).$$

It now only remains to prove (6.1.69), which we shall leave to Problem 1.16. This completes the proof of Theorem 1.4.   ∎

Example 6.1.2. Wilcoxon two-sample statistic (continued). Theorem 6.1.4 establishes the asymptotic normality of the Wilcoxon two-sample statistic for any distributions F and G for which not both σ_{10}^2 and σ_{01}^2 are 0. It can be shown (see, for example, Lehmann (1998, p. 366)) that $\sigma_{10}^2 = \sigma_{01}^2 = 0$ if and only if (6.1.51) holds; i.e., if the distribution of the Y's lies entirely to the right or to the left of the distribution of the X's in either of which cases W_{XY} degenerates to a constant.

In particular, for any (F, G) not satisfying (6.1.51), this verifies (3.3.29) with $W_{XY} = T_N$, which was required for the asymptotic power of the two-sample Wilcoxon test against the shift alternatives

$$(6.1.77) \qquad G(y) = F(y - \theta)$$

(Example 3.3.7.) However, this power result required asymptotic normality not only against a fixed alternative but also against a sequence of alternatives

$$(6.1.78) \qquad G_k(y) = F(y - \theta_k)$$

with $\theta_k \to 0$.

To extend Theorem 6.1.4 to this case, consider once more the proof of the theorem. This proof has two components:

(a) Application of the central limit theorem to T_N^*;

(b) Showing that $E\left(T_N - T_N^*\right)^2 \to 0$.

Of these, (b) causes no new difficulty since σ_{01}^2 and σ_{10}^2 are both continuous functions of θ. A proof of (a) can be based on the Berry-Esseen theorem by means of an argument already used in Example 3.3.2. The conditions of the Berry-Esseen theorem will be satisfied if the third moments of $\phi_{10}(X)$ and $\phi_{01}(Y)$ are bounded and their variances are bounded away from 0. The first of these conditions clearly holds since the ϕ's are between 0 and 1. The variances in question are σ_{10}^2 and σ_{01}^2, which by (6.1.49) are equal to $1/12$ when $F = G$. Under the alternatives (6.1.78), they are continuous functions of θ and hence bounded away from 0 as $\theta \to 0$.

Theorem 6.1.4 generalizes the central limit theorem for i.i.d. variables to U-statistics. A corresponding generalization of the Berry-Esseen theorem is also available, as is a Poisson limit theorem generalizing that obtained for independent binomial trials in formula (1.2.8). A third problem not discussed here is that raised by the degenerate case $\sigma^2 = 0$ in (6.1.55). The limit result (6.1.53) then only states that $\sqrt{N}(U - \theta) \xrightarrow{P} 0$, and a non-degenerate limit distribution will require a different normalization, in typical cases N instead of $\sqrt{N}$. Detailed treatments of these three topics are given in Lee (1990).

Summary

1. A special class of statistical functionals, the expectation functionals, is defined. Their best unbiased estimators are the U-statistics, which include a number of statistics used in testing and estimation, such as the two-sample Wilcoxon statistic, the s-sample Jonckheere-Terpstra statistic, and Gini's mean difference.

2. It is shown that U-statistics are asymptotically normal, provided certain second moment conditions are satisfied.

6.2 Statistical functionals

As pointed out in Section 2.1 (following Example 2.1.2), one commonly speaks of the asymptotic properties of the sample mean $\bar{X} = (X_1 + \cdots + X_n)/n$, although these properties really refer to the sequence $\bar{X}_n$, $n = 1, 2, \ldots$, and the same remark applies to the sample median, the sample variance, and so on. This shortcut terminology is suggested not only by its convenience but also by a feeling that these statistics are in some sense the "same" function of the observations for all n. The following is one way of making this feeling precise.

Each of these statistics, or sequence of statistics, is a consistent estimator of a corresponding population quantity: the sample mean $\bar{X}$ of the expectation $E(X)$, the kth sample moment $\sum (X_i - \bar{X})^k /n$ of the kth population moment $E[X - E(X)]^k$, the pth sample quantile of the pth population quantile $F^{-1}(p)$, etc. Any such population quantity is a function of the distribution F of the X_i (which we are assuming to be i.i.d.) and can therefore be written as $h(F)$, where h is a real-valued function defined over a collection $\mathcal{F}$ of distributions F. The mean $h(F) = E_F(X)$, for example, is defined over the class $\mathcal{F}$ of all F with finite expectation. The functions $h(F)$ are what in the preceding section were defined as statistical functionals.

To establish the connection between the sequence of sample statistics and the functional $h(F)$ that it estimates, define the *sample cdf* $\hat{F}_n$ by

$$(6.2.1) \qquad \hat{F}_n(x) = \frac{\text{Number of } X_i \leq x}{n}.$$

This is the cdf of a distribution that assigns probability $1/n$ to each of the n sample values $X_1, X_2, \ldots, X_n$. For the examples mentioned so far and many others, it turns out that the standard estimator of $h(F)$ based on n observations is equal to $h\left(\hat{F}_n\right)$, the plug-in estimator of $h(F)$. Suppose, for example, that $h(F) = E_F(X)$. The expectation of a random variable with cdf $\hat{F}_n$ is the sum of the probabilities $(1/n)$ multipled by the values (X_i) taken on by a random variable with cdf $\hat{F}_n$, i.e.,

$$(6.2.2) \qquad h\left(\hat{F}_n\right) = \frac{1}{n}X_1 + \cdots + \frac{1}{n}X_n = \bar{X}.$$

Analogously, when $h(F) = E_F\left(X - E(X)\right)^k$, it is seen that

(6.2.3)
$$h\left(\hat{F}_n\right) = \frac{1}{n}\left(X_1 - \bar{X}\right)^k + \cdots + \frac{1}{n}\left(X_n - \bar{X}\right)^k = \frac{1}{n}\sum\left(X_i - \bar{X}\right)^k = M_k.$$

When viewed not as a function of n variables, but as a function of $\hat{F}_n$, $\bar{X}$ or M_k is then indeed the "same" function h defined by (6.2.2) or (6.2.3) for all n.

For the third example mentioned in which $h(F) = F^{-1}(p)$ is a quantile of F, a slight ambiguity mars the relation between the standard estimator of $h(F)$ and $h\left(\hat{F}_n\right) = \hat{F}_n^{-1}(p)$. Suppose, for example, that $p = 1/2$ and that $n = 2m$ is even. Then $\hat{F}_n^{-1}\left(\frac{1}{2}\right)$ could be defined as any point in the interval $\left[X_{(m)}, X_{(m+1)}\right]$. As was discussed in Section 1.6, it is usually defined to be the left-hand end point $X_{(m)}$. On the other hand, the standard definition of the sample median is the midpoint $\left[X_{(m)} + X_{(m+1)}\right]/2$. As was seen in Example 2.4.9, these two estimators are asymptotically equivalent.

Before discussing additional examples, it is convenient to introduce the integral

$$(6.2.4) \qquad \int a\,dF = \int a(x)dF(x) = E_F\left[a(X)\right]$$

so that in particular

$$(6.2.5) \quad \int a\,dF = \begin{cases} \int a(x)f(x)dx & \text{when } F \text{ has a density } f \\ \sum a\left(x_i\right)P_F\left(X = x_i\right) & \text{when } F \text{ is discrete.} \end{cases}$$

Thus, for example,

$$(6.2.6) \qquad \int a(x)d\hat{F}_n(x) = \frac{1}{n}\sum_{i=1}^{n} a\left(x_i\right).$$

The notation (6.2.5) extends in an obvious way to muliple integrals. For example, if X and Y are independent with distributions F and G, respectively, then

(6.2.7)
$$E\left[a(X,Y)\right] = \int\int a(x,y)dF(x)dG(y)$$

$$= \begin{cases} \int\int a(x,y)f(x)g(y)dxdy \quad \text{when } F,G \text{ have densities } f,g \\ \sum\sum a\left(x_i,y_i\right)P_F\left(X=x_i\right)P_G\left(Y=y_j\right) \\ \qquad\qquad\qquad\qquad \text{when } X \text{ and } Y \text{ are discrete.} \end{cases}$$

Since $\left(\hat{F}_m,\hat{G}_n\right)$ takes on the mn pairs of values (x_i,y_j) with probability $1/mn$ each, we have in particular

(6.2.8) $$\int\int a(x,y)d\hat{F}_m(x)d\hat{G}_n(y) = \frac{1}{mn}\sum\sum a\left(x_i,y_j\right).$$

Example 6.2.1 Goodness-of-fit statistics. For testing the hypothesis H that a distribution F equals some specified distribution F_0 on the basis of a sample $X_1,\ldots,X_n$, many goodness-of-fit statistics were discussed in Section 5.7, among them:

(i) The **Kolmogorov statistic**

(6.2.9) $$\sup_x\left|\hat{F}_n(x)-F_0(x)\right|;$$

(ii) the **Cramér-von Mises statistic**

(6.2.10) $$\int\left[\hat{F}_n(x)-F_0(x)\right]^2 dF_0(x);$$

(iii) the classical **Pearson statistic**

(6.2.11) $$\sum_{j=0}^r\left(\hat{p}_j-p_j^0\right)^2/p_j^0,$$

where p_j^0 denotes the probability of the interval $I_j=(a_j,a_{j+1})$ under F_0 and $\hat{p}_j$ denotes the observed proportion of observations in I_j.

Each of these statistics is of the form $h\left(\hat{F}_n\right)$, where in the first two cases

(6.2.12)
$$h(F) = \sup|F(x)-F_0(x)| \quad \text{and} \quad h(F) = \int\left[F(x)-F_0(x)\right]^2 dF_0(x),$$

respectively, and in the third case,

(6.2.13)
$$h(F) = \sum_{j=0}^{r} \frac{[F(a_{j+1}) - F(a_j) - p_j^0]^2}{p_j^0}.$$

□

Before attempting a general theory of these plug-in estimators, it is useful to consider their large-sample behavior in a few examples.

Example 6.2.2 Estimating the cdf. If $F(a) = P(X \leq a)$ for some fixed a, then

(6.2.14)
$$\hat{F}_n(a) = \frac{\text{Number of } X_i \leq a}{n}.$$

This is just the usual estimator of the unknown probability of an event by the observed frequency of its occurrence.

The number Y of $X's \leq a$ has the binomial distribution

(6.2.15)
$$Y : b(p, n) \text{ with } p = F(a).$$

Since

(6.2.16)
$$E\left(\frac{Y}{n}\right) = p \text{ and Var}\left(\frac{Y}{n}\right) = \frac{pq}{n} \ (q = 1 - p),$$

it follows that

(6.2.17)
$$E\left[\hat{F}_n(a)\right] = F(a) \text{ and Var}\left[\hat{F}_n(a)\right] = \frac{1}{n}F(a)[1 - F(a)]$$

so that $\hat{F}_n(a)$ is unbiased and its variance is of order $1/n$. In addition, it follows from (2.3.9) that

(6.2.18)
$$\sqrt{n}\left[\hat{F}_n(a) - F(a)\right] \xrightarrow{L} N\left(0, F(a)[1 - F(a)]\right).$$

Both (6.2.17) and (6.2.18) imply that $\hat{F}_n(a)$ is a consistent estimator of $F(a)$ for each fixed a. However, a much stronger consistency property can be asserted if the difference between $\hat{F}_n(x)$ and $F(x)$ is considered not only for a fixed x but simultaneously for all x, namely

(6.2.19)
$$D_n = \sup_x \left|\hat{F}_n(x) - F(x)\right| \xrightarrow{P} 0 \text{ as } n \to \infty.$$

For continous distribution F, this result was already pointed out in Section 5.7. It remains valid for general F, but the proof then requires special consideration of the discontinuities. For a still stronger result, see, for example, Serfling (1980, Section 2.1.4) or Billingsley (1986, Theorem 20.6).

The limit (6.2.19) states that if we consider a symmetric band of width 2ϵ about F, then the probability that $\hat{F}_n$ will lie entirely within this band tends to 1 for any given $\epsilon > 0$, no matter how small ϵ is. This provides a justification for thinking of $\hat{F}_n$ as an estimator of the unknown distribution function F, and hence of $h\left(\hat{F}_n\right)$ as an estimator of $h(F)$. $\square$

Example 6.2.3 Estimating the mean. Let $h(F)$ be the expectation of F, i.e.,

$$(6.2.20) \qquad h(F) = \int x\, dF(x) = \xi.$$

Then by (6.2.2)

$$(6.2.21) \qquad h\left(\hat{F}_n\right) = \frac{1}{n}\sum_{i=1}^{n} X_i = \bar{X}.$$

(Typically, unless F is normal, this estimator is no longer a good choice for estimating the expectation of F when the distributional form of F is known.)

Clearly, $\bar{X}$ is unbiased, its variance is σ^2/n where σ^2 is the variance of the X's, and $\sqrt{n}\left(\bar{X} - \xi\right) \xrightarrow{L} N\left(0, \sigma^2\right)$, provided $\sigma^2 < \infty$. $\square$

In both Examples 6.2.2 and 6.2.3, the estimator $h\left(\hat{F}_n\right)$ turned out to be unbiased for estimating $h(F)$. This need not be the case, as is shown by the following example.

Example 6.2.4 Central moments. Let $h(F)$ be the kth central moment of F denoted by μ_k in Example 2.1.3. We have

$$(6.2.22) \qquad h(F) = \mu_k = E\left(X_i - \xi\right)^k,$$

where $\xi = E\left(X_i\right)$, and

$$(6.2.23) \qquad h\left(\hat{F}_n\right) = M_k = \frac{1}{n}\sum\left(X_i - \bar{X}\right)^k.$$

In Section 2.1, M_k was found to be a consistent estimator of μ_k. We shall now study the asymptotic behavior of M_k in more detail.

In the simplest case $k = 2$, μ_2 reduces to $\text{Var}\left(X_i\right) = \sigma^2$ and we find that

$$
\begin{aligned}
E\left(M_2\right) &= \frac{1}{n}E\sum\left[\left(X_i - \xi\right) - \left(\bar{X} - \xi\right)\right]^2 \\
&= \frac{1}{n}\left[E\sum\left(X_i - \xi\right)^2 - nE\left(\bar{X} - \xi\right)^2\right] \\
&= \sigma^2 - \frac{\sigma^2}{n} = \frac{n-1}{n}\sigma^2.
\end{aligned}
$$
$(6.2.24)$

The estimator M_2 of $\mu_2 = \sigma^2$ therefore has a negative bias of order $1/n$. It can, of course, be made unbiased by replacing M_2 by $\dfrac{n}{n-1} M_2$.

As a slightly more complicated case that better illustrates the general approach, consider

$$M_3 = \frac{1}{n} \sum (X_i - \bar{X})^3$$
$$= \frac{1}{n} \sum X_i^3 - \frac{3}{n^2} \sum X_i^2 \sum X_j + \frac{3}{n^3} \sum X_i \left(\sum X_j \right)^2 - \frac{1}{n^3} \left(\sum X_j \right)^3.$$

Since the distribution of M_3 does not depend on ξ, assume without loss of generality that $E(X_i) = \xi = 0$ for all i. (This is equivalent to, but more convenient than, carrying ξ along, as was done in (6.2.24).) Then all the terms $X_i^2 X_j$ and $X_i X_j^2$ with $i \neq j$, and the terms $X_i X_j X_k$ with all three subscripts distinct, have zero expectation and one is left with

$$E(M_3) = \mu_3 \left[1 - \frac{3}{n} + \frac{3}{n^2} - \frac{1}{n^2} \right] = \mu_3 - \frac{3}{n} \mu_3 + \frac{2}{n^2} \mu_3.$$

The estimator M_3 therefore again has a bias which is $0(1/n)$. Since

$$E(M_3) = \frac{(n-1)(n-2)}{n^2} \mu_3,$$

the bias can be eliminated through multiplication of M_3 by $n^2/(n-1)(n-2)$.

The same type of argument shows quite generally that

(6.2.25) $$E(M_k) = \mu_k + \frac{a}{n} + O\left(\frac{1}{n^2} \right)$$

with

(6.2.26) $$a = \binom{k}{2} \mu_{k-2} \mu_2 - k\mu_k.$$

(For details, see Problem 2.4(i) or Serfling (1980).)

The variance of M_k can be calculated in the same way. For $k = 2$, one finds (Cramér (1946))

(6.2.27) $$\mathrm{Var}(M_2) = \frac{1}{n^2} E \left[\sum X_i^2 - n\bar{X}^2 \right]^2 - [E(M_2)]^2$$
$$= \frac{\mu_4 - \mu_2^2}{n} - \frac{2(\mu_4 - 2\mu_2^2)}{n^2} + \frac{(\mu_4 - 3\mu_2^2)}{n^3},$$

and quite generally that (Problem 2.4(ii))

(6.2.28) $$\mathrm{Var}(M_k) = \frac{b}{n} + O\left(\frac{1}{n^2} \right)$$

with

(6.2.29) $$b = \mu_{2k} - 2k\mu_{k-1}\mu_{k+1} - \mu_k^2 + k^2\mu_2\mu_{k-1}^2.$$

Equations (6.2.25) and (6.2.28) show that both bias and variance of M_k tend to 0 as $n \to \infty$ and hence that M_k is consistent for estimating μ_k. This was proved more directly in Example 2.1.3.

Instead of the asymptotic behavior of bias and variance, we may be interested in the asymptotic distribution of

(6.2.30) $$\sqrt{n}\,(M_k - \mu_k).$$

It will be shown in Section 6.3 that (6.2.30) tends in law to a normal distribution with mean 0 and variance equal to the constant b given by (6.2.29). Thus the variance of the limit distribution of (6.2.30) is equal to the limit of $\mathrm{Var}\,[\sqrt{n}\,(M_k - \mu_k)]$. □

Example 6.2.5 U- and V-statistics. An important class of functionals consists of the expectation functionals $\theta = h(F)$ defined by (6.1.1). Since $\hat{F}_n$ assigns probability $1/n$ to each of the values $X_1,\dots,X_n$, a independent variables with distribution $\hat{F}_n$ take on each of the possible a-tuples $X_{i_1},\dots,X_{i_a}$ with probability $1/n^a$. The estimator $h\left(\hat{F}_n\right)$ of $\theta = E_F\phi(X_1,\dots,X_a)$ is therefore

(6.2.31) $$V = h\left(\hat{F}_n\right) = \frac{1}{n^a}\sum_{i_1=1}^{n}\cdots\sum_{i_a=1}^{n}\phi(X_{i_1},\dots,X_{i_a}).$$

The statistics V are closely related to the one-sample U-statistics defined by (6.1.3).

When $a = 1$, (6.2.31) reduces to

(6.2.32) $$V = \frac{1}{n}\sum_{i=1}^{n}\phi(X_i),$$

which agrees exactly with the U-statistic for $a = 1$.

Consider next the case $a = 2$. Then

(6.2.33) $$U = \frac{2}{n(n-1)}\sum\sum_{i<j}\phi(X_i,X_j) = \frac{1}{n(n-1)}\sum\sum_{i\neq j}\phi(X_i,X_j)$$

while

(6.2.34)
$$V = \frac{1}{n^2}\sum_{i=1}^{n}\sum_{j=1}^{n}\phi(X_i,X_j) = \frac{1}{n^2}\sum\sum_{i\neq j}\phi(X_i,X_j) + \frac{1}{n^2}\sum_{i=1}^{n}\phi(X_i,X_i).$$

The statistics (6.2.33) and (6.2.34) differ in two ways: V has an extra term corresponding to the pairs (i, j) with $i = j$, and the factor of the common term is $1/n^2$ for V and $1/n(n-1)$ for U. It is intuitively plausible and will be seen below that asymptotically both these differences are negligible for most purposes.

Since

$$(6.2.35) \qquad\qquad\qquad E(U) = \theta,$$

the estimator U of θ is unbiased. On the other hand,

$$(6.2.36) \quad E(V) = \frac{n-1}{n}\theta + \frac{1}{n}E\phi\left(X_1, X_1\right) = \theta + \frac{1}{n}\left[E\phi\left(X_1, X_1\right) - \theta\right]$$

so that typically V will be biased with the bias tending to 0 as $n \to \infty$.

Theorem 6.2.1

(i) *If the conditions of Theorem 6.1.2 are satisfied with $a = 2$, then*

$$(6.2.37) \qquad\qquad\qquad \sqrt{n}\,(V - \theta) \xrightarrow{L} N\left(0, 4\sigma_1^2\right).$$

(ii) *If in addition*

$$(6.2.38) \qquad\qquad\qquad E\phi^2\left(X_1, X_1\right) < \infty,$$

then also

$$(6.2.39) \qquad\qquad\qquad \mathrm{Var}\left[\sqrt{n}\,(V - \theta)\right] \to 4\sigma_1^2.$$

These results correspond exactly to those of Theorem 6.1.2. However, because of the extra term in (6.2.33), (6.2.39) requires the additional condition (6.2.38).

Proof. From (6.2.33) and (6.2.34), it is seen that

$$(6.2.40) \quad \sqrt{n}\,(V_n - \theta) = \frac{n-1}{n}\sqrt{n}\,(U_n - \theta) + \frac{\sqrt{n}}{n^2}\sum\left[\phi\left(X_i, X_i\right) - \theta\right].$$

The first term on the right side tends in law to $N\left(0, 4\sigma_1^2\right)$ by Theorem 6.1.2, and the second term tends to zero in probability by the law of large numbers. This completes the proof of (6.2.37); for the proof of (6.2.39), see Problem 2.15. ∎

These results extend without much difficulty to general a.

Theorem 6.2.2

(i) *If the assumptions of Theorem 6.1.2(i) hold, then (6.1.20) also holds when U is replaced by V.*

(ii) If in addition, assumption (6.1.21) is replaced by the stronger assumption

$$\text{Var } \phi\left(X_{i_1}, \ldots, X_{i_a}\right) < \infty$$

for all $1 \le i_1 \le i_2 \le \cdots \le i_a \le a$, then also (6.1.22) remains valid when U is replaced by V.

For the proof for the case $a = 3$, see Problem 2.5.   □

Example 6.2.6 Quantiles. For fixed $0 < t < 1$, let $h(F) = F^{-1}(t)$ be defined in the usual way as

$$(6.2.41) \qquad F^{-1}(t) = \inf\{x : F(x) \ge t\}$$

Then $h\left(\hat{F}_n\right)$ is the $[nt]$-th order statistic, which is one version of the $[nt]$-th sample quantile. (In this connection see the discussion at the end of Section 1.6.)

If F has a density which is continuous and positive in a neighborhood of t, it was shown in Example 2.4.9 and Problem 4.8 of Chapter 2 that

$$(6.2.42) \qquad \sqrt{n}\left[h\left(\hat{F}_n\right) - h\left(F\right)\right] \overset{L}{\to} N\left(0, \frac{F(t)[1 - F(t)]}{\{f\left[F^{-1}(t)\right]\}^2}\right)$$

Under suitable regularity conditions it is proved by Bickel (1967) that the bias and variance of $h\left(\hat{F}_n\right)$ is of order $1/n$ and that the limit of the variance of (6.2.42) agrees with the asymptotic variance.   □

Examples 6.2.2–6.2.6 treated particular functionals h for which $\sqrt{n}\left[h\left(\hat{F}_n\right) - h(F)\right]$ was shown to tend to a normal limit, and hence $h\left(\hat{F}_n\right)$ to be a consistent estimator of $h(F)$. However, the representation of the estimators as functionals evaluated at the sample cdf is useful not primarily for the consideration of particular cases as in suggesting the possibility of a general theory for this class of estimators.

To develop such a theory, we must first generalize some of the results of Chapter 1 for functions of real variables, to functions of distributions. For this purpose, it is convenient to think of distributions F as points of a space, and to begin by considering the convergence of a sequence of such points. In the vector-valued case of points $\underline{x} = (x_1, \ldots, x_k)$, the following three definitions of $\underline{x}^{(n)} \to \underline{a}$ are equivalent:

(i) $x_i^{(n)} \to a_i$ for each $i = 1, \ldots, k$;

(ii) $\max\limits_{i=1,\ldots,k} \left|x_i^{(n)} - a_i\right| \to 0$;

(iii) $d\left(\underline{x}^{(n)}, \underline{a}\right) \to 0$, where $d\left(\underline{x}, \underline{y}\right) = \sqrt{\sum (x_i - y_i)^2}$.

In generalizing these definitions to distributions, note that we are comparing the coordinates x_i $(i = 1, \ldots, k)$ with the values $F(a)$, $-\infty < a < \infty$ (where we have switched from $F(x)$ to $F(a)$ to avoid the use of x for two different meanings). The comparison becomes clearer if we write $x(i)$ for x_i and establish the correspondence

$$i \to a, \ x \to F.$$

Generalization of (i)–(iii) now leads to the definition of $G_n \to F$ by

(i') $G_n(x) \to F(x)$ for all x;

(ii') $\sup_x |G_n(x) - F(x)| \to 0$;

(iii') $\int [G_n(x) - F(x)]^2 \, dF(x) \to 0$.

However, unlike (i)–(iii), the definitions (i')–(iii') are no longer equivalent. Of the three, (ii') is the most demanding and implies the other two (Problem 2.7). Many other definitions are possible, such as

$$G_n(x) \to F(x) \text{ at all continuity points of } F$$

or

(6.2.43) $$d\left(G_n, F\right) \to 0,$$

where $d(F, G)$ is any measure of the distance between F and G. Here we shall use (ii'), i.e., (6.2.43), where d is the *Kolmogorov distance*

(6.2.44) $$d(F, G) = \sup_x |G(x) - F(x)|.$$

Definition 6.2.1 A functional h is continuous at F if

(6.2.45) $$\sup_x |G_n(x) - F(x)| \to 0 \text{ implies that } h\left(G_n\right) \to h(F).$$

An immediate consequence of this definition and (6.2.19) is

Theorem 6.2.3 *If $X_1, \ldots, X_n$ are i.i.d. with cdf F and if h is a functional that is continuous at F with respect to Kolmogorov distance, then*

(6.2.46) $$h\left(\hat{F}_n\right) \xrightarrow{P} h(F),$$

i.e., $h\left(\hat{F}_n\right)$ is a consistent estimator of $h(F)$.

Proof. By (6.2.19), $d\left(\hat{F}_n, F\right) \to 0$ in probability. The continuity of h implies that then $h\left(\hat{F}_n\right) \to h(F)$ in probability, as was to be proved. ∎

Let us now consider the continuity properties of some functionals.

Example 6.2.7 Continuity of functionals.

(i) For some fixed value a, let

(6.2.47) $$h(F) = F(a).$$

Since

$$\sup |G_n(x) - F(x)| \to 0 \text{ implies } G_n(a) \to F(a),$$

the functional (6.2.47) is clearly continuous.

(ii) Let $h(F)$ be the distance between F and a fixed distribution F_0 given by

(6.2.48) $$h(F) = \int (F - F_0)^2 \, dF_0.$$

Since

$$|h(G_n) - h(F)| = \left| \int \left[(G_n - F_0)^2 - (F - F_0)^2 \right] dF_0 \right|$$

$$\leq 2 \int |G_n - F| \, dF_0 \leq 2 \sup |G_n(x) - F(x)|,$$

the functional (6.2.48) is continuous at all F.

(iii) Let $h(F)$ be the expectation of F, defined for all F for which this expectation exists. Let F be any such distribution and let

$$G_n = (1 - \epsilon_n) F + \epsilon_n H_n, \ \ 0 < \epsilon_n < 1,$$

where H_n is any distribution with finite expectation. Then

$$d(G_n, F) = \sup |G_n(x) - F(x)| = \epsilon_n \sup |H_n(x) - F(x)|$$
$$\leq \epsilon_n \to 0 \text{ as } \epsilon_n \to 0.$$

On the other hand,

$$h(G_n) = (1 - \epsilon_n) h(F) + \epsilon_n h(H_n).$$

The right side can be made to tend to any value whatever by appropriate choice of H_n. Thus $h(F) = E_F(X)$ is not continuous at any F.

The analogous argument applies to higher moments.

Note: This example shows that continuity, although by Theorem 6.2.3 sufficient for consistency, is far from necessary since $\bar{X}$ and the sample moments are consistent estimators of their population counterparts without being continuous.

(iv) Assume that F^{-1} is continuous in a neighborhood of p, and let $h(F)$ denote the pth quantile of F. To see that h is continuous at any such F, note that

$$F(x) + \epsilon \geq G_n(x) \geq F(x) - \epsilon \text{ for all } x$$

implies

$$\inf\left\{x : F(x) + \epsilon \geq p\right\} \leq \inf\left\{x : G_n(x) \geq p\right\} \leq \inf\left\{x : F(x) - \epsilon \geq p\right\}$$

and hence that

$$(6.2.49) \quad d\left(G_n, F\right) \leq \epsilon \text{ implies } F^{-1}(p - \epsilon) \leq G_n^{-1}(p) \leq F^{-1}(p + \epsilon)$$

for any $\epsilon < \min(p, 1 - p)$. From the continuity of F^{-1} at p, it now follows that $G_n^{-1}(p) \to F^{-1}(p)$ as $\epsilon \to 0$. $\qquad\square$

Summary

1. The natural estimator of a functional $h(F)$ defined over a large (nonparametric) class $\mathcal{F}$ of distributions F is the plug-in estimator $h\left(\hat{F}_n\right)$, where $\hat{F}_n$ is the sample cdf.

2. The statistic $\hat{F}_n(a)$ is a consistent estimator of $F(a)$, and $\sqrt{n}[\hat{F}_n(a) - F(a)]$ has a normal limit. The maximum difference $\max\left|\hat{F}_n(x) - F(x)\right|$ tends to 0 in probability, and hence $\hat{F}_n(x) \xrightarrow{P} F(x)$ not only at every fixed point $x = a$ but simultaneously for all x.

3. Approximate values of the bias and variance, and asymptotic normality are obtained for a number of examples, including central moments and quantiles.

4. If h is an expectation functional, $h\left(\hat{F}_n\right)$ is a V-statistic. The V-statistics are closely related to the U-statistics treated in Section 6.1.

5. Continuity of a functional with respect to a distance $d = d(F, G)$ is defined and is illustrated on some examples for the case of Kolmogorov distance.

6.3   Limit distributions of statistical functionals

Theorem 6.2.3 of the preceding section showed that $h\left(\hat{F}_n\right) \to h(F)$ in probability when h is continuous. This result can be viewed as an analog of Theorem 2.1.4 concerning the convergence in probability of $f(Y_n)$, where $\{Y_n, n = 1, 2, \ldots\}$ is a sequence of random variables. In the present section, we shall be concerned with the corresponding problem regarding convergence in law, more specifically with the convergence in law of

$$(6.3.1) \qquad \sqrt{n}\left[h\left(\hat{F}_n\right) - h(F)\right]$$

when h is differentiable in a suitable sense. We are thus aiming for an analog of Theorem 2.5.2 regarding the convergence in law of $\sqrt{n}\,[f(Y_n) - f(\theta)]$. With such a result, we might hope to accomplish three objectives:

(i) an understanding of why functionals such as the median or V-statistics, which seem so far from being sums of independent random variables, have a normal limit;

(ii) a formula for the asymptotic variance of (6.3.1);

(iii) simple sufficient conditions for the asymptotic normality of (6.3.1).

It turns out that we shall be successful with respect to the first two of these aims but less so with the third. The following purely heuristic remarks have the purpose of motivating the results (6.3.8)–(6.3.11).

The proof of Theorem 2.5.2 was based on a Taylor expansion

$$(6.3.2) \qquad f(b) - f(a) = (b - a)f'(a) + R,$$

where the remainder R tends to zero as $b \to a$. Before considering what a corresponding expansion for functionals might look like, it is helpful to take a look at the intermediate case of a differentiable function f of several variables, for which (6.3.2) generalizes to

(6.3.3)

$$f(b_1, \ldots, b_k) - f(a_1, \ldots, a_k) = \sum_{i=1}^{k} (b_i - a_i)\left.\frac{\partial f(x_1, \ldots, x_k)}{\partial x_i}\right|_{\underline{x}=\underline{a}} + R,$$

where $R \to 0$ as $(b_1, \ldots, b_k) \to (a_1, \ldots, a_k)$. (For discussion and application of (6.3.3), see Section 2 of Chapter 5.)

For a corresponding expansion of a functional h, let us write

$$(6.3.4) \qquad h(G) - h(F) = \int h'_x(F)(dG(x) - dF(x)) + R.$$

Here x replaces i; the integral replaces the sum in (6.3.3); $dG(x) - dF(x)$ replaces the difference $b_i - a_i$; F replaces $(a_1, \ldots, a_k)$; and $h'_x(F)$ as a function of x replaces the partial derivative $\partial f(x_1, \ldots, x_k)/\partial x_i \big|_{x=a}$. The expansion (6.3.4) is meaningful if $R \to 0$ as $d(F, G) \to 0$ for a suitable distance function d, for example, the Kolmogorov distance. Since for any constant c,

$$\int c\,(dG(x) - dF(x)) = 0,$$

equation (6.3.4) can determine h'_x only up to an additive constant, and $h'_x(F)$ is therefore usually standardized so that

$$\text{(6.3.5)} \qquad \int h'_x(F)dF(x) = 0.$$

Then (6.3.4) can be written as

$$\text{(6.3.6)} \qquad h(G) - h(F) = \int h'_x(F)dG(x) + R.$$

Let us now replace G by the sample cdf $\hat{F}_n$ and recall that $d\left(\hat{F}_n, F\right)$ tends to zero in probability by (6.2.19). Then

$$\text{(6.3.7)} \qquad \begin{aligned} h\left(\hat{F}_n\right) - h(F) &= \int h'_x(F)d\hat{F}_n(x) + R_n \\ &= \frac{1}{n}\sum_{i=1}^{n} h'_{x_i}(F) + R_n, \end{aligned}$$

where the main term is a sum of i.i.d. variables, which have expectation 0 by (6.3.5). It follows from the central limit theorem that

$$\text{(6.3.8)} \qquad \sqrt{n}\left[h\left(\hat{F}_n\right) - h(F)\right] \xrightarrow{L} N\left(0, \gamma^2(F)\right)$$

with

$$\text{(6.3.9)} \qquad \gamma^2(F) = E\left\{[h'_x(F)]^2\right\} = \int [h'_x(F)]^2 \, dF(x),$$

provided

$$\text{(6.3.10)} \qquad \sqrt{n}R_n \xrightarrow{P} 0.$$

To give meaning to this result, it remains to define $h'_x(F)$. Let δ_x denote the cdf of the distribution that places probability 1 at the point x, and let

$$\text{(6.3.11)} \qquad F_{\epsilon,x} = (1 - \epsilon)F + \epsilon\delta_x.$$

Then it turns out that in many situations in which (6.3.8) and (6.3.9) hold, $h'_x(F)$ is given by the derivative with respect to ϵ of $h\left(F_{\epsilon,x}\right)$ evaluated at

$\epsilon - 0$. It will be convenient to think of this derivative for fixed F as a function of x, and it is commonly denoted by

$$(6.3.12) \qquad IF_{h,F}(x) = \frac{d}{d\epsilon} h\left[(1-\epsilon)F + \epsilon\delta_x\right]|_{\epsilon=0} = h'_x(F).$$

The vector of partial derivatives $\left.\dfrac{\partial f}{\partial x_1}\right|_{x=a}, \ldots, \left.\dfrac{\partial f}{\partial x_k}\right|_{x=a}$ in (6.3.3) is thus replaced by the infinite set of derivatives (6.3.12) for varying x. With the notation (6.3.12), equations (6.3.5) and (6.3.9) become

$$(6.3.13) \qquad \int IF_{h,F}(x)dF(x) = 0$$

and

$$(6.3.14) \qquad \gamma^2(F) = \int IF^2_{h,F}(x)dF(x).$$

The function IF, the so-called *influence function*, has an interest independent of its role in (6.3.4) and (6.3.9). It measures the rate at which the functional h changes when F is contaminated by a small probability of obtaining an observation x, and thus is a measure of the influence of such a contamination. While continuity of h indicates whether the value of h at a distribution close to F is close to its value at h, the influence function provides a measure of this closeness, in particular the influence of a small proportion of observations at x which do not "belong" to F, so-called gross errors. The maximum (over x) of this influence, the quantity

$$(6.3.15) \qquad \lambda^* = \sup_x |IF_{h,F}(x)|,$$

is called the *gross-error sensitivity* of h at F.

Let us now return to the three objectives (i)–(iii) stated at the beginning of the section. An explanation for (i), the common occurrence of asymptotic normality for very non-linear statistics, is seen to be analogous to that given in Theorem 2.5.2 for functions of random variables. It is the fact that if the Taylor expansion (6.3.7) is valid, then $h\left(\hat{F}_n\right) - h(F)$ can to first order be approximated by a sum of i.i.d. random variables. An answer for (ii), a formula for the asymptotic variance of (6.3.1), is given in (6.3.14), which exhibits the asymptotic variance as the integral of the square of the influence function.

There remains problem (iii): To determine conditions under which the heuristic result (6.3.8) can be proved. This issue can be approached in two ways. One is to obtain general conditions on h under which (6.3.10) will hold. Such conditions involving not only the limit (6.3.12) but the more general limit

$$(6.3.16) \qquad \lim_{\epsilon \to 0} \frac{d}{d\epsilon} h\left[(1-\epsilon)F + \epsilon G\right]$$

are provided by the theory of differentiation of functionals and require (6.3.16) to converge uniformly as G ranges over certain sets of distributions. Discussions of this theory can be found, for example, in Serfling (1980), Huber (1981), Fernholz (1983), and Staudte and Sheaffer (1990).

For specific cases, an alternative and often more convenient approach considers the expansion (6.3.7) as a heuristic device suggesting the right answer, and then validates this answer by checking directly that the remainder in (6.3.7) satisfies (6.3.10). Carrying out this approach in specific cases requires the following steps.

1. Calculation of $h(F_{\epsilon,x})$ and the influence function

$$IF_{h,F}(x) = h'_x(F) = \left.\frac{d}{d\epsilon}h(F_{\epsilon,x})\right|_{\epsilon=0}.$$

2. Checking condition (6.3.13).

3. Checking condition (6.3.10).

4. Determining the asymptotic variance given by (6.3.9).

We shall now carry out this program for some specific functionals h. Some of these were treated earlier by different methods. The results are obtained here by a unified approach which provides additional insights.

Example 6.3.1 The mean. When $h(F) = \xi$ is the mean $E_F(x)$ of X, we have

(6.3.17) $h(F_{\epsilon,x}) = (1-\epsilon)\xi + \epsilon x$

and hence

(6.3.18) $h'_x(F) = IF_{h,F}(x) = \frac{d}{d\epsilon}h(F_{\epsilon,x}) = x - \xi.$

Thus $\lambda^* = \infty$ since the influence function (6.3.17) is unbounded. This corresponds to the fact that a small mass placed sufficiently far out can change the expectation by an arbitrarily large amount.

To check (6.3.13), we calculate

$$\int I_{h,F}(x)dF(x) = \int (x-\xi)dF(x) = 0,$$

as was to be shown.

Asymptotic normality of $\sqrt{n}\left[h\left(\hat{F}_n\right) - h(F)\right]$ requires that the remainder R_n defined by (6.3.7) satisfies (6.3.10). In the present case,

$$h\left(\hat{F}_n\right) - h(F) = \bar{X} - \xi = \frac{1}{n}\sum h'_{x_i}(F).$$

Thus $R_n \equiv 0$ and (6.3.10) is trivially satisfied.

Finally,

$$\gamma^2(F) = E\left[h'_X(F)\right]^2 = E\left(X - \xi\right)^2 = \text{Var}(X).$$

It follows that

$$\sqrt{n}\left[h\left(\hat{F}_n\right) - h(F)\right] = \sqrt{n}\left(\bar{X} - \xi\right) \to N\left(0, \text{Var}(X)\right),$$

which in this case is simply the classical CLT.

Note: The last conclusion of course does not constitute a proof of the CLT since the CLT was used to establish (6.3.8).   □

Example 6.3.2 A goodness-of-fit measure. As a more typical example, consider the measure of goodness of fit

(6.3.19)
$$h(F) = \int (F - F_0)^2 \, dF_0$$

corresponding to the Cramér-von Mises statistic (6.2.10). In order to calculate the influence function, recall that

$$F_{\epsilon, x}(t) = (1 - \epsilon)F(t) + \epsilon \delta_x(t),$$

where $\delta_x(t)$ is the cdf of the distribution assigning probability 1 to the point x so that

(6.3.20)
$$\delta_x(t) = \begin{cases} 0 & \text{if } t < x \\ 1 & \text{if } t \geq x. \end{cases}$$

Thus

$$h\left(F_{\epsilon, x}\right) = \int \left[(1 - \epsilon)F(t) + \epsilon\delta_x(t) - F_0(t)\right]^2 dF_0(t).$$

This is a quadratic in ϵ and its derivative (w.r.t. ϵ) is the coefficient of ϵ, which is

(6.3.21)
$$h'_x(F) = 2 \int [F(t) - F_0(t)]\,[\delta_x(t) - F(t)]\,dF_0(t).$$

Since $|h'_x(F)|$ is clearly less than 2, it is seen that the influence function is bounded and hence $\lambda^* < \infty$.

We leave the checking of (6.3.13) to Problem 3.1 and next prove (6.3.8) by checking condition (6.3.10). The remainder R_n is, by (6.3.7) and (6.3.20),

(6.3.22)
$$R_n = \int \left(\hat{F}_n - F_0\right)^2 dF_0 - \int (F - F_0)^2 \, dF_0$$
$$- \frac{2}{n} \int \sum_{i=1}^{n} (F - F_0)\,(\delta_{x_i} - F)\,dF_0.$$

Now $\sum \delta_{x_i}(t)$ is the number of x's $\leq t$ and hence

(6.3.23)
$$\frac{1}{n} \sum \delta_{x_i}(t) = \hat{F}_n(t).$$

Taking the difference of the first two terms in (6.3.21) and using (6.3.22) we find (Problem 3.1) that

(6.3.24)
$$R_n = \int \left(\hat{F}_n - F \right)^2 dF_0 \leq \sup \left(\hat{F}_n(x) - F(x) \right)^2.$$

Since by (5.7.28), $\sqrt{n}D_n = \sqrt{n} \sup \left| \hat{F}_n(x) - F(x) \right|$ tends to a limit distribution, the same is true for nD_n^2, which is therefore bounded in probability. It follows that

$$\sqrt{n}R_n \leq \frac{1}{\sqrt{n}} nD_n^2 \xrightarrow{P} 0,$$

which proves (6.3.8).

This result was already mentioned in the paragraph following (5.7.34), where it was stated that the asymptotic distribution of

$$W_n = h\left(\hat{F}_n \right) = n \int \left(\hat{F}_n - F \right)^2 dF_0$$

is normal, provided $F \neq F_0$. When $F = F_0$, the influence function given by (6.3.20) is seen to be 0 for all x and hence the asymptotic variance $\gamma^2(F_0)$ is also 0. The limit result (6.3.8) then only states that

$$\sqrt{n}\left(h\left(\hat{F}_n \right) - h\left(F_0 \right) \right) \xrightarrow{P} 0.$$

The correct normalizing factor in this case is no longer $\sqrt{n}$ but n. The situation is completely analogous to that encountered in the delta method in Section 2.5 when contrasting the cases $f'(\theta) \neq 0$ and $f'(\theta) = 0$.

To complete the discussion of (6.3.8), we must consider the asymptotic variance

(6.3.25) $$\gamma^2(F) = E\left[h'_X(F)\right]^2 = 4 \int \left[\int (F - F_0)(\delta_x - F)dF_0 \right]^2 dF(x).$$

This value typically has to be obtained by numerical evaluation of the double integral (6.3.25). For some cases in which (6.3.25) can be evaluated explicity, see Problem 3.4(i).   □

Example 6.3.3 Central moments. In Example 6.2.4, we dealt with the bias and variance of $h\left(\hat{F}_n \right) = M_k$ as estimator of $h(F) = \mu_k$. The asymptotic normality of

$$\sqrt{n}\left(M_k - \mu_k \right) = \sqrt{n}\left[h\left(\hat{F}_n \right) - h(F) \right]$$

was shown in Example 5.2.7. We shall now derive this result from the present point of view.

We begin by calculating the influence function of μ_k. Since $F_{\epsilon,k}$ is the distribution of random variable Y defined by

$$(6.3.26) \qquad Y = \begin{cases} X \text{ with probability } 1 - \epsilon \\ x \text{ with probability } \epsilon, \end{cases}$$

it follows that

$$(6.3.27) \qquad h\left(F_{\epsilon,x}\right) = E\left(Y - \eta\right)^k,$$

where

$$\eta = E(Y) = (1 - \epsilon)\xi + \epsilon x.$$

Thus

(6.3.28)
$$\begin{aligned} h\left(F_{\epsilon,x}\right) &= E\left[(Y - \xi) - \epsilon(x - \xi)\right]^k \\ &= E\left(Y - \xi\right)^k - k\epsilon(x - \xi)E\left(Y - \xi\right)^{k-1} + \cdots \\ &= \left[(1 - \epsilon)E\left(X - \xi\right)^k + \epsilon(x - \xi)^k\right] \\ &\quad - k\epsilon(x - \xi)\left[(1 - \epsilon)E\left(X - \xi\right)^{k-1} + \epsilon(x - \xi)^{k-1}\right] + \cdots. \end{aligned}$$

It is seen that $h\left(F_{\epsilon,x}\right)$ is a polynomial in ϵ, and that $IF_{h,F}(x)$ is therefore equal to the coefficient of ϵ in (6.3.28), and hence is equal to

$$(6.3.29) \qquad IF_{h,F}(x) = (x - \xi)^k - \mu_k - k(x - \xi)\mu_{k-1}.$$

It follows that

$$(6.3.30) \qquad E\left[IF_{h,F}(X)\right] = E(X - \xi)^k - \mu_k - k\mu_{k-1}E(X - \xi) = 0$$

and (Problem 3.4(ii)) that

(6.3.31)
$$\begin{aligned} \gamma^2(F) = E\left[IF_{h,F}^2(X)\right] &= E\left[(X - \xi)^k - \mu_k\right]^2 - 2k\mu_{k-1}E\left(X - \xi\right)^{k+1} \\ &\quad + k^2\mu_{k-1}^2 E(X - \xi)^2 \\ &= \mu_{2k} - \mu_k^2 - 2k\mu_{k-1}\mu_{k+1} + k^2\mu_{k-1}^2\mu_2. \end{aligned}$$

Equation (6.3.7) becomes

$$M_k - \mu_k = \frac{1}{n}\sum_{i=1}^{n}(X_i - \xi)^k - \mu_k - \frac{k}{n}\mu_{k-1}\sum(X_i - \xi) + R_n$$

so that the remainder R_n is given by

$$R_n = \frac{1}{n} \sum (X_i - \bar{X})^k - \mu_k - \frac{1}{n} \sum (X_i - \xi)^k + \mu_k + k (\bar{X} - \xi) \mu_{k-1}.$$

Since the distribution of R_n does not depend on ξ, put $\xi = 0$ without loss of generality. Then

$$\sqrt{n} R_n = \sqrt{n} \left[-k\bar{X} \frac{\sum X_i^{k-1}}{n} + \binom{k}{2} \bar{X}^2 \frac{\sum X_i^{k-2}}{n} \right.$$
$$\left. - \cdots + (-1)^k \bar{X}^k + k\bar{X} \mu_{k-1} \right].$$

Here all the terms of the form

$$\sqrt{n} \bar{X}^r \frac{\sum X_i^{k-r}}{n} \quad \text{with } r \geq 2$$

tend to 0 in probability because $\sqrt{n}\bar{X}$ and hence $(\sqrt{n}\bar{X})^r$ is bounded in probability and therefore

$$\sqrt{n} \bar{X}^r = (\sqrt{n}\bar{X})^r / (\sqrt{n})^{r-1} \xrightarrow{P} 0.$$

The sum of the two remaining terms is

$$-k\sqrt{n}\bar{X} \left[\frac{\sum X_i^{k-1}}{n} - \mu_{k-1} \right],$$

which tends to 0 in probability since $\sqrt{n}\bar{X}$ is bounded in probability and $\sum_i X_i^{k-1}/n$ is a consistent estimator of μ_{k-1}. This completes the proof of asymptotic normality with the asymptotic variance given by (6.3.30). The result agrees with that obtained in Example 5.2.7. $\Box$

Example 6.3.4 Expectation functionals; V-statistics. In Example 6.2.5, it was seen that if $h(F)$ is a one-sample expectation functional, then $h\left(\hat{F}_n\right)$ is the associated V-statistic (6.2.31), which is asymptotically equivalent to the corresponding U-statistic (6.1.3). Asymptotic normality of U-statistics, not only for the one-sample case but for the general case of s samples with U given by (6.1.33), was stated in Theorem 6.1.4. The proof was given for $s = 2$, $a_1 = a_2 = 1$ at the end of Section 6.1 and was sketched for $s = 1$, $a = 2$ in Problem 6.1.17. In these proofs, a crucial part was played by the asymptotically equivalent statistics T_N^*, given in (6.1.74) and in Problem 1.17. The present approach provides an explanation of T_N^*: It is just the linear part of the Taylor expansion of $\sqrt{n} \left[h\left(\hat{F}_n\right) - h(F) \right]$. (An alternative way of arriving at the same result is given in Section A.2 of the Appendix.)

We begin with the one-sample case so that $X_1, \ldots, X_n$ are i.i.d. according to F, and let

$$(6.3.32) \qquad h(F) = E_F \phi(X_1, \ldots, X_a)$$

be an expectation functional and

$$(6.3.33) \qquad V = h\left(\hat{F}_n\right) = \frac{1}{n^a} \sum \cdots \sum \phi(X_{i_1}, \ldots, X_{i_a})$$

its V-estimator (6.2.31).

In order to calculate the influence function of h, note that

$$(6.3.34) \qquad \begin{aligned} h(F_{\epsilon,x}) &= \int \cdots \int \phi(y_1, \ldots, y_a) d[(1-\epsilon)F(y_1) + \epsilon\delta_x(y_1)] \\ &\quad \cdots d[(1-\epsilon)F(y_a) + \epsilon\delta_x(y_a)]. \end{aligned}$$

The right side is a polynomial in ϵ of degree a, and its derivative (with respect to ϵ) at $\epsilon = 0$ is the coefficient of ϵ in (6.3.34). Thus

$$\begin{aligned} &IF_{h,F}(x) \\ &= \int \cdots \int \phi(y_1, \ldots, y_a) [d\delta_x(y_1) - dF(y_1)] dF(y_2) \cdots dF(y_a) \\ &\quad + \int \cdots \int \phi(y_1, \ldots, y_a) dF(y_1) [d\delta_x(y_2) - dF(y_2)] dF(y_3) \cdots dF(y_a) \\ &\quad + \cdots. \end{aligned}$$

By (6.2.5),

$$\int \phi(y_1, \ldots, y_a) d\delta_x(y_i) = \phi(y_1, \ldots, y_{i-1}, x, y_{i+1}, \ldots, y_a)$$

and therefore

$$\begin{aligned} IF_{h,F}(x) &= \sum_{i=1}^{a} \Big[\int \cdots \int \phi(y_1, \ldots, y_{i-1}, x, y_{i+1}, \ldots, y_a) \\ &\qquad dF(y_1) \cdots dF(y_{i-1}) dF(y_{i+1}) \cdots dF(y_a) - h(F) \Big] \\ &= \sum_{i=1}^{a} [E\phi(Y_1, \ldots, Y_{i-1}, x, Y_{i+1}, \ldots, Y_a) - h(F)]. \end{aligned}$$

Since ϕ is symmetric in its a arguments, we have

$$(6.3.35) \qquad I_{h,F}(x) = a[\phi_1(x) - h(F)],$$

where

(6.3.36) $\phi_1(x) = E\phi(x, Y_2, \dots, Y_a)$.

It follows from (6.1.14) with $\theta = h(F)$ that

$$\int I_{h,F}(x)dF(x) = a\left[E\varphi_1(X) - h(F)\right] = 0,$$

which checks (6.3.13). Also by (6.1.15),

(6.3.37) $\gamma^2(F) = a^2 \operatorname{Var} \phi_1(X) = a^2\sigma_1^2$.

By the central limit theorem,

$$\frac{a}{n}\sqrt{n}\sum\left[\phi_1(X_i) - h(F)\right] \xrightarrow{L} N\left(0, a^2\sigma_1^2\right)$$

and this suggests by (6.3.7) that

(6.3.38)
$$\sqrt{n}\left[h\left(\hat{F}_n\right) - h(F)\right] = \sqrt{n}\frac{a}{n}\left[\sum\phi_1(X_i) - h(F)\right] + \sqrt{n}R_n \xrightarrow{L} N\left(0, a^2\sigma_1^2\right).$$

To prove (6.3.38), we need to show that

(6.3.39) $\sqrt{n}R_n \xrightarrow{P} 0$,

which will follow if we can show that

(6.3.40)
$$nE\left(R_n^2\right) = nE\left[h\left(\hat{F}_n\right) - \frac{a}{n}\sum\phi_1(X_i)\right]^2$$
$$= n\operatorname{Var}\left[h\left(\hat{F}_n\right) - \frac{a}{n}\sum\phi_1(X_i)\right] \to 0.$$

The limit (6.3.40) can be proved by showing that the three quantities

(6.3.41)
$$n\operatorname{Var} h\left(\hat{F}_n\right),\ n\operatorname{Var}\frac{a}{n}\sum\phi_1(X_i),\ \text{and}\ \operatorname{Cov}\left[nh\left(\hat{F}\right)n\right), \frac{a}{n}\sum\phi_1(X_i)\right]$$

all tend to $a^2\sigma_1^2$ as $n \to \infty$ (Problem 6.3.14). Checking (6.3.41) (which for $a = 2$, $s_1 = s_2 = 1$ is left to Problem 6.1.17) completes the proof of (6.3.38). Note that throughout this discussion we have tacitly assumed that

(6.3.42) $\sigma_1^2 > 0$ and $\operatorname{Var} \phi(X_1, \dots, X_a) < \infty$.

An interesting special case of a V-statistic with $s = 1$, $a = 2$ is provided by letting

(6.3.43) $h(F) = \phi(x_1, x_2) = \displaystyle\int \left[\delta_{x_1}(t) - F_0(t)\right]\left[\delta_{x_2}(t) - F_0(t)\right]dF_0(t)$, where F_0 is some given cdf. Then

$$
h\left(\hat{F}_n\right) = \frac{1}{n^2} \sum_{i=1}^{n} \sum_{j=1}^{n} \phi(X_i, X_j)
$$

$$
= \int \left[\frac{\sum \delta_{x_i}(t)}{n} - F_0(t) \right] \left[\frac{\sum \delta_{x_j}}{n} - F_0(t) \right] dF_0(t)
$$

and hence, by (6.3.23),

$$
h\left(\hat{F}_n\right) = \int \left(\hat{F}_n - F_0\right)^2 dF_0.
$$

Thus $h\left(\hat{F}_n\right)$ is the Cramér-von Mises statistic for which asymptotic normality was proved in Example 6.3.2.

Let us now extend these considerations to the case of V-statistics with $s = 2$, $a_1 = a_2 = 1$ treated earlier in the proof of Theorem 6.1.4 at the end of Section 1. The two cases together give a good picture of the general situation. Let $X_1, \ldots, X_m$ and $Y_1, \ldots, Y_n$ be i.i.d. according to distributions F and G, respectively, and let

(6.3.44) $$h(F, G) = E_{F,G}\phi(X, Y).$$

Then

(6.3.45) $$h\left(\hat{F}_m, \hat{G}_n\right) = \frac{1}{mn} \sum_{i=1}^{m} \sum_{j=1}^{n} \phi(X_i, Y_j),$$

which is both a U- and a V-statistic.

The univariate expansion (6.3.6) with $h_x'(F)$ given by (6.3.12) is now replaced by the bivariate expansion

(6.3.46)

$$
h\left(F^*, G^*\right) - h(F, G) = \int \frac{d}{d\epsilon} h\left(F_{\epsilon,x}, G\right)|_{\epsilon=0} \left[dF^*(x) - dF(x)\right]
$$

$$
+ \int \frac{d}{d\epsilon} h\left(F, G_{\epsilon,y}\right)|_{\epsilon=0} \left[dG^*(y) - dG(y)\right] + R,
$$

where F has been replaced by (F, G) and G by (F^*, G^*). Let us denote the derivatives in (6.3.46) by

(6.3.47) $$I_{h,F,G}^{(1)}(x) = \frac{d}{d\epsilon} h\left(F_{\epsilon,x}, G\right)|_{\epsilon=0}$$

and

$$
I_{h,F,G}^{(2)}(y) = \frac{d}{d\epsilon} h\left(F, G_{\epsilon,y}\right)|_{\epsilon=0},
$$

and suppose that they have been standardized so that

$$\int I_{h,F,G}^{(1)}(x)dF(x) = \int I_{h,F,G}^{(2)}(y)dG(y) = 0.$$

Then the univariate expansion (6.3.4) with $h'_x(F)$ given by (6.3.12) becomes

$$(6.3.48) \quad h(F^*, G^*) - h(F, G) = \int I_{F,G,x}^{(1)}dF^*(x) + \int I_{F,G,y}^{(2)}dG^*(y) + R.$$

We now substitute $\hat{F}_m$ for F^* and $\hat{G}_n$ for G^* to obtain, in generalization of (6.3.7),

(6.3.49)

$$h\left(\hat{F}_m, \hat{G}_n\right) - h(F, G) = \frac{1}{m}\sum_{i=1}^{m} I_{h,F,G}^{(1)}(X_i) + \frac{1}{n}\sum_{j=1}^{n} I_{h,F,G}^{(2)}(Y_j) + R_{m,n}.$$

Let us next evaluate the first two terms in (6.3.49) for the case that h is given by (6.3.44). Then

$$h(F_{\epsilon,x}, G) = \int\int \phi(u,v) d\left[(1-\epsilon)F(u) + \epsilon\delta_x(u)\right] dG(v)$$

and

$$\frac{d}{d\epsilon} h(F_{\epsilon,x}, G)|_{\epsilon=0} = \int \phi(x, v)dG(v) - h(F, G),$$

and analogously

$$\frac{d}{d\epsilon} h(F, G_{\epsilon,y})|_{\epsilon=0} = \int \phi(u, y)\,dF(u) - h(F, G).$$

In the notation of (6.1.70), we have

$$\int \phi(x, v)dG(v) = \phi_{10}(x) \text{ and } \int \phi(u, y)dF(u) = \phi_{01}(y)$$

and hence

(6.3.50)

$$h\left(\hat{F}_m, \hat{G}_n\right) - h(F, G) = \frac{1}{m}\sum \phi_{10}(X_i) + \frac{1}{n}\sum \phi_{01}(Y_j) + R_{m,n}.$$

The auxiliary random variable T_N^* defined by (6.1.74) is thus exactly equal to the linear part of

(6.3.51) $$\sqrt{N}\left[h\left(\hat{F}_m, \hat{G}_n\right) - h(F, G)\right].$$

It was shown in the proof of Theorem 6.1.4 (and Problem 1.16) that $\sqrt{N} R_{m,n} \xrightarrow{P} 0$ and hence that the V-statistic (6.3.45) is asymptotically normal.

The Taylor expansion (6.3.48) has not brought any new results concerning V-statistics. What it does bring, and what was missing in the earlier proof of Theorem 6.1.4, is an interpretation of the linear function T_N^*. In that proof, the functions ϕ_{10} and ϕ_{01} came out of the blue without any explanation. They have now been identified as the two components $I^{(1)}$ and $I^{(2)}$ of the influence function.   □

Examples 6.3.1–6.3.4 illustrate how the influence function approach can be used to obtain the asymptotic normality result (6.3.8) in particular cases. Let us now consider under what circumstances this result cannot be expected to hold. Since we are dealing with a generalization of the delta method, let us recall Theorem 2.5.2, where the conditions required for the asymptotic normality of $\sqrt{n} \left[f\left(\bar{X} \right) - f(\theta) \right]$ were that

(i) f is differentiable at θ

and

(ii) $f'(\theta) \neq 0$.

The situation is quite analogous in the present case.

(i) <u>Lack of smoothness</u>.

Condition (i) suggests that for (6.3.8) to hold, the funtional h needs to be sufficiently smooth (for sufficient conditions, see, for example, Serfling (1980)). As an example in which this is not the case, suppose that $h(F)$ is the absolute value of the population median under the assumptions of Example 2.4.9. Then it follows from Example 2.5.6 of that chapter that $\sqrt{n} \left[h\left(\hat{F}_n \right) - h(F) \right]$ has a limit distribution which, however, is not normal when the population median is zero. Another example is provided by the Kolmogorov distance $h(F) = \sup_x |F(x) - F_0(x)|$. Here it turns out that $\sqrt{n} \left[h\left(\hat{F}_n \right) - h(F) \right]$ tends to a non-normal limit distribution (depending on F) both when $F = F_0$ and when $F \neq F_0$. For a discussion of this result, see, for example, Serfling (1980, Sections 2.1.6 and 2.8.2).

(ii) <u>Vanishing derivative</u>. It was seen in Section 2.5 that when $f'(\theta) = 0$ in the situation of Theorem 2.5.2 but $f''(\theta) \neq 0$, the Taylor expansion of $f(T_n)$ about $f(\theta)$ needs to be carried a step further. The appropriate normalizing constant for $f(T_n) - f(\theta)$ is then no longer $\sqrt{n}$ but n, and the limit distribution is no longer normal. Similarly, the asymptotic normality (6.3.8) breaks down when, for the given distribution F, the derivative $h_x'(F)$ of h is equal to 0 for all x. This possibility is illustrated in Example 6.3.2.

Summary

1. By making a formal Taylor expansion of $h\left(\hat{F}_n\right)$ about $h(F)$, the difference $h\left(\hat{F}_n\right) - h(F)$ is expressed as the average of n i.i.d. random variables plus a remainder R_n. If one can show that $\sqrt{n}R_n \overset{P}{\to} 0$, it follows that $\sqrt{n}\left[h\left(\hat{F}_n\right) - h(F)\right]$ tends in law to a normal distribution with mean 0 and with a variance that is a function of the derivative $h'_F(x)$.

2. The functional derivative h'_F can frequently be calculated as the ordinary derivative of the mixture of F and a one-point distribution with respect to the mixing proportion. This derivative can be interpreted as the influence function $IF_{h,F}(x)$, which measures the influence on $h(F)$ of a small probability placed at a point x.

3. The influence function is calculated for a number of functionals h. For each of them, it is then checked that the asymptotic variance $\gamma^2(F)$ equals the integral of $IF^2_{h,F}(x)$.

4. Examples are given that show how the asymptotic theory of Points 1–3 fails when (i) h is not sufficiently smooth and (ii) the derivative $h'_x(F)$ is identically 0.

6.4   Density estimation

Among the nonparametric estimation problems discussed in the preceding sections, we considered in particular estimating the cdf F and the quantile function F^{-1}. When F has a density f, the latter provides a visually more informative representation of the distribution, and in the present section, we shall therefore consider the estimation of a probability density. We begin with the estimation of $f(y)$ for a given value y, and suppose that f is continuous in a neighborhood of y and that $f(y) > 0$. Continuity assures that $f(y)$ is completely determined by F (Problem 4.1) and hence that $f(y)$ is a functional $h(F)$ defined over the class $\mathcal{F}$ of all distributions F with density f continuous in a neighborhood of y with $f(y) > 0$.

   To estimate $h(F)$ on the basis of a sample $X_1, \ldots, X_n$ from F, we cannot use $h\left(\hat{F}_n\right)$ since $\hat{F}_n$ does not have a density and $h\left(\hat{F}_n\right)$ is therefore not defined. However, since

$$(6.4.1) \qquad f(y) = \lim_{h \to 0} \frac{F(y+h) - F(y-h)}{2h},$$

one might consider the estimator

$$(6.4.2) \qquad \hat{f}_n(y) = \frac{\hat{F}_n(y+h) - \hat{F}_n(y-h)}{2h}.$$

For large n, one would expect $\hat{f}_n(y)$ to be close to

$$[F(y+h) - F(y-h)]/2h,$$

and for small h, the latter will be close to $f(y)$. Thus one can hope that with $h = h_n$ tending to 0 as $n \to \infty$, the estimator $\hat{f}_n(y)$, the so-called *naive density estimator*, will be a consistent estimator of $f(y)$. For suitable sequences h_n, this conjecture is confirmed by Theorem 6.4.1.

It is interesting to note that $\hat{f}_n(y)$ is itself a probability density. Since it is clearly non-negative, one only needs to show that

$$(6.4.3) \qquad \int_{-\infty}^{\infty} \hat{f}_n(y) dy = 1.$$

To see this, write

$$(6.4.4) \qquad \hat{f}_n(y) = \frac{1}{2nh} \sum_{j=1}^{n} I_j(y),$$

where

$$(6.4.5) \qquad I_j(y) = 1 \text{ if } y - h < x_j < y + h \text{ and } 0 \text{ otherwise.}$$

Then

$$\int \hat{f}_n(y) dy = \frac{1}{2nh} \sum_{j=1}^{n} \int_{-\infty}^{\infty} I_j(y) dy = \frac{1}{2nh} \sum_{j=1}^{n} \int_{x_j - h}^{x_j + h} dy = 1.$$

The basic properties of $\hat{f}_n$ are easily obtained from the fact that $n\left[\hat{F}_n(y+h) - \hat{F}_n(y-h)\right]$ has the binomial distribution $b(p, n)$ with

$$(6.4.6) \qquad p = F(y+h) - F(y-h).$$

It follows that

$$(6.4.7) \qquad E\left[\hat{f}_n(y)\right] = \frac{F(y+h) - F(y-h)}{2h} = \frac{p}{2h}.$$

The bias is therefore

$$(6.4.8) \qquad b(y) = \frac{F(y+h) - F(y-h)}{2h} - f(y),$$

which tends to zero, provided

(6.4.9) $h = h_n \to 0$ as $n \to \infty$.

Similarly, the variance of $\hat{f}_n(y)$ is

(6.4.10) $\text{Var}\left[\hat{f}_n(y)\right] = \dfrac{p(1-p)}{4nh^2}.$

As $h_n \to 0$, the value of p

$$p_n = F\left(y + h_n\right) - F\left(y - h_n\right) \to 0.$$

From (6.4.10), we have

$$\text{Var}\left[\hat{f}_n(y)\right] \sim \frac{p_n}{2h_n} \cdot \frac{1}{2nh_n}.$$

Since the first factor on the right side tends to $f(y) > 0$, $\text{Var}\left[\hat{f}_n(y)\right] \to 0$ as $h_n \to 0$ if in addition

(6.4.11) $nh_n \to \infty,$

that is, if h_n tends to 0 more slowly than $1/n$ or, equivalently, if $\dfrac{1}{n} = o\left(h_n\right)$.

From these results, we immediately obtain sufficient conditions for the consistency of $\hat{f}_n(y)$.

Theorem 6.4.1 *A sufficient condition for $\hat{f}_n(y)$ to be a consistent estimator of $f(y)$ is that both (6.4.9) and (6.4.11) hold.*

Proof. This follows directly from Theorem 2.1.1 and the bias-variance decomposition (1.4.6).   ∎

Although $\hat{f}_n(y)$ is consistent for estimating $f(y)$ when the h's satisfy (6.4.9) and (6.4.11), note that $\hat{f}_n(y)$ as an estimator of $f(y)$ involves two approximations: $f(y)$ by $[F(y + h) - F(y - h)]/2h$ and the latter by $\hat{f}_n(y)$. As a result, the estimator turns out to be less accurate than one might have hoped. However, before analyzing the accuracy of $\hat{f}_n(y)$, let us note another drawback. The estimator $\hat{f}_n(y)$, although a density, is a step function with discontinuities at every point $x_j \pm h$, $j = 1, \ldots, n$. If we assume the true f to be a smooth density, we may prefer its estimator also to be smooth.

To see how to modify $\hat{f}_n$ for this purpose, consider the following description of (6.4.2). To any observation x_j, the empirical distribution assigns probability $1/n$. The contribution of x_j to $\hat{f}_n$ (given by the term $I_j(y)/2nh$), spreads this probability mass over the interval $(x_j - h, x_j + h)$ according to a uniform distribution. The ragged appearance of $\hat{f}_n$ results from the discontinuities of the uniform density at its end points. This suggests a natural remedy: In the construction of $\hat{f}_n(y)$, replace the uniform density $U(x_j - h, x_j + h)$ by some other smoother density also centered on x_j and with a scale factor h. Such a density can be written as

$$(6.4.12) \qquad \frac{1}{h} K\left(\frac{y - x_j}{h}\right).$$

With this change, the estimator of $f(y)$ becomes

$$(6.4.13) \qquad \hat{f}_n(y) = \frac{1}{n} \sum_{j=1}^{n} \frac{1}{h} K\left(\frac{y - X_j}{h}\right),$$

which reduces to (6.4.4) for $K = U(-1, 1)$ (Problem 4.3). The estimator (6.4.13) is called a *kernel estimator with kernel* K; the scale factor h is called the *bandwidth*. We shall assume throughout this section that K is a probability density, so that it is ≥ 0 and satisfies

$$(6.4.14) \qquad \int K(z)dz = 1.$$

In addition, we shall restrict attention to densities K that are symmetric about 0. In generalization of (6.4.3), note that $\hat{f}_n$ is a probability density since it is non-negative and since

$$\int \hat{f}_n(y)dy = \frac{1}{n} \sum_{j=1}^{n} \int K(t - x_j)\, dt = \frac{1}{n} \sum_{j=1}^{n} \int K(z)dz = 1.$$

Let us now consider conditions under which the bias of the estimator (6.4.13) tends to 0. Such conditions can be of two kinds: restrictions on (i) the kernel K and (ii) the true density f. Since K is under our control and f is unknown, conditions on K are preferable. Following (6.4.8), we saw that when K is uniform density, then as $h_n \to 0$, the bias of $\hat{f}_n(y)$ tends to 0 for any f that is continuous at y. No further assumptions on f are needed. The corresponding fact is true for (6.4.13) under very weak conditions on K, as is shown by the following theorem.

Theorem 6.4.2 *Let f be any density that is continuous at y and let K be any kernel which is bounded and satisfies*

$$(6.4.15) \qquad yK(y) \to 0 \text{ as } y \to \pm\infty.$$

Then the bias of $\hat{f}_n(y)$ tends to 0, or, equivalently,

$$(6.4.16) \qquad E\left[\hat{f}_n(y)\right] \to f(y) \text{ as } h = h_n \to 0.$$

For a proof, see Parzen (1962), Rosenblatt (1971), or Scott (1992). We shall instead prove a result which does make some smoothness assumptions about f. The proof parallels that of Theorem 4.2.1 and gives somewhat more detailed results, which will be needed to determine the best choice of the bandwidth h.

Theorem 6.4.3 *Let f be three times differentiable with bounded third derivative in a neighborhood of y and let K be a kernel symmetric about 0, with*

(6.4.17)
$$\int K^2(y)dy < \infty, \quad \int y^2 K(y)dy = \tau^2 < \infty, \quad and \quad \int |y|^3 K(y)dy < \infty.$$

(i) Then for any sequence h_n, $n = 1, 2, \ldots$, satisfying (6.4.9),

(6.4.18)
$$\text{bias of } \hat{f}_n(y) = \frac{1}{2}h_n^2 f''(y)\tau^2 + o\left(h_n^2\right).$$

(ii) If, in addition, h_n satisfies (6.4.11), then

(6.4.19)
$$\text{Var}\left[\hat{f}_n(y)\right] = \frac{1}{nh_n}f(y)\int K^2(y)dy + o\left(\frac{1}{nh_n}\right).$$

Proof.
(i) Suppressing the subscript n, we find for the bias $b(y)$ of $\hat{f}_n(y)$,

$$b(y) = E\left[\frac{1}{n}\sum\frac{1}{h}K\left(\frac{y - X_i}{h}\right)\right] - f(y)$$

(6.4.20)
$$= \int \frac{1}{h}K\left(\frac{y - t}{h}\right)f(t)dt - f(y)$$

$$= \int K(z)\left[f(y - hz) - f(y)\right]dz.$$

Now by Taylor's theorem (Theorem 2.5.1),

$$f(y - hz) = f(y) - hzf'(y) + \frac{1}{2}h^2z^2 f''(y) + \frac{1}{6}h^3z^3 f'''(\xi),$$

where ξ lies between y and $y - hz$. Using the fact that $\int zK(z)dz = 0$, we therefore have

(6.4.21)
$$b(y) = \frac{1}{2}h^2 f''(y)\int z^2 K(z)dz + R_n,$$

where, with $|f'''(z)| \leq M$,

$$|R_n| \leq \frac{Mh^3}{6}\int |z|^3 K(z)dz = o\left(h^2\right),$$

as was to be proved.

(ii) By (6.4.13), $\hat{f}_n(y)$ is the average of n i.i.d. random variables and thus

$$
n \, \mathrm{Var} \left[\hat{f}_n(y) \right] = \mathrm{Var} \left[\frac{1}{h} K \left(\frac{y - X}{h} \right) \right]
$$

$$
= \int \frac{1}{h^2} K^2 \left(\frac{y - t}{h} \right) f(t) dt - \left[E \frac{1}{h} K \left(\frac{y - X}{h} \right) \right]^2 .
$$

Since the second term equals

$$
[E \hat{f}_n(y)]^2 = [f(y) + b(y)]^2,
$$

we have by (6.4.18)

$$
n \, \mathrm{Var} \left[\hat{f}_n(y) \right] = \frac{1}{h} \int K^2(z) f(y - hz) dz - \left[f(y) + o \left(h^2 \right) \right]^2
$$

$$
= \frac{1}{h} \int K^2(z) \left[f(y) - hz f'(y) + \frac{1}{2} h^2 z^2 f''(\xi) \right] dz + O(1).
$$

Dividing by n and using (6.4.9) and (6.4.11), we see that all the terms in $\mathrm{Var} \, \hat{f}_n(y)$ except the first tend to 0 and hence that

(6.4.22) $$\mathrm{Var} \left[\hat{f}_n(y) \right] = \frac{1}{hn} f(y) \int K^2(z) dz + o \left(\frac{1}{hn} \right)$$

as $nh_n \to \infty$ and $h_n \to 0$. ∎

Corollary 6.4.1 *Under the assumptions of Theorem 6.4.3,*

(6.4.23) $$E \left[\hat{f}_n(y) - f(y) \right]^2 \to 0$$

and hence $\hat{f}_n(y)$ is a consistent estimator of $f(y)$, provided

(6.4.24) $$h_n \to 0 \text{ and } nh_n \to \infty \text{ as } n \to \infty.$$

Proof. By (6.4.18) and (6.4.19), both the bias and variance of $\hat{f}_n(y)$ tend to zero and the result follows from the fact that

(6.4.25) $$E \left[\hat{f}_n - f(y) \right]^2 = (\mathrm{Bias})^2 + \mathrm{Variance}.$$

In order to specify the estimator $\hat{f}_n(y)$ completely, it is necessary to choose the bandwidth h. To obtain some insight into the effect of this choice, consider the case of $n = 2$ observations and suppose that K is a unimodal density such as the normal. If h is very small, the mass of $1/2$ for each x_i, when spread out according to (6.4.12), will decrease so rapidly away from x_i that the sum of the two contributions will differ little from the effect of each contribution alone (Problem 4.5).

The same is clearly true for any fixed number of observations as h becomes sufficiently small. In particular, $\hat{f}_n(y)$ will then have one peak for each observation. Consider, on the other hand, what happens for large h. Then each contribution is nearly flat over a large interval centered at its observation. As $h \to \infty$, all individual features of the data disappear (Problem 4.5). These comments can be summarized by saying that the estimator undersmooths or oversmooths when h is too small or too large, respectively.

To see how to find the right middle course for h, let us measure the accuracy with which $\hat{f}_n(y)$ estimates $f(y)$ by the expected squared error (6.4.25). From (6.4.18) and (6.4.19), we have

(6.4.26)
$$E\left[\hat{f}_n(y) - f(y)\right]^2 = \frac{1}{4}h^4 \left[f''(y)\right]^2 \tau^4 + \frac{1}{hn}f(y)\int K^2(z)dz + R_n,$$

where

(6.4.27)
$$R_n = o\left(h^4\right) + o\left(\frac{1}{hn}\right).$$

Asymptotically, the best bandwidth is therefore obtained by minimizing

(6.4.28)
$$g(h) = ah^4 + \frac{b}{h}$$

with

(6.4.29)
$$a = \frac{\tau^4}{4}\left[f''(y)\right]^2 \text{ and } b = \frac{1}{n}f(y)\int K^2(z)dz.$$

Since

$$g'(h) = 4ah^3 - \frac{b}{h^2} = 0$$

when

(6.4.30)
$$h = (b/4a)^{1/5},$$

and since this value corresponds to a minimum of g (Problem 4.6), we see that (6.4.28) is minimized by (6.4.30) and that the corresponding minimum values of the two terms of g are

$$ah^4 = \frac{a^{1/5}b^{4/5}}{4^{4/5}} \text{ and } \frac{b}{h} = a^{1/5}b^{4/5} \cdot 4^{1/5}.$$

Substituting a and b from (6.4.29) gives for the value of h minimizing (6.4.26) when R_n is neglected,

$$(6.4.31) \qquad h = \left\{ \frac{f(y) \int K^2(z)dz}{\tau^2 \, [f''(y)]^2} \right\}^{1/5} \bigg/ \frac{1}{n^{1/5}}.$$

It follows from (6.4.21) and (6.4.22) that the corresponding squared bias and variance are both of the order $1/n^{4/5}$.

These results are quite different from those found in (4.2.8) and (4.2.6), where the variance and the expected squared error tend to 0 at rate $1/n$. In the present case, these quantities tend to 0 at the slower rate $1/n^{4/5}$ (Problem 4.9). There is an additional difference between the present and the earlier situation. The two components of (6.4.25) are now both of order $1/n^{4/5}$, and the square of the bias therefore makes a contribution even asymptotically while it is asymptotically negligible when the rates are those given by (4.2.6) and (4.2.8) of Chapter 4.

So far, we have considered $\hat{f}_n(y)$ as an estimator of $f(y)$ at a given point y. However, more often one is interested in the whole curve $\hat{f}_n(y)$, $-\infty < y < \infty$ as an estimator of the density function $f(y)$, $-\infty < y < \infty$. We shall then measure the closeness of the estimator by the *integrated expected squared error*

$$(6.4.32) \quad \int E\left[\hat{f}_n(y) - f(y)\right]^2 dy = \int \operatorname{Var} \hat{f}_n(y)dy + \int \left[\operatorname{Bias} \hat{f}_n(y)\right]^2 dy.$$

From (6.4.26), we have

(6.4.33)
$$\int E\left[\hat{f}_n(y) - f(y)\right]^2 = \frac{h^4}{4}\tau^4 \int [f''(y)]^2 \, dy + \frac{1}{hn} \int K^2(z)dz + \int R_n(y)dy.$$

Assuming that the integrated remainder is of smaller order than the other terms, we determine the best bandwidth for this global criterion by minimizing (6.4.28) with

$$(6.4.34) \qquad a = \frac{\tau^4}{4} \int [f''(y)]^2 \, dy, \quad b = \frac{1}{n} \int K^2(z)dz.$$

The minimizing value of h is given by (6.4.30) and results in the miniumum integrated expected squared error

$$(6.4.35) \qquad \frac{5\tau^{4/5}}{4n^{4/5}} \left\{ \int [f''(y)]^2 \, dy \right\}^{1/5} \left\{ \int K^2(z)dz \right\}^{4/5} + o\left(\frac{1}{n^{4/5}} \right).$$

A statement and proof of conditions on f and K under which (6.4.35) is valid are given in Rosenblatt (1971). (There are some errors in Rosenblatt's formula (23), but they do not affect his proof.)

Analogous considerations apply to the distributional limit behavior of $\hat{f}_n(y)$. By (6.4.13), $\hat{f}_n(y)$ is the average of the n random variables

(6.4.36)
$$\frac{1}{h_n} K\left(\frac{y - X_i}{h_n}\right).$$

Since the common distribution of these variables depends on n, we cannot use the classical CLT (Theorem 2.4.1) but instead base ourselves on Corollary 2.4.1 to obtain conditions for

(6.4.37)
$$\frac{\hat{f}_n(y) - E\hat{f}_n(y)}{\sqrt{\operatorname{Var}\hat{f}_n(y)}} \xrightarrow{L} N(0, 1)$$

to hold. (Problem 4.10 or Rosenblatt (1971)). Besides conditions on f and K, application of Corollary 2.4.1 requires that (6.4.24) be strengthened to

(6.4.38)
$$h_n \to 0 \text{ and } nh_n^3 \to \infty,$$

i.e., that h_n tends to 0 more slowly than $1/n^3$.

A quantity of greater interest than the numerator of (6.4.37) is the difference $\hat{f}_n(y) - f(y)$. Let us now consider the asymptotic behavior of this difference for the case that h is given by (6.4.30). By the definition of bias, we have

(6.4.39)
$$\frac{\hat{f}_n(y) - f(y)}{\sqrt{\operatorname{Var} \hat{f}_n(y)}} = \frac{\hat{f}_n(y) - E\left[\hat{f}_n(y)\right]}{\sqrt{\operatorname{Var} \hat{f}_n(y)}} + \frac{b(y)}{\sqrt{\operatorname{Var} \hat{f}_n(y)}}.$$

Suppose that both the conditions of Theorem 6.4.3 and those required for (6.4.37) are satisfied. Then the first term of (6.4.39) tends in law to $N(0,1)$ while the second term tends to a constant ξ (Problem 4.10),

(6.4.40)
$$\frac{b(y)}{\sqrt{\operatorname{Var}\hat{f}_n(y)}} \to \xi,$$

which is $\neq 0$ unless $f''(y) = 0$. Thus

(6.4.41)
$$\frac{\hat{f}_n(y) - f(y)}{\sqrt{\operatorname{Var}\hat{f}_n(y)}} \xrightarrow{L} N(\xi, 1),$$

where $\sqrt{\operatorname{Var} \hat{f}_n(y)}$ is of order $n^{-2/5}$. The estimator is therefore asymptotically biased also in the distributional sense.

It is interesting to carry the analysis one step further and ask for the kernel K that minimizes the principal term of (6.4.35) or of (6.4.22) with h given by (6.4.31). In both cases, of the various factors making up this term, only $\tau^{4/5} = \left(\int z^2 K(z) dz \right)^{2/5}$ and $\left(\int K^2(z) dz \right)^{4/5}$ involve K, and the desired kernel is obtained by minimizing

$$(6.4.42) \qquad \int z^2 K(z) dz \left[\int K^2(z) dz \right]^2.$$

It can be shown that among all probability densities K, (6.4.42) is minimized by any kernel of the form

$$(6.4.43) \qquad K(z) = \frac{1}{c} p \left(\frac{z}{c} \right),$$

where

$$(6.4.44) \qquad p(z) = \begin{cases} \frac{3}{4} \left(1 - z^2 \right), |z| \leq 1 \\ 0 \text{ elsewhere,} \end{cases}$$

the so-called Epanechnikov kernel. (For a proof of this result, see, for example, Prakasa Rao (1983) or Hodges and Lehmann (1956).)

To see how much is lost by using a suboptimal kernel, let us consider the asymptotic relative efficiency (ARE) of estimators based on two different kernels. As in Section 4.3, this ARE is defined as the ratio of the numbers of observations required by the two estimators to achieve the same accuracy (6.4.35) [or (6.4.22)]. Equating the limiting values of (6.4.35) based on n_1 and n_2 observations, respectively, we find

$$(6.4.45) \qquad \lim \tau_1 \int K_1^2(z) dz / n_1 = \lim \tau_2 \int K_2^2(z) dz / n_2$$

and hence

$$(6.4.46) \qquad e_{2,1} = \lim \frac{n_1}{n_2} = \frac{\tau_1 \int K_1^2(z) dz}{\tau_2 \int K_2^2(z) dx}$$

as the natural analog of (4.3.5). If we take for K_1 the optimal parabolic kernel (6.4.44), $e_{2,1}$ becomes the absolute efficiency (within the class of kernel estimators)

$$(6.4.47) \qquad e(K_2) = \frac{A}{\tau_2 \int K_2^2(z) dz}.$$

Here

$$A = \tau_1 \int K_1^2(z) dz = 3/5\sqrt{5},$$

with K_1 given by (6.4.44). Typical values of the efficiency (6.4.46) are (to three decimals)

$$.951\,(K_2 = \text{ normal})\,,\ .930\,(\text{uniform})\,,\ .986\,(\text{triangular})\,.$$

On the other hand, if the kernel is sufficiently heavy tailed, the efficiency $e(K)$ can become arbitrary small as $\tau_2 \to \infty$ without $\int K^2(z)dz$ tending to 0 (Problem 4.14).

We have so far restricted attention to kernel estimators and have seen that with optimal choice of bandwidth, these achieve a convergence rate of $n^{-4/5}$. Many other types of estimators are available and it is possible to obtain some improvement (for the discussion of such estimators, see the books by Silverman (1986), Scott (1992), and Wand and Jones (1995)). However, in practice substantial improvements typically are impossible (see, for example, Brown and Farrell (1990)). In any case, no nonparametric density estimator can achieve the rate $1/n$, which is the standard rate for the estimation of most functionals $h(F)$ considered in Sections 6.1–6.3, including that of densities given by a parametric family (Problem 4.15). The lower rate for nonparametric density estimation is explained by the local nature of an unrestricted density. Even if we know $f(x)$ for all x outside an arbitrarily small neighborhood $|x - y| < \epsilon$ of y, this tells us nothing about the value of $f(x)$. Thus only observations very close to y provide information concerning $f(y)$. [For a general theory of achievable rates, see Donoho and Liu (1991).]

The estimation of probability densities extends in a fairly straightforward way to the multivariate case. We shall only sketch this theory here. Let $f(\mathbf{y})$, $\mathbf{y} = (y_1, \ldots, y_s)$, be the density of an s-dimensional distribution F and suppose that f is unknown, subject only to certain smoothness conditions. We wish to estimate $f(\mathbf{y})$ at a given point $\mathbf{y}$ at which $f(\mathbf{y})$ is continuous and positive on the basis of n i.i.d. random vectors $\mathbf{X}^j = \left(X_1^{(j)}, \ldots, X_s^{(j)}\right)$ from F. In generalization of (6.4.13), we shall consider s-dimensional kernel estimators

$$(6.4.48) \qquad \hat{f}_n(\mathbf{y}) = \frac{1}{n} \sum_{j=1}^{n} \frac{1}{h_1 \cdots h_s} K\left(\frac{y_1 - X_1^{(j)}}{h_1}, \ldots, \frac{y_s - X_s^{(j)}}{h_s}\right),$$

where K is an s-dimensional probability density satisfying the symmetry condition

$$(6.4.49) \qquad K\left(-z_1, \ldots, -z_s\right) = K\left(z_1, \ldots, z_s\right).$$

Thus in particular

$$(6.4.50) \qquad \int \cdots \int z_j K(\mathbf{z})d\mathbf{z} = 0 \text{ for all } j,$$

where

(6.4.51) $$d\mathbf{z} = dz_1 \cdots dz_s.$$

The interpretation of $\hat{f}_n(y)$ is exactly analogous to that of (6.4.13) in the univariate case. For each observational point $\mathbf{x}^j = (x_1^j, \ldots, x_s^j)$, the probability mass $1/n$ assigned to it by the empirical distribution is spread out according to the s-dimensional probability density

(6.4.52) $$\frac{1}{h_1 \cdots h_s} K\left(\frac{y_1 - x_1^{(j)}}{h_1}, \ldots, \frac{y_s - x_s^{(j)}}{h_s}\right).$$

As in the univariate case, it is easily seen that (Problem 4.16)

(6.4.53) $$\int \cdots \int \hat{f}_n(\mathbf{y}) d\mathbf{y} = 1,$$

so that $\hat{f}_n(\mathbf{y})$ is a probability density.

In order to study the bias and variance of $\hat{f}_n$, we require an extension of the multivariate Taylor theorem (Theorem 5.2.2).

Theorem 6.4.4 *Let f be a real-valued function of s variables for which the third partial derivatives exist in a neighborhood of a point $\mathbf{a}$, and let f_i, f_{ij}, and f_{ijk} denote the first, second, and third partial derivatives*

(6.4.54)
$$f_i(\mathbf{t}) = \left.\frac{\partial f(\mathbf{x})}{\partial x_i}\right|_{\mathbf{x}=\mathbf{t}}, \; f_{ij}(\mathbf{t}) = \left.\frac{\partial^2 f(\mathbf{x})}{\partial x_i \partial x_j}\right|_{\mathbf{x}=\mathbf{t}}$$
$$and \; f_{ijk}(\mathbf{t}) = \left.\frac{\partial^3 f(\mathbf{x})}{\partial x_i \partial x_j \partial x_k}\right|_{\mathbf{x}=\mathbf{t}}.$$

Then

$$f(a_1 + \Delta_1, \ldots, a_s + \Delta_s) - f(a_1, \ldots, a_s)$$
$$= \sum \Delta_i f_i(\mathbf{a}) + \frac{1}{2}\sum\sum \Delta_i \Delta_j f_{ij}(\mathbf{a}) + \frac{1}{6}\sum\sum\sum \Delta_i \Delta_j \Delta_k f_{ijk}(\boldsymbol{\xi}),$$

where $\boldsymbol{\xi}$ is an intermediate point on the line segment connecting $(a_1, \ldots, a_s)$ and $(a_1 + \Delta_1, \ldots, a_s + \Delta_s)$.

To evaluate the bias $b(\mathbf{y})$ of $\hat{f}_n(\mathbf{y})$, we now proceed as in the derivation of (6.4.18), and find in analogy to (6.4.20) that

(6.4.55) $$b(\mathbf{y}) = \int K(z_1, \ldots, z_s)\left[f(y_1 - h_1 z_1, \ldots, y_s - h_s z_s)\right.$$
$$\left. - f(y_1, \ldots, y_s)\right] dz_1, \ldots, dz_s,$$

where here and below we use a single integral sign to indicate integration with respect to $d\mathbf{z}$. We can expand the integrand according to Theorem 6.4.4 as

(6.4.56)
$$
\begin{aligned}
f\left(y_1 - h_1 z_1, \ldots, y_s - h_s z_s\right) - f\left(y_1, \ldots, y_s\right) &= \sum h_j z_j f_j(\mathbf{y}) \\
+ \frac{1}{2} \sum\sum h_i h_j z_i z_j f_{ij}(\mathbf{y}) &- \frac{1}{6} \sum\sum\sum h_i h_j h_k z_i z_j z_k f_{ijk}(\boldsymbol{\xi}),
\end{aligned}
$$

where $\boldsymbol{\xi}$ is a point on the line segment connecting $\mathbf{y}$ and $(\mathbf{y} + \Delta)$. The bias is obtained by multiplying (6.4.56) by $K(\mathbf{z})$ and integrating the product. Here

$$
\int h_j z_j f_j(\mathbf{y}) K(\mathbf{z}) d(\mathbf{z}) = h_j f_j(\mathbf{y}) \int z_j K(\mathbf{z}) d\mathbf{z} = 0
$$

by (6.4.50), so that

(6.4.57)
$$
b(\mathbf{y}) = \frac{1}{2} \sum\sum h_i h_j f_{ij}(\mathbf{y}) \int z_i z_j K(\mathbf{z}) d\mathbf{z} + R_n.
$$

To obtain an estimate of R_n, let us suppose that

(6.4.58)
$$
h_i = c_i h,
$$

where the c's are fixed and $h \to 0$. This means that we are letting all the h's tend to zero at the same rate. Then $R_n = o\left(h^2\right)$ under conditions generalizing those of Theorem 6.4.3(i) (Problem 4.17) and hence

(6.4.59)
$$
b(\mathbf{y}) = \frac{h^2}{2} \sum\sum c_i c_j f_{ij}(\mathbf{y}) \int z_i z_j K(\mathbf{z}) d\mathbf{z} + o\left(h^2\right).
$$

Since $\hat{f}_n(\mathbf{y})$ is a sum of n i.i.d. random variables, we see as in the proof of Theorem 4.3 (ii) that

$$
\begin{aligned}
n \operatorname{Var}\left[\hat{f}_n(\mathbf{y})\right] &= \operatorname{Var}\left[\frac{1}{h_1 \cdots h_s} K\left(\frac{y_1 - X_1}{h_1}, \ldots, \frac{y_s - X_s}{h_s}\right)\right] \\
&= \int \frac{1}{h_1, \ldots, h_s} K^2(\mathbf{z}) f\left(y_1 - h_1 z_1, \ldots, y_s - h_s z_s\right) d\mathbf{z} - \left[f(\mathbf{y}) + o\left(h^2\right)\right]^2 \\
&= \int \frac{1}{h_1, \ldots, h_s} K^2(\mathbf{z}) \left[f(\mathbf{y}) - \sum h_j z_j f_j(\mathbf{y})\right. \\
&\quad \left. + \frac{1}{2} \sum\sum h_i h_j z_i z_j f_{ij}(\boldsymbol{\xi}) d\mathbf{z} + O(1)\right].
\end{aligned}
$$

Restricting attention to h's satisfying (6.4.58), we find

(6.4.60)
$$
\operatorname{Var}\left[\hat{f}_n(\mathbf{y})\right] = \frac{1}{nh^s} \prod_{i=1}^{s} c_i^{-1} f(\mathbf{y}) \int K^2(\mathbf{z}) d\mathbf{z} + o\left(\frac{1}{nh^s}\right)
$$

under suitable conditions on f and K (Problem 4.16). Sufficient conditions for consistency of $\hat{f}_n(\mathbf{f})$ as an estimator of $f(\mathbf{y})$ are then the validity of (6.4.59) and (6.4.60) together with (in generalization of (6.4.24))

$$(6.4.61) \qquad h = h_n \to 0 \text{ and } nh_n^s \to \infty.$$

Let us now determine the rate at which h should tend to zero so as to minimize the expected squared error (6.4.25). Using the formulas (6.4.59) and (6.4.60) for bias and variance and neglecting the remainder terms which will not affect the rate, we find that

$$(6.4.62) \qquad E\left[\hat{f}_n(\mathbf{y}) - f(\mathbf{y})\right]^2 = ah^4 + \frac{b}{h^s} = g(h),$$

where

$$(6.4.63) \qquad a = \frac{1}{4}\left[\sum\sum c_i c_j f_{ij}(\mathbf{y}) \int z_i z_i K(\mathbf{z}) d\mathbf{z}\right]^2$$

and

$$(6.4.64) \qquad b = \frac{\prod c_i^{-1}}{n} f(\mathbf{y}) \int K^2(\mathbf{z}) d\mathbf{z}.$$

Since

$$g'(h) = 4ah^3 - \frac{bs}{h^{s+1}} = 0$$

when

$$h^{s+4} = \frac{bs}{4a}$$

and since this value of h corresponds to a minimum (Problem 4.18), we see that (6.4.62) is minimized by

$$(6.4.65) \qquad h = \left(\frac{bs}{4a}\right)^{\frac{1}{s+4}}.$$

Substituting this value of h into (6.4.62) we find that the squared bias term $\left(ah^4\right)$ and the variance term (b/h^s) both tend to 0 at the rate $(1/n)^{\frac{4}{s+4}}$, and so therefore does the expected squared error. This agrees with the results for the case $s = 1$ and shows that for large n, the accuracy of $\hat{f}_n(y)$ decreases with s.

A dramatic illustration of this "curse of dimensionality" is provided by Silverman (1986). Table 6.4.1 shows the number n of observations required for the relative expected squared error

$$(6.4.66) \qquad E\left[\hat{f}_n(\mathbf{y}) - f(\mathbf{y})\right]^2 / f^2(\mathbf{y})$$

to be $< .1$, for the case that $Y_1, \ldots, Y_s$ are i.i.d. $N(0,1)$, the kernel K is normal, $c_1 = \cdots = c_s = 1$, $y_1 = \cdots = y_s = 0$, and h is the optimal bandwidth for the given situation.

TABLE 6.4.1. Sample size required for (6.4.66) to be $< .1$ ($f, K = $ normal)

s	2	4	6	8	10
n	19	223	2790	43,700	842,000

Source: Table 4.2 of Silverman (1986).

Summary

1. Kernel estimators $\hat{f}_n(y)$ are defined for estimating a probability density $f(y)$, and approximate formulas for their bias, variance, and expected squared error are derived. Conditions are given under which $\hat{f}_n(y)$ is consistent for estimating $f(y)$.

2. With the kernel K fixed, the expected squared error is minimized when the bandwidth h_n tends to 0 at the rate $1/n^{1/5}$. The resulting expected squared error is only of order $1/n^{4/5}$ instead of the rate $1/n$ obtainable for standard estimation problems.

3. The optimal kernel is given by a parabolic density.

4. Other types of estimators can improve the convergence rate of $1/n^{4/5}$, but none can reach the standard rate of $1/n$. This is a consequence of the local nature of a nonparametric density which makes its estimation more difficult than that of the functionals considered in Section 6.3.

5. Under suitable regularity conditions, the kernel estimator $\hat{f}_n(y)$ with optimal bandwidth satisfies

$$\frac{\hat{f}_n(y) - f(y)}{\sqrt{\text{Var}\left(\hat{f}_n(y)\right)}} \to N(\xi, 1)$$

and is thus asymptotically biased.

6. The theory of kernel estimators is extended from univariate to multivariate densities.

6.5   Bootstrapping

In many applications, the functionals $h(F)$ of interest differ from those considered in the preceding sections by depending not only on F but also on the sample size n. In particular, this situation arises routinely when evaluating the performance of an estimator $\hat{\theta}_n$ of some parameter θ or functional $h(F)$. The following are some examples of estimators $\hat{\theta}_n$ based on a sample $X_1, \ldots, X_n$ from F, and of some measures $\lambda_n(F)$ of their performance.

Estimators:

(i) the sample mean $\hat{\theta}_n = \bar{X}$ as an estimator of the expectation $h(F) = E_F(X)$;

(ii) the sample median $\hat{\theta}_n$ as estimator of the population median;

(iii) the V-statistic (6.2.31) as estimator of the expectation (6.1.1);

(iv) the maximum distance $\hat{\theta}_n = \sup_x \left| \hat{F}_n(x) - F_0(x) \right|$ of the sample cdf from a given cdf F_0, as an estimator of $h(F) = \sup_x |F(x) - F_0(x)|$.

Measures of the performance of $\hat{\theta}_n$:

(A) the error distribution $\lambda_n(F) = P_F \left\{ \sqrt{n} \left[\hat{\theta}_n - h(F) \right] \le a \right\}$

or, for some scaling factor $\tau(F)$,

(B) $\lambda_n(F) = P_F \left\{ \dfrac{\sqrt{n} \left[\hat{\theta}_n - h(F) \right]}{\tau(F)} \le a \right\}$.

Alternatively we might be mainly interested in some aspects of these error distributions such as

(C) $\lambda_n(F) =$ the bias of $\hat{\theta}_n$

or

(D) $\lambda_n(F) =$ the variance of $\sqrt{n}\hat{\theta}_n$.

In the earlier sections of this chapter, we were concerned with the estimation of a functional $h(F)$ by means of $h\left(\hat{F}_n\right)$. Correspondingly, we shall now consider estimating $\lambda_n(F)$ by the plug-in estimator $\lambda_n\left(\hat{F}_n\right)$. This estimator together with some elaborations that it often requires is known as the *bootstrap*. It was proposed in this connection by Efron in 1979 and has been found to be applicable in a great variety of situations. Both theory and applications are treated in an enormous literature, summaries of which can be found in the books by Hall (1992), LePage and Billard (1992), Mammen (1992), Efron and Tibshirani (1993), and Shao and Tu (1995).[†]

Before taking up some theoretical aspects of the estimator $\lambda_n\left(\hat{F}_n\right)$, we must consider a problem that is illustrated by the following examples.

Example 6.5.1 Estimating the error probability (A). Suppose we are concerned with the measure $\lambda_n(F)$ given by (A). The plug-in estimator

---

[†]We are here considering the bootstrap only for point estimation. However, it is equally applicable to confidence intervals and hypothesis testing. These aspects are treated in the books cited.

$\lambda_n\left(\hat{F}_n\right)$ of (A) is obtained by replacing the distribution F of the X's in (A) by the distribution $\hat{F}_n$. Then $h(F)$ becomes $h\left(\hat{F}_n\right)$. In addition, the subscript F, which governs the distribution of $\hat{\theta}_n$, must also be changed to $\hat{F}_n$. To see what this last step means, write

$$(6.5.1) \qquad \hat{\theta}_n = \theta\left(X_1, \ldots, X_n\right),$$

that is, express $\hat{\theta}_n$ not as a function of $\hat{F}_n$ but directly as a function of the sample $(X_1, \ldots, X_n)$. If, for example,

$$(6.5.2) \qquad h(F) = E_F(X),$$

then

$$(6.5.3) \qquad \theta\left(X_1, \ldots, X_n\right) = (X_1 + \cdots + X_n)/n.$$

The function θ on the right side of (6.5.1) should have a subscript n which we suppress for the sake of simplicity. The dependence of the distribution of $\hat{\theta}_n$ on F results from the fact that $X_1, \ldots, X_n$ is a sample from F. To replace F by $\hat{F}_n$ in the distribution governing $\hat{\theta}_n$, we must therefore replace (6.5.1) by

$$(6.5.4) \qquad \theta_n^* = \theta\left(X_1^*, \ldots, X_n^*\right),$$

where $X_1^*, \ldots, X_n^*$ is a sample from $\hat{F}_n$. With this notation, $\lambda_n\left(\hat{F}_n\right)$ can now be written formally as

$$(6.5.5) \qquad \lambda_n\left(\hat{F}_n\right) = P_{\hat{F}_n}\left\{\sqrt{n}\left[\theta_n^* - h\left(\hat{F}_n\right)\right] \le a\right\}.$$

To understand this estimator, note that the sample $X_1^*, \ldots, X_n^*$ from $\hat{F}_n$ is only a conceptual sample from this distribution. A statement such as

$$P\left[(X_1^*, \ldots, X_n^*) \in S\right]$$

is simply a convenient notation for the probability that a sample from $\hat{F}_n$ falls into this set. The distribution $\hat{F}_n$ assigns probability $1/n$ to each of the observed values $x_1, \ldots, x_n$, and $(X_1^*, \ldots, X_n^*)$ is a hypothetical sample from this distribution. In particular, the variables $X_1^*, \ldots, X_n^*$ are conditionally independent, given $X_1 = x_1, \ldots, X_n = x_n$. Let us now illustrate the calculation of the estimator (6.5.5) on an example. $\square$

Example 6.5.2 Estimating the error distribution of the mean. Let

$$\theta = h(F) = E_F(X)$$

so that $h\left(\hat{F}_n\right) = \bar{X}$ and let us calculate (6.5.5) in the unrealistic case $n = 2$. Suppose that the ordered values of X_1 and X_2 are

$$X_{(1)} = c < X_{(2)} = d.$$

Then X_1^* and X_2^* are independently distributed with

$$P\left(X_i^* = c\right) = P\left(X_i^* = d\right) = 1/2, \ i = 1, 2.$$

The pair (X_1^*, X_2^*) therefore takes on the four possible pairs of values

(6.5.6) $(c, c), \ (c, d), \ (d, c), \ (d, d),$

each with probability $1/4$. Thus

$$\theta^* = \frac{1}{2}\left(X_1^* + X_2^*\right)$$

takes on the values c, $\frac{1}{2}(c + d)$, d with probabilities $\frac{1}{4}, \frac{1}{2}, \frac{1}{4}$, respectively, so that

$$\theta^* - h\left(\hat{F}_n\right) = \theta^* - \frac{1}{2}(c + d)$$

takes on the values

(6.5.7) $\frac{1}{2}(c - d), \ 0, \ \frac{1}{2}(d - c)$ with probabilities $\frac{1}{4}, \frac{1}{2}, \frac{1}{4}$, respectively.

From this distribution, we can now evaluate the probabilities (6.5.5) for any given value of a (Problem 5.1). ☐

Example 6.5.3 Estimating the bias of the median. Let $\theta = h(F)$ be the median of F and $\lambda_n(F)$ the bias of the sample median $\tilde{\theta}_n$, i.e.,

(6.5.8) $\lambda_n(F) = E\left(\tilde{\theta}_n\right) - \theta.$

Then

(6.5.9) $\lambda_n\left(\hat{F}_n\right) = E\left(\theta_n^*\right) - \tilde{\theta}_n,$

where $\tilde{\theta}_n$ is the median of the sample $X_1, \ldots, X_n$ and θ_n^* is the median of a (hypothetical) sample $X_1^*, \ldots, X_n^*$ from $\hat{F}_n$. Let us this time consider the case $n = 3$. If $X_{(1)} = b$, $X_{(2)} = c$, and $X_{(3)} = d$ denote the ordered observations, the variables (X_1^*, X_2^*, X_3^*) can take on $3^3 = 27$ possible triples of values $(b, b, b), (b, b, c), (b, c, b), \ldots$ in generalization of (6.5.6). To obtain the probabilities for the corresponding ordered triples $\left(X_{(1)}^*, X_{(2)}^*, X_{(3)}^*\right)$, we must count the number of cases with these values. For example,

$$P\left(X_{(1)}^* = b, X_{(2)}^* = b, X_{(3)}^* = c\right) = 3/27$$

since this probability is the sum of the probabilities of the cases (b, b, c), (b, c, b), (c, b, b) for (X_1^*, X_2^*, X_3^*). The resulting distribution for $\left(X_{(1)}^*, X_{(2)}^*, X_{(3)}^*\right)$ is

(6.5.10)

bbb	bbc	bbd	bcc	bcd	bdd	ccc	ccd	cdd	ddd
1/27	3/27	3/27	3/27	6/27	3/27	1/27	3/27	3/27	1/27

The median $X_{(2)}^*$ of (X_1^*, X_2^*, X_3^*) is b for the cases bbb, bbc, and bbd, and so on, so that the distribution of $X_{(2)}^*$ is

(6.5.11) $P\left(X_{(2)}^* = b\right) = \dfrac{7}{27}$, $P\left(X_{(2)}^* = c\right) = \dfrac{13}{27}$, $P\left(X_{(2)}^* = d\right) = \dfrac{7}{27}$.

Therefore $\lambda_n\left(\hat{F}_n\right)$, the estimator of the bias of $\theta_n^* = X_{(2)}^*$, is by (6.5.9)

$$E\left(X_{(2)}^*\right) - X_{(2)} = \left[\frac{7}{27}X_{(1)} + \frac{13}{27}X_{(2)} + \frac{7}{27}X_{(3)}\right] - X_{(2)}$$
$$= \frac{14}{27}\left[\frac{X_{(1)} + X_{(3)}}{2} - X_{(2)}\right].$$

The estimator of the variance of the median $X_{(2)}$ can be calculated in the same way (Problem 5.6(i)).   □

A general algorithm for such calculations, suitable for small n, is discussed by Fisher and Hall (1991). However, the numbers of distinct cases corresponding to (6.5.10) is shown by Hall (1987b) to be $\binom{2n-1}{n}$, which is 92,378 for $n = 10$ and increases exponentially. Except for small values of n, the exact calculation of $\lambda_n\left(\hat{F}_n\right)$ therefore is not feasible.

For large n, this difficulty creates a strange situation. The estimator $\lambda_n\left(\hat{F}_n\right)$ is a well defined function of the observations, but in practice we are unable to compute it. The seminal paper of Efron (1979) provides a way out.

Example 6.5.4 Bootstrapping a probability. Let us begin with case (A), where $\lambda_n\left(\hat{F}_n\right)$ is of the form

(6.5.12) $\lambda_n\left(\hat{F}_n\right) = P_{\hat{F}_n}\left[(X_1^*, \ldots, X_n^*) \in S\right]$

with $X_1^*, \ldots, X_n^*$ being a sample from $\hat{F}_n$. Since $\hat{F}_n$ is a known distribution, evaluating (6.5.12) is a special case of the problem of calculating the probability

$$(6.5.13) \qquad p = P\left[(Y_1, \ldots Y_n) \in S\right],$$

where $(Y_1, \ldots, Y_n)$ has a known distribution G, but S has a shape that makes the calculation of the n-fold integral (6.5.13) difficult. A standard remedy in such cases is simulation, i.e., generating a large sample of vectors $(Y_{i1}, \ldots, Y_{in})$, $i = 1, \ldots, B$, from G and using the frequency $\hat{p}_B$ with which these vectors fall into S as an approximation of p. By the law of large numbers,

$$(6.5.14) \qquad \hat{p}_B \xrightarrow{P} p \text{ as } B \to \infty,$$

so that for sufficiently large B, we can be nearly certain that $\hat{p}_B$ provides a good approximation for p.

In order to apply this idea to the calculation of (6.5.12), we must draw B samples $X_{i1}^*, \ldots, X_{in}^*$ from $\hat{F}_n$:

$$(6.5.15) \qquad \begin{array}{c} X_{11}^*, \ldots, X_{1n}^* \\ \cdots\cdots\cdots\cdots \\ X_{B1}^*, \ldots, X_{Bn}^*. \end{array}$$

These are Bn independent variables, each taking on the values $x_1, \ldots, x_n$ with probability $1/n$. The situation differs from the general case described after (6.5.13) in that G, the joint distribution of n independent variables from $\hat{F}_n$, depends on the observations $x_1, \ldots, x_n$: In fact, each of the variables X_{ij}^* is drawn at random from the population of n elements $\{x_1, \ldots, x_n\}$. This process of *resampling* can be viewed as drawing a sample of size Bn with replacement from $\{x_1, \ldots, x_n\}$; the variables (6.5.15) are called a *bootstrap sample*.

As in the general case, the estimator (6.5.12) is approximated by the frequency $\lambda_{B,n}^*$ with which the vector $(X_{i1}^*, \ldots, X_{in}^*)$ falls into S. For large B, this frequency will be close to $\lambda_n\left(\hat{F}_n\right)$ with high probability. If $\lambda_n\left(\hat{F}_n\right)$ is a reasonable estimator of $\lambda_n(F)$, so is $\lambda_{B,n}^*$, the *bootstrap estimator*, and the two-stage process of estimating $\lambda_n(F)$ by $\lambda_n\left(\hat{F}_n\right)$ and then approximating the latter by $\lambda_{B,n}^*$ is called bootstrapping. $\qquad\square$

Example 6.5.5 Bootstrapping an expectation. Bootstrapping is not restricted to case (A). Suppose that, as in (C), we are interested in estimating the bias

$$(6.5.16) \qquad \lambda_n(F) = E_F \delta\left(X_1, \ldots, X_n\right) - h(F)$$

for some estimator δ of some functional $\theta = h(F)$. Then

$$(6.5.17) \qquad \lambda_n\left(\hat{F}_n\right) = E_{\hat{F}_n}\delta\left(X_1^*, \ldots, X_n^*\right) - h\left(\hat{F}_n\right)$$

and we saw in Example 6.5.3 that the difficulty of computing the first term on the right side of (6.5.17) will typically be prohibitive except for small n. To overcome this problem, we draw a bootstrap sample (6.5.15) as before. For each i, we determine

$$\delta_i^* = \delta\left(X_{i1}^*, \ldots, X_{in}^*\right)$$

and then approximate $\lambda_n\left(\hat{F}_n\right)$ by

$$(6.5.18) \qquad \lambda_{B,n}^* = \frac{1}{n}\sum_{i=1}^{B}\delta_i^* - h\left(\hat{F}_n\right).$$

By the law of large numbers, the average of the i.i.d. variables δ_i^* tends in probability to

$$E\left(\delta_i^*\right) = E\delta\left(X_1^*, \ldots, X_n^*\right),$$

and hence for large B, $\lambda_{B,n}^*$ will, with high probability, be close to $\lambda_n\left(\hat{F}_n\right)$.

The same approach will provide an approximation to $\lambda_n\left(\hat{F}_n\right)$ when $\lambda_n(F)$ is, for example, one of the cases (B) or (D) (Problem 5.8).

Bootstrapping introduces a new element which at first may be somewhat confusing. The following comments may help to clarify its basic logic.

Note 1: It is tempting to think of $\lambda_{B,n}^*$ as an estimator of $\lambda_n\left(\hat{F}_n\right)$. However, $\lambda_n\left(\hat{F}_n\right)$ is not an unknown parameter but a quantity which, in principle, is known although it may be difficult to calculate. The bootstrap sample avoids the necessity of performing this calculation by providing an approximation $\lambda_{B,n}^*$, which for large B is nearly certain to be very close to $\lambda_n\left(\hat{F}_n\right)$. Thus $\lambda_{B,n}^*$ is an *approximator* rather than an estimator of $\lambda_n\left(\hat{F}_n\right)$.

Note 2: An objection that may be raised against the use of $\lambda_{B,n}^*$ as an estimator of $\lambda_n(F)$ is that it depends on the bootstrap samples (6.5.15) and therefore introduces an extraneous randomization into the inference. A repetition of the process would (with the same data $(x_1, \ldots, x_n)$) generate another bootstrap sample and hence a different value for $\lambda_{B,n}^*$. This argument loses much of its force by the fact that for given $x_1, \ldots, x_n$ and sufficiently large B, the approximate $\hat{\lambda}_{B,n}$, though a random variable, is with high probability close to a constant.

Note 3: On first encountering the bootstrap idea, one may feel that it achieves the impossible: to provide additional information (about $\lambda_n(F)$) without acquiring more data. This is, of course, an illusion. The estimator $\lambda_{B,n}^*$ is not an improvement over $\lambda_n\left(\hat{F}_n\right)$, which remains the basic estimator. What $\lambda_{B,n}^*$ accomplishes—and this is very useful but not a miracle—is to provide a simple, and for large B highly accurate, approximation to $\lambda_n\left(\hat{F}_n\right)$ when the latter is too complicated to compute directly.

Note 4: The two-stage process was described above as

(i) estimating $\lambda_n(F)$ by $\lambda_n\left(\hat{F}_n\right)$

and

(ii) approximating $\lambda_n\left(\hat{F}_n\right)$ by $\lambda_{B,n}^*$.

The method is, in fact, much more general, with many alternatives being available for each of the stages.

Regarding (i), depending on the situation, quite different estimators may be used to estimate $\lambda_n(F)$. The following example provides just one illustration. □

Example 6.5.6 The parametric bootstrap. Suppose that we are estimating the median θ of F by the sample median $\hat{\theta} = \theta\left(X_1, \ldots X_n\right)$ and wish to estimate the error probability

$$(6.5.19) \qquad \lambda_n(F) = P\left[\sqrt{n}\left(\tilde{\theta} - \theta\right) \leq a\right].$$

If nothing is known about F, we would estimate $\lambda_n(F)$ by

$$(6.5.20) \qquad \lambda_n\left(\hat{F}_n\right) = P\left[\sqrt{n}\left(\tilde{\theta}^* - \tilde{\theta}\right) \leq a\right],$$

where $\tilde{\theta}^* = \theta\left(X_1^*, \ldots, X_n^*\right)$ is the median of a hypothetical sample $X_1^*, \ldots, X_n^*$ from $\hat{F}_n$.

Suppose, however, that it is known that $X_1, \ldots, X_n$ is a sample from a Cauchy distribution $C(\theta, b)$ with center θ and scale parameter b. We might then still use $\hat{\theta}$ to estimate θ (for an alternative, more efficient estimator, see Section 7.3). However, instead of estimating (6.5.19) by (6.5.20), we might prefer to estimate it by

$$(6.5.21) \qquad \lambda_n\left(F_{\hat{\theta}}\right) = P\left[\sqrt{n}\left(\tilde{\theta}^{**} - \tilde{\theta}\right) \leq a\right],$$

where $\tilde{\theta}^{**} = \theta\left(X_1^{**}, \ldots, X_n^{**}\right)$ is the median of a hypothetical sample $X_1^{**}, \ldots, X_n^{**}$ from $C(\tilde{\theta}, \hat{b})$, the Cauchy distribution centered on $\tilde{\theta}$ and with a scale parameter $\hat{b}$ which is a consistent estimator (e.g., the maximum likelihood estimator) of b. This estimator utilizes the Cauchy structure of F. At the second stage, the estimator (6.5.21) can then be approximated by a bootstrap sample $(X_{i1}^{**}, \ldots, X_{in}^{**})$, $i = 1, \ldots, B$, from the Cauchy distribution $C(\hat{\theta}, \hat{b})$. $\qquad\square$

Let us now return to the general case. Alternatives are available not only for the first stage of the process but also for stage (ii). In particular, bootstrap samples may be obtained by resampling plans other than simple random sampling. A variety of such plans (for example, antithetic, balanced, and importance sampling) are discussed in the bootstrap literature, together with indications of the type of problems for which each is appropriate.

We shall conclude the section with a fairly brief discussion of some theoretical properties of the plug-in estimator $\lambda_n\left(\hat{F}_n\right)$, such as consistency, asymptotic normality, and comparison with more traditional estimators. These problems concern only the behavior of the estimator itself, not that of the approximation $\lambda_{B,n}^*$ which is quite separate and presents little difficulty.[‡]

In much of the bootstrap literature, the bootstrap is said "to work" if $\lambda_n\left(\hat{F}_n\right)$ is consistent for estimating $\lambda_n(F)$ in the sense that

$$(6.5.22) \qquad \lambda_n\left(\hat{F}_n\right) - \lambda_n(F) \xrightarrow{P} 0.$$

Before taking up the question of consistency it is useful to note that typically the sequence λ_n will tend to a limit λ, i.e., that

$$(6.5.23) \qquad \lambda_n(F) \to \lambda(F)$$

for all F under consideration. If $\sqrt{n}\left(\hat{\theta}_n - h(F)\right)$ has a limit distribution, the limit (6.5.23) clearly exists when $\lambda_n(F)$ is given by (A) or (B). In cases (C) and (D), it will often exist with $\lambda(F) \equiv 0$ and $\lambda(F)$ equal to the asymptotic variance, respectively.

Lemma 6.5.1 *If (6.5.23) holds, then (6.5.22) holds if and only if*

$$(6.5.24) \qquad \lambda_n\left(\hat{F}_n\right) \xrightarrow{P} \lambda(F).$$

Proof. The result is seen by writing

$$(6.5.25) \qquad \lambda_n\left(\hat{F}_n\right) - \lambda_n(F) = \left[\lambda_n\left(\hat{F}_n\right) - \lambda(F)\right] - \left[\lambda_n(F) - \lambda(F)\right]. \qquad\blacksquare$$

[‡]For this reason, it might be preferable to separate the two stages terminologically and restrict the term "bootstrap" to the second stage.

This lemma reduces the generalized consistency definition (6.5.22) to the standard definition of convergence in probability of the estimator to a constant.

Proving consistency in the present setting tends to be difficult and beyond the scope of this book because typically no explicit expression is available for the estimator $\lambda_n \left(\hat{F}_n \right)$ which, instead, is expressed conditionally in terms of the hypothetical variables X_i^*. The following example which illustrates both situations in which $\lambda_n \left(\hat{F}_n \right)$ is and is not consistent is an exception; here $\lambda_n \left(\hat{F}_n \right)$ can be expressed directly in terms of the X's.

Example 6.5.7 U-statistics. Let

$$(6.5.26) \qquad \lambda_n(F) = E \left(\sqrt{n} \hat{\theta}_n \right)^2,$$

where $\hat{\theta}_n$ is a U-statistic

$$(6.5.27) \qquad U = \frac{1}{n(n-1)} \sum \sum_{i \neq j} \phi(X_i, X_j).$$

Then the proof of (6.1.17) shows that (Problem 5.11)

$$(6.5.28) \qquad \lambda_n(F) = \frac{4(n-2)}{n-1} \gamma_1^2 + \frac{2}{n-1} \gamma_2^2,$$

where

$$(6.5.29) \qquad \gamma_1^2 = E \phi(X_1, X_2) \phi(X_1, X_3) \geq 0$$

and

$$(6.5.30) \qquad \gamma_2^2 = E \phi^2(X_1, X_2).$$

Thus

$$(6.5.31) \qquad \lambda_n(F) \to \lambda(F) = 4\gamma_1^2.$$

It follows from (6.5.28) that

$$(6.5.32) \qquad \lambda_n \left(\hat{F}_n \right) = \frac{4(n-2)}{n-1} \gamma_1^{*2} + \frac{2}{n-1} \gamma_2^{*2},$$

where

$$(6.5.33) \qquad \gamma_1^{*2} = E_{\hat{F}_n} \phi(X_1^*, X_2^*) \phi(X_1^*, X_3^*)$$

and

$$(6.5.34) \qquad \gamma_2^{*2} = E_{\hat{F}_n} \phi^2(X_1^*, X_2^*).$$

The quantities γ_2^{*2} and γ_1^{*2} can be expressed in terms of the X's as

$$(6.5.35) \qquad \gamma_2^{*2} = \frac{1}{n^2} \sum_{i=1}^{n} \sum_{j=1}^{n} \phi^2 (X_i, X_j)$$

and

$$(6.5.36) \qquad \gamma_1^{*2} = \frac{1}{n^3} \sum_{i=1}^{n} \sum_{j=1}^{n} \sum_{k=1}^{n} \phi (X_i, X_j) \phi (X_i, X_k).$$

That $\lambda_n \left(\hat{F}_n \right)$ may not be consistent for estimating $\lambda_n(F)$ is a consequence of the fact that U (and hence γ_1^2, γ_2^2, and $\lambda_n(F)$) depend only $\phi(x_i, x_j)$ for $i \neq j$, while γ_1^{*2} and γ_2^{*2} (and hence $\lambda_n(\hat{F}_n)$) depend also on $\phi(x_i, x_i)$.

The difficulty does not arise when γ_1^2, γ_2^2, and $\gamma_3^2 = E\phi^2 (X_i, X_i)$ are finite. For then, by the law of large numbers,

$$\gamma_2^{*2} \overset{P}{\to} \gamma_2^2 \text{ and } \gamma_1^{*2} \overset{P}{\to} \gamma_1^2$$

and therefore by (6.5.28), (6.5.31), and (6.5.32), $\lambda_n \left(\hat{F}_n \right) - \lambda_n(F) \overset{P}{\to} 0$, which proves consistency.

On the other hand, suppose that γ_1^2 and γ_2^2 and hence $\lambda_n(F)$ are finite but

$$\gamma_3^3 = E\phi^2 (X_i, X_i) = \infty.$$

Note that

$$\gamma_2^{*2} = \frac{1}{n^2} \sum \sum_{i \neq j} \phi^2 (X_i, X_j) + \frac{1}{n^2} \sum \phi^2 (X_i, X_i).$$

Here the first term tends in probability to γ_2^2 but what we can say about the second term? Clearly,

$$\frac{1}{n} \sum \phi^2 (X_i, X_i) \overset{P}{\to} \infty,$$

but $\frac{1}{n^2} \sum \phi^2 (X_i, X_i)$ can tend either to a finite limit or also to ∞. In the latter case, γ_2^{*2} and hence $\lambda_n \left(\hat{F}_n \right) \overset{P}{\to} \infty$. Since $\lambda_n(F)$ is finite, $\lambda_n \left(\hat{F}_n \right)$ is then no longer consistent.

The first condition, that $\lambda_n(F)$ is finite, will be satisfied for instance by any function ϕ satisfying

$$|\phi(x, y)| < M \text{ for some } M < \infty \text{ and for all } x \neq y.$$

We shall now give an example in which

(6.5.37) $$\frac{1}{n^2} \sum \phi(X_i, X_i) \xrightarrow{P} \infty$$

and hence $\lambda_n\left(\hat{F}_n\right)$ is not consistent.

Let $X_1, \ldots, X_n$ be i.i.d. according to the uniform distribution $U(0, 1)$ and, noting that the boundedness condition on φ imposes no restriction on $\varphi(x, y)$ when $x = y$, let

(6.5.38) $$\phi^2(x, x) = e^{1/x}.$$

To prove (6.5.37), we must show that

(6.5.39) $$P\left[\frac{1}{n^2} \sum e^{1/X_i} > A\right] \to 1 \text{ for any } A > 0.$$

Since

$$\sum e^{1/X_i} > \max e^{1/X_i}$$

it is enough to show that

$$P\left[\max e^{1/X_i} > An^2\right] \to 1$$

or, equivalently, that

$$P\left[\max e^{1/X_i} < An^2\right] = \left[P\left(e^{1/X_1} < An^2\right)\right]^n \to 0.$$

Now

$$P\left(e^{1/X_1} < An^2\right) = P\left[X_1 > \frac{1}{\log\left(An^2\right)}\right] = 1 - \frac{1}{\log(An^2)}.$$

To show that

$$\left(1 - \frac{1}{\log(An^2)}\right)^n \to 0,$$

note that $\dfrac{1}{\log(An^2)} > \dfrac{1}{\sqrt{n}}$ for n sufficiently large. Since (Problem 5.9)

(6.5.40) $$\left(1 - \frac{1}{\sqrt{n}}\right)^n \to 0$$

this completes the proof of (6.5.37). (This example with $\lambda_n(F)$ given by (B) rather than by (6.5.26) is due to Bickel and Freedman (1981).)   □

As is suggested by this example, consistency typically holds, but cannot be taken for granted. A good summary of both the theory and some of the principal examples can be found in Chapter 3 of Shao and Tu (1995).

Note also that the discussion of Example 6.5.7 can be extended to cover asymptotic normality (Problem 5.10).

Let us now return to the general problem of estimating functionals $\lambda_n(F)$ depending on F and on n, such as the performance measures (A)–(D) listed at the beginning of the section. Consider, for example, $\lambda_n(F)$ given by (B). As we saw in Section 6.3 and in earlier chapters, it will often be the case that

$$(6.5.41) \qquad \lambda_n(F) = P_F \left[\frac{\sqrt{n}\left[\hat{\theta}_n - h(F)\right]}{\tau(F)} \leq a \right] \to \Phi(a),$$

and we would then use $\Phi(a)$ as an approximation for $\lambda_n(F)$ (both when F is known and when it is not). The use of the central limit theorem is a prime example of this approach.

If instead of (B) we consider (A), (6.5.36) implies that

$$(6.5.42) \qquad \lambda_n(F) = P_F \left[\sqrt{n}\left(\hat{\theta}_n - h(F)\right) \leq a \right] \to \Phi(a/\tau(F)).$$

If F, and hence $\tau(F)$, are known, we would use $\Phi(a/\tau(F))$ as our approximation for $\lambda_n(F)$. When F is unknown, we would replace $\tau(F)$ by a consistent estimator $\hat{\tau}$, and approximate or estimate $\lambda_n(F)$ by $\Phi(a/\hat{\tau})$. The corresponding remarks apply to measures (C) and (D). If, for example,

$$\lambda_n(F) = \mathrm{Var}\left(\sqrt{n}\hat{\theta}_n\right) \to \tau^2,$$

we would use $\hat{\tau}^2$ as an approximation for $\lambda_n(F)$.

Quite generally, consider now the case that F is unknown and that

$$(6.5.43) \qquad \lambda_n(F) \to \lambda,$$

and suppose that λ does not depend on F. Then we can approximate $\lambda_n(F)$ by either λ or the bootstrap estimator $\lambda_n\left(\hat{F}_n\right)$. Which of the two provides the better approximation?

Suppose that

$$(6.5.44) \qquad \lambda_n(F) = \lambda + \frac{a(F)}{n} + o\left(\frac{1}{n}\right),$$

and, correspondingly,

$$(6.5.45) \qquad \lambda_n\left(\hat{F}_n\right) = \lambda + \frac{a\left(\hat{F}_n\right)}{n} + o_P\left(\frac{1}{n}\right).$$

If

$$(6.5.46) \qquad a\left(\hat{F}_n\right) = a(F) + o_P(1),$$

as will be the case, for example, when $\sqrt{n}\left[a\left(\hat{F}_n\right) - a(F)\right]$ tends to a limit distribution, it follows that the bootstrap estimator $\lambda_n\left(\hat{F}_n\right)$ satisfies

$$(6.5.47) \qquad \lambda_n\left(\hat{F}_n\right) = \lambda_n(F) + o_P\left(\frac{1}{n}\right).$$

Thus the error of the bootstrap estimator is of smaller order than the error of the classical estimator λ given by (6.5.43).

This argument depends crucially on the assumption that the leading term λ in (6.5.44) is independent of F. Otherwise, (6.5.45) will be replaced by

$$(6.5.48) \qquad \lambda_n\left(\hat{F}_n\right) = \lambda\left(\hat{F}_n\right) + \frac{a\left(\hat{F}_n\right)}{n} + o_P\left(\frac{1}{n}\right).$$

The error of the bootstrap estimator then becomes

$$\lambda_n\left(\hat{F}_n\right) - \lambda_n(F) = \left[\lambda\left(\hat{F}_n\right) - \lambda(F)\right] + \frac{1}{n}\left[a\left(\hat{F}_n\right) - a(F)\right] + o_P\left(\frac{1}{n}\right)$$

This is dominated by the first term which is typically of the same order $1/n$ as the error of the classical estimator $\lambda\left(\hat{F}_n\right)$. A rigorous treatment of these issues with many examples and references is given by Hall (1992). We shall here give only one particularly simple example in which it is easy to carry out the calculations explicitly.

Example 6.5.8 Performance of a variance estimator. Let $X_1, \ldots, X_n$ be i.i.d. with variance $h(F) = \sigma^2$ so that

$$h\left(\hat{F}_n\right) = M_2 = \frac{1}{n}\sum\left(X_i - \bar{X}\right)^2.$$

(i) Consider

$$\lambda_n(F) = \operatorname{Var}\left(\sqrt{n}M_2\right).$$

Then by (6.2.27),

$$\lambda_n(F) = \operatorname{Var}\left(\sqrt{n}M_2\right)$$
$$(6.5.49) \qquad = \left(\mu_4 - \mu_2^2\right) - \frac{2\left(\mu_4 - 2\mu_2^2\right)}{n} + \frac{\left(\mu_4 - 3\mu_2^2\right)}{n^2}$$

and hence

(6.5.50) $$\lambda_n(F) \to \lambda(F) = \mu_4 - \mu_2^2,$$

which depends on F.

The classical estimator of $\lambda_n(F)$ is

(6.5.51) $$\lambda\left(\hat{F}_n\right) = M_4 - M_2^2$$

and the bootstrap estimator $\lambda_n\left(\hat{F}_n\right)$ of $\lambda_n(F)$ is

(6.5.52) $$\lambda_n\left(\hat{F}_n\right) = \left(M_4 - M_2^2\right) - \frac{2\left(M_4 - 2M_2^2\right)}{n} + \frac{M_4 - 3M_2^2}{n^2}.$$

In both estimators, the dominating term is $M_4 - M_2^2$ and the error is

$$\left(M_4 - M_2^2\right) - \left(\mu_4 - \mu_2^2\right) + O_p\left(1/n\right).$$

By Example 6.3.3,

(6.5.53) $$M_k = \mu_k + O_p\left(\frac{1}{\sqrt{n}}\right)$$

and the error of both estimators is therefore of the same order $1/\sqrt{n}$.

(ii) Consider next the bias of the estimator M_2 of σ^2. Since $E\left(M_2\right) = (n-1)\sigma^2/n$, we have

(6.5.54) $$\lambda_n(F) = \text{ bias of } M_2 = \sigma^2/n$$

and hence

(6.5.55) $$\lambda_n(F) \to \lambda = 0.$$

Therefore

(6.5.56) $$\lambda - \lambda_n(F) = \sigma^2/n$$

is of order $1/n$. On the other hand, by (6.5.53),

$$\lambda_n\left(\hat{F}_n\right) = \frac{1}{n}M_2 = \frac{1}{n}\left[\sigma^2 + O_p\left(\frac{1}{\sqrt{n}}\right)\right],$$

so that

(6.5.57) $$\lambda_n\left(\hat{F}_n\right) - \lambda_n(F) = O_p\left(1/n^{3/2}\right).$$

Thus, in this case, the bootstrap estimator provides a better approximation to $\lambda_n(F)$, as predicted by the theory since the limit of $\lambda_n(F)$ is independent of F.   □

Summary

1. The bootstrap is concerned with the estimation of functionals $\lambda_n(F)$ that depend on n as well as on F. This situation arises in particular when $\lambda_n(F)$ measures the performance of an estimator $\hat{\theta}_n$.

2. The bootstrap estimator $\lambda_n\left(\hat{F}_n\right)$ can be represented as a function of the actual sample $X_1, \ldots, X_n$ from F and a conceptual sample $X_1^*, \ldots, X_n^*$ from $\hat{F}_n$. This representation shows that, except for very small n, the calculation of $\lambda_n\left(\hat{F}_n\right)$ often requires a prohibitive amount of labor. The difficulty can be overcome by approximating $\lambda_n\left(\hat{F}_n\right)$ by means of bootstrap samples $(X_{i1}^*, \ldots, X_{in}^*)$, $i = 1, \ldots, B$. The resulting approximating converges in probability to $\lambda_n\left(\hat{F}_n\right)$ as $B \to \infty$.

3. Consistency of $\lambda_n\left(\hat{F}_n\right)$ as an estimator of $\lambda_n(F)$ is less routine and more difficult to prove than consistency of classical estimators. The second moment of a U-statistic provides a simple example in which the bootstrap may or may not work, i.e., consistency may or may not hold.

4. When $\lambda_n(F) \to \lambda$ with λ independent of F, the bootstrap estimator typically provides a better approximation to $\lambda_n(F)$ than does the limit value λ, but this result can no longer be expected when the limit $\lambda(F)$ of $\lambda_n(F)$ depends on F.

6.6   Problems

Section 1

1.1 Show that $h(F) = \mathrm{Var}_F(X)$ is an expectation functional.

[**Hint:** Let $\phi(x_1, x_2) = \sum_{i=1}^{2}(x_i - \bar{x})^2$.]

1.2 If θ_1 and θ_2 are expectation functionals, so is $\theta_1 + \theta_2$.

[**Hint:** Let $\theta_1 = E\phi_1(x_1, \ldots, x_a)$ and $\theta_2 = E\phi_2(X_1, \ldots, X_b)$ and suppose that $a \le b$. Let

$$\phi_1'(x_1, \ldots, x_b) = \phi_1(x_1, \ldots, x_a) \text{ for all } x_1, \ldots, x_b,$$

and let $\phi = \phi_1' + \phi_2$.]

1.3 If θ_1 and θ_2 are expectation functionals, so is $\theta_1 \cdot \theta_2$.

[**Hint**: With θ_1 and θ_2 defined as in Problem 1.2, let $\phi(x_1, \ldots, x_{a+b}) = \phi_1(x_1, \ldots, x_a)\phi_2(x_{a+1}, \ldots, x_{a+b})$.]

1.4 Show that the estimator (6.1.3) is symmetric in its n variables.

1.5 Prove (6.1.17) for the case $a = 2$.

[**Hint**: The total number of terms in (6.1.12) is $\binom{n}{2}^2$; the number of terms with i, j, k, l all distinct is $\binom{n}{2}\binom{n-2}{2}$; the number of terms with $i = k$, $j = l$ is $\binom{n}{2}$. The remaining terms are all equal, and their number is $\binom{n}{2}\left[\binom{n}{2} - \binom{n-2}{2} - 1\right]$.]

1.6 Evaluate θ given by (6.1.8) and σ_1^2 given by (6.1.23)

(i) for the normal mixture
$$F = (1 - \epsilon)N(0, 1) + \epsilon N(\eta, \tau^2);$$

(ii) for the uniform distribution $U(\theta - 1/2, \theta + 1/2)$.

1.7 In Example 6.1.1, use (6.1.28) to show that
$$\frac{V_s - E(V_s)}{\sqrt{\mathrm{Var}\, V_s}} \quad \text{and} \quad \frac{W - E(W)}{\sqrt{\mathrm{Var}\, W}}$$
have the same limit distribution.

1.8 Verify (6.1.40) for the case $s = 2$, $a = b = 1$.

[**Hint**: Use the hint of Problem 1.5 but note that the total number of terms is now $m^2 n^2$.]

1.9 Prove the following extension of Lemma 6.1.1. If $X_1, X_2, \ldots$ and $Y_1, Y_2, \ldots$ are independent with distributions F and G, respectively, and if
$$\varphi_{ij}(x_1, \ldots, x_i; y_1, \ldots, y_j) = E\varphi(x_1, \ldots, x_i, X_{i+1}, \ldots, X_a; y_1, \ldots, y_j, Y_{j+1}, \ldots, Y_b)$$

then

(i) $E\varphi_{ij}(X_1, \ldots, X_i; Y_1, \ldots, Y_j) = \theta = E\varphi(X_1, \ldots, X_a; Y_1, \ldots, Y_b)$

and

(ii)
$$\mathrm{Var}\, \varphi_{ij}(X_1, \ldots, X_i; Y_1, \ldots, Y_j)$$
$$= \mathrm{Cov}[\varphi(X_1, \ldots, X_i, X_{i+1}, \ldots, X_a; Y_1, \ldots, Y_j, Y_{j+1}, \ldots, Y_b),$$
$$\varphi(X_1, \ldots, X_i, X'_{i+1}, \ldots, X'_a; Y_1, \ldots, Y_j, Y'_{j+1}, \ldots, Y'_b)].$$

1.10 In Example 6.1.2, evaluate θ given by (6.1.46) when $G(y) = F(y - \Delta)$ and F is

(i) the uniform distribution $U(0, 1)$;

(ii) the double exponential distribution $DE(0, 1)$,

(iii) the logistic distribution $L(0, 1)$.

1.11 (i) For the three parts of the preceding problem, evaluate σ_{10}^2 and σ_{01}^2.

(ii) If $G(y) = F(y - \Delta)$, show that $\sigma_{10}^2 = \sigma_{01}^2$ when F is symmetric.

[**Hint for (ii)**: Without loss of generality, suppose that F is symmetric about 0. Then, for example,

$$P[X_1 < Y_1, Y_1'] = P[X_1 < X_2 + \Delta, X_3 + \Delta],$$

where the X's are i.i.d. according to F.

1.12 In Example 6.1.2, show

(i) that $P(X < Y) = E[1 - G(X)]$;

(ii) that $P(X_1 < Y_1, Y_1') = E[1 - G(X)]^2$;

(iii) that if F is continuous,

$$P(X_1, X_1' < Y_1) = E[F^2(Y)].$$

1.13 (i) Show that the variances σ_{01}^2 and σ_{10}^2 given by (6.1.47) and (6.1.48) can take on values different from $1/12$ when F and G are two different distributions, both symmetric about 0.

(ii) Under the same assumptions about F and G, show that the variance of the U-statistic (6.1.44) can take on values different from its value when $F = G$.

1.14 Let (X_i, Y_i) be i.i.d. according to a bivariate distribution F. Show that $h(F) = \text{Cov}(X_i, Y_i)$ is an expectation functional.

[**Hint**: Problem 1.1.]

1.15 If X_1 and X_2 are independent and $P(X_i \geq u_\alpha) = \alpha$ for $i = 1, 2$, then

$$P[\max(X_1, X_2) \geq u_\alpha] = 1 - (1 - \alpha)^2.$$

1.16 For T_N and T_N^* given by (6.1.66) and (6.1.74), respectively, prove (6.1.69) and (6.1.65).

1.17 Prove Theorem 6.1.2 for the case $a = 2$.

[**Hint:** Let $\phi_1(x) = E\phi(x, X')$, $\theta = E\phi(X, X') = E\phi_1(X)$, and $\sigma_1^2 = \text{Var } \phi_1(X)$. For

$$T_n = \sqrt{n}(U - \theta) \text{ and } T_n^* = \frac{2}{\sqrt{n}} \sum_{i=1}^{n} [\phi_1(X_i) - \theta],$$

show that $E\left(T_n^* - T_n\right)^2 \to 0$.]

1.18 Let X_1, X_2, and X_3 be i.i.d. $N(0,1)$ and let $\phi(x_1, x_2) = e^{\lambda x_1 x_2}$ with $\frac{1}{2} < \lambda \le \frac{1}{\sqrt{2}}$. Then σ_1^2 and σ_2^2 defined by (6.1.15) with $a = 2$ satisfy $\sigma_1^2 < \infty$ and $\sigma_2^2 = \infty$.

[**Hint:** To evaluate

$$E\phi(X_1, X_2) = \frac{1}{2\pi} \int \left[\int e^{\lambda x y} - \frac{1}{2} y^2 dy \right] e^{-\frac{1}{2}e^2} dx,$$

evaluate first the inner integral by completing the square. $E\phi^2(X_1, X_2)$ follows by replacing λ by 2λ. The covariance requires

$$E\left[\phi(X_1, X_2)\phi(X_1, X_e)\right] = E e^{\lambda X_1(X_2 + X_3)}.$$

This can be handled in the same way, using the fact that $X_2 + X_3$ is $N(0, 2)$.]

Section 2

2.1 Let $F = \delta_x$ denote the distribution which assigns probability 1 to the point x. Then

(6.6.1) $$\int a(y)d\delta_x(y) = a(x)$$

and

(6.6.2) $$\int \cdots \int \sum_{i=1}^{k} a(y_i)\, d\delta_{x_1}(y_1) \cdots d\delta_{x_k}(y_k) = \sum_{i=1}^{k} a(x_i).$$

[**Hint:** Equation (6.6.1) follows immediately from (6.2.5); to see (6.6.2), note that

$$\int \cdots \int a(y_1)\, d\delta_{x_1}(y_1)\, dG_2(y_2) \cdots dG_k(y_k) = \int a(y_1)\, d\delta_{x_1}(y_1)$$
$$= a(x_1)$$

for all $G_2, \ldots, G_k$.]

2.2 Determine a functional h defined for all univariate distributions with finite first moment for which $h\left(\hat{F}_n\right) = \sum_{i=1}^{n} |X_i - \bar{X}|$.

2.3 If X and X' are independently distributed according to F, determine $h\left(\hat{F}_n\right)$ when

(i) $h(F) = E|X' - X|$,

(ii) $h(F) = F^{*-1}(p)$ where F^* is the cdf of $\frac{1}{2}(X + X')$.

2.4 Prove the results stated in

(i) (6.2.25) and (6.2.26);

(ii) (6.2.28) and (6.2.29).

2.5 Prove Theorem 6.2.2 for the case $a = 3$.

[**Hint**: Generalize the identity (6.2.34) to the case $a = 3$, and apply the method of proof used for $a = 2$.]

2.6 Show that Definition 5.1.3 is equivalent to the statement that f is continuous at $\underline{a}$ if

$$f\left(\underline{x}^n\right) \to f(\underline{a})$$

for all sequences $\underline{x}^{(n)}$ tending to $\underline{a}$.

2.7 (i) Show that of the convergence conditions (i')–(iii') above (6.2.43)

(a) (ii') $\Rightarrow$ (i');

(b) (ii') $\Rightarrow$ (iii') .

(ii) Give an example of each:

(a) (i') holds but (ii') does not;

(b) (iii') holds but (ii') does not.

[**Hint**: Let G_n be the distribution assigning probability 1/2 to the point $1 - \frac{1}{n}$ and distributing the remaining probability of 1/2 uniformly over the interval $\left(1 - \frac{1}{n}, 1\right)$.]

2.8 The definition (6.2.45) of continuity of a functional h in (6.2.46) was given in terms of the convergence mode (ii').

(i) Show that if h is continuous with respect to (ii'), it is also continuous with respect to (iii').

(ii) Show that the converse of part (i) is not correct.

[**Hint**: Problem 2.7.]

2.9 (i) Let $g(u_1, \ldots, u_k)$ be a continuous function of k variables, and let $h_1, \ldots, h_k$ be k functionals that are continuous at F_0. Then $g[h_1(F), \ldots, h_k(F)]$ is continuous at F_0.

(ii) Show that the functional (6.2.13) is continuous at all F.

[**Hint for Part (ii)**: Part (i) and Example 6.2.7(i).]

The following set of problems deals with the estimation of the quantile function $Q(y) = F^{-1}(y)$ defined by (6.2.41), without the assumption of a density.

2.10 (i) Evaluate $Q(y)$ when F is the cdf of the discrete distribution that assigns probabilities p, q, and r $(p+q+r = 1)$ to the points $a < b < c$, respectively.

(ii) Show that neither of the equations

$$F[Q(y) = y], \quad Q[F(x)] = x$$

holds for all x and y.

2.11 Prove that quite generally

$$F[Q(y)] \geq y, \quad Q[F(x)] \leq x.$$

2.12 In a sample of n from the distribution F of Problem 2.10, let X and Y denote the numbers of observations equal to a and b, respectively, and let $\hat{Q}_n = \hat{F}_n^{-1}$ be the quantile function of $\hat{F}_n$. Show that

$$\hat{Q}_n(y) = \begin{cases} a & \text{for } 0 < y \leq X/n \\[2mm] b & \text{for } \dfrac{X}{n} < y \leq \dfrac{X+Y}{n} \\[2mm] c & \text{for } \dfrac{X+Y}{n} < y \leq 1. \end{cases}$$

2.13 Under the assumptions of the preceding problem, show that

$$\hat{Q}_n(y) \xrightarrow{P} Q(y) \text{ for } y \neq p, \ p+q,$$

and determine the limit behavior of $\hat{Q}_n(y)$ when $y = p$ and $y = p+q$.

2.14 Prove the following generalization of the preceeding problem.

Theorem 6.6.1 *If $X_1, \ldots, X_n$ are i.i.d. according to F, then*

(6.6.3) $$\hat{Q}_n(y) \xrightarrow{P} Q(y)$$

holds unless there exist constants u and $\Delta > 0$ such that

(6.6.4) $$F(a) = y$$

and

(6.6.5) $$F(x) = y \text{ for all } a < x < a + \Delta.$$

(It can also be shown that if such a and Δ do exist, then (6.6.3) does not hold.)

[**Hint**: Let $k = k(n)$ be such that

(6.6.6) $$\frac{k}{n} \le y < \frac{k+1}{n}.$$

and let $n = Q(y)$, δ be any number > 0, and $p = F(x + \delta)$.

(a) Show that under the assumptions made,

$$P\left[\hat{Q}_n(y) \le x + \delta\right] = P\left[\sqrt{n}\left(\frac{S_n}{n} - p\right) \ge \sqrt{n}\left(\frac{k}{n} - p\right)\right] \to 1,$$

where S_n is the number of $X_i \le x + \delta$.

(b) Show similarly that

$$P\left[\hat{Q}_n(y) \le x - \delta\right] \to 0$$

and hence that

$$P\left[x - \delta < \hat{Q}_n(y) \le x + \delta\right] \to 1.]$$

2.15 Prove (6.2.39).

[**Hint**: Use the identity (6.2.34) to evaluate the covariance of the two terms on the right side of (6.2.40) and use (6.1.57) with $a = 1$, $b = 2$.]

2.16 In generalization of (6.2.31) and in the notation of (6.1.38), the V-statistic for $s = 2$ is

$$V = \frac{1}{m^a n^b} \sum \phi\left(X_{i_1}, \dots, X_{i_a}; Y_{j_1}, \dots Y_{j_b}\right)$$

with the summation extending over all

$$1 \le i_1, i_2, \cdots, i_a \le m; \ 1 \le j_1, j_2, \cdots, j_b \le n.$$

(i) If the X's and Y's are i.i.d. according to F and G, respectively, and if

$$h(F, G) = E\phi\left(X_1, \dots, X_a; Y_1, \dots, Y_b\right),$$

show that $V = h\left(\hat{F}_m, \hat{G}_n\right)$.

(ii) State the generalization of Theorem 6.2.2 for the present case.

(iii) Prove the theorem stated in (ii) for the case $a = b = 2$.

2.17 (i) With V given by (6.2.34) and

$$\text{Var } \phi\left(X_i, X_j\right) < \infty, \text{ Var } \phi\left(X_i, X_i\right) = \infty,$$

the asymptotic variance of $\sqrt{n}(V - \theta)$ is $4\sigma_1^2 < \infty$, but $\text{Var}\left[\sqrt{n}(V - \theta)\right] = \infty$ for all n and hence $\text{Var}\left[\sqrt{n}(V - \theta)\right] \to \infty$.

(ii) In particular, the conclusions of part (i) hold when $\phi(x_1, x_2) = x_1 x_2$ and $E\left(X^2\right) < \infty$ but $E\left(X^4\right) = \infty$.

Section 3

3.1 In Example 6.3.2, check (6.3.13) and (6.3.23).

3.2 For a given functional h and cdf F_0, let

$$h^*(F) = h(F) - h\left(F_0\right).$$

Then $IF_{h^*, F}(x) = IF_{h, F}(x)$ for all x.

3.3 (i) Verify the influence function (6.3.29) of μ_k.

(ii) Graph the influence function of $h(F) = \text{Var}_F(X) = \mu_2$.

3.4 (i) Determine $\gamma^2(F)$ for the cases

(i) $F(t) = F_0^K(t)$; (ii) $F(t) = pF_0(t) + qF_0^2(t)$, $0 < p < 1$.

(ii) Verify $\gamma^2(F)$ given by (6.3.31).

3.5 If $h_2 = k\left(h_1\right)$ where h_1 is a functional with influence function $IF_{h_1, F}$ and k is a differentiable function of a real variable, then the influence function of h_2 is given by

(6.6.7) $IF_{h_2, F}(x) = k'\left[h_1(F)\right] \cdot IF_{h_1, F}(x).$

3.6 Use (6.6.7) and (6.3.28) to determine the influence function of the standard deviation of a random variable X with distribution F.

3.7 Under the assumptions of Problem 3.5:

(i) If the influence function of h_1 satisfies (6.3.13) so does the influence function of h_2.

(ii) If $\gamma_{h_1}^2(F)$ is given by (6.3.14) with $h = h_1$, then

$$\gamma_{h_2}^2(F) = \int IF_{h_2, F}^2(x) dF(x) = \left\{k'\left[h_1(F)\right]\right\}^2 \gamma_{h_1}^2(F).$$

3.8 (i) Use the result of Problem 3.7(ii) to evaluate (6.3.14) when $h(F)$ is the standard deviation of F.

(ii) Use the result of Example 6.3.3 and of Theorem 2.5.2 to show that the functional $h(F)$ of part (i) satisfies (6.3.8) with the asymptotic variance $\gamma^2(F)$ determined in part (i) .

3.9 (i) If $h = h_1/h_2$, the influence function of h is

$$(6.6.8) \qquad IF_{h,F}(x) = \frac{h_2(F)IF_{h_1,F}(x) - h_1(F)IF_{h_2,F}(x)}{h_2^2(F)}.$$

(ii) If the influence functions of h_1 and h_2 satisfy (6.3.13) so does that of $h = h_1/h_2$.

3.10 Use (6.6.8) to obtain the influence function of the standardized third moment of a random variable X with distribution F given by

$$(6.6.9) \qquad h(F) = E(X - \xi)^3 / \left[E(X - \xi)^2\right]^{3/2}.$$

Note: For the asymptotic normality of $h\left(\hat{F}_n\right)$, see Problem 7.4 of Chapter 5.

3.11 The Wilcoxon statistic V_s is compared in Example 6.1.1 with the U-statistic corresponding to the function $\phi(x_1, x_2)$ given by (6.1.6). Compare both of these statistics with the V-statistic corresponding to the same ϕ.

3.12 Under the assumptions of Problem 4.8 of Chapter 2, let $h(F) = F^{-1}(p)$ be the population p-quantile $(0 < p < 1)$.

(i) Show that

$$F_{\epsilon,x}^{-1}(p) = F^{-1}\left(\frac{p}{1-\epsilon}\right) \text{ if } p < (1-\epsilon)F(x)$$

and determine $F_{\epsilon,x}^{-1}(p)$ for the remaining values of p.

(ii) By differentiating $F_{\epsilon,x}^{-1}(p)$, show that

$$IF_{h,F}(x) = (p-1)/f\left[F^{-1}(p)\right] \text{ when } x < F^{-1}(p)$$

and determine $IF_{h,F}(x)$ for $x = F^{-1}(p)$ and $x > F^{-1}(p)$.

(iii) Verify (6.3.13) and show that

$$\gamma^2(F) = \int IF_{h,F}^2(x), dF(x) = pq/ \left\{f\left[F^{-1}(p)\right]\right\}^2.$$

(iv) Graph the influence function of part (ii) when F is the standard normal distribution and $p = .2, .4, .6$.

Note: The normal limit $\sqrt{n}\left[h\left(\hat{F}_n\right) - h(F)\right] \xrightarrow{L} N\left(0, \gamma^2(F)\right)$ was shown in Problem 4.8 of Chapter 2.

3.13 With the notation of Section 5.7 and Example 6.2.1, let

$$h(F) = \sum \left(p_j - p_j^0\right)^2 / p_j^0$$

so that $h\left(\hat{F}_n\right)$ is the Pearson statistic (6.2.11) for testing $H : F = F_0$.

(i) Show that

$$IF_{h,F}(x) = 2\frac{p_j - p_j^0}{p_j^0} + A \text{ when } x \in I_j,$$

where $A = -2\sum \left(p_i - p_i^0\right) / p_i^0$.

(ii) Check that (6.3.13) holds.

(iii) Show that

$$\gamma^2(F) = \int I_{h,F}^2(x) dF(x) = 4\sum \left[\frac{p_j - p_j^0}{p_j^0} - \sum \frac{p_i - p_i^0}{p_i^0}\right]^2.$$

(iv) Show that

$$\sqrt{n}\left[h\left(\hat{F}_n\right) - h(F) - \frac{1}{n}\sum I_{h,F}(X_i)\right] \xrightarrow{P} 0.$$

(v) It follows from (i)–(iv) that

$$\sqrt{n}\left[h\left(\hat{F}_n\right) - h(F)\right] \xrightarrow{L} N\left(0, \gamma^2(F)\right).$$

Explain why this does not contradict Theorem 5.5.2 according to which $nh\left(\hat{F}_n\right) \xrightarrow{L} \chi_k^2$.

Note: The normal limit distribution of Problem 3.13 under the assumption of a fixed distribution $F \neq F_0$ and the non-central χ^2 distribution corresponding to $p_i = p_i^0 + \dfrac{\Delta}{\sqrt{n}}$ obtained in Theorem 5.5.3 lead to different approximations for the distribution of Pearson's statistic (6.2.11) for a given n and given values p_i. Which approximation is better depends on the values of the parameters. If all the differences $p_i - p_i^0$ are small, one would expect better results from the non-central χ^2 approximation while the normal approximation would be favored when none of the differences $p_i - p_i^0$ are small. This suggestion is supported by closely related numerical comparisons carried out by Brofitt and Randles (1977).

3.14 Show that each of the three quantities (6.3.41) tends to $a^2\sigma_1^2$ as $n \to \infty$.

Section 4

4.1 Suppose that f_1 and f_2 are densities of F, i.e., satisfy

$$F(b) - F(a) = \int_a^b f_1(x)dx = \int_a^b f_2(x)dx \text{ for all } a < b,$$

and that both are continuous at y. Then $f_2(y) = f_1(y)$.

[**Hint**: Suppose that $f_1(y) < f_2(y)$. Then there exists $\epsilon > 0$ such that $f_1(x) < f_2(x)$ for all x satisfying $y - \epsilon < x < y + \epsilon$.]

4.2 If f is a density satisfying $f(x) \leq M$ for all x, then $\int f^2(x)dx \leq M$.

4.3 Show that (6.4.13) reduces to (6.4.4) when K is the uniform density $U\left(-\frac{1}{2}, \frac{1}{2}\right)$.

4.4 (i) Verify the conditions (6.4.17) of Theorem 6.4.3 when K is the normal density $N(0, 1)$.

(ii) For $K = N(0, 1)$, calculate the principal terms in (6.4.18) and (6.4.19).

4.5 (i) With $n = 2$, $x_1 = -1$, and $x_2 = +1$, graph the estimators (a) (6.4.2) and (b) (6.4.13) with $K = N(0, 1)$ for some representative values of h (small, medium, and large).

(ii) Carry out (i) with $n = 3$; $x_1 = -1$, $x_2 = 1$, $x_3 = 2$.

4.6 Prove that the value of h given by (6.4.30) minimizes (6.4.28).

4.7 Graph $\hat{f}_n(y)$ given by (6.4.13) with $K = N(0, 1)$ as a function of y if $X_1, \ldots, X_n$ is a random sample of $n = 5, 10, 20$ from the following densities f:

(i) $f = N(0, 1)$;

(ii) $f = U(0, 1)$;

(iii) $f = E(0, 1)$;

(iv) $f(x) = \frac{1}{2}$ when $-2 < x < -1$ and $1 < x < 2$, and $= 0$ elsewhere, for a number of values of h.

4.8 (i) The normal mixture density $f = \frac{1}{2}N(a, 1) + \frac{1}{2}N(-a, 1)$ has two modes when a is sufficiently large, and is unimodal for sufficiently small a.

(ii) Work Problem 4.7 for the density f of part (i) for a number of different values of a including both unimodal and bimodal cases.

[**Hint for (i)**: By differentiation of $f(x)$, show that $f'(x) < 0$ for all $x > 0$ (and hence that f is unimodal) when

$$e^{2ax} < \frac{a + x}{a - x} = \left(1 + \frac{x}{a}\right)\left(1 - \frac{x}{a}\right)^{-1}.$$

Expand both sides as a power series and compare the coefficients of x^k for $a < 1/\sqrt{2}$.]

4.9 Obtain expressions for c_1, c_2, and c_3 if the quantities (1) (6.4.21), (2) (6.4.22), and (3) (6.4.25) are expressed as $c_i n^{-4/5} + o(n^{-4/5})$ when h is given by (6.4.30).

4.10 (i) Verify (6.4.40).

(ii) Use Corollary 2.4.1 to obtain conditions for (6.4.37) to hold.

4.11 Determine the limit behavior of (6.4.39) when $h = h_n$ is given not by (6.4.30) but more generally by

(i) $h_n = c/n^{1/5}$;

(ii) $h_n = c/n^\alpha$, $\alpha < 1/5$;

(iii) $h_n = c/n^\alpha$, $\alpha > 1/5$,

where c is any positive constant and $0 < \alpha < 1/3$.

[**Hint**: In each case, evaluate the standardized bias $b(y)/\sqrt{\text{Var } \hat{f}_n(y)}$.]

Note: The result of (ii) shows that it is possible to eliminate the asymptotic bias found in (6.4.39) at the cost of a slower convergence rate for $\hat{f}_n(y)$.

4.12 Show that the quantity (6.4.42) is unchanged when $K(z)$ is replaced by $\frac{1}{c}K\left(\frac{z}{c}\right)$, $c > 0$.

4.13 Verify the value of A given following (6.4.47).

4.14 Show that the efficiency (6.4.47) tends to 0 if

(i) K_2 is a Cauchy distribution truncated at T when $T \to \infty$;

(ii) K_2 is a suitable mixture of normal distributions.

4.15 Let $X_1, \ldots, X_n$ be i.i.d. with density $f(x - \theta)$, and suppose that f' exists and that $\hat{\theta}_n$ is an estimator of θ satisfying $\sqrt{n}\left(\hat{\theta}_n - \theta\right) \to N(0, v^2)$. Find the limit distribution of $\sqrt{n}\left[f\left(y - \hat{\theta}_n\right) - f(y - \theta_n)\right]$ when $f'(\theta) \neq 0$.

4.16 Verify (6.4.53).

4.17 Generalize Theorem 6.4.3(i) to find conditions for the validity of
(i) (6.4.59);
(ii) (6.4.60).

4.18 Check that (6.4.62) is minimized by (6.4.65).

Section 5

5.1 In Example 6.5.2, evaluate the probability (6.5.5) for all a.

5.2 Extend the calculations of Example 6.5.2 to the cases
(i) $n = 3$;
(ii) $n = 4$.

5.3 In Example 6.5.2, suppose $\lambda_n(F)$ is given by (B) instead of (A) so that

$$\lambda_n(F) = P_F\left[\frac{\sqrt{n}\left[\bar{X}_n - E(X)\right]}{\tau(F)} \leq a\right]$$

and

$$\lambda_n\left(\hat{F}_n\right) = P_{\hat{F}_n}\left[\frac{\sqrt{n}\left(\bar{X}_n^* - \bar{X}_n\right)}{\tau\left(\hat{F}_n\right)} \leq a\right].$$

Suppose $\tau^2(F) = \text{Var } X$, so that $\tau^2\left(\hat{F}_n\right) = \frac{1}{n}\sum(X_i - \bar{X})^2$. For this situation calculate $\lambda_n\left(\hat{F}_n\right)$ when
(i) $n = 3$;
(ii) $n = 4$.

5.4 Calculate 10 values of the estimator $\lambda_n(\hat{F}_n)$ of the preceding problem for each of the normal mixture distributions of Problem 4.5 of Chapter 3 corresponding to the four combinations $\epsilon = .1, .2, \tau = 2, 3$ by drawing a sample $X_1, \ldots, X_n$ of $n = 50$ observations from the distribution in question, and then drawing 10 bootstrap samples of $B = 100$ from $\hat{F}_n$.

5.5 Carry out the preceding problem when the 10 values of $\lambda_n(\hat{F}_n)$ are obtained by drawing a new sample $(X_{i1}, \ldots, X_{in})$, $i = 1, \ldots, 10$, for each of the 10 cases and a single bootstrap sample of $B = 100$ for each sample $x_{i1}, \ldots, x_{in}$.

5.6 (i) In Example 6.5.3, determine $\lambda_n(\hat{F}_n)$ when $\lambda_n(F)$ is the variance of $X_{(2)}$.

(ii) Extend the calculations of Example 6.5.3 to the case $n = 4$.

5.7 Let $X_1, \ldots, X_n$ be a sample from F, and let $\theta = h(F)$ and $\tilde{\theta}_n$ be the population and sample median, respectively. Determine the bootstrap estimator $\lambda_n\left(\hat{F}_n\right)$ when

(a) $\lambda_n(F) = P\left[\sqrt{n}\left(\tilde{\theta}_n - \theta\right) \leq a\right]$

and

(b) $\lambda_n(F) = \text{Var}\left[\sqrt{n}\left(\tilde{\theta}_n - \theta\right)\right]$

for (i) $n = 3$ and (ii) $n = 4$.

Note: If n is odd and $x_{(1)} < \cdots < x_{(n)}$, denote the ordered observations, then the median of $X_1^*, \ldots, X_n^*$ is equal to $x_{(i)}$ if and only if at least half of the X_j^* are $\leq x_{(i)}$ and fewer than half are $\leq x_{(i-1)}$. Since the number of $X_j^* \leq x_{(i)}$ has the binomial distribution $b\left(\frac{1}{n}, n\right)$, it is easy to write down a formal expression for the bootstrap estimator. Using this expression, Ghosh et al. (1984) give conditions in case (b) for consistency of $\lambda_n\left(\hat{F}_n\right)$ and give an example in which consistency does not hold. Conditions for consistency in case (a) are given by Bickel and Freedman (1981) and Singh (1981).

5.8 Show how bootstrapping can be used to determine an approximation of $\lambda_n\left(\hat{F}_n\right)$ which converges to $\lambda_n\left(\hat{F}_n\right)$ as $B \to \infty$ for $\lambda_n(F)$ given in Problem 5.3.

5.9 Show that $\left(1 - \dfrac{1}{\sqrt{n}}\right)^n \to 0$ as $n \to \infty$.

[**Hint:** Write $\left(1 - \dfrac{1}{\sqrt{n}}\right)^n = \left[\left(1 - \dfrac{1}{\sqrt{n}}\right)^{\sqrt{n}}\right]^{\sqrt{n}}$].

5.10 Discuss the asymptotic normality of

$$\sqrt{n}\left[\lambda_n\left(\hat{F}_n\right) - \lambda_n(F)\right]$$

under the assumptions of Example 6.5.7 when γ_1^2, γ_2^2, and γ_3^2 are all finite.

[**Hint**: Use the results of Example 6.2.5.]

5.11 Verify (6.5.28).

Bibliographic Notes

The asymptotic theory of statistical functionals was initiated by von Mises (1936, 1937, 1947), who related the asymptotic distribution of $h\left(\hat{F}_n\right)$ to the order of the first non-vanishing term of its Taylor expansion, showing, in particular, a normal limit if the first term does not vanish. To overcome certain technical difficulties, variations of his approach were developed by Filippova (1962), Reeds (1976), Serfling (1980), and Fernholz (1983). The introduction of the influence function by Hampel (1968, 1974) established this theory as an important tool for robust estimation detailed treatments of which are provided by Huber (1981), Hampel et al. (1986), and Staudte and Sheather (1990). U-statistics were introduced by Hoeffding (1948a); comprehensive accounts are given by Lee (1990) and by Koroljuk and Borovskich.

Nonparametric density estimation goes back to a report (1951) by Fix and Hodges, which remained unpublished until 1989 when it appeared with comments by Silverman and Jones. Systematic studies of density estimation were undertaken by Rosenblatt (1956), Cencov (1962), and Parzen (1962). Book-length treatments are given in Prakasa Rao (1983), Silverman (1986), Scott (1992), and Wand and Jones (1995). See also Müller (1997).

The bootstrap methodology was introduced by Efron (1979, 1982a) and has since then undergone an explosive development. Surveys of some aspects of this work can be found in Hall (1992), Mammen (1992), Efron and Tibshirani (1993), and Davison and Hinkley (1997).

7
Efficient Estimators and Tests

Preview

Chapters 3–6 were concerned with the asymptotic performance of statistical estimation and testing procedures, and with the comparison of alternative procedures. In this last chapter, we discuss methods for constructing procedures that are in some sense efficient, that is, asymptotically optimal.

The most widely used approach for such a construction is based on Fisher's maximum likelihood estimators (MLE) and some of its modifications. To avoid certain difficulties, we follow Cramér (1946) and replace the MLE by a consistent root $\hat{\theta}_n$ of the likelihood equation. In Sections 1, 3, and 4, sufficient conditions are given in the one-parameter case for such a root to exist and to be asymptotically normal with variance $1/I(\theta)$. Here $I(\theta)$ is the Fisher information, the properties of which are studied in Section 2. A convenient first order approximation to $\hat{\theta}_n$, which shares its asymptotic properties, is developed in Section 3. The theory is generalized to the multiparameter case in Sections 5 and 6.

Section 7 treats the corresponding testing problem and discusses three classes of test procedures: the likelihood ratio, Wald, and Rao tests. In Section 8, the results of the earlier sections are applied to some multinomial models of contingency tables.

7.1   Maximum likelihood

In this last chapter, we shall consider the construction of estimators which are asymptotically efficient in a sense that will be made precise in Section 7.4. A starting point for the generation of such estimators is the method of maximum likelihood. Suppose for a moment that X is a discrete random quantity taking on values x with probabilities

$$(7.1.1) \qquad P_\theta(x) = P_\theta(X = x)$$

depending on a parameter θ taking on values in Ω. For fixed x, viewed as a function of θ, the probability (7.1.1) is called the *likelihood of θ* and is denoted by

$$(7.1.2a) \qquad L(\theta) = P_\theta(x).$$

The value $\hat\theta(x)$ which assigns the largest possible probability to θ, i.e., which maximizes the likelihood (7.1.2a) in a sense, provides the best "explanation" of the observation x and thus is a natural estimator of θ.

In the same way, if X has probability density $p_\theta(x)$, the likelihood is defined as

$$(7.1.2b) \qquad L(\theta) = p_\theta(x).$$

In both cases, the value of θ maximizing $L(\theta)$ (if it exists) is called the *maximum likelihood estimator* (MLE) of θ.

Example 7.1.1 Poisson. If $Y_1, \ldots, Y_n$ are i.i.d. according to the Poisson distribution $P(\lambda)$, the likelihood is

$$(7.1.3) \qquad L(\lambda) = \lambda^{\Sigma y_i} e^{-n\lambda} / \Pi y_i!$$

This is maximized by (Problem 1.1(i))

$$(7.1.4) \qquad \hat\lambda = \sum y_i / n,$$

which is therefore the MLE of λ.

A difficulty that may be encountered in maximum likelihood estimation is that an MLE may not exist.   □

Example 7.1.2 Non-existence of MLE. Suppose in Example 7.1.1 that for each i we observe only whether Y_i is 0 or positive and let

$$X_i = \begin{cases} 0 & \text{if } Y_i = 0 \\ 1 & \text{if } Y_i > 0. \end{cases}$$

Then

$$P(X_i = 0) = e^{-\lambda}, \; P(X_i = 1) = 1 - e^{-\lambda},$$

and the likelihood is

$$(7.1.5) \qquad L(\lambda) = (1 - e^{-\lambda})^{\Sigma x_i} e^{-\lambda \sum (1 - x_i)}.$$

This is maximized by (Problem 1.1(ii))

$$(7.1.6) \qquad \hat{\lambda} = \log \frac{n}{\sum (1 - x_i)} = -\log(1 - \bar{x}),$$

provided $\sum (1 - x_i) > 0$, i.e., provided not all the x_i are $= 1$.

When all the x's are $= 1$, the likelihood becomes

$$L(\lambda) = \left(1 - e^{-\lambda}\right)^n,$$

which is an increasing function of λ. In this case, the likelihood does not take on its maximum for any finite λ and the MLE does not exist. How serious this problem is depends on how frequently the case $X_1 = \cdots = X_n = 1$ occurs. For any fixed n, the probability

$$(7.1.7) \qquad P(X_1 = \cdots = X_n = 1) = \left(1 - e^{-\lambda}\right)^n$$

tends to 1 as $\lambda \to \infty$. Thus there will exist values of λ for which the probability is close to 1 that the MLE is undefined. (Whether the MLE exists when all the x's are 0 depends on whether 0 is considered a possible value of λ.)

On the other hand, for any fixed λ, the probability of (7.1.7) tends to 0 as $n \to \infty$. If we define a modified MLE $\hat{\lambda}$ by (7.1.6) when at least one of the x_i is $\neq 1$ and arbitrary (for example, equal to 1) otherwise, it is seen that $\hat{\lambda}$ is consistent and in fact satisfies (Problem 1.3)

$$(7.1.8) \qquad \sqrt{n}\left(\hat{\lambda} - \lambda\right) \xrightarrow{L} N\left(0, e^{\lambda} - 1\right).$$

□

The finding of this example, that difficulties encountered with the MLE for finite sample sizes may disappear in the limit as the sample size becomes large, is typical for i.i.d. samples from a suitably smooth parametric family of distributions (but see Example 7.1.3). We shall in the present section consider the asymptotic theory for the case of a single real-valued parameter; the treatment of vector-valued parameters will be taken up in Section 7.5.

A first property expected of any good estimator in the i.i.d. case is that it be consistent. This holds, for example, for the MLE's (7.1.4) and (7.1.6) (Problem 1.4). However, even this rather weak property will not always be satisfied as is shown by the following example due to Ferguson (1982, 1996).

Example 7.1.3 A counterexample. Let $X_1, \ldots, X_n$ be i.i.d. with a distribution which with probability θ is the uniform distribution on $(-1, 1)$, and with probability $(1 - \theta)$ is equal to the triangular distribution with density (Problem 1.5)

$$(7.1.9) \qquad \frac{1}{c(\theta)}\left[1 - \frac{|x - \theta|}{c(\theta)}\right], \quad \theta - c(\theta) < x < \theta + c(\theta),$$

where $c(\theta)$ is a continuous decreasing function of θ with $c(0) = 1$ and $0 < c(\theta) \le 1 - \theta$ for $0 < \theta < 1$, and with $c(\theta) \to 0$ as $\theta \to 1$. Thus, for θ close to 1, the triangular component of the density is concentrated in a narrow interval centered on θ, and its value $1/c(\theta)$ at $x = \theta$ tends to infinity as $\theta \to 1$. The situation is illustrated in Figure 7.1.1, which shows both the uniform density and triangular density corresponding to a value of θ close to 1. In this situation, it turns out that the MLE $\hat{\theta}_n$ exists for all n but that instead of being consistent, i.e., of converging in probability to θ, it converges to 1 as $n \to \infty$ no matter what the true value of θ.

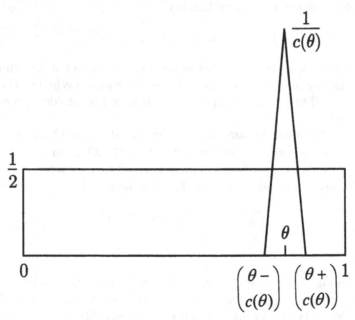

FIGURE 7.1.1. Ferguson's example

The example provides some insight into the working of maximum likelihood. Given a sample $x_1, \ldots, x_n$, we seek the distribution (and hence the value of θ characterizing it), which assigns the greatest likelihood to this sample. Because of the extremely high peak of the density when θ is very close to 1, the occurrence of some x's close to 1 (which for any θ will happen with near certainty when n gets large) is most likely when θ is near 1. Thus, regardless of the true value of θ, for large n the MLE $\hat{\theta}_n$ will be close to 1.

The following is a more formal sketch of the main steps of the argument; the details can be found in Ferguson (1982).

Instead of maximizing the likelihood $L_n(\theta)$, it is more convenient (and equivalent) to maximize the *log likelihood*

$$(7.1.10) \qquad l_n(\theta) = \log L_n(\theta).$$

It can be shown that for any $0 < \alpha < 1$, there exists a constant $K(\alpha)$ such that

$$(7.1.11) \qquad \max_{0 \le \theta \le \alpha} \frac{1}{n} l_n(\theta) \le K(\alpha) \text{ for all } n.$$

On the other hand, it turns out that (Problem 1.6)

$$(7.1.12) \qquad X_{(n)} = \max(X_1, \ldots, X_n) \xrightarrow{P} 1$$

and that

$$(7.1.13) \qquad \frac{1}{n} l_n\left(X_{(n)}\right) \xrightarrow{P} \infty,$$

provided $c(\theta) \to 0$ sufficiently fast as $\theta \to 1$. Since (7.1.13) implies that

$$(7.1.14) \qquad \max_{0 \le \theta \le 1} \frac{1}{n} l_n(\theta) \xrightarrow{P} \infty,$$

it then follows from (7.1.11) that

$$P\left(\hat{\theta}_n > \alpha\right) \to 1$$

for any $0 < \alpha < 1$ and hence that $\hat{\theta}_n \xrightarrow{P} 1$.

The densities $f_\theta(x)$ are quite smooth except for the discontinuity of the derivative at θ caused by the sharp peak of the triangular distribution. This discontinuity can be removed by replacing the family of triangular distributions (7.1.9) by a suitable family of beta distributions (see Ferguson (1982)). $\qquad \square$

Examples 7.1.2 and 7.1.3 illustrate two quite different difficulties that can arise with the MLE. In the first example, the problem is minor. The sample point $X_1 = \cdots = X_n = 1$, for which none of the Y's are equal to zero, suggests a very large value of θ but runs into the technical problem that the range of θ is unbounded. Even for this sample, the MLE, if taken to be ∞, points in the right direction; in addition, the probability of this special case tends to zero as $n \to \infty$ for all values of θ. Thus, for large n, with a slight modification needed only in rare cases, the MLE will provide a good idea of the position of θ.

In contrast, no such technical difficulty arises in Example 7.1.3. Here the MLE always exists, but for large n, it gives a totally misleading picture, suggesting that θ is close to 1, regardless of the true value of θ.

Despite this example, in standard situations the MLE tends to be consistent. Sufficient conditions for this to be the case are given in Wald (1949). Unfortunately, they are often difficult to check and exclude many standard problems. A simpler theory with conditions that are easier to check becomes possible when consistency of the *global* maximum is replaced by that of a suitable sequence of *local* maxima. This modification of the problem is due to Cramér (1946), who gave the following conditions for the existence of a consistent sequence $\hat{\theta}_n$ of local maxima of the likelihood function (satisfied, for instance, by the smoothed version of Example 7.1.3). Such a sequence turns out to be not only consistent but also efficient (see Section 7.4).

For the existence of such a consistent $\hat{\theta}_n$, we make the following assumptions:

(C1) The distributions P_θ of the observations are distinct, i.e., $P_{\theta_1} = P_{\theta_2}$ implies $\theta_1 = \theta_2$.

(C2) The parameter space Ω is an open interval $(\underline{\theta}, \bar{\theta})$,

$$\Omega : -\infty \le \underline{\theta} < \theta < \bar{\theta} \le \infty.$$

(C3) The observations are $X = (X_1, \dots, X_n)$, where the X_i are i.i.d. either with probability density $f_\theta(x)$ which will be assumed to be continuous in x (*the continuous case*), or discrete with probabilities

$$P_\theta(x) = P_\theta(X_i = x)$$

(*the discrete case*).

(C4) The set

$$A = \{x : f_\theta(x) > 0\} \text{ or } A = \{x : P_\theta(x) > 0\}$$

is independent of θ.

(C5) For all x in A, $f_\theta(x)$ or $P_\theta(x)$ is differentiable with respect to θ. Its derivative will be denoted by $f'_\theta(x)$ or $P'_\theta(x)$.

Notes

1. When (C1) holds, the parameter θ is said to be *identifiable*. To illustrate non-identifiability, suppose that $Y_1, \dots, Y_n$ are i.i.d. $N(\eta, 1)$ and let $X_i = |Y_i|$. Then the distribution of the X's depends only on $|\eta|$ and is therefore the same for η and $-\eta$. If only the X's are observed, the parameter η is unidentifiable since two different parameter values (η and $-\eta$) lead to the same distribution of the observations. In this

situation, η clearly cannot be estimated consistently since the X's provide no information as to whether it is positive or negative. (For a general discussion of identifiability with references, see Basu (1983).)

2. Assumption (C4) is satisfied for the mixture distributions of Example 7.1.3 with $A = (-1, 1)$; however, it does not hold for the family of triangular distributions with densities (7.1.9).

3. We shall in the following state results for the continuous case. They remain valid in the discrete case if all integrals are replaced by the corresponding sums.

The following result asserts the consistency not of the MLE but of a suitable sequence of local maxima of the likelihood.

Theorem 7.1.1 Let $X_1, \ldots, X_n$ be i.i.d. with probability density satisfying (C1)–(C5). Then there exists a sequence $\hat{\theta}_n = \hat{\theta}_n(X_1, \ldots, X_n)$ of local maxima of the likelihood function $L_n(\theta)$ which is consistent, i.e., which satisfies

(7.1.15) $$\hat{\theta}_n \xrightarrow{P_\theta} \theta \text{ for all } \theta \in \Omega.$$

For a proof of this result, see, for example, Cramér (1946) or Lehmann and Casella (1998).

The local maxima are determined by setting the derivative of the likelihood equal to 0,

(7.1.16) $$\frac{\partial}{\partial \theta} [f_\theta(x_1) \cdots f_\theta(x_n)] = 0.$$

It is equivalent and, since the likelihood is a product, often more convenient to find instead the local maxima of the log likelihood

(7.1.17) $$l_n(\theta) = \sum \log f_\theta(x_i)$$

by solving the *likelihood equation*

(7.1.18) $$l'_n(\theta) = \sum \frac{f'_\theta(x_i)}{f_\theta(x_i)} = 0.$$

Here, as in $L_n(\theta)$, the notation suppresses the fact that $l_n(\theta)$ and $l'_n(\theta)$ depend also on the x's.

Theorem 7.1.1 does not guarantee the existence of a local maximum for all $(x_1, \ldots, x_n)$ or, for a given n, even for any sample point. (See Example 7.1.2 and Problem 1.8.) It states only that with probability tending to 1 as $n \to \infty$, the likelihood (or equivalently the log likelihood) has a local maximum $\hat{\theta}_n = \hat{\theta}_n(X_1, \ldots, X_n)$ such that $\hat{\theta}_n$ tends to the true value θ in probability. A difficulty which will be discussed in Section 7.3 arises when there are several local maxima since we then do not know which one to choose so as to obtain a consistent sequence.

Corollary 7.1.1 *Under the assumptions of Theorem 7.1.1, if the likelihood equation (7.1.18) has a unique root $\hat{\theta}_n$ for each n and all $(x_1, \dots, x_n)$, then*

(i) *$\hat{\theta}_n$ is a consistent estimator of θ*

and

(ii) *with probability tending to 1 as $n \to \infty$, $\hat{\theta}_n$ is the MLE.*

Proof. Part (i) is an immediate consequence of Theorem 7.1.1.

(ii) If the likelihood equation has a unique root, this may correspond to a local maximum, a local minimum, or an inflection point of $l_n(\theta)$. In the second and third case, there can then exist no local maximum and, by Theorem 7.1.1, the probability of these possibilities therefore tends to 0. It is thus enough to show that if $\hat{\theta}_n$ is a local maximum, it is the MLE. Suppose this is not the case so that at some point $\theta' \neq \hat{\theta}_n$ we have $l_n(\theta') > l_n\left(\hat{\theta}_n\right)$. Then $l_n(\theta)$ must have a local minimum between $\hat{\theta}_n$ and θ', and this contradicts the uniqueness assumption. ∎

Example 7.1.4 Normal. Let $X_1, \dots, X_n$ be i.i.d. according to the normal distribution $N\left(\xi, \sigma^2\right)$. The log likelihood is then

$$(7.1.19) \qquad -n \log \sqrt{2\pi} - n \log \sigma - \frac{1}{2\sigma^2} \sum (x_i - \xi)^2.$$

(i) Consider first the case that σ^2 is known and ξ is the parameter to be estimated. Differentiating (7.1.19) with respect to ξ, we see that the likelihood equation is

$$\frac{1}{\sigma^2} \sum (x_i - \xi) = 0,$$

which has the unique solution $\hat{\xi} = \bar{X}$. Since the log likelihood (7.1.19) is maximized by minimizing

$$(7.1.20) \qquad \sum (x_i - \xi)^2 = \sum (x_i - \bar{x})^2 + n (\bar{x} - \xi)^2,$$

it follows that $\hat{\xi}$ is also the MLE. Of course, that $\bar{X}$ is a consistent estimator of ξ was already seen in Example 2.1.2.

(ii) Suppose next that ξ is known and we wish to estimate σ^2. The likelihood equation then becomes

$$(7.1.21) \qquad -\frac{n}{\sigma} + \frac{1}{\sigma^3} \sum (x_i - \xi)^2 = 0,$$

which has the unique solution

$$(7.1.22) \qquad \hat{\sigma}^2 = \frac{1}{n} \sum (x_i - \xi)^2.$$

Since the log likelihood tends to $-\infty$ both as $\sigma \to 0$ and as $\sigma \to \infty$ (Problem 1.14), it follows that $\hat{\sigma}$ is both a local maximum and the MLE. An exception is the case $x_1 = \cdots = x_n = \xi$ in which the likelihood is a strictly decreasing function of σ, but this event has probability 0.

(iii) As a third possibility, suppose that ξ/σ is known, say

$$(7.1.23) \qquad\qquad \xi = a\sigma, \; 0 < \sigma < \infty,$$

and that we wish to estimate σ. Then the likelihood equation becomes (Problem 1.15)

$$(7.1.24) \qquad\qquad \gamma^2 - \gamma \frac{a \sum x_i}{\sum x_i^2} = \frac{n}{\sum x_i^2},$$

where $\gamma = 1/\sigma$. If $d = a \sum x_i / \left[2 \sum x_i^2 \right]$, (7.1.24) has the two solutions

$$(7.1.25) \qquad\qquad \gamma = d \pm \sqrt{\frac{n}{\sum x_i^2} + d^2}.$$

Only one of these is positive, so that in the range of possible values of σ, the likelihood equation (7.1.24) has the unique solution

$$(7.1.26) \qquad \frac{1}{\hat{\sigma}} = \hat{\gamma} = \frac{a \sum x_i}{2 \sum x_i^2} + \sqrt{\frac{n}{\sum x_i^2} + \frac{a^2 \left(\sum x_i \right)^2}{4 \left(\sum x_i^2 \right)^2}}.$$

Since the log likelihood tends to $-\infty$ as $\sigma \to 0$ or $\sigma \to \infty$ (Problem 1.14(ii)), it follows that $\hat{\sigma}$ is both a local maximum and the MLE. $\square$

Let us now return to the general situation considered in Theorem 7.1.1. Consistency is a rather weak property. However, the assumptions (C1)–(C5), together with some additional smoothness assumptions on the densities $f_\theta(x)$, permit the much stronger conclusion that any consistent sequence $\hat{\theta}_n = \hat{\theta}_n (X_1, \ldots, X_n)$ of roots of the likelihood equation satisfies

$$(7.1.27) \qquad\qquad \sqrt{n} \left(\hat{\theta}_n - \theta \right) \xrightarrow{L} N \left(0, \frac{1}{I(\theta)} \right),$$

where

$$(7.1.28) \qquad I(\theta) = E \left[\frac{\partial}{\partial \theta} \log f_\theta(X) \right]^2 = \int \left[\frac{f_\theta'(x)}{f_\theta(x)} \right]^2 f_\theta(x) dx,$$

which is assumed to satisfy

(7.1.29) $$0 < I(\theta) < \infty.$$

An exact statement and proof of this basic result will be given in Section 7.3. Let us now give some simple illustrations of (7.1.27).

Example 7.1.1 Poisson (continued). If X has the Poisson distribution $P(\lambda)$ with

$$P_\lambda(x) = \frac{\lambda^x}{x!} e^{-\lambda},$$

then

$$\frac{\partial}{\partial \lambda} [\log P_\lambda(x)] = \frac{x}{\lambda} - 1 = \frac{x - \lambda}{\lambda}$$

and hence

$$I(\lambda) = \frac{1}{\lambda^2} E (X - \lambda)^2 = \frac{1}{\lambda}.$$

Suppose now that $X_1, \dots, X_n$ are i.i.d. according to $P(\lambda)$. Then it was stated in Example 7.1.1 that the MLE of λ is $\hat{\lambda} = \bar{X}$. If (7.1.27) applies, it follows that

(7.1.30) $$\sqrt{n} (\bar{X} - \lambda) \to N(0, \lambda).$$

In the present case, (7.1.30) is an immediate consequence of the central limit theorem. $\square$

Example 7.1.4 Normal (continued). Let $X_1, \dots, X_n$ be i.i.d. according to the normal distribution $N(\xi, \sigma^2)$. It is then easily seen that when σ is known, $I(\xi) = 1/\sigma^2$, and when ξ is known, $I(\sigma^2) = 1/2\sigma^4$. In both cases, (7.1.27) follows directly from the CLT (Problem 1.16).

Let us next turn to case (iii) of Example 7.1.3 in which $\xi = a\sigma$, where a is a known positive constant. The log likelihood of a single observation is now equal to

$$\log f_\sigma(x) = -\frac{1}{2\sigma^2} (x - a\sigma)^2 - \log \sigma - \log \sqrt{2\pi},$$

and its derivative with respect to σ is therefore

(7.1.31) $$\frac{\partial}{\partial \sigma} [\log f_\sigma(x)] = \frac{1}{\sigma^3} (x - a\sigma)^2 + \frac{a}{\sigma^2} (x - a\sigma) - \frac{1}{\sigma}.$$

Thus, since $E(X - a\sigma) = E(X - a\sigma)^3 = 0$,

(7.1.32)
$$I(\sigma) = \frac{1}{\sigma^6} E (X - a\sigma)^4 + \frac{a^2}{\sigma^4} E (X - a\sigma)^2 + \frac{1}{\sigma^2} - \frac{2}{\sigma^4} E (X - a\sigma)^2$$
$$= \frac{1}{\sigma^2} [3 + (a^2 - 2) + 1] = \frac{1}{\sigma^2} (a^2 + 2).$$

If (7.1.27) holds (which in Section 7.3 will be seen to be the case), it follows that

(7.1.33) $$\sqrt{n}\,(\hat{\sigma} - \sigma) \xrightarrow{L} N\left(0, \frac{\sigma^2}{a^2 + 2}\right).$$

$\square$

It is often of interest to estimate not θ itself but some function of θ. Suppose that

(7.1.34) $$\eta = g(\theta)$$

is strictly increasing (or decreasing) so that is has an inverse

(7.1.35) $$\theta = g^{-1}(\eta).$$

Then we can use η to label the distribution by writing

(7.1.36) $$f_\theta(x) = f_{g^{-1}(\eta)}(x).$$

If $\hat{\theta}$ maximizes the left side of (7.1.36), the right side is maximized by the value $\hat{\eta}$ of η for which $g^{-1}(\hat{\eta}) = \hat{\theta}$, i.e., by

(7.1.37) $$\hat{\eta} = g\left(\hat{\theta}\right).$$

It follows that the MLE of $g(\theta)$ is $g(\hat{\theta})$.

Let us now drop the assumption that g is strictly monotone. Suppose, for example, that $g(\theta) = \theta^2$. Then we can no longer use $\eta = g(\theta)$ to label the distributions since a given value of η corresponds to more than one distribution. As a consequence, the likelihood of η is undefined and the original definition of an MLE no longer applies. It turns out to be convenient in this case to define the MLE of η by (7.1.37). As we shall see later with this definition, $\hat{\eta}$ will continue to possess the properties which in the monotone case follow from its maximizing the likelihood. The following is an example of such a result.

Corollary 7.1.2 *Under the assumptions of Corollary 7.1.1, suppose that g is a continuous function of θ. Then $\hat{\eta}_n = g\left(\hat{\theta}_n\right)$ is a consistent estimator of η.*

Proof. This follows immediately from Theorem 2.1.4. $\blacksquare$

Summary

1. The maximum likelihood estimator (MLE) is defined as the global maximum of the likelihood function. An example is given in which the MLE does not always exist, but with the probability of this happening tending to 0 as $n \to \infty$. Another example shows that the MLE need not be consistent even in very regular situations.

2. A more convenient theory is obtained by shifting attention from the global maximum to local maxima and hence to roots of the likelihood equation. If $\hat{\theta}_n$ is a consistent sequence of roots of the likelihood equation, it follows under fairly weak regularity conditions that $\sqrt{n}\left(\hat{\theta}_n - \theta\right)$ is asymptotically normal with mean 0 and variance $1/I(\theta)$, where $I(\theta)$ is the Fisher information.

7.2 Fisher information

The quantity $I(\theta)$ defined by (7.1.28) is frequently called "Fisher information" or "the amount of information about θ contained in X." However, the suggestive term "information" should not be taken too seriously. It is justified mainly by the fact that in many cases, under assumptions and with restrictions that will be made precise in Section 7.5, $1/I(\theta)$ is the smallest asymptotic variance obtainable by an estimator δ_n satisfying a limit relation of the form

$$(7.2.1) \qquad \sqrt{n}\left(\delta_n - \theta\right) \xrightarrow{L} N\left(0, v(\theta)\right).$$

Although we speak of the information contained in a single observation X, the justification is thus primarily based on the assumption of a large sample. To see what $I(\theta)$ has to say about the informativeness of a single observation, note that the logarithmic derivative

$$(7.2.2) \qquad \frac{\partial}{\partial \theta}\left[\log f_\theta(x)\right] = \frac{f'_\theta(x)}{f_\theta(x)}$$

is the relative rate* at which the density $f_\theta(x)$ changes (as a function of θ) at the point x. The quantity $I(\theta)$ is just the expected square of this rate. It is plausible that the greater this expectation is at a given value θ_0, the easier it is to distinguish θ_0 from nearby values of θ, and the more accurately therefore θ can be estimated at $\theta = \theta_0$. In this sense, the larger $I(\theta)$ is, the more informative is X. However, this argument leaves many other aspects of the relationship between X and θ out of consideration and provides no basis for thinking of $I(\theta)$ as a comprehensive measure of the amount of information contained in a single observation. As an illustration, consider the following example.

Example 7.2.1 An informative observation with $I(\theta) = 0$. Let X be normally distributed with $E(X) = \theta^3$ and known variance σ_0^2, so that

$$f_\theta(x) = \frac{1}{\sqrt{2\pi}\sigma_0} e^{-\frac{1}{2\sigma_0^2}(x-\theta^3)^2}.$$

---

*An interesting discussion of this rate is given in Pitman (1979, Chapter 4).

Then

$$\frac{\partial}{\partial \theta} \log f_\theta(x) = 3\theta^2 \left(x - \theta^3\right)/\sigma_0^2$$

and hence

$$I(\theta) = 9\theta^4/\sigma_0^2.$$

For $\theta = 0$, we therefore have $I(\theta) = 0$ despite the fact that if σ_0^2 is small, X provides a very accurate estimator of θ^3 and thus of θ . $\square$

Even the asymptotic justification of $I(\theta)$ works only in the regular kind of situation in which (7.1.27) holds. To see the difficulties that arise without this restriction consider the following non-regular example which violates (C4) of Section 1.

Example 7.2.2 Uniform. Let $X_1, \dots, X_n$ be i.i.d. according to the uniform distribution $U(0, \theta)$, so that the likelihood is

(7.2.3)
$$L_n(\theta) = \begin{cases} 1/\theta^n & \text{if } 0 \le x_i \le \theta \text{ for all } i \left(\text{i.e. if } 0 \le x_{(1)} < x_{(n)} \le \theta\right) \\ 0 & \text{otherwise.} \end{cases}$$

It follows that the MLE is equal to $x_{(n)}$, the largest of the x's, i.e.,

(7.2.4)
$$\hat\theta_n = X_{(n)}.$$

As was seen in Example 2.3.7, $n(\hat\theta_n - \theta)$ tends to a non-degenerate limit distribution. Thus

$$\sqrt{n}(\hat\theta_n - \theta) \xrightarrow{P} 0,$$

which can be expressed by saying that $\sqrt{n}(\hat\theta_n - \theta)$ tends in law to the (degenerate) normal distribution $N(0, 0)$. According to (7.1.27), we should therefore have $I(\theta) = \infty$. However,

$$\log f_\theta(x) = \log(1/\theta) = -\log\theta \quad \text{if } x < \theta$$

so that

(7.2.5)
$$\frac{\partial}{\partial \theta} \log f_\theta(x) = \frac{-1}{\theta} \quad \text{if } x < \theta$$

and hence $I(\theta)$ defined by (7.1.28) exists and is equal to

(7.2.6)
$$I(\theta) = \frac{1}{\theta^3} \int_0^\theta dx = \frac{1}{\theta^2} < \infty.$$

Consideration of (7.2.2) indicates why, in the present case, $I(\theta)$ so inadequately reflects the information supplied by even a large sample about θ. As pointed out, the logarithmic derivative (7.2.2) on which $I(\theta)$ is based measures the change of the density from $1/\theta_0$ at θ_0 to $1/\theta$ at a nearby θ. However, it disregards the additional (and in fact much more important) change at the end points θ_0 and θ, where the density changes abruptly from a positive value to 0.   □

In line with these comments, we shall restrict the use of the measure $I(\theta)$ to regular situations in which (7.1.27) holds and we shall interpret the statement that the Fisher information (7.1.28) is the amount of information about θ contained in X simply in the sense of (7.1.27).

Let us next consider some properties of Fisher information.

1. **Alternative expressions for $I(\theta)$.** The following result provides two alternative expressions for Fisher information.

Theorem 7.2.1

(i) Suppose that (C1)–(C5) hold and that, in addition,

(C6) the derivative with respect to θ of the left side of the equation

$$(7.2.7) \qquad \int f_\theta(x)dx = 1$$

can be obtained by differentiating under the integral sign.
Then

$$(7.2.8) \qquad E_\theta \left[\frac{\partial}{\partial \theta} \log f_\theta(X) \right] = 0$$

and hence

$$(7.2.9) \qquad I(\theta) = \text{Var}_\theta \left[\frac{\partial}{\partial \theta} \log f_\theta(X) \right].$$

(ii) Suppose that also

(C6)′ The first two derivatives with respect to θ of $f_\theta(x)$ exist for all $x \in A$ and all θ, and the corresponding derivatives with respect to θ of the left side of (7.2.7) can be obtained by differentiation under the integral sign.
Then also

$$(7.2.10) \qquad I(\theta) = -E_\theta \left[\frac{\partial^2}{\partial \theta^2} \log f_\theta(X) \right].$$

Proof.

(i) Differentiation of both sides of (7.2.7) gives (7.2.8), and (7.2.9) then follows from (7.1.28).

(ii) Differentiation of

$$\frac{\partial}{\partial \theta} \left[\log f_\theta(x) \right] = \left[\frac{\partial}{\partial \theta} f_\theta(x) \right] / f_\theta(x)$$

yields

(7.2.11)
$$\frac{\partial^2 \left[\log f_\theta(x) \right]}{\partial \theta^2} = \frac{\frac{\partial^2}{\partial \theta^2} \left[f_\theta(x) \right]}{f_\theta(x)} - \left[\frac{\frac{\partial}{\partial \theta} f_\theta(x)}{f_\theta(x)} \right]^2,$$

and the result follows from (7.2.7) by taking the expectation of (7.2.11). ∎

As an illustration of (7.2.10) consider once more Example 7.1.4 (continued). From (7.1.31), it is seen that

$$\frac{\partial^2}{\partial \sigma^2} \left[\log f_\sigma(x) \right] = -\frac{3}{\sigma^4} (x - a\sigma)^2$$
$$- \frac{2a}{\sigma^3} (x - a\sigma) - \frac{2a}{\sigma^3} (x - a\sigma) - \frac{a^2}{\sigma^2} + \frac{1}{\sigma^2},$$

and hence that

(7.2.12) $\quad -E \left[\frac{\partial^2}{\partial \sigma^2} \log f_\sigma(X) \right] = \frac{3\sigma^2}{\sigma^4} + \frac{a^2}{\sigma^2} - \frac{1}{\sigma^2} = \frac{1}{\sigma^2} \left(a^2 + 2 \right),$

which agrees with the right side of (7.1.32).

2. **Additivity.** As one might expect, the information supplied by two independent observations is the sum of the information supplied by them separately.

Theorem 7.2.2

(i) Let X and Y be independently distributed with densities f_θ and g_θ satisfying (C1)–(C6). If $I_1(\theta)$, $I_2(\theta)$, and $I(\theta)$ denote the amounts of information about θ contained respectively in X, Y, and (X, Y), then

(7.2.13) $\quad I(\theta) = I_1(\theta) + I_2(\theta).$

(ii) If $X_1, \ldots, X_n$ are i.i.d. with density f_θ satisfying (C1)–(C6), then the information about θ contained in the sample $(X_1, \ldots, X_n)$ equals $nJ(\theta)$, where $J(\theta)$ is the common information contained in the individual X_i.

Proof. (i) By definition,

$$I(\theta) = E\left[\frac{\partial}{\partial\theta}\log f_\theta(X) + \frac{\partial}{\partial\theta}\log g_\theta(Y)\right]^2$$

(7.2.14)

$$= I_1(\theta) + I_2(\theta) + 2E\left[\frac{\partial}{\partial\theta}\log f_\theta(X)\frac{\partial}{\partial\theta}\log f_\theta(Y)\right].$$

Since X and Y are independent, the expected product in the third term is equal to the product of the expectations, and the result follows from (7.2.8).

(ii) This is an immediate consequence of (i). ∎

Note: From (7.2.14), it follows that (7.2.13) fails when

$$E\left[\frac{\partial}{\partial\theta}\log f_\theta(X)\frac{\partial}{\partial\theta}\log g_\theta(Y)\right] = E\left[\frac{\partial}{\partial\theta}\log f_\theta(X)\right] E\left[\frac{\partial}{\partial\theta}\log g_\theta(Y)\right]$$

is different from 0, that is, when both f_θ and g_θ fail to satisfy (7.2.8), Consider, in particular, the case of n i.i.d. random variables $X_1, \ldots, X_n$. In the notation of Theorem 7.2.2(ii), we then have

$$
\begin{aligned}
I(\theta) &= E\left[\sum_{i=1}^n \frac{\partial}{\partial\theta}\log f_\theta(X_i)\right]^2 \\
&= nJ(\theta) + n(n-1)\left[E_\theta\frac{\partial}{\partial\theta}\log f_\theta(X_1)\right]^2.
\end{aligned}
$$

(7.2.15)

Thus if (7.2.8) does not hold, as is the case in Example 7.2.2, $I(\theta)$ is of order n^2 rather than the usual order n.

3. **Information about a function of θ.** In Example 7.1.4 (continued), it was stated that the information $I\left(\sigma^2\right)$ contained in a normal observation X with known mean about σ^2 is $1/2\sigma^4$. Would the same answer obtain for the information that X contains about σ? Let us consider more generally the information $I\left(\sigma^k\right)$ that X contains about $\eta = \sigma^k$. The log likelihood

(7.2.16) $$\log f_\sigma(x) = -\frac{1}{2\sigma^2}(x-\xi)^2 - \log\sigma - \log\sqrt{2\pi}$$

can be expressed in terms of η as

(7.2.17) $$\log f_\eta^*(x) = -\frac{1}{2\eta^{2/k}}(x-\xi)^2 - \log\eta^{1/k} - \log\sqrt{2\pi}.$$

Then

$$\frac{\partial}{\partial\eta}\log f_\eta^*(x) = \frac{1}{k\eta^{\frac{2}{k}+1}}(x-\xi)^2 - \frac{1}{k\eta} = \frac{1}{k\eta^{\frac{2}{k}+1}}\left[(x-\xi)^2 - \eta^{2/k}\right]$$

and it follows that

$$(7.2.18) \qquad I^*(\eta) = \frac{1}{k^2\eta^{(4+2k)/k}}\,\mathrm{Var}\left[(X-\xi)^2\right] = \frac{2}{k^2\sigma^{2k}}.$$

For $k = 2$, this reduces to the earlier $1/2\sigma^4$, as it should. For $k = 1$, we find $I(\sigma) = 2/\sigma^2$, and so on. Quite generally, the information X contains about $g(\theta)$ depends not only on θ but also on g.

The argument leading to (7.2.18) can be used to evaluate $I\,[g(\theta)]$ more generally. Suppose that $\eta = g(\theta)$ is a strictly increasing differentiable function with derivative $g'(\theta)$. Expressed in terms of η, the density $f_\theta(x)$ then becomes

$$f_\eta^*(x) = f_{g^{-1}(\eta)}(x)$$

so that

$$\frac{\partial}{\partial\eta}\left[\log f_\eta^*(x)\right] = \frac{\partial}{\partial\eta}\left[\log f_\theta(x)\right]\frac{dg^{-1}(\eta)}{d\eta}$$

$$= \frac{f_\theta'(x)}{f_\theta(x)}\frac{1}{g'(\theta)},$$

and hence

$$(7.2.19) \qquad I^*(\eta) = I(\theta)/\left[g'(\theta)\right]^2.$$

It is seen from (7.2.19) that a more rapid change in $g(\theta)$ (i.e., an increase in $|g'(\theta)|$) will decrease $I\,[g(\theta)]$. To see why this is reasonable, suppose that $I(\theta)$, and hence the accuracy with which θ can be estimated, is fixed. Then $g(\theta)$ can be estimated more accurately the flatter the function g is at θ since a value θ' at a given small distance from θ will lead to a smaller deviation of $g(\theta')$ from $g(\theta)$ the smaller the value of $|g'(\theta)|$ is.

The most extreme case of this phenomenon occurs when $g'(\theta) = 0$, in which case $I^*[g(\theta)] = \infty$. In this case, the function g is essentially constant in a sufficiently small neighborhood of θ so that only a rough idea of θ is needed to get very precise knowledge of $g(\theta)$. A more quantitative interpretation of the phenomenon is obtained by recalling formula (2.5.14). This shows that if $\hat\theta_n$ satisfies (7.1.27) and $g'(\theta) = 0$ but $g''(\theta) \neq 0$, then $n\left[g\left(\hat\theta_n\right) - g(\theta)\right]$ tends to a limit distribution. Thus the error made when estimating $g(\theta)$ is then of order $1/n$, smaller than the $1/\sqrt{n}$ order which in (7.1.27) provided the motivation for the information criterion $I(\theta)$.

So far, we have assumed that $\eta = g(\theta)$ is a strictly monotone function of θ. Suppose now that this is not the case, that we wish to estimate, for example, $g(\theta) = \theta^2$ or $g(\theta) = \sin\theta$. The argument leading to (7.2.19) then requires some modification since the inverse g^{-1} is no longer uniquely defined.

To see that, except for a small change in notation, formula (7.2.19) continues to apply when the assumption of monotonicity is dropped, suppose that the derivative g' exists and is continuous. Then if $g'(\theta) > 0$, g' will be positive in some neighborhood $(\theta - \epsilon, \theta + \epsilon)$ of θ. In this neighborhood, $\eta = g(\theta)$ is strictly increasing and therefore can be inverted, so that the earlier argument applies. However, the inverse function $g^{-1}(\eta)$ now depends not only on η but also on θ (different values of θ can lead to the same value of η) and therefore will be denoted by $g_\theta^{-1}(\eta)$. To illustrate, let $g(\theta) = \theta^2$. Then

$$(7.2.20) \qquad g_\theta^{-1}(\eta) = \begin{cases} \sqrt{\eta} & \text{if } \theta > 0 \\ -\sqrt{\eta} & \text{if } \theta < 0. \end{cases}$$

To each of these branches of the inverse function, we can now apply without any change the calculation that led to (7.2.19) to obtain

$$(7.2.21) \qquad I_\theta^*(\eta) = I\left[g_\theta^{-1}(\eta)\right]\left[\frac{dg_\theta^{-1}(\eta)}{d\eta}\right]^2.$$

Expressed in terms of θ, this reduces to

$$(7.2.22) \qquad I_\theta^*\left[g(\theta)\right] = I(\theta)/\left[g'(\theta)\right]^2,$$

which is therefore the amount of information supplied by X about $g(\theta)$.

Summary

1. The Fisher information $I(\theta)$ contained in an observation X about θ measures the average relative rate of change in the density $f_\theta(x)$ as a function of θ. The term is misleading since $I(\theta)$ is a reasonable measure of information only asymptotically as $n \to \infty$, and even then only in regular cases.

2. Several properties of $I(\theta)$ are discussed:

(i) Alternative expressions are given which are sometimes more convenient;

(ii) $I(\theta)$ is shown to be additive when two or more independent sets of observations are combined;

(iii) a formula is obtained for the change of $I(\theta)$ under reparametrization.

7.3   Asymptotic normality and multiple roots

The present section will further develop the theory of likelihood-based estimation in two directions. First, we shall discuss conditions under which the asymptotic normality result (7.1.27) holds and give a proof of the resulting theorem. This will be followed by consideration of the problem arising when the likelihood equation has multiple roots. The guaranteed existence of a root that is consistent and hence satisfies (7.1.27) is then inadequate since it does not tell us how to find this root. We shall discuss some ways of getting around this difficulty.

The following theorem gives sufficient conditions for the validity of the limit result (7.1.27) which has been illustrated in the preceding section.

Theorem 7.3.1 *Let $X_1, \ldots, X_n$ be i.i.d. with density $f_\theta(x)$ satisfying (C1)-(C5) of Section 7.1, and suppose that in addition the following conditions hold:*

(C6)'' For all $x \in A$ (specified in (C4)), the density $f_\theta(x)$ is three times differentiable with respect to θ and the third derivative is continuous. The corresponding derivatives of the integral $\int f_\theta(x) dx$ can be obtained by differentiating under the integral sign.

(C7) If θ_0 denotes the true value of θ, there exists a positive number $c(\theta_0)$ and a function $M_{\theta_0}(x)$ such that

$$(7.3.1) \qquad \left| \frac{\partial^3}{\partial \theta^3} \log f_\theta(x) \right| \le M_{\theta_0}(x) \text{ for all } x \in A, \ |\theta - \theta_0| < c(\theta_0)$$

and

$$(7.3.2) \qquad E_{\theta_0}[M_{\theta_0}(X)] < \infty.$$

Then any consistent sequence $\hat{\theta}_n = \hat{\theta}_n(X_1, \ldots, X_n)$ of roots of the likelihood equation satisfies

$$(7.3.3) \qquad \sqrt{n}\left(\hat{\theta}_n - \theta_0\right) \xrightarrow{L} N\left(0, \frac{1}{I(\theta_0)}\right),$$

where it is assumed that

$$(7.3.4) \qquad 0 < I(\theta_0) < \infty.$$

Proof. Let $l_n(\theta)$ denote the log likelihood and $l_n'(\theta)$, $l_n''(\theta), \ldots$ its derivatives with respect to θ. The proof is based on the facts that, on the one hand, $\hat{\theta}_n$ satisfies $l_n'\left(\hat{\theta}_n\right) = 0$ and, on the other, $\hat{\theta}_n$ is known with high probability to be close to θ_0. Expansion of $l_n'\left(\hat{\theta}_n\right)$ about $l_n'(\theta_0)$ will then give us an equation for $\hat{\theta}_n - \theta_0$ that will lead to the desired result.

For any fixed x, expand $l_n'\left(\hat{\theta}_n\right)$ about $l_n'\left(\theta_0\right)$ as

$$(7.3.5) \quad l_n'\left(\hat{\theta}_n\right) = l_n'\left(\theta_0\right) + \left(\hat{\theta}_n - \theta_0\right) l_n''\left(\theta_0\right) + \frac{1}{2}\left(\hat{\theta}_n - \theta_0\right)^2 l_n'''\left(\theta_n^*\right),$$

where θ_n^* lies between θ_0 and $\hat{\theta}_n$. Since the left side is zero, we have

$$(7.3.6) \quad \sqrt{n}\left(\hat{\theta}_n - \theta_0\right) = \frac{l_n'\left(\theta_0\right)/\sqrt{n}}{-(1/n)l_n''\left(\theta_0\right) - (1/2n)\left(\hat{\theta}_n - \theta_0\right)l_n'''\left(\theta_n^*\right)}.$$

(Of course the quantities $l_n'(\theta)$, $l_n''(\theta), \ldots$ depend not only on θ but also on the X's.)

Of the three quantities on the right side, we shall show that

$$(7.3.7) \quad \frac{1}{\sqrt{n}}l_n'\left(\theta_0\right) \xrightarrow{L} N\left(0, I\left(\theta_0\right)\right),$$

that

$$(7.3.8) \quad -\frac{1}{n}l_n''\left(\theta_0\right) \xrightarrow{P} I\left(\theta_0\right),$$

and that

$$(7.3.9) \quad \frac{1}{n}l_n'''\left(\theta_n^*\right) \text{ is bounded in probability.}$$

Since $\hat{\theta}_n \xrightarrow{P} \theta_0$, (7.3.8) and (7.3.9) imply that the denominator on the right side of (7.3.6) tends in probability to $I\left(\theta_0\right)$, and (7.3.3) then follows.

To prove (7.3.7), note that by (7.1.18)

$$(7.3.10) \quad \frac{1}{\sqrt{n}}l_n'\left(\theta_0\right) = \sqrt{n}\frac{1}{n}\sum \frac{f_{\theta_0}'\left(X_i\right)}{f_{\theta_0}\left(X_i\right)}.$$

This is the normalized average of the i.i.d. random variables

$$\frac{f_\theta'\left(X_i\right)}{f_\theta\left(X_i\right)} = \frac{\partial}{\partial\theta}\log f_\theta\left(X_i\right).$$

The expectation of these variables is zero by (7.2.8), their variance is $I\left(\theta_0\right)$ by (7.2.9), and (7.3.7) thus follows from the CLT.

Consider next

$$-\frac{1}{n}l''_n\left(\theta_0\right) = \frac{1}{n}\sum \frac{f_{\theta_0}'^2\left(X_i\right) - f_{\theta_0}\left(X_i\right)f''_{\theta_0}\left(X_i\right)}{f_{\theta_0}^2\left(X_i\right)}$$

$$= \frac{1}{n}\sum\left[\frac{f'_{\theta_0}\left(X_i\right)}{f_{\theta_0}\left(X_i\right)}\right]^2 - \frac{1}{n}\sum \frac{f''_{\theta_0}\left(X_i\right)}{f_{\theta_0}\left(X_i\right)}.$$

By the law of large numbers, the first term tends to

(7.3.11)
$$E_{\theta_0} \left[\frac{f'_{\theta_0}(X_i)}{f_{\theta_0}(X_i)} \right]^2 = I(\theta_0),$$

and the second term to

(7.3.12)
$$E_{\theta_0} \left[\frac{f''_{\theta_0}(X_i)}{f_{\theta_0}(X_i)} \right] = \int f''_{\theta_0}(x)dx.$$

The latter is zero by (7.2.8), which proves (7.3.8).
  Finally,

$$\frac{1}{n} l'''_n(\theta) = \frac{1}{n} \sum \frac{\partial^3}{\partial \theta^3} \log f_\theta(X_i).$$

It follows from (C7) that

$$\left| \frac{1}{n} l'''_n(\theta^*_n) \right| < \frac{1}{n} \sum M_{\theta_0}(X_i)$$

when $\left| \hat{\theta}_n - \theta_0 \right| < c(\theta_0)$ and hence with probability tending to 1. Since the right side tends in probability to $E_{\theta_0}[M(X_1)]$, this completes the proof.■

  Application of the theorem requires checking its various conditions. Differentiation under an integral required for (C6)–(C6)″ is treated in books on calculus. The following conditions typically are easy to apply.

Lemma 7.3.1

(i) A sufficient condition for

$$h(\theta) = \int_a^b g_\theta(x)dx$$

to be differentiable under the integral sign, i.e., for $h'(\theta)$ to exist and to be given by

$$h'(\theta) = \int_a^b \frac{\partial}{\partial \theta} g_\theta(x)dx$$

at a point $\theta = \theta_0$ is that for some $\theta_1 < \theta_0 < \theta_2$, the function g has a continuous derivative with respect to θ in the closed rectangle $a \le x \le b$, $\theta_1 \le \theta \le \theta_2$.

(ii) If the range of integration is infinite, it is required in addition that

(7.3.13)

$$\int_a^b \frac{\partial}{\partial \theta} g_\theta(x) dx \to \int_{-\infty}^\infty \frac{\partial}{\partial \theta} g_\theta(x) dx \qquad \begin{array}{l} \textit{uniformly for } \theta \textit{ in some} \\ \textit{neighborhood of } \theta_0 \end{array}$$

as $a \to -\infty$, $b \to \infty$.

To illustrate the checking of the conditions, let us consider once more the examples of the preceding section.

Example 7.3.1 Poisson. We here replace θ by λ and $f_\theta(x)$ by

(7.3.14)
$$P_\lambda(x) = \frac{\lambda^x}{x!} e^{-\lambda}.$$

Conditions (C1)–(C5) are clearly satisfied with the parameter space Ω being the interval $0 < \lambda < \infty$ and A being the set $\{0, 1, 2, \cdots\}$ of non-negative integers.

To check (C6) in the present case, replace the integral of the density by the sum of the probabilities (7.3.14). That

$$\frac{d}{d\lambda}\left(\sum_{x=0}^\infty \frac{\lambda^x}{x!}e^{-\lambda}\right) = \sum_{x=0}^\infty \left[\frac{\partial}{\partial\lambda}\left(\frac{\lambda^x}{x!}e^{-\lambda}\right)\right]$$

is obvious in the present case since the left side is zero and the right side is

$$\sum_{x=1}^\infty \frac{\lambda^{x-1}}{(x-1)!}e^{-\lambda} - \sum_{x=0}^\infty \frac{\lambda^x}{x!}e^{-\lambda} = 0;$$

the argument for the second and third derivative is completely analogous.

To check (C7), note that

(7.3.15)
$$\frac{\partial^3}{\partial\lambda^3}\left[\log P_\lambda(x)\right] = \frac{\partial^3}{\partial\lambda^3}\left[x\log\lambda - \lambda\right] = \frac{2x}{\lambda^3}.$$

If we required (7.3.1) to hold for all λ, we would be in trouble since $1/\lambda^3 \to \infty$ as $\lambda \to 0$. However, it is required to hold only for λ in some interval $\lambda_0 - c(\lambda_0) < \lambda < \lambda_0 + c(\lambda_0)$. If we take $c(\lambda_0) = \lambda_0/2$, the right side of (7.3.15) is $\leq M_0(x) = \dfrac{2x}{(\lambda_0/2)^3}$, and $E[M_0(X)] = \dfrac{16}{\lambda_0^2} < \infty$.   □

As was pointed out in Example 7.1.1 (continued), in the present case the conclusion of Theorem 7.3.1 is immediate from the CLT. Let us now consider the situation of Example 7.1.4(iii) where the conclusion does not follow so directly.

Example 7.3.2 Normal with fixed coefficient of variation. Let $X_1, \ldots, X_n$ be i.i.d. according to $N\left(a\sigma, \sigma^2\right)$, where a is a given positive number. Then it was seen in Example 7.1.4(iii) that the likelihood equation has a unique solution $\hat{\sigma}$ given by (7.1.26), and in (7.1.32) that $I(\sigma) = \left(a^2 + 2\right)/\sigma^2$. It thus follows from Theorem 7.3.1 that $\sqrt{n}\,(\hat{\sigma} - \sigma) \to N(0, 1/I(\sigma))$, provided we can check conditions (C1)–(C5), (C6)'', and (C7). The first five of the conditions cause no difficulty. Let us therefore consider (C6)'', and for this purpose apply Lemma 7.3.1. The conditions of part (i) of the lemma are obviously satisfied. It remains to check (7.3.13), i.e., that

$$\int_a^b \frac{\partial^k}{\partial\sigma^k} f_\sigma(x)dx \to \int_{-\infty}^{\infty} \frac{\partial^k}{\partial\sigma^k} f_\sigma(x)dx \text{ for } k = 1, 2, 3$$

as $a \to -\infty$, $b \to \infty$, uniformly for σ in some neighborhood of σ_0, where $f_\sigma(x)$ is the density of $N(a\sigma, \sigma^2)$. We shall in fact show that for all k

$$(7.3.16) \qquad \int_b^{\infty} \frac{\partial^k}{\partial\sigma^k} f_\sigma(x) \to 0 \text{ as } b \to \infty$$

uniformly for σ in any sufficiently small interval $0 < \sigma_1 < \sigma < \sigma_2 < \infty$, and the corresponding result holds for the lower limit of the integral.

To prove (7.3.16) note that $\dfrac{\partial^k}{\partial\sigma^k} f_\sigma(x)$ is of the form $P_\sigma(x)f_\sigma(x)$, where $P_\sigma(x)$ is a polynomial in x of degree $2k$ with coefficients depending on σ (Problem 3.1). It is therefore enough to show that

$$(7.3.17) \qquad \int_b^{\infty} x^k e^{-\frac{1}{2\sigma^2}(x-a\sigma)^2} dx \to 0 \text{ uniformly for } \sigma_1 < \sigma < \sigma_2, \text{ as } b \to \infty.$$

Since $(x - a\sigma)^2/\sigma^2$ is a minimum at $\sigma = x/a$ and increases as σ tends away from x/a in either direction, $e^{-\frac{1}{2\sigma^2}(x-a\sigma)^2}$ is an increasing function of σ in the interval (σ_1, σ_2), provided $\sigma_2 < x/a$. Therefore if $b > \sigma_2$, we have

$$e^{-\frac{1}{2\sigma^2}(x-a\sigma)^2} < e^{-\frac{1}{2\sigma_2^2}(x-a\sigma_2)^2} \text{ for all } x < b \text{ and } \sigma_1 < \sigma < \sigma_2,$$

and hence

$$\int_b^{\infty} x^k e^{-\frac{1}{2\sigma^2}(x-a\sigma)^2} dx < \int_b^{\infty} x^k e^{-\frac{1}{2\sigma_2^2}(x-a\sigma_2)^2} dx.$$

Since normal distributions have moments of all orders, it follows that the right side (which no longer depends on σ) tends to zero as $b \to \infty$, and this completes the proof of (7.3.16).

Condition (C7) is easy to check in the present case and is left as a problem.   □

In the regular case, i.e., when conditions (C1)–(C5), (C6)″, and (C7) hold, Theorem 7.3.1 describes the asymptotic behavior of *any* consistent sequence of roots of the likelihood equation. The following result, due to Huzurbazar (1948), shows that there is essentially only one consistent sequence of roots.

Theorem 7.3.2 *Assume that the conditions of Theorem 7.3.1 hold. Then:*

(i) *If $\hat{\theta}_n$ is a consistent sequence of roots of the likelihood equation, the probability tends to 1 as $n \to \infty$ that $\hat{\theta}_n$ is a local maximum of the log likelihood $l_n(\theta)$.*

(ii) *If $\hat{\theta}_{1n}$ and $\hat{\theta}_{2n}$ are two consistent sequences of roots of the likelihood equation, then*

(7.3.18) $$P\left(\hat{\theta}_{1n} = \hat{\theta}_{2n}\right) \to 1 \text{ as } n \to \infty.$$

Proof. (i) We have assumed in (7.3.4) that $I(\theta_0) > 0$ and it follows from (7.3.8) that $\frac{1}{n} l_n''\left(\hat{\theta}_n\right) \xrightarrow{P} -I(\theta_0)$ and hence that

$$P\left[\frac{1}{n} l_n''\left(\hat{\theta}_n\right) < -\frac{1}{2} I(\theta_0)\right] \to 1.$$

However, when at some point the first derivative of a twice-differentiable function is zero and the second derivative is negative, the function has a local maximum at that point. Thus

$$P\left[l_n(\theta) \text{ has a local max. at } \hat{\theta}_n\right]$$
$$\geq P\left[\frac{1}{n} l_n''\left(\hat{\theta}_n\right) < 0\right] \geq P\left[\frac{1}{n} l_n''\left(\hat{\theta}_n\right) < -\frac{1}{2} I(\theta_0)\right] \to 1.$$

(ii) Suppose the probability that $\hat{\theta}_{1n} \neq \hat{\theta}_{2n}$ does not tend to zero as $n \to \infty$. Then, by (i) and Lemma 2.1.2, the probability also does not tend to zero that

(7.3.19) $\hat{\theta}_{1n} \neq \hat{\theta}_{2n}$ and both $\hat{\theta}_{1n}$ and $\hat{\theta}_{2n}$ are local maxima.

When (7.3.19) holds, there exists between the local maxima $\hat{\theta}_{1n}$ and $\hat{\theta}_{2n}$ a local minimum, say θ_n^*, which is then also a root of the likelihood equation. If, in addition, for all sample points with $\hat{\theta}_{1n} = \hat{\theta}_{2n}$ we define θ_n^* to be this common value, it is seen that θ_n^* is a consistent root of the likelihood equation. However, the probability that θ_n^* is a local minimum does not tend to zero, and this contradicts part (i).   ■

Unfortunately, this essential uniqueness does not help us when there are multiple roots since we do not know which of them provide a consistent sequence. Part (i) of Theorem 7.3.2 suggests that we should choose a local maximum, but if there are several of these, Example 7.1.2 (in Ferguson's smoothed version) shows that the largest may not be the right choice unless additional restrictions are imposed on the distributions. One way out of this impasse is available when we know some consistent estimator $\tilde{\theta}_n$ of θ. Then the root of the likelihood equation closest to $\tilde{\theta}_n$ (which can be shown to exist) will also be consistent, and hence constitutes a solution to our problem. A drawback of this approach is that it requires determining all the roots, which may not be an easy task.

The following alternative approach avoids these difficulties. It leads to estimators which are no longer exact roots of the likelihood equation but which have the same asymptotic behavior as $\hat{\theta}_n$.

If, as is frequently the case, the likelihood equation $l'_n(\theta_0) = 0$ has to be solved iteratively, the first step of the Newton-Raphson method for doing so replaces $l'_n(\theta)$ by the linear terms of its Taylor expansion about a starting value $\tilde{\theta}_n$, and therefore replaces the likelihood equation with the equation

$$(7.3.20) \qquad l'_n\left(\tilde{\theta}_n\right) + \left(\theta - \tilde{\theta}_n\right) l''_n\left(\tilde{\theta}_n\right) = 0.$$

This suggests the solution of the equation (7.3.20) for θ,

$$(7.3.21) \qquad \delta_n = \tilde{\theta}_n - \frac{l'_n\left(\tilde{\theta}_n\right)}{l''_n\left(\tilde{\theta}_n\right)},$$

as a first approximation to the solution of the likelihood equation. The procedure is then iterated by replacing $\tilde{\theta}_n$ by δ_n and so on. (For more details on this method, see, for example, Reddien (1985) and Barnett (1966).) To implement the procedure, one must specify the starting point and the stopping rule.

We shall here consider only the first step, that is, we propose (7.3.21) as our estimator of θ. As a starting point, we can take any estimator $\tilde{\theta}_n$ which is not only consistent but has the stronger property of being $\sqrt{n}$-consistent, i.e., which satisfies

$$(7.3.22) \qquad \sqrt{n}\left(\tilde{\theta}_n - \theta\right) \text{ is bounded in probability,}$$

so that $\tilde{\theta}_n$ tends to θ at least at rate $1/\sqrt{n}$. This will be satisfied in particular by any estimator $\tilde{\theta}_n$ for which

$$(7.3.23)$$
$$\sqrt{n}\left(\tilde{\theta}_n - \theta\right) \text{ tends in law to a non} - \text{degenerate limit distribution.}$$

Such estimators, since they are not required to be efficient, are often easy to find.

Theorem 7.3.3 *Under the assumptions of Theorem 7.3.1, if $\tilde{\theta}_n$ is a $\sqrt{n}$-consistent estimator of θ, the estimator (7.3.21) satisfies*

(7.3.24) $$\sqrt{n}\,(\delta_n - \theta) \xrightarrow{L} N(0, 1/I(\theta)).$$

Proof. As in (7.3.5), expand $l'_n\left(\tilde{\theta}_n\right)$ about $l'_n(\theta_0)$ to obtain

(7.3.25) $$l'_n\left(\tilde{\theta}_n\right) = l'_n(\theta_0) + \left(\tilde{\theta}_n - \theta_0\right) l''_n(\theta_0) + \frac{1}{2}\left(\tilde{\theta}_n - \theta_0\right)^2 l'''_n(\theta_n^*),$$

where θ_n^* lies between θ_0 and $\tilde{\theta}_n$. It follows from (7.3.21) that

$$\delta_n = \tilde{\theta}_n - \frac{1}{l''_n\left(\tilde{\theta}_n\right)}\left[l'_n(\theta_0) + \left(\tilde{\theta}_n - \theta_0\right) l''_n(\theta_0) + \frac{1}{2}\left(\tilde{\theta}_n - \theta_0\right)^2 l'''_n(\theta_n^*)\right]$$

and hence that

(7.3.26)
$$\sqrt{n}\,(\delta_n - \theta_0) =$$

$$-\frac{\sqrt{n}\,l'_n(\theta_0)}{l''_n\left(\tilde{\theta}_n\right)} + \sqrt{n}\left(\tilde{\theta}_n - \theta_0\right)\left[1 - \frac{l''_n(\theta_0)}{l''_n\left(\tilde{\theta}_n\right)} - \frac{1}{2}\left(\tilde{\theta}_n - \theta_0\right)\frac{l'''_n(\theta_n^*)}{l''_n\left(\tilde{\theta}_n\right)}\right].$$

To prove (7.3.24), we shall show that

(7.3.27) $$\frac{\sqrt{n}\,l'_n(\theta_0)}{l''_n\left(\tilde{\theta}_n\right)} \xrightarrow{L} N\left(0, 1/I(\theta_0)\right)$$

and

(7.3.28) $$1 - \frac{l''_n(\theta_0)}{l''_n\left(\tilde{\theta}_n\right)} - \frac{1}{2}\left(\tilde{\theta}_n - \theta_0\right)\frac{l'''_n(\theta_n^*)}{l''_n\left(\tilde{\theta}_n\right)} \xrightarrow{P} 0,$$

which together establish the derived result.

The left side of (7.3.27) equals

$$\frac{l'_n(\theta_0)/\sqrt{n}}{l''_n(\theta_0)/n} \cdot \frac{l''_n(\theta_0)/n}{l''_n\left(\tilde{\theta}_n\right)/n},$$

of which the first factor tends in law to $N\left(0, 1/I(\theta_0)\right)$ by (7.3.7) and (7.3.8). The limit result (7.3.27) will therefore follow if we can show that

(7.3.29) $$\frac{l''_n\left(\tilde{\theta}_n\right)/n}{l''_n(\theta_0)/n} \xrightarrow{P} 1.$$

For this purpose, we use the expansion

$$(7.3.30) \qquad l_n'' \left(\tilde{\theta}_n \right) = l_n'' \left(\theta_0 \right) + \left(\tilde{\theta}_n - \theta_0 \right) l_n''' \left(\theta_n^* \right),$$

where θ_n^* is between θ_0 and $\tilde{\theta}_n$. After division by $l_n'' \left(\theta_0 \right)$, the second term on the right side tends to 0 in probability by (C7) and (7.3.8)—the same argument also proves (7.3.28). ∎

Note: The alternative estimator (7.3.21) can be useful not only in the case of multiple roots but also when the likelihood equation has only a single root but cannot be solved explicitly. An iterative method of solution may then be required and such methods do not always converge. (For a discussion of various iterative methods for solving likelihood equations, see, for example, Stuart and Ord (1991, Section 18.21). Theorem 7.3.3 provides asymptotically equivalent estimators which avoid this difficulty.

On the basis of Theorem 7.3.3, it is often easy to construct estimators δ_n satisfying (7.3.24). This only requires checking the conditions of the theorem and knowing or constructing a $\sqrt{n}$-consistent estimator.

Example 7.3.3 Logistic. Let $X_1, \ldots, X_n$ be i.i.d. according to the logistic distribution with cdf and density

$$F_\theta(x) = \frac{1}{1 + e^{-(x-\theta)}} \quad \text{and} \quad f_\theta(x) = \frac{e^{-(x-\theta)}}{\left[1 + e^{-(x-\theta)} \right]^2}, \quad -\infty < x < \infty,$$

and consider the problem of estimating θ. The loglikelihood is

$$l_n(\theta) = \sum \log f_\theta(x_i) = n\theta - \sum x_i - 2 \sum \log \left[1 + e^{\theta - x_i} \right],$$

and the likelihood equation therefore becomes

$$(7.3.31) \qquad \sum_{i=1}^{n} \frac{e^{\theta - x_i}}{1 + e^{\theta - x_i}} = \frac{n}{2}.$$

Since for any sample point $(x_1, \ldots, x_n)$ the left side is a strictly increasing function of θ which goes from 0 at $\theta = -\infty$ to n at $\theta = \infty$, this equation has a unique solution $\hat{\theta}_n$. The conditions of Theorem 7.3.1 and Corollary 7.1.1 are satisfied (Problem 3.2) and by these results $\hat{\theta}_n$ therefore satisfies (7.3.3). A straightforward calculation shows that (Problem 2.4)

$$(7.3.32) \qquad I(\theta) = 1/3$$

so that

$$(7.3.33) \qquad \sqrt{n} \left(\hat{\theta}_n - \theta \right) \xrightarrow{L} N(0,3).$$

To determine $\hat{\theta}_n$, we can solve (7.3.31) numerically. Alternatively, we can obtain explicit estimators δ_n with the same limit behavior as $\hat{\theta}_n$ on the basis of Theorem 7.3.3. For this purpose, we require a $\sqrt{n}$-consistent estimator $\tilde{\theta}_n$ of θ. Since the X_i are symmetrically distributed about θ, we have $E(X_i) = \theta$, and by the CLT, $\tilde{\theta}_n = \bar{X}_n$ satisfies (7.3.23) and thus is $\sqrt{n}$-consistent. The estimator

$$(7.3.34) \qquad \delta_n = \bar{X}_n - \frac{l'_n(\bar{X}_n)}{l''_n(\bar{X}_n)}$$

therefore has the same asymptotic distribution as $\hat{\theta}_n$. By differentiating $l_n(\theta)$, we find that the needed derivatives l'_n and l''_n are given by

$$(7.3.35) \qquad l'_n(\theta) = n - 2\sum \frac{e^{\theta - x_i}}{1 + e^{\theta - x_i}}$$

and

$$(7.3.36) \qquad l''_n(\theta) = -2\sum \frac{e^{x_i - \theta}}{[1 + e^{x_i - \theta}]^2}.$$

Instead of taking $\tilde{\theta}_n$ to be $\bar{X}_n$, we can, of course, take as the starting point any other $\sqrt{n}$-consistent estimator of θ; for example, any of the estimators of the center of a symmetric distribution considered in Example 4.3.4 such as the median, a trimmed mean, the Hodges-Lehmann estimator, etc.   □

In the present case, the mean, median and Hodges-Lehmann estimator will typically be fairly close together, and one may then expect the same to be true for the resulting estimators δ_n. In fact, the following result shows that the estimators $\hat{\theta}_n$ and δ_n, and hence any two δ_n's given by (7.3.21), are quite generally likely to be close together.

Theorem 7.3.4 Let $\hat{\theta}_n$ be a consistent root of the likelihood equation, and let δ_n be given by (7.3.21), where $\tilde{\theta}_n$ is any $\sqrt{n}$-consistent estimator of θ. Then under the assumptions of Theorem 7.3.1, we have

$$(7.3.37) \qquad \sqrt{n}\left(\hat{\theta}_n - \theta_0\right) = -\frac{\sqrt{n}\,l'_n(\theta_0)}{l''_n(\theta_0)} + R_n$$

and

$$(7.3.38) \qquad \sqrt{n}\left(\delta_n - \theta_0\right) = \frac{-\sqrt{n}\,l'_n(\theta_0)}{l''_n(\theta_0)} + R'_n,$$

where R_n and R'_n tend to 0 in probability as $n \to \infty$, and hence

$$(7.3.39) \qquad \sqrt{n}\left(\delta_n - \hat{\theta}_n\right) \xrightarrow{P} 0.$$

Proof. To see (7.3.37), rewrite (7.3.6) as

$$\sqrt{n}\left(\hat{\theta}_n - \theta_0\right) + \frac{l_n'(\theta_0)/\sqrt{n}}{l_n''(\theta_0)/n}$$

$$= \frac{l_n'(\theta_0)/\sqrt{n}}{-l_n''(\theta_0)/n}\left[\frac{1}{1 + \frac{n}{l_n''(\theta_0)}\frac{1}{2n}\left(\hat{\theta} - \theta_0\right)l_n'''(\theta_n^*)} - 1\right].$$

It follows from (7.3.7)–(7.3.9) that the first factor on the right side is bounded in probability and the second factor tends in probability to 0, which verifies (7.3.37). Analogously, (7.3.38) follows from (7.3.27) and (7.3.28) (Problem 3.6). ∎

It can in fact be shown that in a suitably regular situation, if $\delta_n^{(1)}$ and $\delta_n^{(2)}$ are any estimators satisfying

$$(7.3.40) \qquad \sqrt{n}\left(\delta_n^{(i)} - \theta\right) \xrightarrow{L} N(0, 1/I(\theta)),$$

then

$$(7.3.41) \qquad \sqrt{n}\left[\delta_n^{(2)} - \delta_n^{(1)}\right] \xrightarrow{P} 0.$$

Example 7.3.4 Mixtures. Let $X_1, \ldots, X_n$ be i.i.d. according to a distribution

$$F_\theta = \theta G + (1 - \theta) H, \ 0 < \theta < 1,$$

where G and H are two specified distributions with densities g and h, respectively. Conditions (C1)–(C7) will not hold in general but often can be verified for particular choices of g and h and will be assumed in the following. Since

$$(7.3.42) \quad l_n'(\theta) = \sum \frac{g(x_i) - h(x_i)}{\theta g(x_i) + (1 - \theta) h(x_i)} = \sum_{i=1}^{n} \frac{1}{\theta + \dfrac{h(x_i)}{g(x_i) - h(x_i)}}$$

is a strictly decreasing function of θ, the likelihood equation has at most one root $\hat{\theta}_n$ and this satisfies (7.3.3) with (Problem 3.7)

$$(7.3.43) \qquad I(\theta) = \frac{1}{\theta(1 - \theta)}\left[1 - \int_{-\infty}^{\infty} \frac{g(x)h(x)}{\theta g(x) + (1 - \theta)h(x)}dx\right].$$

Two extreme situations clarify the meaning of (7.3.43). If $G = H$, the parameter θ becomes unidentifiable. To this corresponds the fact that the integral on the right side of (7.3.43) is $= 1$, so that $I(\theta) = 0$, suggesting that the observations tell us nothing about θ. At the other extreme is the case in which G and H assign probability 1 to non-overlapping sets. Then $g(x)h(x) = 0$ for all x, and $I(\theta)$ attains its maximum possible value $1/\theta(1-\theta)$. We can then tell from which of the distributions G and H each X_i comes and beyond this the observations supply no additional information about θ. The situation thus reduces to n binomial trials, each with possible outcomes G and H having probabilities θ and $1 - \theta$, respectively, and $1/\theta(1 - \theta)$ is just the information provided by such a trial (Problem 2.1).

When the two distributions have some overlap but are vastly disparate, the observations may provide nearly as much information even when their densities are positive on the same set. Suppose, for example, that G and H are the normal distributions $N(-\xi, 1)$ and $N(\xi, 1)$, respectively, with ξ large. Then the observations are likely to take on large negative or positive values and thereby give strong indications as to from which distribution each comes. Correspondingly, $I(\theta)$ given by (7.3.43) tends to its maximum value $1/\theta(1-\theta)$ as $\xi \to \infty$ (Problem 3.8). (For approximations to $I(\theta)$, see Hill (1963).)

As in Example 7.3.3, the likelihood equation can be solved only numerically. Explicit estimators δ_n satisfying (7.3.24) can be found using (7.3.21). For this purpose, we require a $\sqrt{n}$-consistent estimator of θ. If G and H have finite expectations ξ and η, respectively, we have

$$E(X_i) = \theta\xi + (1 - \theta)\eta.$$

Then the sample mean $\bar{X}_n$ tends in probability to $\theta\xi + (1 - \theta)\eta$ and hence with high probability we have

$$\bar{X}_n \doteq \theta\xi + (1 - \theta)\eta \text{ and thus } \theta \doteq \left(\bar{X}_n - \eta\right) / \left(\xi - \eta\right),$$

provided $\eta \neq \xi$. This suggests the estimator

$$(7.3.44) \qquad \tilde{\theta}_n = \frac{\bar{X}_n - \eta}{\xi - \eta},$$

which is seen to be $\sqrt{n}$-consistent if not only the expectation but also the variance of the X's is finite. (Problem 3.9(i).)

If $E(X_i)$ does not exist (for example, if G and H are Cauchy distributions), let a be a point for which $G(a) \neq H(a)$. Then

$$(7.3.45) \qquad p = P(X_i \leq a) = \theta G(a) + (1 - \theta)H(a).$$

Since the frequency

$$(7.3.46) \qquad \hat{p}_n = (\#X_i \leq a)/n$$

tends in probability to p, we will have with high probability

$$\hat{p}_n \text{ close to } \theta G(a) + (1 - \theta)H(a)$$

and hence

$$\theta \text{ close to } \frac{\hat{p}_n - H(a)}{G(a) - H(a)}.$$

This suggests the estimator

(7.3.47)
$$\tilde{\theta}_n = \frac{\hat{p}_n - H(a)}{G(a) - H(a)},$$

which is $\sqrt{n}$-consistent (Problem 3.9(ii)). The estimator (7.3.47) also can be used when $\eta - \xi$. $\qquad\square$

In Theorems 7.3.1–7.3.3, it was assumed that both x and θ are real-valued. The case of vector-valued parameters involves additional complications which will be taken up in Section 7.5. On the other hand, the generalization to vector-valued x is trivial. An examination of the proofs of Theorems 7.3.1–7.3.3 shows that nothing requires the X's in the i.i.d. sequence $X_1, X_2, \ldots$ to be real-valued, except that integrals such as $\int f_\theta(x)dx$ will become $\int\int f_\theta(x, y)dxdy$ or $\int\int\int f_\theta(x, y, z)dxdydz$, etc. Let us illustrate the bivariate case with an example.

Example 7.3.5 Correlation coefficient. Let $(X_1, Y_1), \ldots, (X_n, Y_n)$ be i.i.d. according to a bivariate normal distribution with means $E(X_i) = E(Y_i) = 0$, variances $\text{Var}(X_i) = \text{Var}(Y_i) = 1$, and unknown correlation coefficient ρ. We have

(7.3.48)
$$f_\rho(x, y) = \frac{1}{2\pi\sqrt{1 - \rho^2}} e^{-\frac{1}{2(1-\rho^2)}(x^2 - 2\rho xy + y^2)}, \quad -1 < \rho < 1,$$

and the log likelihood is therefore

(7.3.49)
$$l_n(\rho) = -\frac{1}{2(1 - \rho^2)} \sum (x_i^2 - 2\rho x_i y_i + y_i^2) - n\log\left(2\pi\sqrt{1 - \rho^2}\right).$$

The likelihood equation, after some simplification, becomes (Problem 3.10(i))

(7.3.50)
$$\rho\left(1 - \rho^2\right) + \left(1 + \rho^2\right)\frac{\sum x_i y_i}{n} - \rho\frac{\sum x_i^2 + \sum y_i^2}{n} = 0.$$

This cubic equation has at least one solution in the interval $-1 < \rho < 1$ corresponding to a local maximum of the likelihood and at most two such solutions (Problem 3.10(ii).) (A more detailed analysis of the roots is given in Stuart and Ord (1991, Section 8.9), where it is shown that the probability of there being more than one root tends to 0 as $n \to \infty$.) When there are two maxima, the theory does not tell us which is closer to the true value. We may then instead utilize Theorem 7.3.3. In the present case, the estimator

$$\tilde{\theta}_n = \sum X_i Y_i / n$$

is a $\sqrt{n}$-consistent estimator of $E(X_i Y_i) = \rho$, and thus can be used in (7.3.21).

To determine the asymptotic distribution of $\sqrt{n}(\delta_n - \rho)$, it remains to evaluate $I(\rho)$. From (7.3.49) with $n = 1$, it follows that

$$\frac{\partial}{\partial \rho} \log f_\rho(x, y) = \frac{-\rho}{(1 - \rho^2)^2} \left(x^2 - 2\rho xy + y^2\right) + \frac{xy}{1 - \rho^2} + \frac{\rho}{1 - \rho^2},$$

and hence that

$$(7.3.51) \quad I(\rho) = \frac{1}{(1 - \rho^2)^4} E\left[\rho\left(X^2 + Y^2\right) - \left(1 + \rho^2\right) XY\right]^2 - \frac{\rho^2}{(1 - \rho^2)^2}.$$

Equations (5.4.33) and (5.4.39) show that

$$(7.3.52) \quad \begin{array}{l} E\left(X^2 Y^2\right) = 1 + 2\rho^2, \ E\left(XY^2\right) = E\left(X^2 Y\right) = 0, \\ E\left(XY^3\right) = E\left(X^3 Y\right) = 3\rho, \ E\left(X^4\right) = E\left(Y^4\right) = 3. \end{array}$$

Therefore, after some simplification (Problem 3.10(iii)),

$$(7.3.53) \quad I(\rho) = \left(1 + \rho^2\right) / \left(1 - \rho^2\right)^2.$$

□

In the examples considered so far, we found that in Examples 7.3.1–7.3.4 the likelihood equation has at most one root; in Example 7.3.5, there are three roots but the probability of more than one (real) root tends to 0 as $n \to \infty$. The following example shows that, in general, the situation is not so simple.

Example 7.3.6 Cauchy. Let $X_1, \ldots, X_n$ be i.i.d. according to a Cauchy distribution with density

$$(7.3.54) \quad f_\theta(x) = \frac{1}{\pi} \frac{1}{1 + (x - \theta)^2}.$$

Then the likelihood equation becomes

$$(7.3.55) \quad \frac{1}{2} l'_n(\theta) = \sum_{i=1}^n \frac{\theta - x_i}{1 + (x_i - \theta)^2} = 0.$$

By putting the middle expression on a common denominator, (7.3.55) is seen to be equivalent to an equation of the form $P_n(\theta) = 0$, where P_n is a polynomial of degree $2n - 1$ with coefficients depending on the x's. For any $(x_1, \dots, x_n)$, the likelihood equation can therefore have at most $2n-1$ roots.

As $\theta \to \infty$, each term $(\theta - x_i) \Big/ \left[1 + (x_i - \theta)^2\right]$ tends to zero through positive values, and the same is therefore true of the sum. Analogously, $l'_n(\theta)$ tends to zero through negative values as $\theta \to -\infty$. This shows that (7.3.55) always has at least one root and that the number R_n of roots is odd, say

$$R_n = 2K_n - 1,\ 1 \leq K_n \leq n.$$

The limiting behavior of the number of roots is determined by the following remarkable result due to Reeds (1985) which we shall not prove: The probability $P(K_n = k)$ that the likelihood equation (7.3.55) has exactly $2k - 1$ roots satisfies

$$\text{(7.3.56)} \qquad P(K_n = k) \to \frac{1}{k!}\left(\frac{1}{\pi}\right) e^{-1/\pi} \text{ as } n \to \infty,$$

that is, the random variable K_n has a limiting Poisson distribution $P(\lambda)$ with $\lambda = 1/\pi$.

Table 7.3.1 compares some Monte Carlo results for sample sizes $n = 5, 7, 11, 15$ with the limit distribution (7.3.56). (The Monte Carlo sample sizes are 3,000 for $n = 5, 7$ and 1,000 for $n = 11, 15$.) Since K_n can take

TABLE 7.3.1. Monte Carlo and asymptotic distribution of K_n

	K				
n	1	2	3	4	1 + 2 + 3 + 4
5	.652	.768	.069	.011	1.0000
7	.670	.261	.058	.009	.998
11	.706	.245	.036	.012	.999
15	.696	.262	.039	.002	.999
∞	.727	.232	.037	.004	1.000

Source: Barnett (1966) and Reeds (1985).

on values as high as n, it is interesting to note that with probability close to 1, its actual value is ≤ 4 (and hence the number R_n of the roots of the likelihood equation ≤ 7) for all n. (On the other hand, it can be shown that $P(K_n = n)$ is positive for every n.)

In the present case, we cannot take the sample mean $\bar{X}_n$ for $\tilde{\theta}_n$ in (7.3.21) since it is not consistent, but by Example 2.4.9, the sample median is $\sqrt{n}$-consistent and therefore provides a possible choice for $\tilde{\theta}_n$. The asymptotic distribution of the resulting estimator δ_n or of the consistent root of the likelihood equation is then given by (7.3.24) with $I(\theta) = 1/2$ (Problem 2.4).

TABLE 7.4.1. Efficiency of the trimmed mean for a Cauchy distribution

α	.5	.45	.4	.38	.35	.3	.25	.2	.15	.1	.5	0
Effic. of $\bar{X}_\alpha$	.811	.855	.876	.878	.873	.844	.786	.696	.575	.419	.228	0

Adapted from Rothenberg, Fisher and Tilenius (1964).

Note: Barnett (1966 p. 154) finds that the local maximum closest to the median is not always the absolute maximum although the proportion of such cases is very small.   □

Summary

1. Regularity conditions are given under which any consistent sequence of roots of the likelihood equation is aymptotically normal, and it is shown how to check these conditions in specific cases.

2. The first Newton-Raphson approximation for solving the likelihood equation provides an estimator which is asymptotically equivalent to the consistent solution. This estimator circumvents the difficulty of selecting the correct root of the likelihood equation. Implementation of the estimator only requires a $\sqrt{n}$-consistent starting point. The procedure is illustrated on a number of examples.

7.4   Efficiency

In small sample theory, an estimator δ is often considered optimal for estimating θ if it minimizes $E_\theta (\delta - \theta)^2$ for all θ. This definition runs into the difficulty that for any given θ_0 there exists an estimator δ_0 for which $E_{\theta_0} (\delta_0 - \theta_0)^2 = 0$, namely the constant estimator $\delta_0 \equiv \theta_0$ which estimates θ to be θ_0 regardless of the observations. Thus an estimator δ minimizing $E_\theta (\delta - \theta)^2$ simultaneously for all θ does not exist. Two principal approaches to overcome this difficulty are:

(a) to restrict the class of competing estimators by some suitable desirable condition; for example, unbiasedness;

(b) to replace the local optimality criterion which is required to hold for every θ by a global criterion such as minimizing the maximum expected squared error (minimax) or some average of the expected squared error (Bayes).

The corresponding problem arises when trying to define efficient large-sample estimators, and we can approach it in the same way. We shall here follow route (a) and begin by restricting attention to consistent estimators. This at least eliminates constant estimators but turns out not to includes most of the estimators encountered in this and previous chapters. The most desirable estimators in this class are those for which $v(\theta)$ is as small as possible.

For a long time it was believed that in the i.i.d. case, under suitable regularity assumptions on the distributions of the X's, (7.4.1) implies[†]

$$(7.4.2) \qquad\qquad v(\theta) \geq 1/I(\theta).$$

Since under the assumptions of Theorem 7.3.1 the estimator $\delta_n = \hat{\theta}_n$ satisfies (7.4.1) with $v(\theta) = 1/I(\theta)$, $\hat{\theta}_n$ would then be efficient in the sense of having uniformly minimum asymptotic variance among all estimators satisfying (7.4.1). This conjecture was disproved by Hodges by means of the following construction (see Le Cam (1953)).

Example 7.4.1 The Hodges counterexample. Let δ_n be a sequence of estimators satisfying (7.4.1) with $0 < v(\theta) < \infty$ for all θ. We shall now construct an estimator δ'_n which satisfies (7.4.1) with $v(\theta)$ replaced by

$$(7.4.3) \qquad v'(\theta) = \begin{cases} a^2 v(\theta) & \text{when } \theta = \theta_0 \\ v(\theta) & \text{when } \theta \neq \theta_0, \end{cases}$$

where θ_0 is any given value of θ and a is any constant. For $a < 1$, we then have $v'(\theta) \leq v(\theta)$ for all θ and $v'(\theta_0) < v(\theta_0)$ so that δ'_n is an improvement over δ_n.

Without loss of generality, let $\theta_0 = 0$ and let

$$(7.4.4) \qquad\qquad \delta'_n = \begin{cases} a\delta_n & \text{if } |\delta_n| \leq 1/\sqrt[4]{n} \\ \delta_n & \text{otherwise.} \end{cases}$$

Then

$$P_\theta\left[\sqrt{n}\,|\delta'_n - \theta| \leq c\right] = P_\theta\left[\sqrt{n}\,|\delta_n| \leq c\sqrt[4]{n} \text{ and } \sqrt{n}\,|a\delta_n - \theta| \leq c\right]$$
$$+ P_\theta\left[\sqrt{n}\,|\delta_n| > c\sqrt[4]{n} \text{ and } \sqrt{n}\,|\delta_n - \theta| \leq c\right].$$

If $\theta = 0$, the second term on the right side tends to zero and the first term has the same limit as

$$P_0\left[\sqrt{n}\,|a\delta_n| \leq c\right],$$

which is $\Phi\left(\dfrac{c}{av(\theta)}\right) - \Phi\left(\dfrac{-c}{av(\theta)}\right)$, i.e.,

$$(7.4.5) \qquad\qquad \sqrt{n}\delta'_n \xrightarrow{L} N\left(0, a^2 v^2(\theta)\right).$$

[†]The information inequality (7.4.2) holds for the exact variance $v(\theta)$ of any unbiased estimator of θ. However, this does not imply that (7.4.2) must hold in the present situation since an estimator satisfying (7.4.1) need not be unbiased and since by Section 4.2 the asymptotic variance may be smaller than the actual variance.

Analogously, if $\theta \neq 0$, the first term tends to zero while the second term has the same limit as $P\left(\sqrt{n}\,|\delta_n - \theta| \leq c\right)$ (Problem 4.1). Thus $\sqrt{n}\left(\delta_n' - \theta\right) \to N\left(0, v'(\theta)\right)$ with $v'(\theta)$ given by (7.4.3).   □

An easy extension of the argument shows that we can improve $v(\theta)$ (and by putting $a = 0$ even reduce it to 0) not only at any given point θ_0 but at any given finite number of points (Problem 4.2). However it has been shown (Le Cam (1953), Bahadur (1964)) that in the regular cases with which we are dealing here, improvement over $v(\theta) = 1/I(\theta)$ is possible only on relatively "small" sets (technically, sets of Lebesgue measure 0) and in particular not for all values θ in any interval, no matter how small.

Example 7.4.1 shows that even within the class of estimators satisfying (7.4.1), no uniformly best estimator exists. However, the desired result (7.4.2) does follow if we impose one further restriction on the estimators, namely require $v(\theta)$ to be continuous (as it is for all the usual estimators).[‡] This result is a consequence of the fact that if $I(\theta)$ and $v(\theta)$ are both continuous and if $v(\theta) < 1/I(\theta)$ for any θ_0, then strict inequality also holds for some interval containing θ_0, which would violate Le Cam's theorem.

Within the class of estimators δ_n satisfying (7.4.1) with continuous $v(\theta)$, estimators satisfying (7.1.27) are therefore *efficient* in the sense that their asymptotic variance cannot be improved anywhere. An important difference from the small-sample situation of unbiased estimators with uniformly minimum variance is that, in the present situation, efficient estimators are not unique. The consistent root of the likelihood equation is one such estimator; the class of one-step estimators (7.3.21) provides another example; a further large and important class of efficient estimators will be considered later in this section and in Section 7.5. Efficient estimators provide a standard with which other estimators can be compared; the ARE e_{δ_n', δ_n} of an estimator δ_n' with respect to an estimator δ_n satisfying (7.3.24) will be called the (absolute) efficiency of δ_n'.

Example 7.4.2 Normal. In Example 7.1.4(iii) and its continuations, we considered the estimation of σ based on a sample $X_1, \ldots, X_n$ from $N\left(a\sigma, \sigma^2\right)$. A consistent root $\hat{\sigma}_n$ of the likelihood equation given by (7.1.26) provided an efficient estimator of σ and we found that $I(\sigma) = \left(a^2 + 2\right)/\sigma^2$. Other obvious estimators of σ are

$$(7.4.6) \qquad \delta_{1n} = \bar{X}/a \text{ with } \sqrt{n}\left(\delta_{1n} - \sigma\right) \xrightarrow{L} N\left(0, \sigma^2/a^2\right)$$

[‡]It can be shown that estimators which are superefficient in the sense of satisfying (7.4.1) but not (7.4.2) for all θ necessarily have some undesirable properties (see, for example, Bahadur (1983) and Lehmann (1983)).

and

$$(7.4.7) \qquad \delta_{2n} = \sqrt{\frac{\sum (X_i - \bar{X})^2}{n-1}} \quad \text{with } \sqrt{n}\,[\delta_{2n} - \sigma] \xrightarrow{L} N\left(0, \sigma^2/2\right).$$

We notice that the asymptotic variances of (7.4.6) and (7.4.7) are both greater than $1/I(\sigma) = \sigma^2/(a^2+2)$ for all a. They must of course be at least as large since $\hat{\sigma}_n$ is efficient. For large a, δ_{1n} is nearly as efficient as $\hat{\sigma}_n$, and for small a, the corresponding fact is true for δ_{2n}.

The estimators (7.4.6) and (7.4.7) suggest the more general class of estimators

$$(7.4.8) \qquad \delta_{3n} = \alpha \delta_{1n} + (1-\alpha)\,\delta_{2n}.$$

Since δ_{1n} and δ_{2n} are independent, the asymptotic variance of δ_{3n} is

$$(7.4.9) \qquad \left[\alpha^2 \cdot \frac{1}{a^2} + (1-\alpha)^2 \cdot \frac{1}{2}\right]\sigma^2.$$

The value of α minimizing (7.4.9) is (Problem 4.3)

$$(7.4.10) \qquad \alpha = \frac{a^2}{a^2+2}$$

and the corresponding value of (7.4.9) is $\sigma^2/(a^2+2)$. The associated estimator is therefore efficient. $\qquad\qquad\square$

Example 7.4.3 Cauchy. In Example 7.3.6, we discussed efficient estimators of the center of the Cauchy distribution (7.3.54), and suggested the median $\tilde{\theta}_n$ as a starting point for the one-step estimator (7.3.21). One may wonder how much efficiency is lost if the median itself is used as the estimator of θ rather than the adjusted estimator (7.3.21). It follows from (2.4.19) that

$$(7.4.11) \qquad \sqrt{n}\left(\tilde{\theta}_n - \theta\right) \xrightarrow{L} N\left(0, \frac{\pi^2}{4}\right).$$

On the other hand, the efficient estimator δ_n given by (7.3.21) satisfies

$$(7.4.12) \qquad \sqrt{n}\,(\delta_n - \theta) \xrightarrow{L} N(0, 2).$$

Thus the ARE of $\tilde{\theta}_n$ with respect to δ_n, which is also the absolute efficiency of $\tilde{\theta}_n$, is $\frac{8}{\pi^2} \doteq .81$.

We can improve this efficiency if instead of the median, we consider the class of trimmed means defined by (4.3.19), the asymptotic variance of which is given by (4.3.21). In the notation of Theorem 4.3.1, a proportion

We can improve this efficiency if instead of the median, we consider the class of trimmed means defined by (4.3.19), the asymptotic variance of which is given by (4.3.21). In the notation of Theorem 4.3.1, a proportion α of the observations is discarded at each end and the trimmed mean $\bar{X}_\alpha$ is the average of the remaining central portion of the sample. For the Cauchy case, the asymptotic variance of this estimator has been obtained by Rothenberg et al. (1964) and the resulting efficiency is shown in Table 7.4.1. (The value $\alpha = .5$ corresponds to the median.) The maximum efficiency of .878 obtains for $\alpha = .38$, i.e., when 24% of the sample is retained. One can do better by using other linear functions of the order statistics or quantiles. Bloch (1966), for example, finds that a linear function of five quantiles (defined in Section 2.1) corresponding to $p = .13, .40, .50, .60, .87$ with weights $-.052, .3485, .407, .3485, -.052$ has an efficiency of .95.   □

The following result shows that efficiency of an estimator is preserved under smooth transformations.

Theorem 7.4.1 *Suppose that δ_n is an efficient estimator of θ satisfying (7.4.1) with $v(\theta) = 1/I(\theta)$ and that g is a differentiable function of θ. Then $g(\delta_n)$ is an efficient estimator of $g(\theta)$ at all points θ for which $g'(\theta) \neq 0$.*

Proof. By Theorem 2.5.2,

$$\sqrt{n}\left[g(\delta_n) - g(\theta)\right] \xrightarrow{L} N\left(0, \left[g'(\theta)\right]^2 / I(\theta)\right).$$

By (7.2.22), the asymptotic variance $\left[g'(\theta)\right]^2 / I(\theta)$ is the reciprocal of the amount of information $I^*\left[g(\theta)\right]$ that an observation X_i contains about $g(\theta)$. The result therefore follows from the definition of efficiency.   ∎

The estimators (7.3.21) provide a large class of efficient estimators. The remainder of this section will be concerned with another such class.

Example 7.4.4 Binomial. In Examples 1.4.1 and 4.3.2, we considered the Bayes estimation of a probability p when X is binomial $b(p, n)$ and p has a prior beta distribution $B(a, b)$. The Bayes estimator is then

(7.4.13)
$$\delta_n(X) = \frac{a + X}{a + b + n}.$$

Writing

$$\sqrt{n}\left(\delta_n(X) - p\right) = \sqrt{n}\left(\frac{X}{n} - p\right) + \frac{\sqrt{n}}{a + b + n}\left[a - (a + b)\frac{X}{n}\right],$$

we see that $\sqrt{n}\left[\delta_n(X) - p\right]$ has the same limit distribution $N(0, pq)$ as the efficient estimator X/n. This shows that the estimators (7.4.13) are themselves efficient for all a and b.   □

What is found in this special case is, in fact, a quite general phenomenon. Let λ be a prior density for θ and let $\delta_\lambda = \delta_\lambda(X_1, \ldots, X_n)$ be the *Bayes estimator* which minimizes the average squared error

$$(7.4.14) \qquad \int E(\delta - \theta)^2 \lambda(\theta)d\theta.$$

Then δ_λ is equal to the conditional expectation (under λ) of θ given $(x_1, \ldots, x_n)$,

$$(7.4.15) \qquad \delta_\lambda = E_\lambda(\theta|x_1, \ldots, x_n)$$

(see, for example, Berger (1985) or Robert (1994)). For i.i.d. X's, it turns out under fairly general conditions that δ_λ satisfies (7.3.24) and hence is asymptotically efficient.

For this result, conditions are needed not only on $l_n(\theta)$ but also on λ. For suppose that λ assigns probability 0 to some subinterval ω of the parameter space Ω. Then the posterior (i.e., conditional) distribution of θ given $(x_1, \ldots, x_n)$ will also assign probability 0 to ω and we then cannot expect the Bayes estimator to be consistent for values of θ in ω. The following example illustrates this phenomenon. For a statement of sufficient conditions and a proof of the result, see, for example, Lehmann and Casella (1998).

Example 7.4.5 Discrete prior. Let $X_1, \ldots, X_n$ be i.i.d. $N(\theta, 1)$, $-\infty < \theta < \infty$, and consider the Bayes estimator for the prior λ that assigns probabilities p and $q = 1 - p$ to the two points $a < b$. Then the posterior distribution of θ given $x_1, \ldots, x_n$ assigns to the point a the probability (Problem 4.4(i))

$$p_a = P(\theta = a|x_1, \ldots, x_n) = \frac{pe^{-\frac{1}{2}\sum(a-x_i)^2}}{pe^{-\frac{1}{2}\sum(a-x_i)^2} + qe^{-\frac{1}{2}\sum(b-x_i)^2}}$$

and to $\theta = b$ the probability $p_b = 1 - p_a$. The Bayes estimator by (7.4.15) is then

$$(7.4.16) \qquad \delta_\lambda = \frac{ape^{-\frac{1}{2}\sum(a-x_i)^2} + bqe^{-\frac{1}{2}\sum(b-x_i)^2}}{pe^{-\frac{1}{2}\sum(a-x_i)^2} + qe^{-\frac{1}{2}\sum(b-x_i)^2}}.$$

We shall now consider the limit behavior of this estimator for fixed θ as $n \to \infty$.

Write δ_λ as

$$(7.4.17) \qquad \delta_\lambda = \frac{a + b\frac{q}{p}e^{-\frac{1}{2}\sum[(b-x_i)^2-(a-x_i)^2]}}{1 + \frac{q}{p}e^{-\frac{1}{2}\sum[(b-x_i)^2-(a-x_i)^2]}}.$$

Since

$$\frac{1}{2}\sum\left[(b-x_i)^2-(a-x_i)^2\right]=(b-a)\sum\left(\frac{b+a}{2}-x_i\right)$$
$$=n(b-a)\left(\frac{b+a}{2}-\bar{x}\right)$$

and $\bar{X}\xrightarrow{P}\theta$, it follows that this exponent tends in probability to ∞ for any $\theta<\dfrac{a+b}{2}$ and hence that

(7.4.18a) $\delta_\lambda\xrightarrow{P}a$ for any $\theta<\dfrac{a+b}{2}$,

and, analogously,

(7.4.18b) $\delta_\lambda\xrightarrow{P}b$ for any $\theta>\dfrac{a+b}{2}$.

$\square$

The Bayes estimator is therefore not consistent. (For the behavior of δ_λ when $\theta=(a+b)/2$, see Problem 4.4(ii).)

That, subject to some general conditions, all Bayes estimators are asymptotically efficient leaves the investigator free to select which prior distribution λ to use. If there is prior information concerning the likelihood of different values of θ, this would play an important role in the choice. If no such prior information is available, one may want to model this lack of knowledge by using a *non-informative* prior.

Example 7.4.4 Binomial (continued). If X is binomial $b(p,n)$, the classical choice to model ignorance is the uniform distribution for p on $(0,1)$ which assigns the same probability density to all possible values of p. This approach was used extensively by Laplace and throughout the 19th century. Since the uniform distribution is the special case of the beta distribution $B(a,b)$ with $a=b=1$, it follows from (7.4.13) that the corresponding Bayes estimator is

(7.4.19) $$\delta_n(X)=\frac{X+1}{n+2}.$$

Although the uniform distribution may at first glance seem the natural way to model ignorance, two objections can be raised:

(i) The choice of p as the parameter is somewhat arbitrary; one could equally well parametrize the distribution by $\sqrt{p},\,p^2,\dots$ and a uniform distribution for any of these choices would lead to a non-uniform distribution for the others.

(ii) When it comes to the estimation of p, the situation varies widely with the true value of p: Values close to 0 and 1 are much easier to estimate accurately (in terms of expected square error or asymptotic variance) than values in the middle of the range. In order to take this asymmetry into account, one may wish to determine a function $\eta = g(p)$, all values of which are equally easy to estimate, and then assign a uniform distribution to η. Instead of determining $g(p)$ for this particular case, let us consider the problem more generally.   □

Let $X_1, \ldots, X_n$ be i.i.d. according to a density $f_\theta(x)$ satisfying the conditions of Theorem 7.3.1. An efficient estimator of θ is the consistent root $\hat{\theta}_n$ of the likelihood equation and its asymptotic distribution is given by

$$\sqrt{n} \left(\hat{\theta}_n - \theta \right) \to N(0, 1/I(\theta)).$$

Unless $I(\theta)$ is constant, some values of θ can thus be estimated more accurately than others. Let us therefore seek out a function $g(\theta)$ for which the asymptotic variance of $\sqrt{n} \left[g \left(\hat{\theta}_n \right) - g(\theta) \right]$ is constant. This function is just the variance-stabilizing transformation which was found in (2.5.7) to be given by (in the present notation)

(7.4.20) $$g'(\theta) = c\sqrt{I(\theta)},$$

where c can be any constant, which we will take to be positive. For $\eta = g(\theta)$, we might then consider a uniform distribution as uninformative in the sense of not providing more information about some values than about others.

A qualitative explanation of (7.4.20) can be obtained from the considerations above (7.2.20). When $I(\theta)$ is small so that θ is difficult to estimate accurately, the derivative $g'(\theta)$ given by (7.4.20) is also small so that $g(\theta)$ changes slowly, thereby making it relatively easy to estimate it accurately. The transformation g given by (7.4.20) balances these two effects so that the resulting $g(\theta)$ can be estimated with the same accuracy for all θ.

Let us now determine the prior distribution for θ induced by the uniform distribution for $g(\theta)$. For the moment, let us assume that the range of g is finite, and then without essential loss of generality that it is the interval $(0, 1)$. If $g(\theta)$ is uniformly distributed on $(0, 1)$, we have

$$P[g(\theta) < y] = y \text{ for any } 0 < y < 1$$

and hence

(7.4.21) $$P(\theta < z) = P[g(\theta) < g(z)] = g(z).$$

Here we are using the fact that g is strictly increasing since $g'(\theta) > 0$ for all θ by (7.4.20). It follows from (7.4.21) that the resulting non-informative prior for θ has density $g'(\theta)$ given by (7.4.20), where c must now be chosen so that the integral of (7.4.20) is equal to 1. It was proposed for this purpose by Jeffreys (1939) and is also known as the Jeffreys prior.

Let us now return to the first objection raised above against the uniform prior for p and see how the Jeffreys prior fares under transformations of the parameter. Let $\zeta = h(\theta)$ be an alternative parametrization and, for simplicity, suppose that h is strictly increasing. We saw earlier that the information about $h(\theta)$ supplied by an observation that contains information $I(\theta)$ about θ is $I(\theta)/[h'(\theta)]^2$. The Jeffreys distribution for ζ therefore has density proportional to $\sqrt{I(\theta)}/h'(\theta)$, which assigns to $\theta = h^{-1}(\zeta)$ the density

$$(7.4.22) \qquad \frac{c\sqrt{I(\theta)}}{h'(\theta)} \frac{d\zeta}{d\theta} = c\sqrt{I(\theta)}.$$

This is the Jeffreys prior for θ, which is therefore unchanged under reparametrization. This choice of prior thus takes care of both the objections raised against the uniform prior for p in Example 7.4.4.

The invariance of the prior under transformations of the parameter has an important consequence. Since the Bayes estimator corresponding to this prior is the same for all parametrizations, we can choose as our basic parameter whichever is most convenient.

Example 7.4.4 Binomial (concluded). If $X_i = 1$ or 0 as the ith trial in a sequence of binomial trials is a success or failure, the information contained in an observation about p is (Problem 2.1)

$$I(p) = 1/pq.$$

The Jeffreys prior for p is therefore proportional to

$$(7.4.23) \qquad\qquad 1/\sqrt{pq},$$

and hence is the beta distribution $B(1/2, 1/2)$. The Bayes estimator corresponding to this non-informative prior is by (7.4.13)

$$(7.4.24) \qquad\qquad \delta_n(X) = \frac{X + 1/2}{n + 1}.$$

$\square$

Example 7.4.6 Location parameter. If $X_1, \dots, X_n$ are i.i.d. with density $f(x-\theta)$, $-\infty < \theta < \infty$, the information $I(\theta)$ is a constant, independent of θ (Problem 2.3). The Jeffreys prior (7.4.20) is therefore also constant. One might visualize it as the uniform distribution on $(-\infty, \infty)$. Unfortunately,

$$\int_{-\infty}^{\infty} c\, d\theta = \infty,$$

so that we are no longer dealing with a probability distribution. Prior densities whose integral is infinite are said to be *improper*. Their use can be justified by considering them as approximations to a sequence of proper prior distributions. In the present example, the improper density which is constant on $(-\infty, \infty)$ can, for example, be viewed as an approximation to the uniform distribution on $(-A, A)$ for large A or to a normal distribution with large variance, in the following sense.

The Bayes estimator (7.4.15) is the conditional expectation of θ given the observations; what matters therefore is the conditional distribution of θ. As an example, suppose that $f(x - \theta)$ is the normal distribution with mean θ and variance 1 and that θ is uniformly distributed on $(-A, A)$. Then the conditional density of θ given $(x_1, \ldots, x_n)$ is (Problem 4.8(i))

$$(7.4.25) \qquad \frac{e^{-\frac{1}{2}\sum(x_i-\theta)^2}}{\displaystyle\int_{-A}^{A} e^{-\frac{1}{2}\sum(x_i-\xi)^2}\,d\xi} \qquad \text{for } -A < \theta < A.$$

By writing $\sum (x_i - \theta)^2 = \sum (x_i - \bar{x})^2 + n(\bar{x} - \theta)^2$ and expanding the denominator analogously, this ratio is seen to be equal to

$$(7.4.26) \qquad \frac{e^{-\frac{n}{2}(\theta-\bar{x})^2}}{\displaystyle\int_{-A}^{A} e^{-\frac{n}{2}(\xi-\bar{x})^2}\,d\xi}.$$

Since the denominator tends to

$$(7.4.27) \qquad \int_{-\infty}^{\infty} e^{-\frac{n}{2}(\xi-\bar{x})^2}\,d\xi = \sqrt{\frac{2\pi}{n}}$$

as $A \to \infty$, it follows that the conditional density (7.4.25) tends to the normal density $N(\bar{x}, 1/n)$.

Consider now the corresponding formal calculation in which the prior $U(-A, A)$ is replaced by the improper prior $c \cdot d\theta$, $-\infty < \theta < \infty$. If we ignore the fact that this is not a density since its integral is infinite, and write the joint (improper) density of θ and the x's as

$$ce^{-\frac{1}{2}\sum(x_i-\theta)^2}/(2\pi)^{n/2},$$

the posterior density corresponding to (7.4.25) and (7.4.26) becomes

$$\frac{e^{-\frac{1}{2}\sum(x_i-\theta)^2}}{\displaystyle\int_{-\infty}^{\infty} e^{-\frac{1}{2}\sum(x_i-\xi)^2}\,d\xi} = \frac{e^{-\frac{n}{2}(\theta-\bar{x})^2}}{\displaystyle\int_{-\infty}^{\infty} e^{-\frac{n}{2}(\xi-\bar{x})^2}\,d\xi} = \sqrt{\frac{n}{2\pi}}\,e^{-\frac{n}{2}(\theta-\bar{x})^2},$$

which is the density of the normal distribution $N(\bar{x}, 1/n)$. Thus, although we started with an improper prior, we end up with a proper posterior distribution. This is intuitively plausible. Since the X's are normally distributed about θ with variance 1, even a single observation provides a good idea of the position of θ. The Bayes estimator for the improper prior $c \cdot d\theta$ is the expectation of $N(\bar{x}, 1/n)$ and is therefore $\bar{x}$.

The posterior distribution $N(\bar{x}, 1/n)$ obtained in this way is the limit of the distributions (7.4.26) as $A \to \infty$ and the Jeffreys Bayes estimator $\bar{X}$ is the limit of the proper Bayes estimators when θ is $U(-A, A)$ (Problems 4.8 and 4.9). It is in this sense that the improper prior can be viewed as a limiting case of the priors $U(-A, A)$.   □

Returning now to the efficiency of improper Bayes estimators, we can say quite generally that if for some n_0 the conditional distribution of θ given $x_1, \ldots, x_{n_0}$ is proper for all $x_1, \ldots, x_{n_0}$ (even though the prior is improper), the efficiency of the Bayes estimator continues to hold under the conditions required for proper priors.

Example 7.4.7 Scale parameter. Suppose that $X_1, \ldots, X_n$ are i.i.d. with density

$$(7.4.28) \qquad \theta^{-1} f(x/\theta), \ \theta > 0.$$

Then it was seen in Problem 2.6 that $I(\theta)$ is proportional to $1/\theta^2$ and hence the Jeffreys prior has a density of the form

$$g'(\theta) = c/\theta.$$

Since $\displaystyle\int_a^b 1/\theta = \log b - \log a$, we are again dealing with an improper prior.

This case can be handled in complete analogy to that of a location family (Problem 4.12).   □

Example 7.4.8 Poisson process. In a Poisson process, let the number of events occurring in a time interval of length l have the Poisson distribution $P(\lambda l)$, with the numbers occurring in non-overlapping intervals being independent. In such a process, the *waiting time* T required from the starting point, say $t = 0$, until the first event occurs has the exponential density

$$(7.4.29) \qquad f_\lambda(t) = \lambda e^{-\lambda t}, \ t > 0.$$

We shall now consider the estimation of λ for two sampling schemes.

(i) *Direct sampling.* We observe the numbers of events $X_1, \ldots, X_n$ occurring in non-overlapping intervals of length 1, so that

$$P_\lambda(X_1 = x_1, \ldots, X_n = x_n) = \frac{\lambda^{\sum x_i}}{\Pi x_i!} e^{-n\lambda}.$$

It was seen in Example 7.1.1 (continued) that, in this case, $I(\lambda) = 1/\lambda$. The Jeffreys prior is therefore improper with density $1/\sqrt{\lambda}$. The corresponding posterior density of λ given the x's is of the form

$$(7.4.30) \qquad C(x_1, \ldots, x_n) \lambda^{x-1/2} e^{-n\lambda},$$

where $x = \sum x_i$ and C is determined so that the integral of (7.4.30) from 0 to ∞ is equal to 1. The distribution (7.4.30) is a gamma distribution, the expectation of which is

$$(7.4.31) \qquad \delta(X) = \frac{1}{n} X + \frac{1}{2n},$$

the Bayes estimator for the Jeffreys prior. The consistent root of the likelihood equation was earlier seen to be $\hat{\theta}_n = X/n$.

(ii) *Inverse sampling.* Continue the process until n events have occurred, and let $T_1, \ldots, T_n$ denote the waiting times it takes for the occurrence of the first event, from the first to the second, and so on. The joint density of the T's is then

$$(7.4.32) \qquad \Pi f_\lambda(t_i) = \lambda^n e^{-\lambda \Sigma t_i} \text{ for all } t_i > 0.$$

The information a variable T_i contains about λ is (Problem 4.13)

$$(7.4.33) \qquad I(\lambda) = 1/\lambda^2,$$

and the Jeffreys prior is therefore improper with density proportional to λ. The posterior density of λ is

$$(7.4.34) \qquad \lambda^{n+1} e^{-\lambda \sum t_i},$$

which is proper for all $\sum t_i > 0$ and is in fact the density of a gamma distribution. The expectation of (7.4.34) is

$$(7.4.35) \qquad \frac{n+2}{\sum T_i},$$

and this is therefore the Bayes estimator in the present case.

It is interesting to note that the non-informative priors are different in cases (i) and (ii). It may seem awkward to have the prior depend on the sampling scheme but this is of course a consequence of the definition of non-informativeness given in and following Example 7.4.4 (continued). $\square$

The present section has exhibited large classes of asymptotically efficient estimators, an embarrassment of riches. First order asymptotic theory provides no guidance for choosing among these estimators. They all have the same asymptotic distribution and in fact any pair δ_{1n}, δ_{2n} of these estimators satisfies

$$\sqrt{n}\,(\delta_{2n} - \delta_{1n}) \xrightarrow{P} 0.$$

The most widely applied method of estimation at present is maximum likelihood, which in regular cases has the advantages of

(i) providing typically, although not always, a consistent and hence efficient root of the likelihood equation;

(ii) having a certain intuitive appeal as the value of the parameter making the observed data most plausible;

(iii) leading to a unique estimator, although this uniqueness is marred in practice by the fact that different algorithms for calculating the maximum will lead to different answers.

On the other hand, writing from a Bayesian point of view, Berger (1985, p. 90) states:

> We would argue that non-informative prior Bayesian analysis is the *single most powerful analysis* (Berger's italics) in the sense of being the *ad hoc* method most likely to yield a sensible answer for a given investment effort.

And he adds that

> the answers so attained have the added feature of being, in some sense, the most "objective" statistical answers obtainable (which is attractive to those who feel objectivity is possible).

For further discussion of this choice of prior, see, for example, Bernardo and Smith (1994), Robert (1994), and Kass and Wasserman (1996).

The difference between the MLE and the Jeffreys Bayes estimator is typically of order $1/n$ and, except for very small sample sizes the choice between these two representative estimators, is therefore not too important.

Summary

1. An asymptotically normal estimator is called efficient if its asymptotic variance is $1/I(\theta)$. A theorem of Le Cam implies that such an estimator under suitable regularity conditions minimizes the asymptotic variance among all asymptotically normal estimators whose asymptotic variance is a continuous function of the parameter. Efficient estimators thus provide a standard with which other estimators can be compared.

2. Not only consistent roots of the likelihood equations and the one-step estimators (7.3.21) are asymptotically efficient but so are all Bayes estimators that satisfy some not very stringent conditions, including the requirement that the prior density is positive and continuous for all parameter values.

3. To obtain a Bayes estimator of θ, one must first specify a prior distribution for θ. A popular choice is the Jeffreys prior which assigns a uniform distribution to the function $g(\theta)$ for which the information $I^*[g(\theta)]$ is constant.

7.5 The multiparameter case I. Asymptotic normality

The previous sections developed the theory of efficient estimation for regular models containing only one parameter. However, except in the very simplest cases, parametric models typically involve several parameters and we shall now extend the results of Sections 7.1–7.4 to this more general situation.

We begin by generalizing the concept of Fisher information under assumptions corresponding to (C1)–(C5) of Section 7.1 which will be stated in Theorem 7.5.1.

Definition 7.5.1 If X is a real- (or vector-) valued random variable with density $f_\theta(x)$ where $\theta = (\theta_1, \dots, \theta_k)$, the information matrix $I(\theta)$ is the $k \times k$ matrix with elements

$$(7.5.1) \qquad I_{ij}(\theta) = E\left[\frac{\partial}{\partial \theta_i} \log f_\theta(X) \frac{\partial}{\partial \theta_j} \log f_\theta(X)\right].$$

Example 7.5.1 Normal. Let X have the normal distribution $N\left(\xi, \sigma^2\right)$ with both parameters unknown. Then

$$\log f_{\xi,\sigma^2}(x) = -\frac{1}{2} \log \sigma^2 - \frac{1}{2\sigma^2}(x - \xi)^2 - \log\left(\sqrt{2\pi}\right),$$

so that

$$\frac{\partial}{\partial \xi} \log f_{\xi,\sigma^2}(x) = \frac{1}{\sigma^2}(x - \xi) \text{ and } \frac{\partial}{\partial \sigma^2} \log f_{\xi,\sigma^2}(x) = -\frac{1}{2\sigma^2} + \frac{1}{2\sigma^4}(x - \xi)^2.$$

It follows that

$$(7.5.2) \qquad I\left(\xi, \sigma^2\right) = \begin{pmatrix} 1/\sigma^2 & 0 \\ 0 & 1/2\sigma^4 \end{pmatrix}.$$

$\square$

As will be stated more precisely in Theorem 7.5.2, a consistent root $\left(\hat{\theta}_1, \ldots, \hat{\theta}_k\right)$ of the likelihood equations, under suitable regularity conditions, satisfies

$$(7.5.3) \qquad \left(\sqrt{n}\left(\hat{\theta}_1 - \theta_1\right), \ldots, \sqrt{n}\left(\hat{\theta}_k - \theta_k\right)\right) \xrightarrow{L} N\left(\mathbf{0}, I^{-1}(\theta)\right),$$

where $\mathbf{0} = (0, \ldots, 0)$. If $I(\theta)$ is diagonal, as is the case in (7.5.2), so is $I^{-1}(\theta)$ by Problem 3.14(ii) of Chapter 5, and it then follows that the variables $\sqrt{n}\left(\hat{\theta}_1 - \theta_1\right), \ldots, \sqrt{n}\left(\hat{\theta}_k - \theta_k\right)$ are asymptotically independent.

Example 7.5.2 Location-scale families. In generalization of Example 7.5.1, suppose that

$$(7.5.4) \qquad f_{\xi,\sigma}(x) = \frac{1}{\sigma} f\left(\frac{x - \xi}{\sigma}\right).$$

Then the off-diagonal element I_{12} of $I(\xi, \sigma)$ is

$$(7.5.5) \qquad I_{12} = \frac{1}{\sigma^2} \int y \left[\frac{f'(y)}{f(y)}\right]^2 f(y) dy$$

where f' denotes the derivative of f. This is 0 whenever f is symmetric about the origin (Problem 5.1). $\qquad\qquad\square$

Example 7.5.3 Multinomial. In the multinomial situation of Example 5.4.1, the joint distribution of the variables $Y_1, \ldots, Y_k$ is

$$(7.5.6) \qquad \begin{aligned} &P\left(y_1, \ldots, y_k\right) = \\ &\frac{n!}{y_1!, \ldots, y_{k+1}!} p_1^{y_1} \cdots p_k^{y_k} \left(1 - p_1 - \cdots - p_k\right)^{n-y_1-\cdots-y_k}. \end{aligned}$$

Thus,

$$\frac{\partial}{\partial p_i} \log P\left(y_1, \ldots, y_k\right) = \frac{y_i}{p_i} - \frac{y_{k+1}}{p_{k+1}}$$

and

$$I_{ij} = E\left[\left(\left(\frac{Y_i}{p_i} - n\right) - \left(\frac{Y_{k+1}}{p_{k+1}} - n\right)\right)\left(\left(\frac{Y_j}{p_j} - n\right) - \left(\frac{Y_{k+1}}{p_{k+1}} - n\right)\right)\right].$$

It follows from expansion of the right side and (5.4.15) (Problem 5.1(ii)) that

$$(7.5.7) \qquad I_{ij} = \begin{cases} n\left[\dfrac{1}{p_i} + \dfrac{1}{p_{k+1}}\right] & \text{if } i = j \\ n/p_{k+1} & \text{if } i \neq j. \end{cases}$$

$$\square$$

In order to generalize Theorem 7.2.1, let us next state the multivariate analog of assumptions (C1)–(C6) .

(M1) The distributions P_θ are distinct.

(M2) There exists a neighborhood[§] N of the true parameter point θ^0 which lies entirely within (i.e., is a subset of) Ω.

Note: Since the true θ^0 is unknown, this condition can typically be verified only by requiring that (M2) hold for every possible point θ^0 of Ω. A set Ω in R^k with this property is called *open*. Typical examples of open sets are the intervals of points $(\theta_1, \ldots, \theta_k)$ given by

$$|\theta_i - a_i| < c \text{ for all } i = 1, \ldots, k$$

and the balls

$$\sum_{i=1}^k (\theta_i - a_i)^2 < c.$$

Here it is crucial in both cases that the inequalities are strict (Problem 5.2). In terms of this definition, there is little lost in replacing (M2) by the stronger assumption

(M2′) The parameter space Ω is open.

(M3) The observations $X_1, \ldots, X_n$ are n i.i.d. random variables or vectors with continuous density f_θ or discrete probability distribution $P_\theta(x) = P_\theta(X_i = x)$.

(M4) The set A on which $f_\theta(x)$ or $P_\theta(x)$ is positive is independent of θ.

(M5) For all x in A, the partial derivatives $\dfrac{\partial}{\partial \theta_i} f_\theta(x)$ or $\dfrac{\partial}{\partial \theta_i} P_\theta(x)$ exist for all $i = 1, \ldots, k$.

(M6) The partial derivatives of the left side of the equation

(7.5.8) $$\int f_{\theta_1, \ldots, \theta_k}(x) dx = 1$$

exist and can be obtained by differentiating under the integral sign, and the corresponding fact holds for $P_\theta(x)$.

In the remainder of the section we shall state results and assumptions only for the continuous case.

[§]Defined in (5.1.2).

Theorem 7.5.1

(i) If (M1)–(M6) hold, then

(7.5.9)
$$E_\theta \left[\frac{\partial}{\partial \theta_i} f_{\theta_1,\ldots,\theta_k}(X) \right] = 0$$

and hence

(7.5.10)
$$I_{ij}(\theta) = \mathrm{Cov} \left[\frac{\partial}{\partial \theta_i} \log f_\theta(X), \frac{\partial}{\partial \theta_j} \log f_\theta(X) \right].$$

(ii) If in addition to (M1)–(M5) we have

(M6)′ The first two partial derivatives $\dfrac{\partial^2}{\partial \theta_i \partial \theta_j}$ of $f_\theta(x)$ exist for all $x \in A$ and all $\theta \in \Omega$, and if the corresponding derivatives of the left side of (7.5.8) can be obtained by differentiating under the integral sign, then also

(7.5.11)
$$I_{ij}(\theta) = -E_\theta \left[\frac{\partial^2}{\partial \theta_i \partial \theta_j} \log f_\theta(X) \right].$$

Proof. The argument is completely analogous to that of Theorem 7.2.1 (Problem 5.3). ∎

Corollary 7.5.1 *Under the assumptions of Theorem 7.5.1(i), the matrix $I(\theta)$ is positive semi-definite.*

Proof. Example 5.3.3. ∎

The additivity results stated for the one-parameter case in Theorem 7.2.2 remain valid under the assumptions of Theorem 7.5.1(i) with $I(\theta)$, $I_1(\theta)$, $\ldots$ now denoting information matrices (Problem 5.4). The transformation formula (7.2.20) also extends in a natural way to the multiparameter case. More specifically, let

(7.5.12)
$$\eta_i = g_i(\theta_1,\ldots,\theta_k), \quad i = 1,\ldots,k,$$

and let J be the Jacobian matrix with $(i,j)^{\mathrm{th}}$ element $\partial \eta_i / \partial \theta_j$. Then the information matrix for η will be (Problem 5.5)

(7.5.13)
$$I^*(\eta) = (J')^{-1} I(\theta) J^{-1}.$$

Let us next consider the multiparameter versions of Theorems 7.1.1 and 7.3.1, with the likelihood equation (7.1.18) now replaced by the set of equations

(7.5.14)
$$\frac{\partial}{\partial \theta_j} [f(x_1,\theta),\ldots,f(x_n,\theta)] = 0, \quad j = 1,\ldots,k,$$

or, equivalently,

$$(7.5.15) \qquad l'_j(\theta) = \frac{\partial}{\partial\theta_j} l_n(\theta) = \sum_{i=1}^{n} \frac{\partial}{\partial\theta_j} \log f(x_i, \theta) = 0, \ j = 1, \ldots, k,$$

where $l_n(\theta)$ as before denotes the log of the likelihood and $l'_j(\theta)$ its derivative with respect to θ_j.

In addition to assumptions (M1)–(M6)′, we require:

(M6)″ For all $x \in A$, all third derivatives $\dfrac{\partial^3}{\partial\theta_i\partial\theta_j\partial\theta_r} f_\theta(x)$ exist and are

continuous, and the corresponding derivatives of the integral (7.5.8) exist and can be obtained by differentiating under the integral sign,

and

(M7) If $\theta_0 = \left(\theta_1^{(0)}, \ldots, \theta_k^{(0)}\right)$ denotes the true value of θ, there exist functions $M_{ijr}(x)$ and a positive number $c(\theta_0)$ such that

$$\left| \frac{\partial^3}{\partial\theta_i\partial\theta_j\partial\theta_r} \log f_\theta(x) \right| \le M_{ijr}(x)$$

for all θ with $\sum \left(\theta_i - \theta_i^{(0)}\right)^2 < c(\theta_0)$, where $E_{\theta_0}[M_{ijr}(X)] < \infty$ for all i, j, r.

Finally, in generalization of (7.3.4), we shall assume that

(M8) the elements $I_{ij}(\theta)$ of the information matrix are finite and the positive semidefinite matrix $I(\theta)$ is positive definite.

Theorem 7.5.2 *Under assumptions (M1)–(M5), (M6)″, (M7), and (M8), there exists a solution $\hat\theta_n = \left(\hat\theta_{1n}, \ldots, \hat\theta_{kn}\right)$ of the likelihood equations which is consistent, and any such solution satisfies*

$$(7.5.16) \qquad \left(\sqrt{n}\left(\hat\theta_{1n} - \theta_1^{(0)}\right), \ldots, \sqrt{n}\left(\hat\theta_{kn} - \theta_k^{(0)}\right)\right) \xrightarrow{L} N\left(0, I^{-1}(\theta_0)\right).$$

For a proof, see, for example, Lehmann and Casella (1998).

In particular, if the solution of the likelihood equations is unique, it will then satisfy (7.5.16). On the other hand, it turns out that part (ii) of Corollary 7.1.1 does not generalize to the multiparameter case. Even if the unique root of the likelihood equations corresponds to a local maximum, it need not be the global maximum of the likelihood function.

Example 5.1 Normal (continued). For a sample $X_1, \ldots, X_n$ from $N\left(\xi, \sigma^2\right)$, the log likelihood is

$$(7.5.17) \qquad l_n(\xi, \sigma) = -n \log\sigma - \frac{1}{2\sigma^2} \sum (x_i - \xi)^2 - n \log\left(\sqrt{2\pi}\right).$$

The likelihood equations are therefore

$$\frac{\partial}{\partial \xi} l_n(\xi, \sigma) = \frac{1}{\sigma^2} \sum (x_i - \xi) = 0,$$

$$\frac{\partial}{\partial \sigma} l_n(\xi, \sigma) = -\frac{n}{\sigma} + \frac{1}{\sigma^3} \sum (x_i - \xi)^2 = 0,$$

which have the unique solution

$$(7.5.18) \qquad \hat{\xi} = \bar{x}, \quad \hat{\sigma}^2 = \frac{1}{n} \sum (x_i - \bar{x})^2.$$

It follows from Theorem 7.5.2 and Example 7.5.1 that

$$(7.5.19) \qquad \left(\sqrt{n}(\bar{X} - \xi), \ \sqrt{n} \left[\frac{1}{n} \sum (X_i - \bar{X})^2 - \sigma^2 \right] \right) \xrightarrow{L} N(0, \Sigma),$$

with $\Sigma = (\sigma_{ij}) = I^{-1}$ given by

$$(7.5.20) \qquad \sigma_{11} = \sigma^2, \ \sigma_{12} = 0, \ \sigma_{22} = 2\sigma^4.$$

This can, of course, also be checked directly from the bivariate central limit theorem (Problem 5.6).   □

Example 5.3 Multinomial (continued). From (7.5.6), it is seen that the likelihood equations are

$$(7.5.21) \qquad \frac{y_i}{p_i} = \frac{y_{k+1}}{p_{k+1}} \quad (i = 1, \dots, k),$$

from which we get

$$n - y_{k+1} = \sum_{i=1}^{k} y_i = \frac{y_{k+1}}{p_{k+1}} \sum_{i=1}^{k} p_i = \frac{y_{k+1}}{p_{k+1}} (1 - p_{k+1})$$

and hence $y_{k+1} = n p_{k+1}$. Substitution in (7.5.21) gives

$$(7.5.22) \qquad \hat{p}_i = \frac{y_i}{n}, \ i = 1, \dots, k+1.$$

It follows from Theorem 7.5.2 that

$$\left(\sqrt{n} \left(\frac{Y_i}{n} = p_1 \right), \dots, \sqrt{n} \left(\frac{Y_k}{n} - p_k \right) \right) \xrightarrow{L} N(0, \Sigma),$$

with $\Sigma = (\sigma_{ij}) = I^{-1}$, where I is given by (7.5.7). Comparison with (5.4.13) shows that σ_{ij} is given by (5.4.14). This result was seen directly from the CLT in Example 5.4.1.   □

Example 7.5.4 Bivariate normal. Consider a sample from the bivariate normal distribution (5.2.1), and let $\theta = (\xi, \eta, \sigma^2, \tau^2, \rho)$, $\sigma > 0$, $\tau > 0$, $|\rho| < 1$. Then the log likelihood is

(7.5.23)
$$l_n(\theta) = -n \left[\log \sigma + \log \tau + \tfrac{1}{2} \log \left(1 - \rho^2 \right) \right]$$
$$- \frac{1}{2 \left(1 - \rho^2 \right)} \left[\frac{1}{\sigma^2} \sum (x_i - \xi)^2 - \frac{2\rho}{\sigma \tau} \sum (x_i - \xi)(y_i - \eta) + \frac{1}{\tau^2} \sum (y_i - \eta)^2 \right].$$

and the likelihood equations become, after some simplification,

$$\frac{\partial}{\partial \xi} : \quad \frac{1}{\sigma^2} \sum (x_i - \xi) = \frac{\rho}{\sigma \tau} \sum (y_i - \eta),$$

$$\frac{\partial}{\partial \eta} : \quad \frac{1}{\tau^2} \sum (y_i - \eta) = \frac{\rho}{\sigma \tau} \sum (x_i - \xi),$$

$$\frac{\partial}{\partial \sigma} : \quad \frac{1}{\left(1 - \rho^2 \right)} \left[\frac{1}{\sigma^3} \sum (x_i - \xi)^2 - \frac{\rho}{\sigma^2 \tau} \sum (x_i - \xi)(y_i - \eta) \right] = \frac{n}{\sigma},$$

$$\frac{\partial}{\partial \tau} : \quad \frac{1}{\left(1 - \rho^2 \right)} \left[\frac{1}{\tau^3} \sum (y_i - \eta)^2 - \frac{\rho}{\sigma \tau^2} \sum (x_i - \xi)(y_i - \eta) \right] = \frac{n}{\tau},$$

$$\frac{\partial}{\partial \rho} : \quad \rho \left[\frac{\sum (x_i - \xi)^2}{\sigma^2} + \frac{\sum (y_i - \eta)^2}{\tau^2} \right] - \left(1 + \rho^2 \right) \frac{\sum (x_i - \xi)(y_i - \eta)}{\sigma \tau}$$
$$= n\rho \left(1 - \rho^2 \right).$$

The first two equations have the solution

(7.5.24)
$$\hat{\xi} = \bar{x}, \quad \hat{\eta} = \bar{y}$$

or $\rho^2 = 1$; the latter is ruled out if the distribution is assumed to be non-degenerate. The solution of the last three equations is

$$\hat{\sigma}^2 = \frac{1}{n} \sum (x_i - \xi)^2, \quad \hat{\tau}^2 = \frac{1}{n} \sum (y_i - \eta)^2, \quad \hat{\rho}\hat{\sigma}\hat{\tau} = \frac{1}{n} \sum (x_i - \xi)(y_i - \eta).$$

Thus the likelihood equations have the unique solution (7.5.24) and

(7.5.25)
$$\hat{\sigma}^2 = \frac{1}{n} \sum (x_i - \bar{x})^2, \quad \hat{\tau}^2 = \frac{1}{n} \sum (y_i - \bar{y})^2,$$
$$\hat{\rho} = \frac{\sum (x_i - \bar{x})(y_i - \bar{y})}{\sqrt{\sum (x_i - \bar{x})^2 \sum (y_i - \bar{y})^2}}.$$

If we put

(7.5.26) $\quad \gamma = \rho \sigma \tau = E \left(X_i - \xi \right) \left(Y_i - \eta \right), \quad \hat{\gamma} = \frac{1}{n} \sum \left(X_i - \bar{X} \right) \left(Y_i - \bar{Y} \right),$

we can use the multivariate CLT to show that the joint limit distribution of

(7.5.27)
$$\sqrt{n} \left(\hat{\xi} - \xi \right), \quad \sqrt{n} \left(\hat{\eta} - \eta \right), \quad \sqrt{n} \left(\hat{\sigma}^2 - \sigma^2 \right), \quad \sqrt{n} \left(\hat{\tau}^2 - \tau^2 \right), \quad \sqrt{n} \left(\hat{\gamma} - \gamma \right)$$

is the 5-variate normal distribution with zero means and covariance matrix (Problem 5.7)

$$(7.5.28) \qquad \Sigma = \begin{pmatrix} \Sigma_1 & 0 \\ 0 & \Sigma_2 \end{pmatrix},$$

where

$$(7.5.29) \qquad \Sigma_1 = \begin{pmatrix} \sigma^2 & \rho\sigma\tau \\ \rho\sigma\tau & \tau^2 \end{pmatrix}$$

and Σ_2 is the 3×3 matrix with elements σ_{ij} given by (5.4.33). These latter covariances must then be evaluated for the case that the (X_i, Y_i) are bivariate normal when they become (see, for example, Anderson (1984, p. 49)

$$(7.5.30) \qquad \begin{array}{c} \\ \hat{\sigma}^2 \\ \hat{\tau}^2 \\ \hat{\gamma} \end{array} \begin{array}{ccc} \hat{\sigma}^2 & \hat{\tau}^2 & \hat{\gamma} \\ \begin{pmatrix} 2\sigma^4 & 2\rho^2\sigma^2\tau^2 & 2\rho\sigma^3\tau \\ 2\rho^2\sigma^2\tau^2 & 2\tau^4 & 2\rho\sigma\tau^3 \\ 2\rho\sigma^3\tau & 2\rho\sigma\tau^3 & 2\rho^2\sigma^2\tau^2 \end{pmatrix} \end{array}$$

Instead of the variables $\hat{\sigma}^2$, $\hat{\tau}^2$, and $\hat{\gamma}$, we may prefer to work with $\hat{\sigma}^2$, $\hat{\tau}^2$, and $\hat{\rho} = \hat{\gamma}/\hat{\sigma}\hat{\tau}$ as estimators of σ^2, τ^2, and ρ, respectively. The joint limit distribution of these variables is again normal with a covariance matrix that can be determined from (5.4.21), and which then reduces to (Problem 5.8)

$$(7.5.31) \qquad \begin{array}{c} \hat{\sigma}^2 \\ \hat{\tau}^2 \\ \hat{\rho} \end{array} \begin{pmatrix} 2\sigma^4 & 2\rho^2\sigma^2\tau^2 & \rho\left(1-\rho^2\right)\sigma^2 \\ 2\rho^2\sigma^2\tau^2 & 2\tau^4 & \rho\left(1-\rho^2\right)\tau^2 \\ \rho\left(1-\rho^2\right)\sigma^2 & \rho\left(1-\rho^2\right)\tau^2 & \left(1-\rho^2\right)^2 \end{pmatrix}.$$

$\square$

Example 7.5.5 Multivariate normal. Let us now generalize Examples 7.5.1 ($k = 1$) and 7.5.4 ($k = 2$) by considering the general case of an i.i.d. sample $X_1, \ldots, X_n$ from the multivariate normal density (5.4.1) of Chapter 5, with both the mean vector $(\xi_1, \ldots, \xi_k)$ and the covariance matrix $\Sigma = A^{-1}$ unknown.

The log likelihood is equal to

$$\frac{n}{2} \log |A| - \frac{1}{2} \sum_{\nu=1}^{n} \sum_{i=1}^{k} \sum_{j=1}^{k} a_{ij} (x_{i\nu} - \xi_i)(x_{j\nu} - \xi_j)$$

(7.5.32)
$$= \frac{n}{2} \log |A| - \frac{1}{2} \sum_{\nu=1}^{n} \sum_{i=1}^{k} \sum_{j=1}^{k} a_{ij} (x_{i\nu} - \bar{x}_i)(x_{j\nu} - \bar{x}_j)$$

$$+ n \sum_{i=1}^{k} \sum_{j=1}^{k} a_{ij} (\bar{x}_i - \xi_i)(\bar{x}_j - \xi_j).$$

The derivative with respect to ξ_i of (7.5.32) is

(7.5.33)
$$\frac{\partial}{\partial \xi_i} (5.32) = -2n \sum_{j=1}^{k} a_{ij} (\bar{x}_j - \xi_j), \quad i = 1, \dots, k.$$

Since the matrix $A = (a_{ij})$ is assumed to be nonsingular, the system of linear equations $\frac{\partial}{\partial \xi_i}(5.32) = 0$ in the k variables $(\bar{x}_j - \xi_j)$, $j = 1, \dots, k$, has the unique solution $\bar{x}_j - \xi_j = 0$, so that the likelihood equations have the unique solution

(7.5.34)
$$\hat{\xi}_i = \bar{X}_i.$$

Consider next the derivative of (7.5.32) with respect to a_{ij}. For this purpose, we need to know how to differentiate a symmetric determinant with respect to its elements.

Lemma 7.5.1 Let $A = (a_{ij})$ be a symmetric matrix with inverse $A^{-1} = \Sigma = (\sigma_{ij})$. Then

(7.5.35)
$$\frac{\partial \log |A|}{\partial a_{ij}} = \begin{cases} 2a_{ij} & \text{if } i \neq j \\ a_{ij} & \text{if } i = j. \end{cases}$$

For a proof, see, for example, Graybill (1983) or Harville (1997).
From this lemma, it follows that

$$\frac{\partial}{\partial a_{ij}} (7.5.32) = \begin{cases} n\sigma_{ij} - nS_{ij} + 2n (\bar{x}_i - \xi_i)(\bar{x}_j - \xi_j) & \text{when } j \neq i \\ \frac{n}{2}\sigma_{ii} - \frac{n}{2}S_{ii} + n(\bar{x}_i - \xi_i)^2 & \text{when } j = i, \end{cases}$$

where

(7.5.36)
$$S_{ij} = \sum (X_{i\nu} - \bar{X}_i)(X_{j\nu} - \bar{X}_j)/n.$$

Combined with (7.5.34), these equations have the unique solution

(7.5.37)
$$\hat{\sigma}_{ij} = S_{ij}.$$

The estimators $\left\{ \hat{\xi}_i = \bar{X}_i, \ \hat{\sigma}_{ij} = S_{ij} \right\}$ are in fact the MLEs, as is shown, for example, in Anderson (1984) and Seber (1984).

The joint limit distribution of the variables

$$(7.5.38) \qquad \sqrt{n} \left(\bar{X}_i - \xi_i \right), \ i = 1, \dots, k, \text{ and } \sqrt{n} \left(S_{ij} - \sigma_{ij} \right), \ i \le j,$$

is now easily obtained from the CLT since it is unchanged if the S_{ij} are replaced by (Problem 5.9)

$$(7.5.39) \qquad S'_{ij} = \sum \left(X_{i\nu} - \xi_i \right) \left(X_{j\nu} - \xi_j \right) / n.$$

By the multivariate CLT (Theorem 5.4.4), the joint limit distribution of the variables (7.5.38) is therefore multivariate normal with means zero and with covariance matrix equal to the covariance matrix of the variables

$$(7.5.40) \qquad X_i - \xi_i \ (i = 1, \dots, k) \text{ and } T_{jl} = \left(X_j - \xi_j \right) \left(X_l - \xi_l \right), \ j \le l,$$

where $(X_1, \dots, X_k)$ has the distribution (5.4.1). The required covariances are

$$(7.5.41) \qquad \mathrm{Cov} \left(X_i, \ X_j \right) = \sigma_{ij}, \ \mathrm{Cov} \left(X_i, \ T_{jl} \right) = 0$$

and

$$(7.5.42) \qquad \mathrm{Cov} \left(T_{ij}, \ T_{rl} \right) = \sigma_{ir} \sigma_{jl} + \sigma_{il} \sigma_{jr}.$$

For a proof of (7.5.42), see, for example, Anderson (1984, pp. 81–82).

Note: The asymptotic normality of the joint limit distribution of the variables (7.5.38) does not require the assumption of normality made at the beginning of Example 7.5.5. Let $X_1, \dots, X_n$ be an i.i.d. sample from any k-variate distribution with mean $(\xi_1, \dots, \xi_k)$ and nonsingular covariance matrix Σ. Then the joint limit distribution of the variables (7.5.38) is multivariate normal with mean zero regardless of the underlying distribution of the X's. This follows from the proof given in the normal case since the multivariate CLT does not require normality of the parent distribution. However, the asymptotic covariance matrix will no longer be given by (7.5.41) and (7.5.42), which depend on the assumption of normality. Even the asymptotic independence of the variables $\sqrt{n} \left(\bar{X}_i - \xi_i \right)$ and $\sqrt{n} \left(S_{jl} - \sigma_{jl} \right)$ no longer holds without some symmetry assumption. (Note that when the X's are normal, this independence holds not only asymptotically but also exactly for any sample size $n \ge 2$.) □

The difficulties in determining the efficient root of the likelihood equation which were discussed in Section 7.3 for the case of a single parameter tend to be even more severe when several parameters are involved. The following extension of Theorem 7.3.3 will then frequently provide a convenient alternative.

Theorem 7.5.3 *Suppose the assumptions of Theorem 7.5.2 hold and that $\tilde{\theta}_{jn}$ is a $\sqrt{n}$-consistent estimator of θ_j for $j = 1, \ldots, k$. Let δ_{in} $(i = 1, \ldots, k)$ be the solution of the set of linear equations*

$$(7.5.43) \qquad \sum_{i=1}^{k} \left(\delta_{in} - \tilde{\theta}_{in} \right) l''_{ij} \left(\tilde{\theta}_n \right) = -l'_j \left(\tilde{\theta}_n \right), \quad j = 1, \ldots, k.$$

Then $\delta_n = (\delta_{1n}, \ldots, \delta_{kn})$ satisfies (7.5.3).

For a proof, see, for example, Lehmann and Casella (1998).

Note that (7.5.43) is just a set of linear equations in the unknowns δ_{in}. For $k = 1$, the solution of (7.5.43) reduces to (7.3.21).

Example 7.5.6 Normal mixtures. Consider an industrial setting with a production process in control, so that the outcome follows a known distribution, which we shall take to be the standard normal distribution. However, it is suspected that the production process has become contaminated, with the contaminating portion following some other, unknown normal distribution $N\left(\eta, \tau^2\right)$. To get an idea of what is going on, a sample $X_1, \ldots, X_n$ of the output is drawn. The X's are therefore assumed to be i.i.d. according to the distribution

$$(7.5.44) \qquad pN(0,1) + qN\left(\eta, \tau^2\right).$$

As a first step, it is desired to estimate the unknown parameters p, η, and τ^2.

Let us begin with the simpler two-parameter problem in which it is assumed that $\tau^2 = 1$. Then the joint density of the X's is

$$(7.5.45) \qquad \prod_{i=1}^{n} \left[pe^{-\frac{1}{2}x_i^2} + qe^{-\frac{1}{2}(x_i - \eta)^2} \right] / (2\pi)^{n/2}.$$

The loglikelihood is therefore

$$(7.5.46) \qquad l_n(p, \eta) = \sum_{i=1}^{n} \log \left[pe^{-\frac{1}{2}x_i^2} + qe^{-\frac{1}{2}(x_i - \eta)^2} \right] - \frac{n}{2} \log(2\pi)$$

and its derivatives are

$$(7.5.47) \qquad \frac{\partial}{\partial p} l_n(p, \eta) = \sum_{i=1}^{n} \frac{e^{-\frac{1}{2}x_i^2} - e^{-\frac{1}{2}(x_i - \eta)^2}}{pe^{-\frac{1}{2}x_i^2} + qe^{-\frac{1}{2}(x_i - \eta)^2}}$$

and

$$(7.5.48) \qquad \frac{\partial}{\partial \eta} l_n(p, \eta) = \sum_{i=1}^{n} \frac{-q\left(x_i - \eta\right) e^{-\frac{1}{2}(x_i - \eta)^2}}{pe^{-\frac{1}{2}x_i^2} + qe^{-\frac{1}{2}(x_i - \eta)^2}}.$$

Rather than solving the likelihood equations, we shall find estimators satisfying (7.5.3) by means of Theorem 7.5.3.

To obtain the needed starting $\sqrt{n}$-consistent estimators for p and η, we shall employ the *method of moments* pioneered by Karl Pearson (1894). If there are k unknown parameters, this method consists in calculating k moments of X_i—typically the first k—and then determining the parameter values for which the k population moments equal the corresponding sample moments. In the present case, $k = 2$ and we find (Problem 5.12(i))

$$(7.5.49) \qquad E(X_i) = q\eta, \ E(X_i^2) = p + q(\eta^2 + 1) = 1 + q\eta^2.$$

The estimating equations for q and η are therefore

$$(7.5.50) \qquad q\eta = \bar{X} \text{ and } q\eta^2 = \frac{1}{n}\sum X_i^2 - 1,$$

and the resulting estimators of q and η are

$$(7.5.51) \qquad \tilde{\eta} = \frac{\frac{1}{n}\sum X_i^2 - 1}{\bar{X}} \text{ and } \tilde{q} = \frac{\bar{X}^2}{\frac{1}{n}\sum X_i^2 - 1}.$$

That these estimators are $\sqrt{n}$-consistent follows from the fact that by the bivariate central limit theorem,

$$(7.5.52) \qquad \sqrt{n}(\bar{X} - q\eta), \ \sqrt{n}\left(\frac{1}{n}\sum X_i^2 - q\eta^2\right)$$

have a bivariate normal limit distribution. By Theorem 5.4.6,

$$(7.5.53) \qquad \sqrt{n}(\tilde{\eta} - \eta), \ \sqrt{n}(\tilde{q} - q)$$

then also have a normal limit distribution and $\sqrt{n}$-consistency of $\tilde{\eta}$ and $\tilde{q}$ follows (Problem 5.12(ii)).

Note: For finite n, there is positive probability that $\tilde{q}$ will not be between 0 and 1 and hence will not provide a satisfactory estimate of q. However, the probability of this event tends to 0 as $n \to \infty$ (Problem 5.13).

Let us now return to the original three-parameter model (7.5.44). It is interesting to note that the MLE does not exist for this model. To see this, note that the joint density of the X's is a sum of non-negative terms, one of which is proportional to

$$\frac{1}{\tau}p^{n-1}qe^{-\frac{1}{2}\sum\limits_{i=2}^{n}x_i^2 - \frac{1}{2\tau^2}(y_1-\eta)^2},$$

which tends to infinity for $y_1 = \eta$ as $\tau \to 0$. The likelihood is therefore unbounded. However, the assumptions of Theorems 7.5.2 and 7.5.3 are satisfied. To apply the latter, we use the method of moments and calculate (Problem 5.14(i))

(7.5.54)
$$E\left(X_i\right) = q\eta, \ E\left(X_i^2\right) = p + q\left(\eta^2 + \tau^2\right), \ E\left(X_i^3\right) = q\eta\left(\eta^2 + 3\tau^2\right)$$

and thus obtain the estimating equations

(7.5.55)
$$q\eta = \bar{X}, \ q\left(\eta^2 + \tau^2 - 1\right) = \frac{1}{n}\sum X_i^2 - 1,$$
$$q\eta\left(\eta^2 + 3\tau^2\right) = \frac{1}{n}\sum X_i^3,$$

which again can be solved explicitly and provide the needed $\sqrt{n}$-consistent starting estimators (Problem 5.14(ii)).

In the same way, the method of moments can, in principle, be used to obtain starting estimators for the full five-parameter mixture model which replaces (7.5.44) by

(7.5.56)
$$pN\left(\xi, \sigma^2\right) + qN\left(\eta, \tau^2\right).$$

However, the likelihood equations then become more complicated. In addition, for the higher moments the convergence of the sample moments to the true ones tends to be much slower and the sensitivity of the estimators to the model assumptions much greater. For these reasons, alternative starting estimators have been explored. Mixture models and the estimation and testing of their parameters are discussed in the books by Everitt and Hand (1981), Titterington, Smith, and Makov (1985), McLachlan and Basford (1988), and Lindsay (1995). See also Titterington (1997).   □

Summary

1. In the multiparameter case, the Fisher information $I(\theta)$ becomes a matrix with properties generalizing those in the one-parameter situation.

2. Regularity conditions are given under which the likelihood equations have a consistent, asymptotically normal solution.

3. The difficulty presented by multiple roots of the likelihood equations can be circumvented by a one-step estimator which only requires a $\sqrt{n}$-consistent starting point and the solution of a set of linear equations.

7.6   The multiparameter case II. Efficiency

In the one-parameter case, two asymptotically normal estimators are compared in terms of the variances of the limit distribution. The natural k-

variate generalization is in terms of the covariance matrices of the estimators. More specifically, suppose that $\delta_n = (\delta_{1n}, \ldots, \delta_{kn})$ and $\delta'_n = (\delta'_{1n}, \ldots, \delta'_{kn})$ are two estimators of $\theta = (\theta_1, \ldots, \theta_k)$ such that

(7.6.1)
$$\sqrt{n}\,(\delta_{1n} - \theta_1), \ldots, \sqrt{n}\,(\delta_{kn} - \theta_k) \text{ and } \sqrt{n}\,(\delta'_{1n} - \theta_1), \ldots, \sqrt{n}\,(\delta'_{kn} - \theta_k)$$

are both asymptotically normal with means 0 and with covariance matrices $\Sigma(\theta)$ and $\Sigma'(\theta)$. Then δ_n will be considered more efficient (in the sense of $\geq$) than δ'_n if

(7.6.2) $\Sigma'(\theta) - \Sigma(\theta)$ is positive semidefinite for all θ.

This definition is justified by the following result.

Theorem 7.6.1 *If δ_n and δ'_n are k-variate normal with common mean θ and covariance matrices $\Sigma(\theta)$ and $\Sigma'(\theta)$, respectively, if C is any symmetric (about the origin) convex set in R^k and if (7.6.2) holds, then*

(7.6.3)
$$P\left[\left(\sqrt{n}\,(\delta_{1n} - \theta_1), \ldots, \sqrt{n}\,(\delta_{kn} - \theta_k)\right) \in C\right]$$
$$\geq P\left[\left(\sqrt{n}\,(\delta'_{1n} - \theta_1), \ldots, \sqrt{n}\,(\delta'_{kn} - \theta_k)\right) \in C\right].$$

For a proof, see, for example, Tong (1990, p. 73).
A related property is provided by the following result.

Theorem 7.6.2 *Let δ_n and δ'_n satisfy (7.6.1) and let g be any function of k variables for which the k first partial derivatives exist in a neighborhood of $\theta = (\theta_1, \ldots, \theta_k)$. Then*

$$\sqrt{n}\,[g\,(\delta_{1n}, \ldots, \delta_{kn}) - g\,(\theta_1, \ldots, \theta_k)]$$

and

$$\sqrt{n}\,[g\,(\delta'_{1n}, \ldots, \delta'_{kn}) - g\,(\theta_1, \ldots, \theta_k)]$$

are asymptotically normal with means 0 and variances

(7.6.4) $v(\theta) = \sum\sum \dfrac{\partial g}{\partial \theta_i} \dfrac{\partial g}{\partial \theta_j} \sigma_{ij}(\theta)$ *and* $v'(\theta) = \sum\sum \dfrac{\partial g}{\partial \theta_i} \dfrac{\partial g}{\partial \theta_j} \sigma'_{ij}(\theta),$

where $\sigma_{ij}(\theta)$ and $\sigma'_{ij}(\theta)$ are the (ij)-th element of Σ and Σ', respectively. If, in addition, (7.6.2) holds, then

(7.6.5) $v(\theta) \leq v'(\theta)$ *for all θ.*

Proof. Asymptotic normality with variances (7.6.4) follows from Theorem 5.4.6. If $\Sigma' - \Sigma$ is positive semidefinite, the inequality (7.6.5) is an immediate consequence of Definition 5.3.3.   ∎

A special case of Theorem 7.6.2 is the fact that $\sqrt{n}\,(\delta_{in} - \theta_i)$ and $\sqrt{n}\,(\delta'_{in} - \theta_i)$ are asymptotically normal with zero means and variances $v_i(\theta)$ and $v'_i(\theta)$, respectively, satisfying

$$(7.6.6) \qquad\qquad v_i(\theta) \le v'_i(\theta).$$

In seeking an efficient estimator of $(\theta_1, \ldots, \theta_k)$, we shall, in generalization of the one-parameter case, restrict attention to estimators δ_n satisfying

$$(7.6.7) \qquad \left(\sqrt{n}\,(\delta_{1n} - \theta_1), \ldots, \sqrt{n}\,(\delta_{kn} - \theta_k)\right) \xrightarrow{L} N\left(0, \Sigma(\theta)\right)$$

for some nonsingular covariance matrix $\Sigma(\theta)$, and within this class, consider an estimator efficient if it minimizes $\Sigma(\theta)$ in the sense of (7.6.2). Under the assumptions of Theorem 7.5.2, the estimators $\left(\hat{\theta}_{1n}, \ldots, \hat{\theta}_{kn}\right)$ satisfying (7.5.16) are then efficient, that is, satisfy

$$(7.6.8) \qquad \begin{array}{l} \Sigma(\theta) - I^{-1}(\theta) \text{ is positive semidefinite for any estimator } \delta_n \\ \text{satisfying (7.6.7),} \end{array}$$

provided we add the restriction that all elements of the matrices $\Sigma(\theta)$ and $I^{-1}(\theta)$ are continuous functions of θ. (For a proof, see Bahadur (1964)).

It was seen in Section 7.4 that (7.6.8) does not hold without some restrictions on the class of estimators δ_n. The significance of these restrictions, asymptotic normality, and continuity of the covariances raises issues that are beyond the scope of this book. The following comments are included to give a more complete picture. However, they will provide only a rather vague idea of the situation.

In the multivariate case, well-known counterexamples to the efficiency of the MLE without such restriction are Stein-type shrinkage estimators. (See, for example, Lehmann and Casella (1998).) These show the MLE to be inadmissible when estimating a normal vector mean with $k \ge 3$. This small-sample result implies the corresponding result asymptotically. In these examples, the asymptotic covariances are continuous. However, the limit distribution of the more efficient estimator is not normal; in addition, it is biased, so that the comparison has to be made in terms of expected squared error rather than variance. Analogous examples become available without the restriction $k \ge 3$ when one adopts a local point of view similar to that taken in Theorem 3.3.3 in a testing framework. (See, for example, Bickel (1984).) At the same time, the local approach permits the introduction of appropriate regularity conditions (Bickel et al., 1993).

General classes of estimators that are efficient (i.e., satisfy (7.5.16)) are:

(i) consistent solutions $\hat{\theta}_n$ of the likelihood equations under the assumptions of Theorem 7.5.2;

(ii) the one-step estimators (7.5.43) under the assumptions of Theorem 7.5.3;

(iii) Bayes estimators under assumptions stated for example in Le Cam (1953) and Ferguson (1996).

The estimators (i) and (ii) are illustrated in the preceding section. We shall now consider an example of multivariate Bayes estimation.

Example 7.6.1 Multinomial. In generalization of the binomial, Example 1.3.1 and of Example 7.4.4 of the present chapter, suppose that $(Y_1, \ldots, Y_{k+1})$ have the multinomial distribution (5.5.29), and that the vector $(p_1, \ldots, p_{k+1})$ is distributed with probability density

$$(7.6.9) \qquad C p_1^{a_1-1} \cdots p_{k+1}^{a_{k+1}-1}, \; 0 < p_i, \; \sum p_j = 1, \; 0 < a_i,$$

where

$$(7.6.10) \qquad C = \Gamma(a_1 + \cdots + a_{k+1}) / \Gamma(a_1) \cdots \Gamma(a_{k+1}).$$

The *Dirichlet distribution* (7.6.9) is a generalization of the beta distribution that was assumed in the binomial example, and under it we have

$$(7.6.11) \qquad E(p_i) = a_i / \sum_{j=1}^{k+1} a_j.$$

The conditional joint distribution of the p's given $(y_1, \ldots, y_{k+1})$ is proportional to (Problem 6.1)

$$p_1^{y_1+a_1-1} \cdots p_{k+1}^{y_{k+1}+a_{k+1}-1}$$

and is therefore again a Dirichlet distribution. The Bayes estimator of $(p_1, \ldots, p_{k+1})$ under squared error is, in generalization of (7.4.13), the conditional expectation of $(p_1, \ldots, p_{k+1})$ given the y's, and by (7.6.11) is thus

$$(7.6.12) \qquad \left(\frac{a_1 + y_1}{n + \sum a_j}, \ldots, \frac{a_{k+1} + y_{k+1}}{n + \sum a_j} \right).$$

The joint limit distribution of

$$(7.6.13) \qquad \sqrt{n} \left(\frac{Y_1 + a_1}{n + \sum a_j} - p_1 \right), \ldots, \sqrt{n} \left(\frac{Y_{k+1} + a_{k+1}}{n + \sum a_j} - p_{k+1} \right)$$

is the same as that of the MLE

$$(7.6.14) \qquad \sqrt{n} \left(\frac{Y_1}{n} - p_1 \right), \ldots, \sqrt{n} \left(\frac{Y_{k+1}}{n} - p_{k+1} \right)$$

which was derived in Example 5.4.1 and in Example 7.5.3 of the present chapter.

As was the case in the binomial Example 7.4.4, the asymptotic distribution of the Bayes estimators (7.6.12) is independent of the choice of $a_1, \ldots, a_{k+1}$, and all these estimators are asymptotically efficient. This independence of the asymptotic distribution of Bayes estimators of the prior distribution is a quite general phenomenon. For an exact statement and proof, see Ferguson (1996) or Lehmann and Casella (1998).   □

The definition of a prior distribution that could be considered noninformative is more complicated in the multivariate than in the univariate case treated in Section 7.4, and we shall not consider it here. For references, see, for example, Bernardo and Smith (1994), Robert (1994), and Kass and Wasserman (1996).

Instead of considering the estimation of the vector parameter $(\theta_1, \ldots, \theta_k)$, let us now suppose that interest centers on one of the parameters, say θ_1, with the others playing the role of nuisance parameters. If past experience provides a considerable amount of information concerning $(\theta_2, \ldots, \theta_k)$, one might consider treating them as known. On the other hand, such reliance on past experience always carries risks. The decision of whether to treat them as known might depend on the gain in efficiency resulting from this assumption. If $\hat{\theta}_1$ and $\hat{\hat{\theta}}_1$ denote efficient estimators when $\theta_2, \ldots, \theta_k$ are respectively unknown or known, we shall then be interested in the ARE of $\hat{\theta}_1$ to $\hat{\hat{\theta}}_1$. The following example illustrates the calculation of such an ARE.

Example 7.6.2 Correlation coefficient. Consider the estimation of the correlation coefficient ρ on the basis of a sample $(X_1, Y_1), \ldots, (X_n, Y_n)$ from a bivariate normal distribution. If the means ξ and η and variances σ^2 and τ^2 of X and Y are known, it was seen in Example 7.3.5 that an efficient estimator $\hat{\hat{\rho}}$ of ρ satisifies

$$(7.6.15) \qquad \sqrt{n}\left(\hat{\hat{\rho}} - \rho\right) \to N\left(0, \frac{(1-\rho^2)^2}{1+\rho^2}\right).$$

On the other hand, if $\hat{\rho}$ is an efficient estimator of ρ when ξ, η, σ^2, and τ^2 are unknown, it follows from Example 7.5.4 that

$$(7.6.16) \qquad \sqrt{n}\left(\hat{\rho} - \rho\right) \xrightarrow{L} N\left(0, \left(1-\rho^2\right)^2\right).$$

By the definition of the ARE and (4.3.13), we therefore have

$$(7.6.17) \qquad e_{\hat{\rho}, \hat{\hat{\rho}}} = \frac{1}{1+\rho^2}.$$

This is always ≤ 1, as it must be since $\hat{\hat{\rho}}$ is efficient when the other parameters are known and $\hat{\rho}$ is a competing estimator in this situation. The maximum gain in efficiency occurs when ρ is close to ± 1. Knowledge of the remaining parameters then nearly doubles the efficiency. On the other hand, asymptotically this knowledge brings no gain when $\rho = 0$. $\square$

Example 7.6.3 Normal mixtures. In Example 7.5.6, we considered one-step estimators for a normal mixture model with the starting estimators obtained by the method of moments. If, instead, these moment estimators are used to estimate the parameters without the adjustment (7.3.21), one must expect a loss of efficiency. The magnitude of this loss has been studied by Tan and Chang (1972) for the mixture model

$$(7.6.18) \qquad pN\left(\xi, \sigma^2\right) + qN\left(\eta, \sigma^2\right)$$

in which p, ξ, η, and σ^2 are unknown but the variances of the two normal distributions are assumed to be equal. The authors found that the ARE of the moment estimator to the one-step estimator (or equivalently the MLE) depends only on

$$(7.6.19) \qquad p \text{ and } \Delta = (\eta - \xi)/\sigma.$$

Because of the symmetry of (7.6.18) with respect to the two component distributions, attention can further be restricted to the values $0 < p \leq .5$. Table 7.6.1 shows some of the ARE's for the estimators of p and $\theta = (\eta + \xi)/2$ (which are also the absolute efficiencies of the moment estimators).

TABLE 7.6.1. Efficiency of moment estimators

Δ	p					
	.05	.2	.4	.05	.2	.4
.5	.001	.010	.071	.009	.029	.133
1	.190	.565	.808	.212	.608	.914
2	.572	.771	.781	.645	.802	.783
4	.717	.832	.846	.623	.647	.640
5	.849	.904	.911	.630	.649	.647
	ARE for p			ARE for $\theta = (\eta + \xi)/2$		

Source: Tan and Chang (1972).

The table clearly shows the dramatic improvement of efficient estimators over the moment estimators for small values of Δ (i.e., when the two normal distributions are relatively close), particularly when $|p - 1/2|$ is large. Table 7.6.2 shows the values of the asymptotic variance of $\sqrt{n}\left(\hat{\theta}_n - \theta\right)$ for the efficient estimator $\hat{\theta}_n$ of θ for $p = .2$ and $.4$ and a number of values of Δ. This table indicates the large sample sizes that would be required for accurate estimation of θ when Δ is small. $\square$

In general, the presence of unknown nuisance parameters decreases the efficiency with which the parameters of interest can be estimated. This is,

TABLE 7.6.2. Asymptotic variance of $\hat{\theta}_n$

Δ	.5	1	2	3	4	5
$p = .2$	185.5	105.0	10.9	3.8	2.3	1.8
$p = .4$	101.3	41.1	5.8	2.4	1.5	1.2

Source: Tan and Chang (1972).

however, not always the case, as can be seen from the estimation of the mean ξ and variance σ^2 of a normal distribution treated in Example 7.5.1 (continued). Here the estimator of ξ is $\bar{X}$, regardless of whether or not σ^2 is known so that there is no efficiency loss due to not knowing σ^2. The MLE of σ^2 is $\hat{\sigma}^2 = \sum (X_i - \bar{X}^2)/n$ when ξ is unknown and $\hat{\hat{\sigma}}^2 = \sum (X_i - \xi)^2/n$ when ξ is known, but as is seen from Example 2.4.4,

$$\sqrt{n}\left(\hat{\sigma}^2 - \sigma^2\right) \text{ and } \sqrt{n}\left(\hat{\hat{\sigma}}^2 - \sigma^2\right)$$

have the same limit distribution. Thus ignorance of ξ causes no loss of asymptotic efficiency in the estimation of σ^2.

To see just when there is or is not a loss of efficiency, consider a model depending on the k parameters $\theta_1, \ldots, \theta_k$, with positive definite information matrix

$$(7.6.20) \qquad I(\theta) = (I_{ij}) \text{ and } I^{-1}(\theta) = J(\theta) = (J_{ij}).$$

Then, under the assumptions of Theorem 7.5.2, there exist efficient estimators $\left(\hat{\theta}_{1n}, \ldots, \hat{\theta}_{kn}\right)$ of $(\theta_1, \ldots, \theta_k)$ such that

$$(7.6.21) \qquad \sqrt{n}\left(\hat{\theta}_{1n} - \theta_1\right), \ldots, \sqrt{n}\left(\hat{\theta}_{kn} - \theta_k\right)$$

has a joint multivariate limit distribution with mean $(0, \ldots, 0)$ and covariance matrix $J(\theta)$. In particular therefore,

$$(7.6.22) \qquad \sqrt{n}\left(\hat{\theta}_{jn} - \theta_j\right) \to N\left(0, J_{jj}\right).$$

On the other hand, if $\theta_1, \ldots, \theta_{j-1}, \theta_{j+1}, \ldots, \theta_k$ are known, it follows from the definition of $I(\theta)$ and Theorem 7.3.1 with $k = 1$ that an efficient estimator $\hat{\hat{\theta}}_{jn}$ of θ_j satisfies

$$(7.6.23) \qquad \sqrt{n}\left(\hat{\hat{\theta}}_{jn} - \theta_j\right) \to (0, 1/I_{jj}).$$

The argument of Example 7.6.2 then shows that

$$(7.6.24) \qquad \frac{1}{I_{jj}} \leq J_{jj},$$

and we are looking for conditions under which equality holds in (7.6.24).

For this purpose, let us give a direct algebraic proof of (7.6.24) for the case $k = 2$, $j = 1$, using the fact that by (5.3.6),

$$(7.6.25) \qquad J_{11} = I_{22}/\Delta, \quad J_{12} = -I_{12}/\Delta, \quad J_{22} = I_{11}/\Delta,$$

where

$$\Delta = I_{11}I_{22} - I_{12}^2.$$

Here Δ is positive since I is positive definite (Problem 6.2), and we therefore have $I_{11}I_{22} \leq \Delta$, which is equivalent to (7.6.24). We furthermore see that equality holds in (7.6.24) if and only if

$$(7.6.26) \qquad I_{12} = 0,$$

which is therefore a necessary and sufficient condition for $\hat{\theta}_{1n}$ to be as efficient as $\hat{\hat{\theta}}_{1n}$. By (7.6.25), condition (7.6.26) is equivalent to

$$(7.6.27) \qquad J_{12} = 0$$

and since J is the asymptotic covariance matrix of $\sqrt{n}\left(\hat{\theta}_{1n} - \theta_1\right)$, $\sqrt{n}\left(\hat{\theta}_{2n} - \theta_2\right)$, another equivalent condition is that

$$(7.6.28) \qquad \sqrt{n}\left(\hat{\theta}_{1n} - \theta_1\right) \text{ and } \sqrt{n}\left(\hat{\theta}_{2n} - \theta_2\right)$$

are asymptotically independent. Still another equivalent condition is that (Problem 6.3)

$$(7.6.29) \qquad \sqrt{n}\left(\hat{\hat{\theta}}_{1n} - \theta_1\right) \text{ and } \sqrt{n}\left(\hat{\hat{\theta}}_{2n} - \theta_2\right)$$

are asymptotically independent.

These results are illustrated by the estimation of ξ and σ^2 in a normal distribution discussed above. Formula (7.5.2) shows that $I_{12} = 0$; the estimators $\bar{X}$ and $\hat{\sigma}^2$ are asymptotically independent, as are $\bar{X}$ and $\hat{\hat{\sigma}}^2$, and the efficiency with which either parameter can be estimated is unaffected by whether or not the other parameter is known.

For more general location-scale families (7.5.4), it follows from Example 7.5.2 that the efficiency with which the location or scale parameter can be estimated is not impaired by not knowing the other parameter, provided the probability density is symmetric about the origin.

It is interesting to note that the conditions (7.6.26)–(7.6.29) are symmetric in the two parameters. As a result, if the efficiency with which θ_1 can be estimated does not depend on whether θ_2 is known, the corresponding fact holds for the estimation of θ_2.

For analogous results concerning the estimation of k parameters, see Problems 6.4–6.6.

The theory of efficient estimation developed in this and the preceding section has so far been restricted to the case of a single sample of i.i.d. variables or vectors. However, the theory extends with only minor changes to the case of two or more samples.

Let us consider first the case that θ is real-valued and suppose that $X_{\gamma 1}, \dots, X_{\gamma n_\gamma}$ ($\gamma = 1, \dots, r$) are i.i.d. according to a distribution with density $f_{\gamma,\theta}$ and that the r samples are independent. We shall take r as fixed and assume that the sample sizes n_γ all tend to infinity at the same rate. More specifically, we shall consider a sequence of sample sizes $\left(n_1^{(\nu)}, \dots, n_r^{(\nu)}\right)$, $\nu = 1, 2, \dots$, with total sample size

$$N_\nu = n_1^{(\nu)} + \cdots + n_r^{(\nu)}$$

such that

(7.6.30)
$$\frac{n_\gamma^{(\nu)}}{N_\nu} \to \lambda_\gamma \text{ as } \nu \to \infty,$$

where $\sum \lambda_\gamma = 1$ and the λ_γ are > 0.

A central concept in the i.i.d. case is the amount of information $I(\theta)$ contained in a single observation. To see how to generalize this idea, recall that the total amount of information contained in n i.i.d. observations is $T(\theta) = nI(\theta)$. In the present situation, assuming the information contained in independent observations to be additive, the total amount of information is

$$T(\theta) = \sum_{\gamma=1}^r n_\gamma I_\gamma(\theta),$$

where $I_\gamma(\theta)$ denotes the information provided by an observation from $f_{\gamma,\theta}$. The average information per observation is therefore

$$\frac{1}{N_\nu} T(\theta) = \sum_{\gamma=1}^r \frac{n_\gamma}{N_\nu} I_\gamma(\theta),$$

which tends to

(7.6.31)
$$I(\theta) = \sum_{\gamma=1}^r \lambda_\gamma I_\gamma(\theta)$$

as $N_\nu \to \infty$. We shall see below that this average amount of information plays the role of the information measure $I(\theta)$ defined by (7.1.28) in the i.i.d. case.

If

$$(7.6.32) \qquad L_\gamma(\theta) = \prod_{i=1}^{n_\gamma} f_{\gamma,\theta}(x_{\gamma,i}) \text{ and } l_\gamma(\theta) = \sum_{i=1}^{n_\gamma} \log f_{\gamma,\theta}(x_{\gamma,i})$$

denote the likelihood and log likelihood of θ based on the γth sample, the likelihood and log likelihood based on all N observations is respectively

$$(7.6.33) \qquad L(\theta) = \prod_{\gamma=1}^{r} L_\gamma(\theta) \text{ and } l(\theta) = \sum_{\gamma=1}^{r} l_\gamma(\theta),$$

and the likelihood equation becomes

$$l'(\theta) = \sum_{\gamma=1}^{r} l'_\gamma(\theta) = 0.$$

These considerations extend to the multiparameter case, where, however, the notation becomes rather complicated. In addition, it will be at variance with that used in the one-sample case, with $I_\gamma(\theta)$ and $I(\theta)$ now denoting the information matrices and with the likelihood equation replaced by the set of equations

$$(7.6.34) \qquad l^{(j)}(\theta) = \sum_{\gamma=1}^{r} l_\gamma^{(j)}(\theta) = 0, \; j = 1, \ldots, k,$$

where for each γ, the derivative $l_\gamma^{(j)}(\theta)$ is given by

$$l_\gamma^{(j)}(\theta) = \sum_{i=1}^{n_\gamma} \frac{\partial}{\partial \theta_j} \log f_{\gamma,\theta}(x_{\gamma,i}).$$

In generalization of Theorem 7.3.1, we then have

Theorem 7.6.3 *For each $\gamma = 1, \ldots, r$, let $X_{\gamma 1}, \ldots, X_{\gamma n_\gamma}$ be i.i.d. with density $f_{\gamma,\theta}(x)$ satisfying the assumptions of Theorem 7.5.2, and suppose that all $N = \sum n_\gamma$ observations are independent. Let $\left(n_1^{(\nu)}, \ldots, n_r^{(\nu)}\right)$ be a sequence of sample sizes satisfying (7.6.30) where, in the following, we shall suppress the index ν. Then any consistent sequence*

$$\hat{\theta}_N = \hat{\theta}_N(X_{11}, \ldots, X_{1n_1}, ; \cdots ; X_{r1}, \ldots, X_{rn_r})$$

of the likelihood equations (7.6.34) satisfies

$$(7.6.35) \qquad \sqrt{N}\left(\hat{\theta}_N - \theta_0\right) \xrightarrow{L} N\left(0, I^{-1}(\theta_0)\right),$$

where θ_0 denotes the true value of θ and $I(\theta)$ is given by (7.6.31).

Proof. We shall give the proof only for the case $k - 1$, i.e., that θ is real-valued. Then equations (7.3.5) and (7.3.6) hold exactly as in the proof of Theorem 7.3.1. The proof will be completed by verifying (7.3.7)–(7.3.9), with n replaced by N throughout.

To prove the result corresponding to (7.3.7), write

$$(7.6.36) \qquad \frac{1}{\sqrt{N}} l'(\theta) = \sum_{\gamma=1}^{r} \left(\sqrt{\frac{n_\gamma}{N}} \frac{1}{\sqrt{n_\gamma}} l'_\gamma(\theta) \right).$$

By (7.3.7) with n_γ in place of n, we have

$$(7.6.37) \qquad \frac{1}{\sqrt{n_\gamma}} l'_\gamma(\theta) \xrightarrow{L} N\left(0, I_\gamma(\theta_0)\right)$$

and the result now follows from (7.3.36), (7.6.30), and the fact that $l'_1(\theta)$, $\ldots, l'_r(\theta)$ are independent.

Analogously

$$(7.6.38) \qquad -\frac{1}{N} l''(\theta_0) = -\frac{1}{N} \sum_{\gamma=1}^{r} l''_\gamma(\theta_0) = \sum \left[\frac{n_\gamma}{N} \cdot \frac{-1}{n_\gamma} l''_\gamma(\theta_0) \right].$$

Since $n_\gamma/N \to \lambda_\gamma$ by (7.6.30) and $-l''_\gamma(\theta_0)/n_\gamma \xrightarrow{P} I_\gamma(\theta_0)$ by (7.3.8), it is seen that (7.6.38) tends in probability to $I(\theta_0)$, as was to be proved—the extension of (7.3.9) follows in the same way (Problem 6.7).

Le Cam's result, stated following Example 7.4.1, continues to apply in the multisample case and shows that an estimator $\hat{\theta}_N$ satisfying (7.6.35) is efficient in the class of all estimators satisfying (7.4.1) (with N in place of n) with continuous $v(\theta)$. Other results that remain in force with only the obvious changes are Theorem 7.3.3 concerning one-step estimators (Problem 6.8) and multivariate generalizations such as Theorems 7.5.2, 7.5.3, and 7.6.2 (Problem 6.9). ∎

Example 7.6.4 The normal two-sample problem. Let $X_1, \ldots, X_m$ and $Y_1, \ldots, Y_n$ be independent samples from the normal distributions $N\left(\xi, \sigma^2\right)$ and $N\left(\eta, \tau^2\right)$, respectively. If all four parameters are unknown, it is seen as in Example 7.5.1 (continued) that the likelihood equations lead to the estimators of ξ and σ^2 given by (7.5.18) for the one-sample case and to the corresponding one-sample estimators for η and τ^2. Let us next consider the situations in which it is assumed that either $\tau^2 = \sigma^2$ or $\eta = \xi$.

(i) *Common variance.*

The log likelihood is now

$$l\left(\xi, \eta, \sigma^2\right) = -(m+n) \log \sigma - \frac{1}{2\sigma^2} \left[\sum (x_i - \xi)^2 + \sum (y_j - \eta)^2 \right] + C$$

and the likelihood equations become

$$\frac{\partial}{\partial \xi} : \frac{m}{\sigma^2} (\bar{x} - \xi) = 0, \quad \frac{\partial}{\partial \eta} : \frac{n}{\sigma^2} (\bar{y} - \eta) = 0,$$

and

$$\frac{\partial}{\partial \sigma} : -\frac{N}{\sigma} + \frac{1}{\sigma^3} \left[\sum (x_i - \xi)^2 + \sum (y_j - \eta)^2 \right],$$

where $N = m + n$. These have the unique solution

$$\hat{\xi} = \bar{x}, \quad \hat{\eta} = \bar{y}, \quad \hat{\sigma}^2 = \frac{1}{N} \left[\sum (x_i - \bar{x})^2 + \sum (y_i - \bar{y})^2 \right],$$

which are therefore the efficient estimators of ξ, η, and σ^2, respectively.

(ii) *Common mean*

To estimate the common mean $\eta = \xi$, let us suppose first that σ^2 and τ^2 are known. The log likelihood is then

$$l(\xi) = -m \log \sigma - n \log \tau - \frac{1}{2\sigma^2} \sum (x_i - \xi)^2 - \frac{1}{2\tau^2} \sum (y_j - \xi)^2 + C,$$

and the likelihood equation has the unique solution (Problem 6.10(i))

$$(7.6.39) \qquad \hat{\xi} = \left(\frac{m\bar{X}}{\sigma^2} + \frac{n\bar{Y}}{\tau^2} \right) \Big/ \left(\frac{m}{\sigma^2} + \frac{n}{\tau^2} \right),$$

which is a special case of the estimator given in Example 2.2.2 with the weights (2.2.8).

If σ^2 and τ^2 are unknown, the efficiency of an efficient estimator $\tilde{\xi}$ of ξ cannot exceed that of $\hat{\xi}$. However, if $\hat{\sigma}^2$ and $\hat{\tau}^2$ are any consistent estimators of σ^2 and τ^2, respectively, in particular, for example, if $\hat{\sigma}^2$ is given by (7.5.18) and $\hat{\tau}^2$ is defined analogously, and if

$$(7.6.40) \qquad \tilde{\xi} = \left(\frac{m\bar{X}}{\hat{\sigma}^2} + \frac{n\bar{Y}}{\hat{\tau}^2} \right) \Big/ \left(\frac{m}{\hat{\sigma}^2} + \frac{n}{\hat{\tau}^2} \right),$$

it is tempting to claim that the asymptotic distribution of $\sqrt{N} \left(\tilde{\xi} - \xi \right)$ is the same as that of $\sqrt{N} \left(\hat{\xi} - \xi \right)$. However, this conclusion is false (Problem 6.10(ii)).   □

This argument shows that $\hat{\xi}, \hat{\sigma}^2$, and $\hat{\tau}^2$ minimize respectively the asymptotic variances of $\sqrt{N} \left(\hat{\xi} - \xi \right)$, $\sqrt{N} \left(\hat{\sigma}^2 - \sigma^2 \right)$, and $\sqrt{N} \left(\hat{\tau}^2 - \tau^2 \right)$ when the other parameters are unknown. However, it is not clear that this proves that the vector $\left(\hat{\xi}, \hat{\sigma}^2, \hat{\tau}^2 \right)$ is efficient for estimating (ξ, σ^2, τ^2) in the sense of (7.6.8). The following theorem shows that this conclusion is in fact justified.

Theorem 7.6.4 *For each $i = 1, \ldots, k$, let $\hat{\theta}_i$ be an efficient estimator of θ_i when the remaining parameters are unknown. Then under the assumptions of Theorem 7.5.2, the vector $\left(\hat{\theta}_1, \ldots, \hat{\theta}_k \right)$ is efficient for estimating $(\theta_1, \ldots, \theta_k)$ in the sense of (7.6.8).*

Proof. Under the assumptions of Theorem 7.5.2, there exists an efficient vector estimator $(\theta_1^*, \ldots, \theta_k^*)$ of $(\theta_1, \ldots, \theta_k)$. Thus

(7.6.41) $\hat{\Sigma} - \Sigma^*$ is positive semidefinite,

where $\hat{\Sigma}$ and Σ^* denote the asymptotic covariance matrices of $\sqrt{n}(\hat{\theta}_1 - \theta_1)$, $\ldots, \sqrt{n}(\hat{\theta}_k - \theta_k)$ and $\sqrt{n}(\theta_1^* - \theta_1), \ldots, \sqrt{n}(\theta_k^* - \theta_k)$, respectively. Therefore

(7.6.42) $$\sum_{i=1}^{k} \sum_{j=1}^{k} \left(\hat{\sigma}_{ij} - \sigma_{ij}^* \right) u_i u_j \geq 0 \text{ for all vectors } (u_1, \ldots, u_k),$$

where $\hat{\sigma}_{ij}$ and σ_{ij}^* denote the (i,j)-th element of $\hat{\Sigma}$ and Σ^*. Putting $u_1 = \cdots = u_{i-1} = u_{i+1} = \cdots = u_k = 0$, we have

$$\sigma_{ii}^* \leq \hat{\sigma}_{ii}$$

and hence, by the efficiency of $\hat{\theta}_i$,

(7.6.43) $$\sigma_{ii}^* = \hat{\sigma}_{ii}.$$

It follows that

$$\sum_{i \neq j} \sum \left(\hat{\sigma}_{ij} - \sigma_{ij}^* \right) u_i u_j \geq 0 \text{ for all } (u_1, \ldots, u_k).$$

Let all u's except u_i and u_j be 0. Then

$$\left(\hat{\sigma}_{ij} - \sigma_{ij}^* \right) u_i u_j \geq 0 \text{ for all } u_i, u_j.$$

Since $u_i u_j$ can be both positive and negative, this implies that $\hat{\sigma}_{ij} = \sigma_{ij}^*$, so that $\hat{\Sigma} = \Sigma^*$, as was to be proved. ∎

Example 7.6.5 Regression. Let X_{ij}, $j = 1, \ldots, n_i$ and $i = 1, \ldots, r$ be independently normally distributed with common variance σ^2 and with means

(7.6.44) $$E(X_{ij}) = \alpha + \beta v_i.$$

The likelihood equations are (Problem 6.11(i))

(7.6.45) $$\alpha \sum n_i v_i + \beta \sum n_i v_i^2 = \sum n_i v_i x_{i\cdot},$$
$$\alpha \sum n_i + \beta \sum n_i v_i = \sum n_i x_{i\cdot},$$
$$\sigma^2 = \frac{1}{n} \sum \sum (x_{ij} - \alpha - \beta v_i)^2,$$

where $x_{i.} = \sum x_{ij}/n_i$ and $n = \sum n_{i.}$. They have a unique solution $\left(\hat{\alpha}, \hat{\beta}, \hat{\sigma}^2\right)$, which is efficient since the assumptions of Theorem 7.5.2 hold for each of the r samples. The estimators agree with the least squares estimators given for a more general situation in (2.7.11) (Problem 6.11(ii)). $\square$

In the regression model (7.6.44), we assumed that the observations were taken at a fixed number of levels, many observations at each level. In contrast, the model considered in Examples 2.7.4 and 2.7.6 permits the number of levels to tend to infinity as the number of observations increases and, in particular, includes the case in which there is only one observation at each level, so that the number of observations equals the number of levels. This is an example of the general situation of independent nonidentical observations. This case requires more changes in the theory than the extension from one sample to a fixed number of samples greater than one, and we shall begin by considering a simple one-parameter example.

Example 7.6.6 Independent binomials with common p. Let X_i $(i = 1, \ldots, n)$ be independent binomial $b(p, k_i)$. Then

$$L(p) = \prod \binom{k_i}{x_i} p^{x_i} q^{k_i - x_i}$$

and an easy calculation shows that the maximum likelihood estimator is

(7.6.46)
$$\hat{p} = \frac{\sum X_i}{\sum k_i}.$$

Since $\sum X_i$ has the binomial distribution $b(p, N)$ with $N = \sum k_i$, we have

(7.6.47)
$$\sqrt{N}(\hat{p} - p) \xrightarrow{L} N(0, pq).$$

Here the normalizing factor $\sqrt{N}$ may have order very different from $\sqrt{n}$. Suppose, for example, that $k_i = i^\alpha$ $(\alpha > 0)$. Then by (1.3.6),

$$N = \sum_{i=1}^{n} i^\alpha \sim \frac{n^{1+\alpha}}{1+\alpha}.$$

Therefore $\sqrt{N}$ is of order $n^{(1+\alpha)/2}$ which can be of any order $\geq \sqrt{n}$. (For an example in which also orders less than $\sqrt{n}$ are possible, see Problem 6.13.)

The normalizing factor $\sqrt{N}$ can be given an interpretation in terms of information. The amount of information that X_j contains about p is

(7.6.48)
$$I_j(p) = k_j/pq$$

and the total information the sample contains about p by Theorem 7.2.2 is therefore

(7.6.49) $$T_n(p) = \sum I_j(p) = \sum k_j/pq = N/pq.$$

Thus the normalizing factor C_n required for

(7.6.50) $$\sqrt{C_n}(\hat{p} - p) \to N(0, 1)$$

is just $\sqrt{T_n(p)}$.

Quite generally, the total amount of information in a set of independent variables $X_1, \ldots, X_n$ is

(7.6.51) $$T_n(\theta) = \sum_{j=1}^{n} I_j(\theta),$$

where $I_j(\theta)$ is the amount of information in X_j, and it turns out that under suitable regularity conditions

(7.6.52) $$\sqrt{T_n(\theta_0)}(\hat{\theta}_n - \theta_0) \xrightarrow{L} N(0, 1).$$

To see how to obtain conditions under which (7.6.52) holds, consider the proof of Theorem 7.3.1. The expansion (7.3.6) remains valid with

$$l_n(\theta_0) = \sum_{j=1}^{n} \log f_{j,\theta}(X_j),$$

but (7.3.6) has to be replaced by

(7.6.53)

$$\sqrt{T_n(\theta_0)}\left(\hat{\theta}_n - \theta_0\right) = \frac{l_n'(\theta_0)/\sqrt{T_n(\theta_0)}}{-\dfrac{1}{T_n(\theta_0)}l_n''(\theta_0) - \dfrac{1}{2T_n(\theta_0)}\left(\hat{\theta}_n - \theta_0\right)l_n'''(\theta_n^*)}.$$

Then

$$l_n'(\theta_0) = \sum \frac{\partial}{\partial \theta} \log f_{j,\theta}(X_j)$$

is a sum of n independent variables with mean 0 and variance $I_j(\theta_0)$, and conditions for

(7.6.54) $$\frac{l_n'(\theta_0)}{\sqrt{T_n(\theta_0)}} \text{ to tend to } N(0, 1)$$

are given, for example, in Theorem 2.7.1. Sufficient conditions for

(7.6.55) $$-l_n''(\theta_0)/T_n(\theta_0) \xrightarrow{P} 1$$

can be obtained from Example 2.2.2 (Problem 6.15), and the remainder term can be handled as in the proof of Theorem 7.3.1.

A common feature of the extensions considered so far is that the number of parameters remains fixed, independent of sample size. When this is not the case, the situation can be very different, as is seen in the following example due to Neyman and Scott (1948).   □

Example 7.6.7 Common variance. Let $(X_{\alpha 1}, \ldots, X_{\alpha r})$, $\alpha = 1, 2, \ldots, n$, be independently distributed as $N\left(\theta_\alpha, \sigma^2\right)$. The situation is similar to that of Theorem 7.6.3, but the sample sizes (previously denoted by n_α and assumed to $\to \infty$) are now fixed and it is the number n of samples (previously assumed fixed and denoted by r) that now tends to infinity. The parameter being estimated is σ^2, while the means $\theta_1, \ldots, \theta_n$ are nuisance parameters, the number of which tends to infinity with the number of samples. If $N = rn$, the log likelihood is

$$l_n\left(\sigma^2; \theta_1, \ldots, \theta_n\right) = -N \log \sigma - \frac{1}{2\sigma^2} \sum \sum (x_{\alpha j} - \theta_\alpha)^2 + c,$$

and the likelihood equations have the unique solution

$$\hat{\theta}_\alpha = X_{\alpha\cdot}, \quad \hat{\sigma}^2 = \frac{1}{rn} \sum_{\alpha=1}^n \sum_{j=1}^r (X_{\alpha j} - X_{\alpha\cdot})^2.$$

Consider now the n statistics

$$S_\alpha^2 = \frac{1}{r} \sum_{j=1}^r (X_{\alpha j} - X_{\alpha\cdot})^2, \quad \alpha = 1, \ldots, n.$$

They are i.i.d. and, since

$$E\left(S_\alpha^2\right) = \frac{r-1}{r}\sigma^2,$$

we see that

$$\hat{\sigma}^2 = \frac{1}{n} \sum_{\alpha=1}^n S_\alpha^2 \xrightarrow{P} \frac{r-1}{r}\sigma^2,$$

so that $\hat{\sigma}^2$ is not consistent. While a simple modification (multiplication by $r/(r-1)$) remedies this failure, the point of the example is that the unique solution of the likelihood equations itself is not consistent.   □

For some recent discussions of this example, see Bickel et al. (1993) and Barndorff-Nielsen and Cox (1994).

Summary

The asymptotic performance of an asymptotically normal estimator of a vector parameter is characterized by the covariance matrix $\Sigma(\theta)$ of the limit distribution. An estimator δ_n is considered more efficient than a competitor δ'_n if $\Sigma'(\theta) - \Sigma(\theta)$ is positive semidefinite for all θ. An asymptotically normal estimator is called efficient if $\Sigma(\theta) = I^{-1}(\theta)$.

When estimating a single parameter θ_1, the presence of nuisance parameters $(\theta_2, \ldots, \theta_k)$ generally decreases the efficiency with which θ_1 can be estimated. A necessary and sufficient condition for this not to be the case is that the elements $I_{12}(\theta), \ldots, I_{1k}(\theta)$ of the information matrix are all 0.

The theory of efficient estimation is extended from the i.i.d. case to that of a fixed number r of large samples. In this extension, the average amount of information (averaged over the different samples) plays the role of $I(\theta)$ in the i.i.d. case. An example shows that this theory no longer applies when r is not fixed but is allowed to tend to ∞.

7.7   Tests and confidence intervals

So far in this chapter we have been concerned with efficient methods of point estimation, the one-parameter case being treated in Sections 1–4, and the multiparameter case in Sections 5 and 6. We shall now discuss tests and confidence procedures based on these efficient estimators and begin with the case of a single parameter θ.

A. The Wald Test. Suppose that $X_1, \ldots, X_n$ are i.i.d. and consider an estimator $\hat{\theta}_n$ which is efficient in the sense of satisfying

$$(7.7.1) \qquad \sqrt{n}\left(\hat{\theta}_n - \theta\right) \xrightarrow{L} N\left(0, 1/I(\theta)\right),$$

for example, a consistent root of the likelihood equation under the assumptions of Theorem 7.3.1. If $\hat{I}_n$ is any consistent estimator of $I(\theta)$, it follows that

$$(7.7.2) \qquad \sqrt{n}\left(\hat{\theta}_n - \theta\right)\sqrt{\hat{I}_n} \xrightarrow{L} N(0,1)$$

and hence that

$$(7.7.3) \qquad \hat{\theta}_n - \frac{u_{\alpha/2}}{\sqrt{n\hat{I}_n}} < \theta < \hat{\theta}_n + \frac{u_{\alpha/2}}{\sqrt{n\hat{I}_n}}$$

are confidence intervals for θ with asymptotic confidence coefficient $1 - \alpha$. Here $u_{\alpha/2}$ is the upper $\alpha/2$ point of the standard normal distribution. If $I(\theta)$ is a continuous function of θ, as is typically the case, it follows from Theorem 2.1.4 that $\hat{I}_n = I\left(\hat{\theta}_n\right)$ is a consistent estimator of $I(\theta)$. An alternative estimator is suggested by (7.2.10). Since under the assumptions of Theorem 7.2.1(ii) and 7.3.1, the proof of Theorem 7.3.2 shows that

$$(7.7.4) \qquad -\frac{1}{n}l_n''\left(\hat{\theta}_n\right) \xrightarrow{P} I(\theta),$$

the left side of (7.7.4) provides a consistent estimator of $I(\theta)$. For a comparison of the estimators $I\left(\hat{\theta}_n\right)$ and $-l_n''\left(\hat{\theta}_n\right)/n$, see Efron and Hinkley (1978) and Runger (1980).

For testing the hypothesis

$$(7.7.5) \qquad H : \theta = \theta_0$$

against the two-sided alternatives $\theta \neq \theta_0$ at asymptotic level α, (7.7.3) leads to the *Wald test* with acceptance region

$$(7.7.6) \qquad \theta_0 - \frac{u_{\alpha/2}}{\sqrt{n\hat{I}_n}} < \hat{\theta}_n < \theta_0 + \frac{u_{\alpha/2}}{\sqrt{n\hat{I}_n}}$$

or, equivalently, with rejection region

$$(7.7.7) \qquad \left|\hat{\theta}_n - \theta_0\right| \geq \frac{u_{\alpha/2}}{\sqrt{n\hat{I}_n}}.$$

Since the calculation of the level of this test is based solely on the distribution of $\hat{\theta}_n$ at θ_0, we can in (7.7.6) and (7.7.7) replace $\hat{I}_n$ by $I(\theta_0)$ and reject H when

$$(7.7.8) \qquad \left|\hat{\theta}_n - \theta_0\right| \geq \frac{u_{\alpha/2}}{\sqrt{nI(\theta_0)}}.$$

If H is to be tested against the one-sided alternatives $\theta > \theta_0$ rather than against $\theta \neq \theta_0$, the rejection region (7.7.8) is replaced by

$$(7.7.9) \qquad \hat{\theta}_n - \theta_0 \geq \frac{u_{\alpha}}{\sqrt{nI(\theta_0)}}.$$

B. The Likelihood Ratio Test. A standard approach to testing (7.7.5) is provided by the likelihood ratio test, which rejects when the maximum of the likelihood divided by the likelihood under H,

$$(7.7.10) \qquad \frac{L_x\left(\hat{\theta}_n\right)}{L_x(\theta_0)},$$

is sufficiently large, where $\hat{\theta}_n$ is the MLE. Taking logarithms, this is equivalent to rejecting when

$$(7.7.11) \qquad \Delta_n = l_n\left(\hat{\theta}_n\right) - l_n(\theta_0)$$

is sufficiently large, where, as earlier, l_n denotes the log likelihood. The following theorem provides the asymptotic null-distribution of (7.7.11) when $\hat{\theta}_n$ is a consistent root of the likelihood equation. Typically, but not always, $\hat{\theta}_n$ will be the MLE. With this understanding, we shall call the test based on (7.7.10) or (7.7.11) the *likelihood ratio test* without insisting that $\hat{\theta}_n$ is actually the MLE.

Theorem 7.7.1 *Under the assumptions of Theorem 7.3.1, if $\hat{\theta}_n$ is a consistent root of the likelihood equation, the null-distribution of $2\Delta_n$ tends to a χ^2-distribution with 1 degree of freedom.*

Proof. For any $(x_1, \dots, x_n)$, we can expand $l_n\left(\hat{\theta}_n\right)$ and $l'_n(\theta_0)$ in analogy with (7.3.5) to obtain

$$(7.7.12) \quad l_n\left(\hat{\theta}_n\right) - l_n(\theta_0) = \left(\hat{\theta}_n - \theta_0\right) l'_n(\theta_0) + \frac{1}{2}\left(\hat{\theta}_n - \theta_0\right)^2 l''_n(\theta_n^*).$$

Also, by (7.3.5) we have, since $l'_n\left(\hat{\theta}_n\right) = 0$,

$$(7.7.13) \quad -l'_n(\theta_0) = \left(\hat{\theta}_n - \theta_0\right) l''_n(\theta_0) + \frac{1}{2}\left(\hat{\theta}_n - \theta_0\right)^2 l'''_n(\theta_n^{**}),$$

where θ_n^* and θ_n^{**} lie between θ_0 and $\hat{\theta}_n$. Substitution of (7.7.13) into (7.7.12) leads to

(7.7.14)

$$\Delta_n = -n\left(\hat{\theta}_n - \theta_0\right)^2 \left[\frac{l''_n(\theta_0)}{n} - \frac{1}{2}\frac{l''_n(\theta_n^*)}{n} + \frac{1}{2}\left(\hat{\theta}_n - \theta_0\right)\frac{l'''_n(\theta_n^{**})}{n}\right].$$

The third term tends to 0 in probability by (7.3.9), while $l''_n(\theta_0)/n$ and $l''_n(\theta_n^*)/n$ both tend to $-I(\theta_0)$ in probability by (7.3.8) and the argument of (7.3.29), respectively. It follows that $2\Delta_n$ has the same limit distribution as

$$(7.7.15) \qquad\qquad n\left(\hat{\theta}_n - \theta_0\right)^2 I(\theta_0)$$

and, in view of (7.7.1), this completes the proof. ∎

Actually, we have proved slightly more, namely that the test statistics

$$(7.7.16) \qquad\qquad 2\Delta_n \text{ and } n\left(\hat{\theta}_n - \theta_0\right)^2 I(\theta_0)$$

are asymptotically equivalent under θ_0 in the sense that their difference tends to 0 in probability (Problem 7.4). It follows from the following lemma that the Wald and likelihood ratio tests are asymptotically equivalent under H in the sense of Definition 3.1.1.

Lemma 7.7.1 *Suppose that V_n and V_n' are two sequences of test statistics satisfying $V_n' - V_n \xrightarrow{P} 0$. If V_n has a continuous limit distribution F_0, this is also the limit distribution of V_n', and for any constant c, the tests with rejection regions $V_n \geq c$ and $V_n' \geq c$ are asymptotically equivalent.*

Proof. That F_0 is the limit distribution of V_n' follows from Corollary 2.3.1. To prove asymptotic equivalence of the two tests, let $V_n' = V_n + W_n$. Then

$$P[V_n < c \text{ and } V_n' \geq c] = P[c - W_n \leq V_n < c],$$

and this tends to zero by Problem 3.16(iii) of Chapter 2. That $P[V_n \geq c$ and $V_n' < c] \to 0$ is seen analogously. ∎

Example 7.7.1 Normal mean. Let $X_1, \ldots, X_n$ be i.i.d. $N(\theta, 1)$. Then $\hat{\theta}_n = \bar{X}$ by Example 7.1.4. To test the hypothesis $H : \theta = 0$, note that

$$l_n(\theta) = -\frac{1}{2} \sum (x_i - \theta)^2 - n \log\left(\sqrt{2\pi}\right),$$

and hence

$$2\Delta_n = \sum x_i^2 - \sum (x_i - \bar{x})^2 = n\bar{x}^2.$$

The likelihood ratio test therefore rejects when

$$n\bar{x}^2 \geq v_\alpha,$$

where v_α is the upper α point of χ_1^2. Since $I(\theta_0) = 1$, the Wald Test (7.7.8) rejects when

$$\sqrt{n}\,|\bar{x}| \geq u_{\alpha/2}.$$

In this case, these two rejection regions are not only asymptotically equivalent, but they are exactly the same and their level is exactly equal to α. □

Adaptation of the likelihood ratio test to the one-sided alternatives $\theta > \theta_0$ is less simple than it was for the Wald test. Rather than giving a general theory of the one-sided problem, we shall illustrate it by Example 7.7.1 as a prototype of the general case.

Example 7.7.1. Normal mean (continued). In Example 7.7.1, suppose that only values $\theta \geq 0$ are possible and that we wish to test $H : \theta = 0$ against $\theta > 0$. Then the results of this chapter concerning asymptotic normality and efficiency of a consistent root of the likelihood equation no longer apply because the parameter space $\theta \geq 0$ is not open. The MLE of θ is now (Problem 7.7.1(i))

$$(7.7.17) \qquad \hat{\theta}_n = \begin{cases} \bar{X} & \text{if } \bar{X} > 0 \\ 0 & \text{if } \bar{X} \leq 0. \end{cases}$$

For $\theta > 0$, the asymptotic behavior of $\hat{\theta}_n$ is as before, but for $\theta = 0$ there is a change (Problem 7.7.1(ii) and 7.7.1(iii)). Since the equation $l'_n(\theta) = 0$ has no solution when $\bar{X} < 0$, let us consider the likelihood ratio test based on (7.7.11) with $\hat{\theta}_n$ given by (7.7.17), so that (for this example only)

$$(7.7.18) \qquad 2\Delta_n = \sum x_i^2 - \sum (x_i - \bar{\theta}_n)^2 = \begin{cases} n\bar{x}^2 & \text{if } \bar{x} > 0 \\ 0 & \text{if } \bar{x} \leq 0. \end{cases}$$

Since we are rejecting for large values of $2\Delta_n$ and since

$$P_H\left(\bar{X} > 0\right) = 1/2,$$

the rejection for $\alpha < 1/2$ will be a subset of the set $\bar{x} > 0$. By (7.7.18), the likelihood ratio test therefore rejects when

$$(7.7.19) \qquad \sqrt{n}\bar{X} \geq u_\alpha.$$

For generalizations of this example, see Chernoff (1954), Feder (1968), and the related Problem 7.22. $\qquad\square$

C. The Rao Scores Test. Both the Wald and likelihood ratio tests require evaluation of $\hat{\theta}_n$. Let us now consider a third test for which this is not necessary. When dealing with large samples, interest tends to focus on distinguishing the hypothetical value θ_0 from nearby values of θ. (If the true value is at some distance from θ_0, a large sample will typically reveal this so strikingly that a formal test may be deemed unnecessary.) The test of $H : \theta = \theta_0$ against $\theta > \theta_0$ is locally most powerful if it maximizes the slope $\beta'(\theta_0)$ of the power function $\beta(\theta)$ at $\theta = \theta_0$. Standard small-sample theory shows that this test rejects H for large values of

$$(7.7.20) \qquad \frac{\frac{\partial}{\partial\theta}[f_\theta(x_1)\cdots f_\theta(x_n)]|_{\theta=\theta_0}}{f_{\theta_0}(x_1)\cdots f_{\theta_0}(x_n)} = \sum_{i=1}^n \frac{f'_{\theta_0}(x_i)}{f_{\theta_0}(x_i)} = l'_n(\theta_0).$$

(See, for example, Casella and Berger (1990), p. 377.) We saw in (7.3.7) that

$$\frac{1}{\sqrt{n}}l'_n(\theta_0) \xrightarrow{L} N(0, I(\theta_0));$$

it follows that the locally most powerful rejection region

$$(7.7.21) \qquad \frac{l'_n(\theta_0)}{\sqrt{nI(\theta_0)}} \geq u_\alpha$$

has asymptotic level α.

The corresponding two-sided rejection region given by

$$(7.7.22) \qquad \frac{|l'_n(\theta_0)|}{\sqrt{nI(\theta_0)}} \geq u_{\alpha/2}$$

is also locally best in a suitable sense. The quantity $l'_n(\theta)$ is called the *score function*; the test (7.7.22) is the *Rao score test*.

It is interesting to note that the Rao test (7.7.22) and the Wald test (7.7.7) are asymptotically equivalent. This follows from the fact that

$$(7.7.23) \qquad -\frac{1}{\sqrt{n}} l'_n(\theta_0) - \sqrt{n}\left(\hat{\theta}_n - \theta_0\right) I(\theta_0) \xrightarrow{P} 0,$$

which, in turn, is a consequence of (7.3.6) and (7.3.8) (Problem 7.10).

Example 7.7.2 Logistic. Consider a sample $X_1, \ldots, X_n$ from the logistic distribution with density

$$f_\theta(x) = \frac{e^{(x-\theta)}}{\left[1 + e^{(x-\theta)}\right]^2}.$$

Then

$$l_n(\theta) = \sum (x_i - \theta) - 2\sum \log\left(\left[1 + e^{(x_i - \theta)}\right]\right)$$

and

$$l'_n(\theta) = -n + \sum \frac{2e^{(x_i - \theta)}}{1 + e^{(x_i - \theta)}}.$$

We also require $I(\theta)$, which by (7.3.32) is equal to $1/3$ for all θ. The Rao scores test of $H : \theta = \theta_0$ vs. $K : \theta > \theta_0$ therefore rejects H when

$$\sqrt{\frac{3}{n}} \sum \frac{e^{(x_i - \theta_0)} - 1}{e^{(x_i - \theta_0)} + 1} \geq u_\alpha.$$

In this case, the MLE does not have an explicit expression and therefore the Wald and likelihood ratio tests are less convenient. However, we can get alternative, asymptotically equivalent tests by replacing $\hat{\theta}_n$ by the more convenient one-step estimator δ_n given by (7.3.21) or by other efficient estimators (see, for example, Stroud (1971) and Problem 7.7).   □

The results obtained for the Wald, Rao, and likelihood ratio tests are summarized in the following theorem.

Theorem 7.7.2 *Under the assumptions of Theorem 7.7.1, the Wald test (7.7.7), the Rao test (7.7.22), and the likelihood ratio test which for Δ_n defined by (7.7.11) rejects when*

$$(7.7.24) \qquad \sqrt{2\Delta_n} \geq u_{\alpha/2},$$

are asymptotically equivalent under H and all have asymptotic level α.

The three tests are asymptotically equivalent not only under H but, under slightly stronger assumptions than those made for the case $\theta = \theta_0$, also under a sequence of alternatives

$$(7.7.25) \qquad\qquad \theta_n = \theta_0 + \frac{c}{\sqrt{n}},$$

which in Theorem 3.3.3 was seen to be appropriate for power calculations. In particular, the asymptotic power against the alternatives (7.7.25) is thus the same for all three tests. In terms of their performance, we can therefore not distinguish between them to the degree of accuracy considered here. They do differ, however, in convenience and ease of interpretation.

Both the likelihood ratio test and the Wald test require calculating an efficient estimator $\hat{\theta}_n$, while the Rao test does not and is therefore the most convenient from this point of view. On the other hand, the Wald test, being based on the studentized difference

$$\left(\hat{\theta}_n - \theta_0 \right) \sqrt{nI\left(\theta_0\right)},$$

is more easily interpretable and has the advantage that, after replacement of $I\left(\theta_0\right)$ by $I\left(\hat{\theta}_n\right)$, it immediately yields confidence intervals for θ.

The Wald test has the drawback, not shared by the other two, that it is only asymptotically but not exactly invariant under reparametrization. To see this, let $\eta = g(\theta)$ and suppose that g is differentiable and strictly increasing. Then $\hat{\eta}_n = g\left(\hat{\theta}_n\right)$ and by (7.2.20)

$$(7.7.26) \qquad\qquad \sqrt{I^*(\eta)} = \sqrt{I(\theta)}/g'(\theta).$$

Thus the Wald statistic for testing $\eta = \eta_0$ is

$$\left[g\left(\hat{\theta}_n\right) - g\left(\theta_0\right) \right] \sqrt{nI^*\left(\eta_0\right)} = \sqrt{nI\left(\theta_0\right)} \left(\hat{\theta}_n - \theta_0 \right) \frac{g\left(\hat{\theta}_n\right) - g\left(\theta_0\right)}{\left(\hat{\theta}_n - \theta_0 \right)} \cdot \frac{1}{g'\left(\theta_0\right)}.$$

The product of the third and fourth factor tends to 1 as $\hat{\theta}_n \to \theta_0$ but typically will differ from 1 for finite n.

For a careful comparison of the three tests, these considerations need to be supplemented by simulations and second order calculations. It seems that the best choice depends on the situation, but various investigations so far have not led to any clear conclusions. For further discussion, see, for example, Cox and Hinkley (1974, Section 9.3), Tarone (1981), and Rayner and Best (1989).

The Wald, Rao, and likelihood ratio tests, both one- and two-sided, are consistent in the sense that for any fixed alternative, the probability of rejection tends to 1 as the sample size tends to infinity. For the test (7.7.9), for example, the power against an alternative $\theta > \theta_0$ is

$$\beta_n(\theta) = P_\theta \left[\sqrt{n} \left(\hat{\theta}_n - \theta_0 \right) \geq u_\alpha / \sqrt{I(\theta_0)} \right]$$

(7.7.27)

$$= P_\theta \left[\sqrt{n} \left(\hat{\theta}_n - \theta \right) \geq \frac{u_\alpha}{\sqrt{I(\theta_0)}} - \sqrt{n}(\theta - \theta_0) \right],$$

which tends to 1 as $n \to \infty$ by (7.7.1).

In the light of this result, the following example is surprising and somewhat disconcerting.

Example 7.7.3 Cauchy. Let $X_1, \ldots, X_n$ be a sample from the Cauchy distribution with density (7.3.54) so that

$$l_n'(\theta) = 2 \sum_{i=1}^n \frac{x_i - \theta}{1 + (x_i - \theta)^2}.$$

Since $I(\theta) = 1/2$ (Problem 2.4), the Rao test (7.7.21) rejects when

(7.7.28)
$$2\sqrt{\frac{2}{n}} \sum_{i=1}^n \frac{x_i - \theta}{1 + (x_i - \theta)^2} \geq u_\alpha.$$

As $\theta \to \infty$ with n remaining fixed,

$$\min(X_i - \theta_0) \xrightarrow{P} \infty$$

so that the left side of (7.7.28) tends in probability to 0 (Problem 7.11). Since $u_\alpha > 0$ (for $\alpha < 1/2$), it follows that the power of the test (7.7.28) tends to 0 as $\theta \to \infty$. The test was, of course, designed to maximize the power near θ_0; nevertheless, a test with such low power at a distance from H is unsatisfactory.   □

Examples in which an analogous unsatisfactory behavior occurs for the Wald test are discussed, for example, by Vaeth (1985), Mantel (1987), and by Le Cam (1990a), who proposes modifications to avoid the difficulty.

There is, of course, no contradiction between stating that the power function $\beta_n(\theta)$ satisfies

(7.7.29) $\beta_n(\theta) \to 1$ as $n \to \infty$ for any $\theta > \theta_0$

and

(7.7.30) $\beta_n(\theta) \to 0$ as $\theta \to \infty$ for any fixed n.

The limit (7.7.30) shows that the consistency asserted by (7.7.29) is not uniform in θ even when θ is bounded away from θ_0. For any given θ and any ϵ, there exists $n(\theta)$ such that

$$\beta_n(\theta) > 1 - \epsilon \text{ for all } n > n(\theta).$$

However, no n will work for all θ; as $\theta \to \infty$, so will $n(\theta)$.

Example 7.7.1. Cauchy (continued). Instead of the Rao test (7.7.21), let us consider the Wald test (7.7.9), which against a fixed alternative $\theta > \theta_0$ has power

$$\begin{aligned}
\beta_n(\theta) &= P_\theta \left[\sqrt{n} \left(\hat{\theta}_n - \theta_0 \right) \geq \sqrt{2} u_\alpha \right] \\
&= P_\theta \left[\sqrt{n} \left(\hat{\theta}_n - \theta \right) \geq \sqrt{2} u_\alpha - \sqrt{n} \left(\theta - \theta_0 \right) \right].
\end{aligned}$$

This time $\beta_n(\theta) \to 1$ as $\theta \to \infty$. The two tests (7.7.21) and (7.7.9) therefore behave very differently as $\theta \to \infty$, which may seem puzzling in view of the earlier claim of their asymptotic equivalence. However, this claim was asserted only under or near the hypothesis. It was based on (7.3.6)–(7.3.9). Of these, (7.3.6) is an identity which is always valid. On the other hand, (7.3.7)–(7.3.9) are only proved under H, that is, assuming θ_0 to be the true value of θ. If instead some value $\theta \neq \theta_0$ is correct, these limit results need no longer hold.

So far, we have assumed that the model depends on a single real-valued parameter θ. Suppose next that θ is vector-valued, $\theta = (\theta_1, \ldots, \theta_k)$ say, as in Section 7.5 and that the hypothesis

(7.7.31) $$H : \theta = \theta_0 = (\theta_1^o, \ldots, \theta_k^o)$$

is to be tested against the alternatives $\theta \neq \theta_0$.

The considerations leading to the Wald test (7.7.7) easily generalize. The starting point is an estimator $\hat{\theta}_n$ which is efficient, i.e., which satisfies (7.5.16). Then it follows from Theorem 5.4.2 that the quadratic form

(7.7.32)
$$n \left(\hat{\theta}_n - \theta \right)' I(\theta) \left(\hat{\theta}_n - \theta \right) = n \sum \sum I_{ij}(\theta) \left(\hat{\theta}_{in} - \theta_i \right) \left(\hat{\theta}_{jn} - \theta_j \right)$$

has a limiting χ^2-distribution with k degrees of freedom. If v_α is the upper α-point of this distribution and if $\hat{I}_{ij}$ is any consistent estimator of $I_{ij}(\theta)$, the ellipsoids

(7.7.33) $$n \sum \sum \hat{I}_{ij} \left(\hat{\theta}_{in} - \theta_i \right) \left(\hat{\theta}_{jn} - \theta_j \right) < v_\alpha$$

thus constitute confidence sets for the vector θ with asymptotic confidence coefficient $1 - \alpha$.

The joint distribution (7.5.16) of the variables $\sqrt{n} \left(\hat{\theta}_{1n} - \theta_1 \right), \ldots,$ $\sqrt{n} \left(\hat{\theta}_{kn} - \theta_k \right)$ can also be used to obtain simultaneous confidence intervals for the k parameters $\theta_1, \ldots, \theta_k$. The method is analogous to that illustrated in Example 5.2.5.

An acceptance region for testing the hypothesis (7.7.31) can be obtained by replacing θ by θ_0 in (7.7.33), as was done for (7.7.6). However, since the calculation of the level is based on the distribution of $\hat{\theta}_n$ at θ_0, we can now replace $\hat{I}_{ij}$ by $I_{ij}(\theta_0)$ in (7.7.33) and thus, in generalization of (7.7.7), obtain the rejection region of the *Wald test*

$$(7.7.34) \qquad W_n = n \sum\sum I_{ij}(\theta_0)\left(\hat{\theta}_{in} - \theta_i^o\right)\left(\hat{\theta}_{jn} - \theta_j^o\right) \geq v_\alpha$$

at asymptotic level α.

To obtain the Rao scores test of H, note that, in generalization of (7.3.7), the joint distribution of the score statistics

$$l_i'(\theta_0) = \left.\frac{\partial}{\partial\theta_i}l_n(\theta)\right|_{\theta=\theta_0}$$

under the assumptions of Theorem 7.5.1 satisfies (Problem 7.12)

$$(7.7.35) \qquad \left(\frac{1}{\sqrt{n}}l_1'(\theta_0),\ldots,\frac{1}{\sqrt{n}}l_k'(\theta_0)\right) \to N(0, I(\theta_0)).$$

From this limit result and Theorem 5.4.2, it follows that the scores test

$$(7.7.36) \qquad R_n = \frac{1}{n}\left(l_1'(\theta_0),\ldots,l_k'(\theta_0)\right)I^{-1}(\theta_0)\begin{pmatrix} l_1'(\theta_0) \\ \vdots \\ l_k'(\theta_0) \end{pmatrix} \geq v_\alpha$$

has asymptotic level α. Since $\frac{1}{n}I^{-1}(\theta.)$ is the asymptotic covariance matrix of $\left(\hat{\theta}_{1n} - \theta_1^o,\ldots,\hat{\theta}_{kn} - \theta_k^o\right)$, (7.7.36) can also be written as

$$(7.7.37) \qquad R_n = \sum\sum \sigma_{ij}(\theta_0)\,l_i'(\theta_0)\,l_j'(\theta_0) \geq v_\alpha,$$

where $\sigma_{ij}(\theta_0)$ denotes the asymptotic covariance of $\left(\hat{\theta}_{in} - \theta_i^o\right)$ and $\left(\hat{\theta}_{jn} - \theta_j^o\right)$.

Consider finally the likelihood ratio test. If Δ_n is defined as in (7.7.11), then Theorem 7.7.1 generalizes in the way one would expect. Specifically, under the assumptions of Theorem 7.5.2, if $\hat{\theta}_n$ is a consistent root of the likelihood equations, the distribution of $2\Delta_n$ tends to a χ^2-distribution with k degrees of freedom. The proof parallels that of Theorem 7.7.1 with the difference that the Taylor expansions of $l_n\left(\hat{\theta}_n\right)$ and $l_n'\left(\hat{\theta}_n\right)$ given by (7.7.12) and (7.7.13) are replaced by the corresponding multivariate Taylor expansions (Problem 7.13).

These results are summarized and slightly strengthened in the following extension of Theorem 7.7.2.

Theorem 7.7.3 *Under the assumptions of Theorem 7.5.2, the Wald test (7.7.34), the Rao test (7.7.37), and the likelihood ratio test which rejects when*

$$(7.7.38) \qquad 2\Delta_n \geq v_\alpha$$

are asymptotically equivalent under H and all have asymptotic level α.

Example 7.7.4 Normal one-sample problem. Let $X_1, \dots, X_n$ be i.i.d. $N(\xi, \sigma^2)$. The problem of testing

$$(7.7.39) \qquad H : \xi = \xi_0, \ \sigma = \sigma_0$$

was treated on an ad hoc basis in Example 5.2.4. Let us now consider it from the present point of view. The unique roots of the likelihood equations are by Example 7.5.1 (continued)

$$(7.7.40) \qquad \hat{\xi} = \bar{X}, \quad \hat{\sigma}^2 = \frac{1}{n} \sum (X_i - \bar{X})^2,$$

and the information matrix and its inverse are, by (7.5.2),

$$(7.7.41) \qquad I(\xi, \sigma^2) = \begin{pmatrix} 1/\sigma^2 & 0 \\ 0 & 1/2\sigma^4 \end{pmatrix} \text{ and } I^{-1}(\xi, \sigma^2) = \begin{pmatrix} \sigma^2 & 0 \\ 0 & 2\sigma^4 \end{pmatrix}.$$

It follows that the Wald statistic is

$$(7.7.42) \qquad W_n = \frac{n}{\sigma_0^2}(\bar{X} - \xi_0)^2 + \frac{n}{2\sigma_0^4}(\hat{\sigma}^2 - \sigma_0^2)^2$$

and the score statistic R_n can be determined analogously (Problem 7.14(i)).

From (7.7.40), it is seen that the log likelihood ratio statistic is (Problem 7.14(ii))

$$(7.7.43) \qquad 2\Delta_n = \frac{n(\bar{X} - \xi_0)^2}{\sigma_0^2} + \frac{n\hat{\sigma}^2}{\sigma_0^2} - n - n\left[\log\hat{\sigma}^2 - \log\sigma_0^2\right].$$

The asymptotic equivalence of these two test statistics under H can be checked directly by a Taylor expansion of $\log\hat{\sigma}^2 - \log\sigma_0^2$ (Problem 7.14(iii)). □

Example 7.7.5 The multinomial one-sample problem. Let $Y_1, \dots, Y_{k+1}$ have the multinomial distribution $M(p_1, \dots, p_{k+1}; n)$ given by (5.5.29), and consider the hypothesis

$$(7.7.44) \qquad H : p_i = p_i^o, \ i = 1, \dots, k+1.$$

For testing H, Pearson's χ^2-test based on the statistic

$$(7.7.45) \qquad Q_n = n \sum \left(\frac{Y_i}{n} - p_i^o\right)^2 / p_i^o$$

was proposed in Example 5.5.4.

To obtain the Wald, Rao, and likelihood ratio tests, we need the following results obtained in Examples 5.4.1, 5.5.4, 7.5.3, and 7.5.3 (continued):

(i) The likelihood equations have the unique solution

(7.7.46)
$$\hat{p}_i = Y_i/n.$$

(ii) The (i, j)th element of the information matrix $I(p)$ is

(7.7.47)
$$I_{ij} = \begin{cases} n\left[\dfrac{1}{p_i} + \dfrac{1}{p_{k+1}}\right] & \text{if } i = j \\ n/p_{k+1} & \text{if } i \neq j. \end{cases}$$

(iii) The conditions of Theorem 7.5.2 are satisfied and therefore

(7.7.48)
$$\sqrt{n}\left(\frac{Y_1}{n} - p_1\right), \ldots, \sqrt{n}\left(\frac{Y_k}{n} - p_k\right) \xrightarrow{L} N(0, \Sigma)$$

where the (i, j)th element σ_{ij} of $\Sigma = I^{-1}(p)$ is

(7.7.49)
$$\sigma_{ij} = \begin{cases} p_i(1 - p_i) & \text{if } j = i \\ -p_i p_j & \text{if } j \neq i. \end{cases}$$

By (7.7.46) and (7.7.47), the Wald statistic is therefore

(7.7.50)
$$W_n = n\sum_{i=1}^{k}\sum_{j=1}^{k} \frac{1}{p^o_{k+1}}\left(\frac{Y_i}{n} - p^o_i\right)\left(\frac{Y_i}{n} - p^o_i\right) + n\sum_{i=1}^{k} \frac{1}{p^o_i}\left(\frac{Y_i}{n} - p^o_i\right)^2.$$

It was seen in the proof of Theorem 5.5.2 that $W_n = Q_n$, so that the Wald test in the present case reduces to Pearson's χ^2-test.

Using (7.7.46), it is seen (Problem 7.15 (i)) that the logarithm of the likelihood ratio statistic is

(7.7.51)
$$\Delta_n = \sum_{i=1}^{k+1} Y_i \log \frac{Y_i}{np^o_i}.$$

In the present case, it is easy to check directly (Problem 7.15(ii)) that under H,

(7.7.52)
$$W_n - 2\Delta_n \xrightarrow{P} 0 \text{ as } n \to \infty.$$

That W_n has a limiting χ^2_k-distribution was shown in Example 5.5.4 and the same limit distribution for $2\Delta_n$ thus follows from (7.7.52). Alternatively, we can prove this limit for $2\Delta_n$ from Theorem 7.7.1. For the Rao score test, see Problem 7.8.   □

The hypotheses considered so far are simple, that is, they completely specify the distribution being tested. We shall now turn to the case, more frequently met in applications, in which the hypothesis is composite. Assuming again that $\theta = (\theta_1, \ldots, \theta_k)$, the hypothesis is then typically of the form

$$(7.7.53) \qquad H : g_1(\theta) = a_1, \ldots, g_r(\theta) = a_r, \quad 1 \le r < k.$$

We shall instead treat the standard hypothesis

$$(7.7.54) \qquad H : \theta_1 = \theta_1^o, \ldots, \theta_r = \theta_r^o, \quad 1 \le r < k,$$

which can be obtained from (7.7.53) through reparametrization. For testing H against the alternatives that $\theta_i \ne \theta_i^o$ for at least some i, we have the following generalization of the likelihood ratio part of Theorem 7.7.2.

Theorem 7.7.4 *Suppose the assumptions of Theorem 7.5.2 hold and that $\left(\hat{\theta}_{1n}, \ldots, \hat{\theta}_{kn}\right)$ are consistent roots of the likelihood equations for $\theta = (\theta_1, \ldots, \theta_k)$. Suppose, in addition, that the corresponding assumptions hold for the parameter vector $(\theta_{r+1}, \ldots, \theta_k)$ when $\theta_i = \theta_i^o$ for $i = 1, \ldots, r$, and that $\hat{\hat{\theta}}_{r+1,n}, \ldots, \hat{\hat{\theta}}_{kn}$ are consistent roots of the likelihood equations for $(\theta_{r+1}, \ldots, \theta_k)$ under H. In generalization of (7.7.10), consider the likelihood ratio statistic*

$$(7.7.55) \qquad \frac{L_x\left(\hat{\theta}_n\right)}{L_x\left(\hat{\hat{\theta}}_n\right)},$$

where $\hat{\hat{\theta}}_n = \left(\theta_1^o, \ldots, \theta_r^o, \hat{\hat{\theta}}_{r+1,n}, \ldots, \hat{\hat{\theta}}_{kn}\right)$. Then under H, if

$$(7.7.56) \qquad \Delta_n = l_n\left(\hat{\theta}_n\right) - l_n\left(\hat{\hat{\theta}}_n\right),$$

the statistic $2\Delta_n$ has a limiting χ_r^2-distribution (Problem 7.16).

The conclusions of Theorems 7.7.2–7.7.4 hold not only for the i.i.d. case but, under the assumptions and in the notation of Theorem 7.6.3, also for the multisample case when each component distribution $f_{\gamma,\theta}(x)$ satisfies the assumptions of Theorem 7.5.2. (Problem 7.17.)

Example 7.7.6 Multinomial two-sample problem. Let $(X_1, \ldots, X_{r+1})$ and $(Y_1, \ldots, Y_{r+1})$ be independent with multinomial distributions $M(p_1, \ldots, p_{r+1}; m)$ and $M(q_1, \ldots, q_{r+1}, n)$, respectively, and consider the hypothesis

$$(7.7.57) \qquad H : p_i = q_i \text{ for all } i = 1, \ldots, r+1.$$

Since $\sum_{i=1}^{r+1} p_i = \sum_{i=1}^{r+1} q_i = 1$, we shall as in Example 7.5.3 (continued) take $(p_1, \ldots, p_r, q_1, \ldots, q_r)$ as our parameters. In order to reduce the hypothesis to (7.7.54), we make the further transformation to the parameters

(7.7.58)
$$\begin{aligned} \theta_i &= q_i - p_i, & i = 1, \ldots, r, \\ \theta_{i+r} &= q_i, & i = 1, \ldots, r. \end{aligned}$$

The hypothesis then becomes

(7.7.59)
$$H : \theta_1 = \cdots = \theta_r = 0.$$

The likelihood is

(7.7.60)
$$\frac{m!}{\prod_{i=1}^{r+1} x_i!} \prod_{i=1}^{r+1} p_i^{x_i} \cdot \frac{n!}{\prod_{j=1}^{r+1} y_j!} \prod_{j=1}^{r+1} q_j^{y_j}$$

and the likelihood equations have the unique solution

$$\hat{p}_i = \frac{X_i}{m}, \quad \hat{q}_i = \frac{Y_i}{n}$$

and hence

$$\hat{\theta}_i = \frac{Y_i}{n} - \frac{X_i}{m}, \quad i = 1, \ldots, r,$$

$$\hat{\theta}_{i+r} = \frac{Y_i}{n}, \quad i = 1, \ldots, r.$$

Under the hypothesis, the likelihood reduces to

(7.7.61)
$$\frac{m! n!}{\prod x_i! \prod y_j!} \prod_{i=1}^{r} q_i^{x_i + y_i} (1 - q_1 - \cdots - q_r)^{x_{k+1} + y_{k+1}}$$

and the likelihood equations have the unique solution

(7.7.62)
$$\hat{\hat{q}}_i = \hat{\hat{\theta}}_{i+r} = \frac{X_i + Y_i}{m + n}, \quad i = 1, \ldots, r,$$

while $\hat{\hat{\theta}}_1 = \cdots = \hat{\hat{\theta}}_r = 0$ (Problem 7.18).

The likelihood ratio statistic thus is

$$\prod_{i=1}^{r+1} \left(\frac{X_i}{m}\right)^{X_i} \left(\frac{Y_i}{n}\right)^{Y_i} \bigg/ \left(\frac{X_i + Y_i}{m + n}\right)^{X_i + Y_i}$$

and its logarithm is

(7.7.63)
$$\Delta_{m,n} = \sum_{i=1}^{r+1} \left[X_i \log\left(\frac{X_i}{m}\right) + Y_i \log\left(\frac{Y_i}{n}\right) - (X_i + Y_i) \log\left(\frac{X_i + Y_i}{m + n}\right) \right].$$

All the conditions of Theorem 7.5.2 are satisfied and Theorem 7.7.4 applies with $k = 2r$ so that $2\Delta_{m,n}$ tends in law to χ_r^2 as m and n tend to infinity while satisfying (7.6.30), i.e. at the same rate.[¶] This is the same limit distribution as that obtained in Example 5.5.5 for the statistics $Q = Q_{m,n}$ given by (5.5.51). It can, in fact, again be shown that

$$(7.7.64) \qquad 2\Delta_{m,n} - Q_{m,n} \xrightarrow{P} 0.$$

$\square$

Let us next consider the extensions of the Wald and Rao tests of the hypothesis (7.7.54), which are based on the limits

$$(7.7.65) \qquad \left(\sqrt{n}\left(\hat{\theta}_{1n} - \theta_1\right), \ldots, \sqrt{n}\left(\hat{\theta}_{rn} - \theta_r\right)\right) \xrightarrow{L} N\left(0, \Sigma^{(r)}(\theta)\right)$$

and

$$(7.7.66) \qquad \left(\frac{1}{\sqrt{n}}\frac{\partial}{\partial\theta_1}l_n(\theta), \ldots, \frac{1}{\sqrt{n}}\frac{\partial}{\partial r}l_n(\theta)\right) \xrightarrow{L} N\left(0, I^{(r)}(\theta)\right),$$

respectively. Here $\Sigma^{(r)}(\theta)$ and $I^{(r)}(\theta)$ are the submatrices consisting of the upper left-hand corner formed by the intersection of the r first rows and columns of the covariance matrix $\Sigma(\theta)$ and the information matrix $I(\theta)$, respectively. From Theorem 5.4.2, it then follows that the quadratic forms

$$(7.7.67) \qquad n\left(\hat{\theta}_n - \theta\right)'\left[\Sigma^{(r)}(\theta)\right]^{-1}\left(\hat{\theta}_n - \theta\right)$$

and

$$(7.7.68) \qquad \frac{1}{n}\left[\frac{\partial}{\partial\theta_1}l_n(\theta), \ldots, \frac{\partial}{\partial\theta_r}l_n(\theta)\right]' I^{(r)}(\theta)^{-1}\left[\frac{\partial}{\partial\theta_1}l_n(\theta), \ldots, \frac{\partial}{\partial\theta_r}l_n(\theta)\right]$$

are distributed in the limit as χ^2 with r degrees of freedom. Unfortunately, the equations

$$[\Sigma(\theta)]^{-1} = I(\theta), \quad [I(\theta)]^{-1} = \Sigma(\theta)$$

which simplified the writing of the Wald and Rao test statistics (7.7.32) and (7.7.36) no longer hold when Σ and I are replaced by $\Sigma^{(r)}$ and $I^{(r)}$. Instead (Problem 7.20),

$$(7.7.69) \qquad \left[\Sigma^{(r)}\right]^{-1} = \left[I^{(r)}\right] - I_{12}I_{22}^{-1}I_{21},$$

[¶]For the role of assumption (7.6.30), see the Note following Lemma 3.1.1.

where

(7.7.70)
$$I(\theta) = \begin{pmatrix} I^{(r)} & I_{12} \\ I_{21} & I_{22} \end{pmatrix}_{k-r}^{r}.$$

Since $I_{12} = I'_{21}$ and since I_{22} is positive definite, (7.7.69) shows that

(7.7.71)
$$\left[\Sigma^{(r)} \right]^{-1} \leq I^{(r)}.$$

In the special case $r = 1$, this inequality was already seen in (7.6.24). As in that case, the inequality (7.7.71) corresponds to the fact that $(\theta_1, \dots, \theta_r)$ can be estimated at least as efficiently when $(\theta_{r+1}, \dots, \theta_k)$ is known as when it is unknown (Problem 7.21).

It follows from (7.7.69) that equality holds in (7.7.71) when $I_{12} = 0$ or, equivalently, by Problem 3.14 of Chapter 5, when the corresponding submatrix Σ_{12} of Σ is 0 and the vectors

$$\left(\sqrt{n} \left(\hat{\theta}_{1n} - \theta_1 \right), \dots, \sqrt{n} \left(\hat{\theta}_{rn} - \theta_r \right) \right) \text{ and}$$
$$\left(\sqrt{n} \left(\hat{\theta}_{r+1,n} - \theta_{r+1} \right), \dots, \sqrt{n} \left(\hat{\theta}_{kn} - \theta_k \right) \right)$$

are therefore asymptotically independent. In that case, the quadratic forms (7.7.67) and (7.7.68) reduce to

(7.7.72)
$$n \sum_{i=1}^{r} \sum_{j=1}^{r} I_{ij} \left(\theta^o \right) \left(\hat{\theta}_{in} - \theta_i^o \right) \left(\hat{\theta}_{jn} - \theta_j^o \right)$$

and

(7.7.73)
$$\frac{1}{n} \sum_{i=1}^{r} \sum_{j=1}^{r} \sigma_{ij} \left(\theta^o \right) \frac{\partial}{\partial \theta_i} l_n(\theta) \frac{\partial}{\partial \theta_j} l_n(\theta)|_{\theta=\theta_0},$$

respectively.

There is another important case in which (7.7.67) and (7.7.68) simplify, although in a different way. When $r = 1$,

$$\Sigma^{(r)}(\theta) = \sigma_{11}(\theta) \text{ and } I^{(r)}(\theta) = I_{11}(\theta)$$

so that (7.7.67) and (7.7.68) reduce to

(7.7.74)
$$\frac{n \left(\hat{\theta}_{1n} - \theta_1 \right)^2}{\sigma_{11}(\theta)} \text{ and } \left[\frac{\partial}{\partial \theta_1} l_n(\theta) \right]^2 / n I_{11}(\theta),$$

respectively, where $\sigma_{11}(\theta)$ is the asymptotic variance of $\sqrt{n} \left(\hat{\theta}_{1n} - \theta_1 \right)$.

In all these cases, the coefficients $I_{ij}(\theta_0)$ and $\sigma_{ij}(\theta_0)$ in the quadratic forms (7.7.72) and (7.7.73), and $1/\sigma_{11}(\theta)$ and $1/I_{11}(\theta)$ in (7.7.74) may even, under the hypothesis, depend on the nuisance parameters $(\theta_{r+1}, \ldots, \theta_k)$. They then have to be replaced by consistent estimators such as $I_{ij}\left(\hat{\theta}_n\right)$ and $\sigma_{ij}\left(\hat{\theta}_n\right)$ for the quadratic forms to become usable as test statistics.

When considering maximum likelihood estimation earlier in this chapter, we obtained not only the asymptotic performance of the estimators $\hat{\theta}_n$ but also established their efficiency within the class of asymptotically normal estimators with continuous asymptotic variance. The corresponding situation for testing is more complicated. A more satisfactory theory, which is also available for estimation and which avoids the restriction to asymptotically normal estimators with continuous asymptotic variance, is based on the concept of asymptotic local minimaxity. This theory, which is beyond the scope of this book, is sketched in Le Cam and Yang (1990); see also Lehmann and Casella (1998, Section 6.8).

Summary

1. Asymptotic tests and confidence sets can be obtained from the results of the preceding sections in three different ways.

   (i) The asymptotic distribution of the efficient estimators $\hat{\theta}_n = \left(\hat{\theta}_{1n}, \ldots, \hat{\theta}_{kn}\right)$ can be inverted in the usual way to obtain asymptotic confidence sets for $\theta = (\theta_1, \ldots, \theta_k)$ and hence also tests of the hypothesis $H : \theta_1 = \theta_{10}, \ldots, \theta_k = \theta_{k0}$.

   (ii) One can use as a starting point the asymptotic distribution not of the estimators $\hat{\theta}_n$ but instead of the score statistics $l_i'(\theta)$, $i = 1, \ldots, k$. The procedure of (i) leads to the Wald test and that of (ii) to Rao's score test.

   (iii) A third possibility is the likelihood ratio test which rejects when $l_n\left(\hat{\theta}_n\right) - l_n\left(\theta_0\right)$ is sufficiently large.

2. The three types of tests are developed, compared, and illustrated first for the simple hypothesis H when $k = 1$, then for the general case of the simple hypothesis H for arbitrary k, and finally in the presence of nuisance parameters.

7.8 Contingency tables

Section 5.6 gave an account of inference in a 2×2 table. We shall now extend some of these results not only to a general $a \times b$ table but for quite general contingency tables.

Suppose that we have N multinomial trials with probabilities $p_1, \ldots, p_{k+1}$ for its $k+1$ possible outcomes. Then the numbers $n_1, \ldots, n_{k+1}$ of trials resulting in these outcomes has the multinomial distribution $N(p_1, \ldots, p_{k+1};$

N) given by (5.5.29). The p's are, of course, constrained to be between 0 and 1; in addition, they must satisfy

$$(7.8.1) \qquad \sum_{i=1}^{k+1} p_i = 1.$$

The vector of probabilities $\underline{p} = (p_1, \dots, p_{k+1})$ therefore lies in a k-dimensional subset of R^{k+1}; correspondingly, the $k+1$ p's can be expressed as functions of k parameters $\theta_1, \dots, \theta_k$—for example, $\theta_1 = p_i$, $i = 1, \dots, k$—which vary over a subset of R^k. General classes of contingency tables are obtained by assuming that the p's are functions

$$(7.8.2) \qquad p_i = p_i(\theta_1, \dots, \theta_s)$$

of $s \leq k$ parameters varying over a set ω in R^k.

Example 7.8.1 Two-way table. In generalization of the situation leading to the 2×2 table given as Table 5.6.1, consider N trials, the possible outcomes of which are classified according to two criteria as $A_1, A_2, \dots,$ or A_a and as $B_1, B_2, \dots,$ or B_b.

For example, a sample of N voters could be classified by age (young, middle aged, old) and by gender (F, M). Let n_{ij} denote the number of trials resulting in outcomes A_i and B_j, let $n_{i+} = n_{i1} + \cdots + n_{ib}$ denote the number resulting in outcome A_i and $n_{+j} = n_{ij} + \cdots + n_{aj}$ the number resulting in outcome B_j. The results can be displayed in the $a \times b$ table shown as Table 7.8.1.

TABLE 7.8.1. $a \times b$ contingency table

	B_1	B_2	$\cdots$	B_b	
A_1	n_{11}	n_{12}	$\cdots$	n_{1b}	n_{1+}
A_2	n_{21}	n_{22}	$\cdots$	n_{2b}	n_{2+}
$\vdots$					
A_a	n_{a1}	n_{a2}	$\cdots$	n_{ab}	n_{a+}
	n_{+1}	n_{+2}	$\cdots$	n_{+b}	N

A submodel that is frequently of interest is obtained by assuming that the categories A and B are independent. If

$$p_{ij} = P(A_i \text{ and } B_j), \quad p_{i+} = P(A_i), \quad \text{and } p_{+j} = P(B_j),$$

this assumption is expressed formally by

$$(7.8.3) \qquad p_{ij} = p_{i+} \cdot p_{+j}.$$

Since the probabilities p_{i+} and p_{j+} must lie between 0 and 1 and satisfy

$$(7.8.4) \qquad \sum_{i=1}^{a} p_{i+} = \sum_{j=1}^{b} p_{+j} = 1$$

but are otherwise unconstrained, the ab probabilities are functions of the $(a-1)+(b-1)$ parameters

$$(7.8.5) \qquad p_{1+}, \dots, p_{a-1,+}; \; p_{+1}, \dots, p_{+,b-1},$$

which can serve as the $s = a + b - 2$ parameters $\theta_1, \dots, \theta_s$ of (7.8.2).

This example extends in the obvious way to three-way and higher tables with various possible independence structures (Problems 8.1–8.5).

As in Example 5.4.1, let $X_i^{(\nu)} = 1$ if the νth trial results in outcome O_i and let

$$(7.8.6) \qquad Y_i = \sum_{\nu=1}^{N} X_i^{(\nu)}, \; i = 1, \dots, k+1,$$

be the total number of trials resulting in outcome O_i. Then the N vectors $\left(X_1^{(\nu)}, \dots, X_{k+1}^{(\nu)}\right)$, $\nu = 1, \dots, N$, are i.i.d., so that the theory of Sections 7.5 and 7.6 can be applied to obtain efficient estimators of the parameters $(\theta_1, \dots, \theta_s)$ of (7.8.2).

The likelihood is

$$(7.8.7) \qquad L(\theta) = \frac{N!}{y_1! \cdots y_{k+1}!} [p_1(\theta)]^{y_1} \cdots [p_{k+1}(\theta)]^{y_{k+1}},$$

and the log likelihood is therefore

$$(7.8.8)$$
$$l_n(\theta) = y_1 \log[p_1(\theta)] + \cdots + y_{k+1} \log[p_{k+1}(\theta)] + \log(N!/y_1! \cdots y_{k+1}!).$$

Differentiating with respect to θ_j, we obtain the likelihood equations

$$(7.8.9) \qquad \sum_{i=1}^{k+1} \frac{y_i}{p_i(\theta)} \cdot \frac{\partial p_i(\theta)}{\partial \theta_j} = 0 \text{ for } j = 1, \dots, s.$$

To conclude the existence of a consistent solution of these equations that satisfies (7.5.16) requires checking the conditions of Theorem 7.5.2. Of these, (M_1), (M_3), and (M_4) are obviously satisfied. Condition $(M2)$ must be stipulated but typically causes no difficulty, and $(M5)$ and $(M6'')$ will be satisfied provided the functions $p_i(\theta)$ have three continuous derivatives. Condition (M8) can be checked once the information matrix has been computed. Alternatively, one can replace the assumption that $I(\theta)$ is positive definite by the assumption that the matrix $\left(\dfrac{\partial p_i(\theta)}{\partial \theta_j}\right)$ has rank s, which implies $(M8)$ (Problem 8.6).

This leaves $(M7)$. Here it is important to recall that (as was pointed out in Example 7.3.1) the bound on the third derivative is required not for all θ but only in some neighborhood of the true value, which in (M7) was denoted by θ^0. To see what is involved, consider

(7.8.10) $$f_\theta(x_1,\dots,x_k) = p_1(\theta)^{x_1} p_2(\theta)^{x_2} \cdots p_{k+1}(\theta)^{x_{k+1}},$$

where one of $x_1,\dots,x_{k+1}$ is 1 and the others are 0. Then

(7.8.11) $$\frac{\partial}{\partial\theta_j} f_\theta(x_1,\dots,x_k) = f_\theta(x_1,\dots,x_k) \sum_{i=1}^{k+1} \frac{x_i}{p_i(\theta)} \cdot \frac{\partial p_i(\theta)}{\partial\theta_j}.$$

Continuing to differentiate, one sees that the third derivatives occurring in (M7) are the product of $f_\theta(x_1,\dots,x_k)$ with the sum of a number of terms, each of which is the product of three factors: (i) one of the x's, (ii) a negative power of $p_i(\theta)$, and (iii) a first, second, and third derivative of $p_i(\theta)$. The factor $f_\theta(x_1,\dots,x_k)$ given by (7.8.10) is between 0 and 1 and the x's only take on the values 0 and 1. Both these factors are therefore bounded.

The factors $[1/p_i(\theta)]^t$ would cause difficulty if they had to be bounded over all $0 < p_i(\theta^0) < 1$. However, a bound is only needed in a neighborhood of the true θ^0. Since $0 < p_i(\theta^0) < 1$, it follows from the continuity of p_i that there then also exists a neighborhood of θ^0 in which $\epsilon < p_i(\theta^0)$ for some $\epsilon > 0$ and in which therefore $[1/p_i(\theta)]^t$ is bounded. Condition (M7) will thus hold, provided the functions $p_i(\theta)$ satisfy

(7.8.12) For each i, the first three derivatives of $p_i(\theta)$ are in absolute value less than some constant M for all θ in some neighborhood of θ^0.

Under the stated conditions it follows that there exists a consistent root $\hat\theta = \left(\hat\theta_1,\dots,\hat\theta_s\right)$ of the likelihood equations satisfying (7.5.16). The $(i,j)^{\text{th}}$ element of the information matrix is

(7.8.13) $$I_{ij}(\theta) = E\left[\frac{\partial}{\partial\theta_i} \log f_\theta(X) \frac{\partial}{\partial\theta_j} \log f_\theta(X)\right],$$

which by (7.8.11) is the expectation of

(7.8.14)
$$\left[\frac{X_1}{p_1}\frac{\partial p_1}{\partial\theta_i} + \cdots + \frac{X_{k+1}}{p_{k+1}}\frac{\partial p_{k+1}}{\partial\theta_i}\right]\left[\frac{X_1}{p_1}\frac{\partial p_1}{\partial\theta_j} + \cdots + \frac{X_{k+1}}{p_{k+1}}\cdot\frac{\partial p_{k+1}}{\partial\theta_j}\right].$$

Here

$$(X_1,\dots,X_{k+1}) = (0,\dots,0,1,0,\dots,0)$$

with probability p_i when the 1 occurs in the ith position. Thus

$$E(X_u X_v) = \begin{cases} 0 & \text{if } u \neq v \\ p_u & \text{if } u = v \end{cases}$$

and (7.8.13) reduces to

$$(7.8.15) \qquad I_{ij} = \sum_{u=1}^{k+1} \frac{p_u}{p_u^2} \frac{\partial p_u}{\partial \theta_i} \frac{\partial p_u}{\partial \theta_j} = \sum_{u=1}^{k+1} \frac{1}{p_u} \frac{\partial p_u}{\partial \theta_i} \frac{\partial p_u}{\partial \theta_j}.$$

We have therefore proved the following result.   □

Theorem 7.8.1 *Let* $(Y_1, \ldots, Y_{k+1})$ *have the multinomial distribution* $M(p_1, \ldots, p_{k+1}; N)$ *with the p's given by (7.8.2). Concerning* θ *and the functions* $p_i(\theta)$, *assume the following:*

(i) The parameter space Ω *of* θ *is an open set in* R^s *and* $0 < p_i(\theta) < 1$ *for all* $\theta \in \Omega$ *and all* i.

(ii) If $\theta \neq \theta'$, *the probability vectors* $(p_1(\theta), \ldots, p_{k+1}(\theta))$ *and* $(p_1(\theta'), \ldots, p_{k+1}(\theta'))$ *are distinct.*

(iii) There exists a neighborhood of the true value θ^0 *in which the functions* $p_i(\theta)$ *have three continuous derivatives satisfying (7.8.12).*

(iv) The information matrix (7.8.15) is nonsingular.

Then the likelihood equations (7.8.9) have a consistent root $\hat{\theta} = (\hat{\theta}_1, \ldots, \hat{\theta}_{k+1})$, *and*

$$(7.8.16) \qquad \left(\sqrt{N} \left(\hat{\theta}_1 - \theta_0 \right), \ldots, \sqrt{N} \left(\hat{\theta}_s - \theta_s \right) \right) \xrightarrow{L} N \left(0, I^{-1}(\theta) \right)$$

where the elements $I_{ij}(\theta)$ *of the matrix* $I(\theta)$ *are given by (7.8.15).*

For similar results with somewhat different conditions, see Rao (1973) and Agresti (1990).

Example 7.8.2 Unrestricted multinomial. This example was already treated in Example 5.4.1, and in Example 7.5.3. To consider it from the present point of view, suppose again that $(Y_1, \ldots, Y_{k+1})$ is distributed as $M(p_1, \ldots, p_{k+1}; N)$. Since the p's must add up to 1, we have $s = k$ and can put, for example, $\theta_i = p_i (i = 1, \ldots, k)$. Then (7.8.2) becomes

$$p_i(\theta) = \theta_i \quad (i = 1, \ldots, s),$$
$$p_{k+1}(\theta) = 1 - (\theta_1 + \cdots + \theta_s)$$

and Ω is the set

$$\Omega = \{\theta : 0 < \theta_i < 1 \text{ for } i = 1, \ldots, s\}.$$

Clearly, conditions (i)–(iii) of Theorem 7.8.1 are satisfied. To check (iv), it is enough by Problem 8.6 to check that the matrix $\left(\dfrac{\partial p_i(\theta)}{\partial \theta_j} \right)$ has rank s.

In the present case, this matrix is

$$
\begin{pmatrix}
1 & 0 & \cdots & 0 \\
0 & 1 & \cdots & 0 \\
 & & & \\
0 & 0 & \cdots & 1 \\
-1 & -1 & \cdots & -1
\end{pmatrix},
$$

the first s rows of which form a nonsingular $s \times s$ submatrix and which is therefore of rank s. The likelihood equations are solved in Example 7.5.3 and have the unique solutions $\hat{\theta}_i = \hat{p}_i = Y_i/n$ $(i = 1, \ldots, k)$. Theorem 7.8.1 then gives the asymptotic distribution of $\left(\hat{\theta}_1, \ldots, \hat{\theta}_k\right)$ which agrees with the result of Example 7.5.3 (continued). $\qquad \square$

Example 7.8.3 Two-way table under independence. Consider the two-way Table 7.8.1 under the assumption (7.8.3) of independence. Since the probabilities p_{i+} and p_{+j} are restricted only by (7.8.4), we have $s = (a - 1) + (b - 1)$ and can take

$$
\theta_1 = p_{1+}, \ldots, \theta_{a-1} = p_{a-1,+},
$$
$$
\theta'_1 = p_{+1}, \ldots, \theta'_{b-1} = p_{+,b-1}
$$

as our parameters. Then

$$
\begin{aligned}
p_{ij}(\theta) &= \theta_i \theta'_j \text{ for } i = 1, \ldots, a-1; \ j = 1, \ldots, b-1; \\
p_{ib}(\theta) &= \theta_i \left(1 - \theta'_1 - \cdots - \theta'_{b-1}\right) \text{ for } i = 1, \ldots, a-1; \\
p_{aj}(\theta) &= \left(1 - \theta_1 - \cdots - \theta_{a-1}\right) \theta'_j \text{ for } j = 1, \ldots, b-1; \\
p_{ab}(\theta) &= 1 - \theta_1 - \cdots - \theta_{a-1} \left(1 - \theta'_1 - \cdots - \theta'_{b-1}\right).
\end{aligned}
$$

The conditions of Theorem 7.8.1 are easily checked as in the previous example (Problem 8.7).

The log likelihood is given by (7.8.8 with the $p_i(\theta)$ replaced by the $p_{ij}(\theta)$ and the y's by the n_{ij}'s, and thus is

$$
\begin{aligned}
l_n(\theta) &= \log\left(N! / \prod_{ij} n_{ij}!\right) + \sum_{i=1}^{a-1} n_{i+} \log \theta_i + \sum_{j=1}^{b-1} n_{+j} \log \theta'_j \\
&\quad + n_{a+} \log\left(1 - \theta_1 - \cdots - \theta_{a-1}\right) + n_{+b} \log\left(1 - \theta'_1 - \cdots - \theta'_{b-1}\right).
\end{aligned}
$$

The likelihood equations for $\theta_1, \ldots, \theta_{a-1}$ therefore become

$$
\frac{n_{i+}}{\theta_i} = \frac{n_{a+}}{\theta_a}, \quad i = 1, \ldots, a-1,
$$

where $\theta_a = 1 - \theta_1 - \cdots - \theta_{a-1}$. Denoting this common ratio by c, we have $\theta_i = cn_{i+}$ for $i = 1, \ldots, a$, and, summing over i, we find the unique solution of these likelihood equations to be

$$\hat{\theta}_i = \frac{n_{i+}}{N}, \quad i = 1, \dots, a.$$

Analogously, the equations for $\theta_1', \dots, \theta_b'$ have the solution

$$\theta_j' = \frac{n_{+j}}{N}, \quad j = 1, \dots, b.$$

The resulting estimators for

$$p_{ij} = p_{i+}p_{+j}$$

are

(7.8.17) $$\hat{p}_{ij} = \frac{n_{i+}n_{+j}}{N^2}.$$

To obtain the limit distribution of

(7.8.18) $$\sqrt{N}\left(\hat{p}_{ij} - p_{i+}p_{+j}\right)$$

under the independence assumption (7.8.3), we begin by obtaining it in the unrestricted multinomial model. $\qquad\square$

Lemma 7.8.1 *In the unrestricted multinomial model, the joint limit distribution of the two variables*

(7.8.19) $$\sqrt{N}\left(\frac{n_{i+}}{N} - p_{i+}\right) \text{ and } \sqrt{N}\left(\frac{n_{+j}}{N} - p_{+j}\right)$$

is the bivariate normal distribution $N(0, \Sigma)$, *where* $\Sigma = (\sigma_{ij})$ *is given by*

(7.8.20)
$$\sigma_{11} = p_{i+}(1 - p_{i+}), \quad \sigma_{22} = p_{+j}(1 - p_{+j}), \quad \text{and } \sigma_{12} = p_{ij} - p_{i+}p_{+j}.$$

Proof. Without loss of generality, let $i = j = 1$, and let

$$p_{11} = \pi_1, \quad p_{12} + \dots + p_{1b} = \pi_2, \quad p_{21} + \dots + p_{a1} = \pi_3,$$

so that

(7.8.21) $$\pi_1 + \pi_2 = p_{1+} \text{ and } \pi_1 + \pi_3 = p_{+1}.$$

It is convenient to begin with the joint distribution of the three variables

$$X = \sqrt{N}\left(\frac{n_{11}}{N} - \pi_1\right), \quad Y = \sqrt{N}\left(\frac{n_{12} + \dots + n_{1b}}{N} - \pi_2\right),$$
$$Z = \sqrt{N}\left(\frac{n_{21} + \dots + n_{a1}}{N} - \pi_3\right).$$

Since n_{11}, $n_{12} + \cdots + n_{1b}$, $n_{21} + \cdots + n_{a1}$ are the numbers of trials resulting in three of the outcomes of N multinomial trials with the four outcomes

$$i = j = 1; \quad i = 1, \ j > 1; \quad i > 1, \ j = 1; \quad i > 1, \ j > 1,$$

it follows from Example 5.4.1 that (X, Y, Z) has a trivariate normal limit distribution with mean $(0, 0, 0)$ and covariance matrix (τ_{ij}) where

$$(7.8.22) \qquad \tau_{ii} = \pi_i (1 - \pi_i) \text{ and } \tau_{ij} = -\pi_i \pi_j \text{ for } i, \ j = 1, 2, 3.$$

The variables $(X + Y, \ X + Z)$ therefore have a bivariate normal limit distribution with mean $(0, 0)$, with variances

$$
\begin{aligned}
\sigma_{11} &= \tau_{11} + 2\tau_{12} + \tau_{22} = \pi_1 (1 - \pi_1) - 2\pi_1 \pi_2 + \pi_2 (1 - \pi_2) \\
&= (\pi_1 + \pi_2)(1 - \pi_1 - \pi_2), \\
\sigma_{22} &= (\pi_1 + \pi_3)(1 - \pi_1 - \pi_3)
\end{aligned}
$$
$(7.8.23)$

and covariance

$$
\begin{aligned}
\sigma_{12} &= \pi_1 (1 - \pi_1) - \pi_1 \pi_2 - \pi_1 \pi_3 - \pi_2 \pi_3 \\
&= \pi_1 - (\pi_1 + \pi_2)(\pi_1 + \pi_3)
\end{aligned}
$$
$(7.8.24)$

and by (7.8.21) this completes the proof of the lemma. ∎

The limit distribution of (7.8.18) can now be obtained from Theorem 5.2.3 with

$$U = \frac{n_{1+}}{N}, \ V = \frac{n_{+1}}{N}, \ \xi = p_{1+}, \ \eta = p_{+1},$$

and $f(u, v) = uv$. We then have

$$\left. \frac{\partial f}{\partial u} \right|_{\xi, \eta} = p_{+1}, \quad \left. \frac{\partial f}{\partial v} \right|_{\xi, \eta} = p_{1+},$$

and it follows that (7.8.18) with $i = j = 1$ has a normal limit distribution with mean 0 and variance

$$\sigma^2 = p_{+1}^2 \sigma_{11} + 2 p_{+1} p_{1+} \sigma_{12} + p_{1+}^2 \sigma_{22},$$

which, in terms of the p's, equals (Problem 8.8)

$$(7.8.25) \quad \sigma^2 = p_{+1} p_{1+} \left[p_{+1} (1 - p_{1+}) + p_{1+} (1 - p_{+1}) + 2 (p_{11} - p_{+1} p_{1+}) \right].$$

Under the assumption of independence, the last term drops out, and the asymptotic variance of (7.8.18) reduces to

$$(7.8.26) \qquad \sigma^2 = p_{+i} p_{j+} \left[p_{+i} (1 - p_{j+}) + p_{j+} (1 - p_{+i}) \right].$$

Let us now compare the estimator $\hat{\hat{p}}_{ij}$ of p_{ij} derived as (7.8.17) under the independence assumption with the estimator

$$(7.8.27) \qquad \hat{p}_{ij} = \frac{n_{ij}}{N},$$

which was obtained in (7.5.22) for the unrestricted multinomial model. When the independence assumption is justified, it follows from (5.4.15) and from (7.8.26) that the ARE of $\hat{p}_{ij}$ to $\hat{\hat{p}}_{ij}$ is

$$(7.8.28) \qquad e = \frac{p_{i+}\left(1 - p_{+j}\right) + p_{+j}\left(1 - p_{i+}\right)}{1 - p_{i+}p_{+j}},$$

which is equal to

$$(7.8.29) \qquad 1 - \frac{\left(1 - p_{i+}\right)\left(1 - p_{+j}\right)}{1 - p_{i+}p_{+j}}.$$

Thus the efficiency is always ≤ 1, as it, of course, must be since under the independence assumption the estimator $\hat{\hat{p}}_{ij}$ is efficient.

As an example, consider an $a \times a$ table with $p_{ij} = 1/a^2$ for all i and j. Then $p_{i+} = p_{+j} = 1/a$ so that independence holds and

$$(7.8.30) \qquad e = 1 - \frac{\left(1 - \frac{1}{a}\right)^2}{1 - \frac{1}{a^2}} = 1 - \frac{a - 1}{a + 1} = \frac{2}{a + 1}.$$

This is a decreasing function of a which is equal to $2/3$ when $a = 2$, $1/3$ when $a = 5$, and tends to 0 as $a \to \infty$. More generally, e given by (7.8.20) tends to 0 as p_{i+} and p_{+j} both tend to 0. When independence holds, the loss of efficiency that results from ignoring this fact can therefore be quite severe.

One must, however, also consider the consequences of using the estimator $\hat{\hat{p}}_{ij}$ in the belief that independence obtains when this is in fact not the case. When $p_{ij} \neq p_{i+}p_{+j}$, it follows from the asymptotic normality of (7.8.18) that

$$(7.8.31) \qquad \sqrt{N}\left(\hat{\hat{p}}_{ij} - p_{ij}\right) = \sqrt{N}\left(\hat{\hat{p}}_{ij} - p_{i+}p_{+j}\right) - \sqrt{N}\left(p_{ij} - p_{i+}p_{+j}\right)$$

tends in probability to $+\infty$ or $-\infty$ as $p_{ij} < p_{i+}p_{+j}$ or $p_{ij} > p_{i+}p_{+j}$, i.e., the estimator $\hat{\hat{p}}_{ij}$ is then highly biased. Despite the resulting efficiency advantage of the estimators $\hat{\hat{p}}_{ij}$ when independence holds, this assumption should thus not be made lightly. □

A natural next problem is to test the hypothesis of independence. We have available the three tests (likelihood ratio, Wald, and Rao) discussed in the preceding section as well as Pearson's χ^2-test considered in Section 5.7. The Wald and Rao tests have the disadvantage that they require evaluation of the matrices $\left[\Sigma^{(r)}(\theta)\right]^{-1}$ and $\left[I^{(r)}(\theta)\right]^{-1}$ and we shall therefore here restrict attention to likelihood ratio and χ^2-tests which in the present situation are much simpler.

It follows from (7.8.17) and (7.8.27) that the likelihood ratio statistic (7.7.54) is equal to

$$(7.8.32) \qquad \frac{\prod\limits_{ij} n_{ij}^{n_{ij}}}{\prod\limits_{ij} (n_{i+}n_{+j})^{n_{ij}} / N^N},$$

and hence that its logarithm is given by

$$(7.8.33) \qquad \Delta_N = \sum_i \sum_j n_{ij} \log \left(\frac{n_{ij}}{(n_{i+}n_{+j})/N} \right).$$

By Theorem 7.7.4, the statistic $2\Delta_N$ has, under the hypothesis, a limiting χ_r^2-distribution where r is the number of parameters specified by the hypothesis. In the present case, the number of parameters in the unrestricted model is $ab - 1$, while under the hypothesis of independence, it is $(a - 1) + (b - 1)$. It follows that

$$(7.8.34) \qquad r = (ab - 1) - (a - 1) - (b - 1) = (a - 1)(b - 1).$$

The likelihood ratio test therefore rejects the hypothesis of independence at asymptotic level α when

$$(7.8.35) \qquad 2\Delta_N > C,$$

where

$$(7.8.36) \qquad \int_C^\infty \chi^2_{(a-1)(b-1)} = \alpha.$$

Similarly, substitution (with the appropriate change of notation) of the estimators (7.8.17) into (5.7.6) shows that Pearson's χ^2-statistic becomes

$$(7.8.37) \qquad X^2 = \sum \sum \frac{\left(n_{ij} - \frac{n_{i+}n_{+j}}{N} \right)^2}{n_{i+}n_{+j}/N},$$

which also has a limiting χ_r^2-distribution with r given by (7.8.34). An alternative to (7.8.35) therefore rejects at asymptotic level α when

$$(7.8.38) \qquad X^2 > C,$$

with C given by (7.8.36), as earlier.

The two tests are asymptotically equivalent as is shown, for example, in Agresti (1990, Section 12.3.4), which also contains references and additional information about these tests. Further results are provided by Loh (1989) and Loh and Yu (1993).

Summary

1. We consider contingency tables which exhibit the results of N multinomial trials with $k+1$ outcomes and with the probabilities of these outcomes being functions $p_i = p_i(\theta_1, \ldots, \theta_s)$ of $s \leq k$ parameters.

2. Conditions on the functions p_i are given for the likelihood equations to have a consistent root $\left(\hat{\theta}_1, \ldots, \hat{\theta}_s\right)$ and the asymptotic covariance matrix of the vector $\left(\sqrt{N}\left(\hat{\theta}_1 - \theta_1\right), \ldots, \sqrt{N}\left(\hat{\theta}_s - \theta_s\right)\right)$ is evaluated.

3. The theory is applied to an $a \times b$ two-way layout both in the unrestricted case and under the hypothesis H of independence, and the likelihood ratio test of H is obtained.

7.9 Problems

Section 1

1.1 Show that

  (i) the likelihood (7.1.3) is maximized by (7.1.4);

  (ii) the likelihood (7.1.5) is maximized by (7.1.6), provided not all the x_i are $= 1$.

1.2 Calculate the value of λ for which the probability (7.1.7) is equal to $1 - \epsilon$ for $\epsilon = .05, .1, .2$ and $n = 1, 2, 3$.

1.3 Prove (7.1.8).

1.4 Show that the MLEs (7.1.4) and (7.1.6) are consistent estimators of λ.

1.5 (i) Show that (7.1.9) defines a probability density.

  (ii) Give an example of a function $c(\theta)$ satisfying the conditions stated following (7.1.9).

1.6 Prove (7.1.12).

1.7 In the context of Example 7.1.3, Ferguson (1982) shows that (7.1.13) holds for

(7.9.1) $$c(\theta) = e(1 - \theta)e^{-1/(1-\theta)^4}.$$

Provide a graph of $c(\theta)$ and the resulting $f_\theta(x)$ for $\theta = .7, .8, .9, .95$.

1.8 Let X take on the values 1 and 0 with probabilities p and $q = 1 - p$, respectively, and suppose that the parameter space Ω is the open interval $0 < p < 1$. Then the likelihood function $L(p)$ has no local maxima for either $x = 0$ or $x = 1$ and in neither case does the MLE exist.

1.9 Let $X_1, \ldots, X_n$ be i.i.d. copies of the random variable X of the preceding problem, so that the likelihood of p is

$$p^{\sum x_i} q^{n - \sum x_i}.$$

(i) Show that the conclusions of the preceding problem continue to hold if either $x_1 = \cdots = x_n = 0$ or $x_1 = \cdots = x_n = 1$.

(ii) If $0 < \sum x_i < n$, the likelihood equation has the unique root $\hat{p} = \sum x_i / n$, and this corresponds to a local maximum.

(iii) If $0 < \sum x_i < n$, the estimator $\hat{p}$ of (ii) is the MLE.

[**Hint for (iii)**: Show that $l_n(p) \to \infty$ as $p \to 0$ or 1.]

1.10 Sketch a function $h(\theta)$, $-\infty < \theta < \infty$, which

(i) has two local maxima but no global maximum;

(ii) has two local maxima, one of which is also the global maximum.

1.11 Let $X_1, \ldots, X_n$ be i.i.d. $N(\xi, 1)$ with $\xi > 0$. Then

(i) the MLE is $\overline{X}$ when $\overline{X} > 0$ and does not exist when $\overline{X} \leq 0$;

(ii) the MLE exists with probability tending to 1 as $n \to \infty$;

(iii) the estimator

$$\hat{\xi}_n = \overline{X} \text{ when } \overline{X} > 0; \quad \hat{\xi}_n = 1 \text{ when } \overline{X} \leq 0$$

is consistent.

1.12 Let $X_1, \ldots, X_n$ be i.i.d. according to a Weibull distribution with density

(7.9.2) $\theta a (\theta x)^{a-1} e^{-(\theta x)^a}, \quad a, \theta, x > 0.$

Show that

(i) the likelihood equation has a unique solution $\hat{\theta}$;

(ii) the solution $\hat{\theta}$ corresponds to a local maximum of $l_n(\theta)$;

(iii) $\hat{\theta}$ is the MLE.

1.13 Let $X_1, \ldots, X_n$ be i.i.d. according to the Cauchy density

$$\frac{a}{\pi} \frac{1}{x^2 + a^2}, \quad -\infty < x < \infty, \quad 0 < a.$$

For the solution $\hat{a}$ of the likelihood equation, show the results of (i)–(iii) of the preceding problem with a in place of θ.

1.14 Show that the log likelihood (7.1.19) tends to $-\infty$ as $\sigma \to 0$ and as $\sigma \to \infty$

(i) for any fixed ξ;

(ii) for any fixed value of $a = \xi/\sigma$.

1.15 Verify the likelihood equation (7.1.24).

1.16 Under the assumptions of Example 7.1.3, use the CLT to show that the MLEs $\hat{\xi} = \overline{X}$ in (i) and $\hat{\sigma}^2$ given by (7.1.22) satisfy (7.1.27).

1.17 The definition of $\hat{\eta}$ given by (7.1.37) as the MLE for non-monotone g acquires additional justification from the following property. Let ω_η be the set of θ-values for which $g(\theta) = \eta$ and let

$$M(\eta) = \sup_{\theta \in \omega_\eta} L(\theta).$$

Then if $\hat{\theta}$ maximizes $L(\theta)$, $\hat{\eta}$ maximizes $M(\eta)$ (Zehna (1966), Berk (1967)).

[**Hint**: Use the facts that

$$L\left(\hat{\theta}\right) = \sup_{\Omega} L(\theta), \quad M(\hat{\eta}) = \sup_{\omega_\eta} L(\theta), \quad \text{and } L\left(\hat{\theta}\right) = M(\hat{\eta}).]$$

Section 2

2.1 If X has the binomial distribution $b(p, n)$,

(i) determine $I(p)$ from (7.1.28);

(ii) show that the MLE $\hat{p} = \sum X_i/n$ derived in Problem 1.9 satisfies (7.1.27).

2.2 In a sequence of binomial trials, let X_i be the number of trials between the $(i-1)^{\text{st}}$ and i^{th} success. Then $Y = \sum_{i=1}^{n} X_i$ has the negative binomial distribution $Nb(p, n)$ given in Table 1.6.2.

(i) Show that the MLE of p is $\hat{p} = n/(Y + n)$, i.e., the proportion of successes when sampling stops.

(ii) Determine $I(p)$ from (7.1.28).

(iii) Verify (7.1.27).

[**Hint for (iii)**: Use the limit result obtained in Problem 4.12 of Chapter 2 and Theorem 2.5.2.]

2.3 For a location family with density $f(x-\theta)$, where $f(x) > 0$ and $f'(x)$ exists for all x, show that

(i) $I(\theta)$ is independent of θ;

(ii) $I(\theta)$ is given by

$$(7.9.3) \qquad\qquad I \equiv \int \frac{[f'(x)]^2}{f(x)}\,dx.$$

2.4 Evaluate (7.9.3) for the cases that f is

(i) normal $N(0,1)$,

(ii) the logistic distribution $L(0,1)$;

(iii) the Cauchy distribution $C(0,1)$.

2.5 Let X have the double exponential distribution with density $\frac{1}{2}e^{-|x-\theta|}$.

(i) Determine $I(\theta) = I$ from (7.1.28).

(ii) Use the fact that $\sum |x_i - \theta|$ is minimized when $\theta = \hat\theta_n$ is the median of the X's to show that $\hat\theta_n$ is the MLE.

(iii) Check that $\hat\theta_n$ satisfies (7.1.27) despite the fact that $f_\theta'(x)$ does not exist when $x = \theta$.

[**Hint for (iii)**: Problem 3.4 of Chapter 2.]

2.6 For the scale family with density $\theta^{-1}f(x/\theta)$, $\theta > 0$, where $f(x) > 0$ and $f'(x)$ exists for all x, show that

$$(7.9.4) \qquad\qquad I(\theta) = \frac{1}{\theta^2}\int\left[\frac{xf'(x)}{f(x)} + 1\right]^2 f(x)\,dx.$$

2.7 Use (7.9.4) to evaluate $I(\theta)$ for the three densities of Problem 2.4.

2.8 Use the alternative expression (7.2.10) to determine $I(\theta)$ and compare it with the value obtained from (7.1.28) for the following cases:

(i) Example 7.1.1;

(ii) Problem 2.1(i);

(iii) Problem 2.2(ii).

2.9 Solve the preceding problem for the following cases:

(i) Problem 2.4 (i)–(iii);

(ii) Problem 2.5.

2.10 In Example 7.2.2, show that both (7.2.9) and (7.2.10) fail.

2.11 Show that (7.2.14) holds quite generally if $I(\theta)$ is defined by (7.2.10) instead of (7.1.28).

2.12 If X has the binomial distribution $b(p, n)$, determine the information X contains about $I(p)$. (See Problem 2.1.)

2.13 Check whether (C6) holds for the following distributions:

(i) the uniform distribution $U\left(\theta - \frac{1}{2},\ \theta + \frac{1}{2}\right)$;

(ii) the triangular distribution with density

$$f_\theta(x) = 1 - |x - \theta|, \quad -\theta < x < \theta;$$

(iii) the asymmetric triangular distribution with density

$$f_\theta(x) = \begin{cases} 2(\theta + 1 - x)/3 & \text{if } \theta < x < \theta + 1 \\ x - 2 - \theta & \text{if } \theta - 2 < x < \theta. \end{cases}$$

2.14 For each of the distributions of the preceding problem

(i) calculate $I(\theta)$ from (7.1.28);

(ii) obtain the limit distribution of a suitably normalized MLE;

(iii) compare the results of (i) and (ii) along the lines of Example 7.2.2.

Section 3

3.1 In Example 7.3.2:

(i) show that $\partial^k f_\sigma(x)/\partial\sigma^k$ is of the form stated there;

(ii) check condition (C7).

3.2 In Examples 7.3.1 and 7.3.3, check that the conditions of Theorem 7.3.1 and Corollary 7.3.1 are satisfied.

3.3 If $\hat{\theta}_n$ is a $\sqrt{n}$-consistent estimator of θ:

(i) determine constants a_n such that $\hat{\theta}_n + a_n$ is consistent but no longer $\sqrt{n}$-consistent;

(ii) determine constants b_n such that the conclusion of (i) holds for estimators $(1 + b_n)\hat{\theta}_n$ when $\theta \neq 0$.

The following problem shows that Theorem 7.3.3 no longer holds if $\hat{\theta}_n$ is only required to be consistent rather than $\sqrt{n}$-consistent.

3.4 Let $X_1, \ldots, X_n$ be i.i.d. $N(\theta, 1)$. Use Problem 3.3(ii) with $\hat{\theta}_n = \overline{X}_n$ to construct a sequence of estimators of the form (7.3.21) with $\tilde{\theta}_n$ consistent (but not $\sqrt{n}$-consistent) for which (7.3.24) does not hold.

3.5 Suppose that the assumptions of Theorem 7.3.3 hold and that $I(\theta)$ is a continuous function of θ. Then the estimator

(7.9.5)
$$\delta_n = \tilde{\theta}_n + \frac{l_n'\left(\tilde{\theta}_n\right)}{nI\left(\tilde{\theta}_n\right)}$$

satisfies (7.3.24).

[**Hint**: Use (7.3.8) and Theorem 2.1.1.]

3.6 Prove (7.3.38).

3.7 Prove (7.3.43).

[**Hint**: To simplify (7.1.28), use the fact that

$$h - g = [\theta g + (1 - \theta)h - g] / (1 - \theta)$$

and

$$g - h = [\theta g + (1 - \theta)h - h] / \theta.$$

3.8 If $G = N(-\xi, 1)$ and $H = N(\xi, 1)$, show that (7.3.43) tends to $1/\theta(1 - \theta)$ as $\xi \to \infty$.

3.9 (i) Show that the estimator (7.3.44) is $\sqrt{n}$-consistent, provided the X's have finite variance.

(ii) Show that (7.3.47) is $\sqrt{n}$-consistent.

(iii) Determine a $\sqrt{n}$-consistent estimator of θ when $\eta = \xi$ but $E_G\left(X_i^2\right) \neq E_H\left(X_i^2\right)$ along the lines of (7.3.44) rather than (7.3.47).

3.10 (i) Verify the likelihood equation (7.3.50).

(ii) Prove the statement concerning the roots of the likelihood equation (7.3.50) made following (7.3.50).

(iii) Verify (7.3.53).

3.11 In Example 7.3.6, check that the conditions of Theorem 7.3.1 are satisfied.

3.12 In Example 7.3.6, discuss the number and nature (i.e., maximum, minimum, or inflection point) of the roots of the likelihood equation when $n = 2$.

4.1 In Example 7.4.1, prove that $\sqrt{n}\,(\delta_n' - \theta) \xrightarrow{L} N\,(0, v'(\theta))$ when $\theta \neq 0$.

4.2 If δ_n satisfies (7.4.1):

(i) construct an estimator δ_n' which satisfies (7.4.3) for any given θ_0 ;

(ii) for any given $\theta_1, \ldots, \theta_k$ construct an estimator δ_n' that satisfies (7.4.1) with $v(\theta)$ replaced by a function $v'(\theta)$ for which

$$v'(\theta) \begin{cases} < v(\theta) & \text{for } \theta = \theta_1, \ldots, \theta_k \\ = v(\theta) & \text{for all other } \theta. \end{cases}$$

4.3 Let δ_{1n} and δ_{2n} be independent and such that $\sqrt{n}\,(\delta_{in} - \theta) \xrightarrow{L} N\,(0, \sigma_i^2)$, $i = 1, 2$, and let $\delta_{3n} = \alpha \delta_{1n} + (1 - \alpha)\,\delta_{2n}$. Then the value of α minimizing the asymptotic variance of $\sqrt{n}\,(\delta_{3n} - \theta)$ is $\alpha = \sigma_2^2/\left(\sigma_1^2 + \sigma_2^2\right)$ and for this value the asymptotic variance of δ_{3n} is $1/\left(\dfrac{1}{\sigma_1^2} + \dfrac{1}{\sigma_2^2}\right)$.

4.4 In Example 7.4.5,

(i) verify the formula for $P\,(\theta = a|x_1, \ldots, x_n)$;

(ii) show that when $\theta = (a + b)/2$, the Bayes estimator δ_λ converges to a or b with probability $1/2$ each.

[**Hint for (ii)**: Problem 3.11 of Chapter 2.]

4.5 In Example 7.4.5, obtain the Bayes estimator if the prior assigns positive probabilities $p_1, \ldots, p_k$ $\left(\sum p_i = 1\right)$ to the points $a_1 < \cdots < a_k$.

4.6 In Example 7.4.5, suppose that the X's are i.i.d. according to the exponential distribution with density $\dfrac{1}{\theta}e^{-x/\theta}, x, \theta > 0$. With λ defined as in the example, show that there exists a value θ_0 (depending on a and b) such that (7.4.18) holds with $(a + b)/2$ replaced by θ_0.

4.7 Solve the preceding problem for the case that the X's are i.i.d. and $X_i = 1$ or 0 with probabilities θ and $1 - \theta$, respectively.

4.8 In Example 7.4.6,

(i) verify the conditional density (7.4.26);

(ii) show that the conditional density formally obtained from the improper Jeffreys prior is the proper distribution $N\,(\bar{x}, 1/n)$.

4.9 In Example 7.4.6, show that for any fixed $(x_1, \ldots, x_n)$ the Bayes estimator corresponding to the prior $U(-A, A)$ tends to $\bar{x}$, the Bayes estimator corresponding to the Jeffreys prior.

4.8 In Example 7.4.6,

(i) verify the conditional density (7.4.26);

(ii) show that the conditional density formally obtained from the improper Jeffreys prior is the proper distribution $N(\bar{x}, 1/n)$.

4.9 In Example 7.4.6, show that for any fixed $(x_1, \ldots, x_n)$ the Bayes estimator corresponding to the prior $U(-A, A)$ tends to $\bar{x}$, the Bayes estimator corresponding to the Jeffreys prior.

4.10 Solve the preceding two problems when the prior for θ is $N(0, A^2)$ instead of $U(-A, A)$.

4.11 Solve Problems 4.8 and 4.9 for the case that $X_1, \ldots, X_n$ are i.i.d. $N(0, \sigma^2)$ and σ has the prior distribution $U(a, A)$, and that $a \to 0$ and $A \to \infty$.

4.12 (i) Let X be a random variable with density $f(x - \theta)$. If θ has the improper uniform prior on $(-\infty, \infty)$, the posterior distribution of θ given x is a proper distribution with density $f(x - \theta)$.

(ii) If X has density $\frac{1}{\theta} f\left(\frac{x}{\theta}\right)$, $\theta > 0$, and θ has the improper prior with density a/θ, $0 < \theta < \infty$, determine the posterior density of θ given x and show that it is a proper density.

4.13 Verify formula (7.4.33).

4.14 Consider a sequence of binomial trials with success probability p which is continued until n successes have been obtained.

(i) Find the Jeffreys prior for p and determine whether it is a proper or improper distribution.

(ii) Obtain the posterior distribution of p given $Y = y$, where $Y + n$ is the number of trials required to achieve n successes.

(iii) Determine the Bayes estimator of p.

4.15 In the following situations, show that the difference between the MLE and the Jeffreys Bayes estimator is $O_P(1/n)$:

(i) Example 7.4.8 (i);

(ii) Example 7.4.8 (ii);

(iii) Example 7.4.4 (concluded);

(iv) Problem 4.12;

(v) Problem 4.10.

Section 5

5.1 (i) Show that (7.5.5) is zero when f is symmetric about 0.

(ii) Verify (7.5.7).

5.2 In the (x, y)-plane, show that the set

$$\left|x - x^0\right| < a, \ \left|y - y^0\right| < b$$

is an open set, but the set

$$\left|x - x^0\right| < a, \ \left|y - y^0\right| \leq b$$

is not.

5.3 Prove Theorem 7.5.1.

5.4 State and prove the multivariate generalizations of Theorem 7.2.2.

5.5 Prove formula (7.5.13).

5.6 Prove (7.5.19) directly from the bivariate CLT.

5.7 Prove that the joint limit distribution of the variables (7.5.27) is multivariate normal with means 0 and covariance matrix given by (7.5.28) and (7.5.29).

[**Hint:** Use the facts that (a) the limit distribution of $(\hat{\sigma}^2, \hat{\tau}^2, \hat{\gamma})$ is unchanged when $(\overline{X}, \overline{Y})$ is replaced by (ξ, η) and (b) the variables $(X_i - \xi, \ Y_i - \eta)$ are uncorrelated with the variables $(X_i - \xi)^2$, $(Y_i - \eta)^2$, $(X_i - \xi)(Y_i - \eta)$.]

5.8 Prove the result corresponding to Problem 5.7 for the variables $\hat{\sigma}^2$, $\hat{\tau}^2$, and $\hat{\rho}$, and, in particular, verify (7.5.31).

5.9 Show that the limit distribution of the variables (7.5.38) is as claimed in the text.

5.10 Let $X_1, \ldots, X_n$ be i.i.d. according to the gamma density

$$(7.9.6) \qquad \frac{1}{\Gamma(a)b^a} x^{a-1} e^{-x/b}, \ a > 1, \ b > 0,$$

so that

$$(7.9.7) \qquad E(X_i) = ab, \ \mathrm{Var}(X_i) = ab^2.$$

(i) Show that the likelihood equations for a and b are

$$(7.9.8) \qquad ab = \overline{X}; \ \log a - \Gamma'(a) = \log \overline{X} - \frac{1}{n}\sum \log X_i.$$

(ii) As an alternative to consistent solutions of (7.9.8), use Theorem 7.5.3 to determine asymptotically efficient estimators with starting point determined by the method of moments, i.e., by solving the equations

$$(7.9.9) \qquad ab = \overline{X}; \; ab^2 = \frac{1}{n} \sum (X_i - \overline{X})^2 .$$

5.11 In generalization of Problem 3.13, show that the result of Theorem 7.5.3 remains valid when the coefficients $l''_{ij}\left(\tilde{\theta}_n\right)$ in (7.5.43) are replaced by $-I_{ij}\left(\tilde{\theta}_n\right)$, provided $I_{ij}(\theta)$ is continuous for all i, j.

5.12 (i) Verify the moments (7.5.49).

(ii) Show that the variables (7.5.53) have a joint normal limit distribution, and determine its covariance matrix.

(iii) For $k = 2$, give explicit expressions for the estimators $(\delta_{1n}, \delta_{2n})$ obtained from (7.5.43).

5.13 Show that $\tilde{q}$ defined by (7.5.51) can take on values both less than 0 and greater than 1, but that the probability of this happening tends to 0 as $n \to \infty$.

5.14 (i) Verify the moments (7.5.54).

(ii) Solve (7.5.55) to obtain explicit estimators $\tilde{q}$, $\tilde{\eta}$, and $\tilde{\tau}$, and show that they are $\sqrt{n}$-consistent.

5.15 Let $X_1, \ldots, X_n$ be i.i.d. with distribution P_θ depending on a real-valued parameter θ, and suppose that

$$E_\theta(X) = g(\theta) \text{ and } \text{Var}_\theta(X) = \tau^2(\theta) < \infty,$$

where g is continuously differentiable function with derivative $g'(\theta) > 0$ for all θ. Then the estimator $\tilde{\theta}$ obtained by the method of moments, i.e., by solving the equation

$$g(\theta) = \overline{X},$$

is $\sqrt{n}$-consistent.

[**Hint:** Since $\tilde{\theta} = g^{-1}\left(\overline{X}\right)$, the result follows from Theorem 2.5.2.]

5.16 A vector $\underline{X}^{(n)} = \left(X_1^{(n)}, \ldots, X_k^{(n)}\right)$ is *bounded in probability* if for any $\epsilon > 0$ there exists a k-dimensional cube

$$C = \{(x_1, \ldots, x_k) : \; |x_i| < K \text{ for all } i\}$$

and a value n_0 such that

$$P\left(\underline{X}^{(n)} \in C\right) > 1 - \varepsilon \text{ for all } n > n_0.$$

A sequence of estimators $\left(\hat{\theta}_{1n}, \ldots, \hat{\theta}_{kn}\right)$ of $(\theta_1, \ldots, \theta_k)$ is $\sqrt{n}$-consistent if the vector

$$\sqrt{n}\left(\hat{\theta}_{1n} - \theta_1\right), \ldots, \sqrt{n}\left(\hat{\theta}_{kn} - \theta_k\right)$$

is bounded in probability.

Show that $\underline{X}^{(n)}$ is bounded in probability if and only if $X_i^{(n)}$ is bounded in probability for each i, and hence that $\left(\hat{\theta}_{1n}, \ldots, \hat{\theta}_{kn}\right)$ is $\sqrt{n}$-consistent if and only if $\hat{\theta}_{in}$ is $\sqrt{n}$-consistent for each i.

5.17 Let $X_1, \ldots, X_n$ be i.i.d. with distribution P_{θ_1, θ_2} depending on two real parameters θ_1 and θ_2. Suppose that

$$E_{\theta_1, \theta_2}(X) = g_1(\theta_1, \theta_2); \quad E_{\theta_1, \theta_2}\left(X^2\right) = g_2(\theta_1, \theta_2),$$

where the functions g_i have continuous partial derivatives $\partial g_i / \partial \theta_j$, that the Jacobian $\| \partial g_i / \partial \theta_j \|$ is $\neq 0$, and that the estimating equations

$$g_1(\theta_1, \theta_2) = \bar{X}, \quad g_2(\theta_1, \theta_2) = \sum_{i=1}^{n} X_i^2 / n$$

have a unique solution $\left(\tilde{\theta}_{1n}, \tilde{\theta}_{2n}\right)$.

(i) Give a heuristic argument indicating why $\left(\tilde{\theta}_{1n}, \tilde{\theta}_{2n}\right)$ will typically be $\sqrt{n}$-consistent.

(ii) State conditions under which the conclusion of (i) will hold.

(iii) Generalize (i) and (ii) to the case of k parameters.

Section 6

6.1 In Example 7.6.1, show that the conditional distribution of $p_1, \ldots, p_{k+1}$ given $y_1, \ldots, y_{k+1}$ is the Dirichlet distribution with parameters $a_i' = a_i + y_i$ for $i = 1, \ldots, k+1$.

6.2 A 2×2 matrix $A = \begin{pmatrix} a & b \\ b & c \end{pmatrix}$ is positive definite if and only if (a) a and c are positive and (b) $ac > b^2$.

[**Hint:** Use the fact that A is positive definite if and only if $ax^2 + 2bx + c > 0$ for all x.]

6.3 Under the assumptions of Theorem 7.5.2, show that (7.6.29) is equivalent to (7.6.28).

6.4 (i) Under the assumptions of Theorem 7.5.2, show that equality holds in (7.6.24) (and hence that the efficiency with which θ_j can be estimated is the same whether or not $(\theta_1, \ldots, \theta_{j-1}, \theta_{j+1}, \ldots, \theta_k)$ is known), provided

$$(7.9.10) \qquad I_{ij}(\theta) = 0 \text{ for all } i \neq j.$$

(ii) Equivalent to (7.9.10) is that $\sqrt{n}\left(\hat{\theta}_{jn} - \theta_j\right)$ is asymptotically independent of $\left\{\sqrt{n}\left(\hat{\theta}_{in} - \theta_i\right), i \neq j\right\}$.

[**Hint**: (i) If (7.9.10) holds, then also $J_{ij}(\theta) = 0$ for all $i \neq j$ by Problem 3.14 of Chapter 5. The result then follows from the fact that the (j,j)th element of IJ is 1.]

6.5 If the information matrix is of the form

$$(7.9.11) \qquad I = \begin{pmatrix} I_1 & 0 \\ 0 & I_2 \end{pmatrix} \begin{matrix} r \\ s \end{matrix}$$

and if the assumptions of Theorem (7.5.2) hold, then

(i) the two sets of variables $\left\{\sqrt{n}\left(\hat{\theta}_i - \theta_i\right), i = 1, \ldots, r\right\}$ and $\left\{\sqrt{n}\left(\hat{\theta}_j - \theta_j\right), j = r+1, \ldots, r+s\right\}$ are mutually independent;

(ii) The parameter vector $(\theta_1, \ldots, \theta_r)$ can be estimated with the same efficiency whether or not $(\theta_{r+1}, \ldots, \theta_{r+s})$ is known.

6.6 Illustrate the results of the preceding problem on the multivariate normal distribution of Example 7.5.5, with $(\theta_1, \ldots, \theta_r)$ corresponding to the vector of means and $(\theta_{r+1}, \ldots, \theta_{r+s})$ to the matrix of covariance.

6.7 Under the assumptions of Theorem 7.6.3, verify that (7.3.9) holds with l_n replaced by l.

6.8 State and prove an analog to Theorem 7.3.3 for the case of several samples treated in Theorem 7.6.3.

6.9 Extend to the case of several samples the results of

(i) Theorem 7.5.2;

(ii) Theorem 7.5.3;

(iii) Theorem 7.6.2.

6.10 In Example 7.6.4(ii):

(i) check that (7.6.39) is the unique solution of the likelihood equation;

(ii) show that $\sqrt{n}(\hat{\hat{\xi}} - \hat{\xi})$ does not tend to 0 in probability and hence that $\sqrt{n}\left(\hat{\xi} - \xi\right)$ and $\sqrt{n}\left(\hat{\hat{\xi}} - \xi\right)$ do not have the same limit distribution.

6.11 (i) Verify the likelihood equations (7.6.45).

(ii) Show that the equations (7.6.45) have a unique solution which agrees with the least squares estimators.

6.12 If A and B are two positive semi-definite symmetric $k \times k$ matrices with the same diagonal elements, then if $B - A$ is positive semi-definite, $A = B$.

[**Hint**: The proof is analogous to that of Theorem 6.4.7.]

6.13 (i) If X has the Poisson distribution $P(a\lambda)$ ($a =$ unknown), determine the amount of information X contains about λ.

(ii) Let X_i ($i = 1, \ldots, n$) be independent $P(a_i\lambda)$. Determine the MLE $\hat{\lambda}$ of λ and show that $k_n(\hat{\lambda}_n - \lambda) \overset{L}{\to} N(0, 1)$ for suitable k_n.

(iii) Show that $k_n = \sqrt{T_n(\lambda)}$, where $T_n(\lambda)$ is the total amount of information, and determine a sequence of a's for which $T_n(\lambda)$ is of order $n^{1/4}$.

6.14 In contrast to the preceding problem show that

(i) if X is $N(0, a^2\sigma^2)$, the amount of information in X about σ^2 is independent of a;

(ii) if X_i ($i = 1, \ldots, n$) are independent $N(0, a_i^2\sigma^2)$, the total amount of information $T_n(\sigma^2)$ is of order $\sqrt{n}$.

[**Hint**: Note that X/a is $N(0, \sigma^2)$.]

6.15 (i) Let $Y_1, Y_2, \ldots$ be a sequence of independent positive random variables with $E(Y_i) = \eta$ and $\text{Var}(Y_i) = \tau_i^2$. Show that if

$$E\left(\overline{Y} - \eta\right)^2 = \frac{1}{n^2}\sum_{i=1}^{n}\tau_i^2 \to 0,$$

then $\overline{Y}/\eta \overset{P}{\to} 1$.

(ii) Use (i) to obtain sufficient conditions for (7.6.55).

(ii) for the second term in the denominator of (7.6.53) to tend to zero in probability.

[**Hint for (i)**: Use Theorem 2.7.1.]

6.17 In Example 7.6.5, determine for all possible values of the v's whether the efficiency with which β can be estimated depends on whether or not α is known.

6.18 Let $X_1, \ldots, X_n$ be i.i.d. according to the exponential distribution $E(\xi, a)$ with density

$$(7.9.12) \qquad f_{\xi, a}(x) = \frac{1}{a} e^{-(x-\xi)/a} \text{ for } x > \xi.$$

(i) When ξ is known, show that the likelihood equation for estimating a has a unique solution which is efficient by Theorem 7.3.1.

(ii) When ξ is unknown, show that the likelihood equations have a unique solution $\left(\hat{\xi}, \hat{a}\right)$ and that $\hat{a}$ is efficient for estimating a despite the fact that the conditions of Theorem 7.5.2 are not satisfied.

[**Hint for (ii)**: Use part (i).]

6.19 (i) In Example 7.6.7, show that $\hat{\sigma}^2$ is consistent for estimating σ^2 if n remains fixed and $r \to \infty$.

(ii) What happens to $\hat{\sigma}^2$ if $r = \sqrt{n}$ and $n \to \infty$? Generalize.

Section 7

7.1 (i) Verify the maximum likelihood estimator $\hat{\theta}_n$ given by (7.7.17).

(ii) If $\theta > 0$, show that $\sqrt{n}\left(\hat{\theta}_n - \theta\right) \xrightarrow{L} N(0, 1)$.

(iii) If $\theta = 0$, the probability is 1/2 that $\hat{\theta}_n = 0$ and 1/2 that $\sqrt{n}\left(\hat{\theta}_n - \theta\right) \xrightarrow{L} N(0, 1)$.

7.2 Determine the rejection region of the Wald, Rao, and likelihood ratio tests of the following hypotheses against two-sided alternatives:

(i) $H : \lambda = \lambda_0$ when the X's are i.i.d. with Poisson distribution $P(\lambda)$;

(ii) $H : p = p_0$ when the X's are independent with $P(X_i = 1) = p$, $P(X_i = 0) = 1 - p$;

(iii) $H : \sigma^2 = \sigma_0^2$ when the X's are i.i.d. $N\left(0, \sigma^2\right)$.

7.3 For each part of the preceding problem, determine the one-sided Wald and likelihood ratio tests.

7.4 Verify the asymptotic equivalence of the statistics (7.7.16).

7.4 Verify the asymptotic equivalence of the statistics (7.7.16).

7.5 Let $X_1, \ldots, X_n$ be i.i.d. according to the Pareto distribution with density

$$f_\theta(x) = \theta c^\theta / x^{\theta+1}, \ 0 < c < x, \ 0 < \theta.$$

(i) Determine the unique solution $\hat{\theta}_n$ of the likelihood equation and find the limit distribution of $\sqrt{n}\left(\hat{\theta}_n - \theta\right)$.

(ii) Determine the Wald, Rao, and likelihood ratio tests of $H : \theta = \theta_0$ against $\theta \neq \theta_0$.

[**Hint for (i)**: Let $\eta = 1/\theta$ and solve the problem first for η.]

7.6 Suppose that $\sqrt{n}\,(T_n - \theta_0)$ and $\sqrt{n}\,(T'_n - \theta_0)$ each tends to $N(0, 1)$ as $n \to \infty$. Then the tests with rejection regions

$$\sqrt{n}\,(T_n - \theta_0) \geq u_\alpha \text{ and } \sqrt{n}\,(T'_n - \theta_0) \geq u_\alpha,$$

respectively, are asymptotically equivalent if

$$\sqrt{n}\,(T'_n - T_n) \xrightarrow{P} 0$$

but not necessarily if $T'_n - T_n \xrightarrow{P} 0$.

[**Hint for Second statement**: Let $T'_n = T_n + n^{-1/4}$.]

7.7 Show that both the test (7.7.8) and that based on (7.7.11) retain their asymptotic level when $\hat{\theta}_n$ is replaced by the one-step estimator $\hat{\delta}_n$ given by (7.3.21).

7.8 Determine the Rao score test for the hypothesis (7.7.44) and show directly that the associated test statistic R_n has the same limit distribution as W_n.

7.9 In the model of Problem 7.2 (ii), suppose that p is restricted to the interval $p_0 \leq p_0 < 1$ where $0 < p_0 < 1$.

(i) Determine the MLE $\hat{p}_n$ of p.

(ii) In analogy with Problem 7.1 determine the limit behavior of $\sqrt{n}\,(\hat{p}_n - p)$ both when $p > p_0$ and when $p = p_0$.

7.10 Prove (7.7.23).

7.11 Show that the left side of (7.7.28) tends in probability to 0 as $\theta \to \infty$ while n remains fixed.

7.12 Prove (7.7.35) under the assumptions of Theorem 7.5.1.

7.13 Prove Theorem 7.7.3.

7.14 For the normal one-sample problem of Example 7.7.4:

(i) determine the Rao score statistic R_n;

(ii) show that $2\Delta_n$ is given by (7.7.43);

(iii) show directly that $2\Delta_n - W_n \overset{P}{\to} 0$ under H but not under a fixed alternative $(\xi, \sigma^2) \neq (\xi_0, \sigma_0^2)$.

7.15 For the multinomial one-sample problem of Example 7.7.5:

(i) show that Δ_n is given by (7.7.51);

(ii) check directly that (7.7.52) holds.

[**Hint for (ii)**: If $Z_i = \left(\dfrac{Y_i}{n} - p_i \right) / p_i$, then $W_n = \hat{\theta} = \sum_{i=1}^{k+1} p_i Z_i^2$ and $\Delta_n = \sum n p_i (Z_i + 1) \log (Z_i + 1)$. To prove (ii), expand $\log (1 + Z_i)$ and use the facts that $\sum p_i Z_i = 0$ and $\sqrt{n} Z_i \overset{L}{\to} N(0, q_i/p_i)$.]

7.16 Under the assumptions of Theorem 7.7.4, prove that $2\Delta_n \overset{L}{\to} \chi_r^2$.

7.17 Prove that the conclusions of Theorems 7.7.2–7.7.4 hold for the multisample case under the assumptions of Theorem 7.6.3.

7.18 In Example 7.7.6, verify (7.7.62).

7.19 Let the square matrices A and B be partitioned as

$$
A = \begin{pmatrix} A_{11} & A_{12} \\ A_{21} & A_{22} \end{pmatrix} \begin{matrix} r \\ s \end{matrix} \quad \text{and} \quad B = \begin{pmatrix} B_{11} & B_{12} \\ B_{21} & B_{22} \end{pmatrix} \begin{matrix} r \\ s \end{matrix} .
$$

Then the product $C = AB$ can be calculated as if the submatrices were elements, i.e., $C = \begin{pmatrix} C_{11} & C_{12} \\ C_{21} & C_{22} \end{pmatrix} \begin{matrix} r \\ s \end{matrix}$, where, for example, $C_{12} = A_{11} B_{12} + A_{12} B_{22}$.

7.20 Let A be partitioned as in the preceding problem and suppose that A, A_{11}, and A_{22} are nonsingular. Then if $B = A^{-1}$, the submatrices A_{ij} and B_{ij} satisfy

$$
A_{11} = \left(B_{11} - B_{12} B_{22}^{-1} B_{21} \right)^{-1}, \quad A_{12} = -B_{11}^{-1} B_{12} A_{22},
$$
$$
A_{21} = -B_{22}^{-1} B_{21} A_{11}, \quad A_{22}^{-1} = \left(B_{22} - B_{21} B_{11}^{-1} B_{12} \right)^{-1}.
$$

[**Hint**: Show that BA is the identity matrix.]

7.21 Under the assumptions of Theorem 7.7.4, derive the inequality (7.7.71) from the fact that $(\theta_1, \ldots, \theta_r)$ can be estimated at least as efficiently when $(\theta_{r+1}, \ldots, \theta_k)$ is known as when it is unknown.

7.22 Prove that (7.7.65) and (7.7.66) hold for the multisample case under the assumptions of Theorem 7.6.3.

7.23 For the two-sample Poisson problem, three tests were proposed in Example 3.1.5 and Problem 1.21 of Chapter 3, given by (3.1.34), (3.1.84), and (3.6.6), respectively. Obtain the likelihood ratio, Wald and Rao, tests for this problem, and determine whether any of these tests coincide.

7.24 Random effects model. Let X_{ij}, $j = 1, \ldots, m$, $i = 1, \ldots, s$ be sm observations satisfying

$$X_{ij} = \xi + A_i + U_{ij},$$

where the unobservable random variables A_i and U_{ij} are independently normally distributed with mean 0 and variances σ_A^2 and σ^2, respectively. The probability density of the X's is

$$\frac{1}{(2\pi)^{sm/2}\, \sigma^{s(m-1)}\, (\sigma^2 + m\sigma_{A^2})^s}\, \exp\left[-\frac{1}{2(\sigma^2 + m\sigma_A^2)}\right] \times$$
$$\left[sm\,(x_{..} - \xi)^2 + S_A^2\right] - \frac{1}{2\sigma^2} S^2,$$

where

$$S_A^2 = m \sum_{i=1}^{s} (X_{i.} - X_{..})^2 \ \text{and} \ S^2 = \sum_{i=1}^{s}\sum_{j=1}^{m} (X_{ij} - X_{i.})^2,$$

so that

$$S_A^2 / \left(\sigma^2 + m\sigma_A^2\right) \ \text{and} \ S^2/\sigma^2$$

are independent χ_{s-1}^2 and $\chi_{s(m-1)}^2$, respectively. The hypothesis H : $\sigma_A^2 = 0$ is to be tested against the alternatives $\sigma_A^2 > 0$. Under these assumptions show the following:

(i) The MLEs of σ_A^2 and σ^2 are $\hat{\sigma}^2 = S^2/s(m-1)$ and

$$\hat{\sigma}_A^2 = \begin{cases} \dfrac{1}{s}S_A^2 - \dfrac{1}{s(m-1)}S^2 & \text{if } S_A^2 > S^2/(m-1) \\ 0 & \text{otherwise.} \end{cases}$$

(ii) Under H, the MLEs of σ^2 and σ_A^2 are

$$\hat{\sigma} = \left(S_A^2 + S^2\right)/sm \ \text{and} \ \hat{\sigma}^2_A = 0.$$

(iii) As $s \to \infty$,

$$\sqrt{s}\left(\frac{S_A^2/\left(\sigma^2 + m\sigma_A^2\right)}{s} - 1\right) \text{ and } \sqrt{s(m-1)}\left[\frac{S^2/\sigma^2}{s(m-1)} - 1\right]$$

both tend in law to $N(0,2)$ and hence

$$P\left[S_A^2 > \frac{S^2}{m-1}\right] \to 1/2.$$

[**Hint for (ii)**: Example 2.3.2.]

7.25 Use the preceding problem to determine the Wald and the likelihood ratio test of H at level $\alpha < 1/2$.

Section 8

8.1 Consider a three-way layout with probabilities for N multinomial trials with possible outcomes O_{ijk} having probabilities p_{ijk} $(i = 1, \ldots, a;$ $j = 1, \ldots, b;$ $k = 1, \ldots, c)$ assumed to satisfy one of the following model assumptions:

(a) unrestricted except for $\sum_i \sum_j \sum_k p_{ijk} = 1$;

(b) outcome k independent of (i, j), i.e.,

(7.9.13) $p_{ijk} = p_{ij+}p_{++k}$,

where $+$ indicates summations over the corresponding subscript;

(c) complete independence, i.e.,

(7.9.14) $p_{ijk} = p_{i++}p_{+j+}p_{++k}.$

For each of the three models find a suitable representation (7.8.2) and the value of s.

[**Hint**: For models (a) and (b), this was already worked out (with a different notation) in Examples 7.8.1 and 7.8.2.]

8.2 For each of the three models of the preceding section, show that the likelihood equations have a unique solution and determine the corresponding estimators

$$\hat{p}_{ijk}, \; \hat{\hat{p}}_{ijk}, \; \text{and} \; \hat{\hat{\hat{p}}}_{ijk}$$

of p_{ijk} in terms of the variables n_{ijk}, n_{ij+}, n_{i++}, etc.

8.3 Obtain the limit distributions

$$\sqrt{N}\left(\hat{\hat{p}}_{ijk} - p_{ijk}\right), \ \sqrt{N}\left(\hat{\hat{p}}_{ijk} - p_{ijk}\right), \text{ and } \sqrt{N}\left(\hat{p}_{ijk} - p_{ijk}\right)$$

under each of the models (a), (b), and (c).

8.4 Obtain and discuss the ARE of

(i) $\hat{p}_{ijk}$ to $\hat{\hat{p}}_{ijk}$ in models (b) and (c);

(ii) $\hat{\hat{p}}_{ijk}$ to $\hat{\hat{p}}_{ijk}$ in model (c).

8.5 What can you say about the asymptotic behavior of

(i) $\hat{\hat{p}}_{ijk}$ in model (a) when (7.9.13) does not hold;

(ii) $\hat{\hat{p}}_{ijk}$ in model (b) when (7.9.14) does not hold?

8.6 Let the likelihood be given by (7.8.7). Then if the Jacobian $\left|\left(\dfrac{\partial p_i(\theta)}{\partial \theta_j}\right)\right|$ is $\neq 0$, the information matrix $I(\theta)$ is positive definite.

8.7 Check the conditions of Theorem 7.8.1 in Example 7.8.3.

8.8 Verify equation (7.8.25).

8.9 Evaluate and discuss the efficiency (7.8.29) for an $a \times b$ table in which $p_{ij} = 1/ab$.

Bibliographic Notes

Maximum likelihood estimation has a long history (see, for example, Edwards (1974), Savage (1976), Pfanzagl (1994), and Stigler (1997). Its modern development as an efficient method of estimation in parametric families was initiated by Edgeworth (1908/9) and Fisher (1922a, 1925) (see Pratt (1976) for an analysis of their respective contributions and Aldrich (1997) for the development of Fisher's ideas). However, Fisher's claim of efficiency of the MLE was disproved by an example of Hodges reported by Le Cam (1953). Accounts of the work of Le Cam, Hajek, and others to rescue Fisher's program can be found in Wong (1992) and Beran (1995).

Conditions for the consistency of MLE's were given by Wald (1949); examples of inconsistency are discussed, for example, by Le Cam (1990b) and Ferguson (1995). Easier and more widely applicable conditions were found by Cramér (1946) for consistent sequences of local instead of global maxima.

Fisher's measure of information plays a central role in his theory of efficient estimation (1922a, 1925). Its relation to other information measures is discussed in Kendall (1973). (For its role for estimation in nonparametric and semiparametric models, see, for example, Ritov and Bickel (1990).

The asymptotic efficiency of Bayes estimators is treated by Le Cam (1953). Accounts within the general framework of large-sample theory can be found, for example, in Ferguson (1995), Ibragimov and Has'minskii (1981), and Lehmann and Casella (1998).

Likelihood ratio tests were introduced by Neyman and Pearson (1928) as small-sample procedures; their large-sample performance was established by Wilks (1938) and Wald (1943). The Wald and Rao tests were proposed in Wald (1943) and Rao (1947), respectively.

For further discussion of these tests, see, for example, Stroud (1971), Rao (1973), Cox and Hinkley (1974), Buse (1982), Strawderman (1983), Tarone (1988), Le Cam (1990a), and Ferguson (1995).

Historical aspects of contingency tables are considered in the bibliographic notes to Chapter 5.

Appendix

This appendix briefly sketches some topics which a reader of large-sample literature is likely to encounter but which, because of their more mathematical nature, are not included in the main text.

Section A.1. The Lindeberg Condition for the central limit theorem

In Theorem 2.7.2, we stated the Liapunov condition (2.7.14) as a simple sufficient condition for the asymptotic normality of a sum of random variables which are independent but not necessarily identically distributed. The following theorem, due to Lindeberg (1922) provides a weaker condition which, under a mild additional assumption, is not only sufficient but also necessary. At first sight, the Lindeberg condition looks rather forbidding, but it turns out to be surprisingly easy to apply.

Theorem A.1.1. *Let X_{nj}, $j = 1, \ldots, n$; $n = 1, 2, \ldots$, form a triangular array of independent variables with*

(A.1.1) $$E(X_{nj}) = \xi_{nj} \text{ and } \text{Var}(X_{nj}) = \sigma_{nj}^2,$$

and let

(A.1.2) $$s_n^2 = \sigma_{n1}^2 + \cdots + \sigma_{nn}^2.$$

If G_{nj} denotes the distribution of $Y_{nj} = X_{nj} - \xi_{nj}$ and

(A.1.3) $$\tau_{nj}^2(t) = \int_{|y| > t s_n} y^2 dG_{nj}(y),$$

then

(A.1.4)
$$\frac{\bar{X}_n - \bar{\xi}_n}{\sqrt{\text{Var}(\bar{X}_n)}} \xrightarrow{L} N(0, 1)$$

provided

(A.1.5)
$$\frac{\tau_{n1}^2(t) + \cdots + \tau_{nn}^2(t)}{\sigma_{n1}^2 + \cdots + \sigma_{nn}^2} \to 0 \text{ for each } t > 0.$$

For further discussion and a proof, see, for example, Feller (1971), Chung (1974), Billingsley (1986), or Petrov (1995).

As a simple example, let us prove the following result, which was already stated and proved in Chapter 2 as Corollary 2.7.1.

Corollary A.1.1. *If there exists a constant A such that $|X_{nj}| \leq A$ for all j and n, then a sufficient condition for (A.1.4) to hold is that*

(A.1.6)
$$s_n^2 \to \infty \text{ as } n \to \infty.$$

Proof. We have $|Y_{nj}| \leq 2A$, and hence $\tau_{nj}^2(t) = 0$ whenever $ts_n > 2A$. Given any $t > 0$, it follows from (A.1.6) that there exists n_0 such that $ts_n > A$ for all $n > n_0$ and hence that the numerator of (A.1.5) is zero for all $n > n_0$. ∎

Under the assumptions of Corollary A.1.1, we have $\sigma_{nj}^2 \leq A^2$ for all $j = 1, \ldots, n$ and hence

(A.1.7)
$$\max_{j=1,\ldots,n} \sigma_{nj}^2 / s_n^2 \to 0 \text{ as } n \to \infty$$

if (A.1.6) holds. Quite generally, (A.1.7) is a consequence of the Lindeberg condition (A.1.5) (Problem 1.1). Condition (A.1.7) states that none of the variances $\sigma_{n1}^2, \ldots, \sigma_{nn}^2$ can make a relatively large contribution to s_n^2. Some such condition is needed to preclude possibilities such as $X_2 = \cdots = X_n = 0$. In that case, the left side of (A.1.4) reduces to $(X_1 - \xi_1)/\sigma_1$, which can have any distribution with mean 0 and variance 1. If (A.1.7) holds, the Lindeberg condition (A.1.5) is not only sufficient but also necessary for the validity of (A.1.4).

Let us next show that the Liapunov condition (2.7.14) implies (A.1.5). To see this, note that

$$\int_{|y|>ts_n} y^2 dG_{nj}(y) \leq \frac{1}{ts_n} \int_{|y|>ts_n} |y|^3 dG_{nj}(y) \leq \frac{1}{ts_n} E|Y_{nj}|^3.$$

Thus

$$\frac{1}{s_n^2} \sum_{j=1}^n \int_{|y|>ts_n} y^2 dG_{nj}(y) \leq \frac{1}{ts_n^3} \sum E|Y_{nj}|^3.$$

Condition (2.7.14) states that the right side tends to 0, and hence implies (A.1.5).

Section A.2. Hajek's projection method

A central fact of first-order, large-sample theory is the asymptotic normality of various estimators and test statistics. For linear functions of independent variables, this result is established by the central limit theorem (for example, Theorem 1.1 of Section A.1), the applicability of which is greatly extended by Slutsky's theorem and the delta method. In more complicated situations, the asymptotic normality of a statistic T_n can often be proved by comparing T_n with a statistic T_n^* known to be asymptotically normal and showing that

$$(A.2.1) \qquad\qquad E(T_n - T_n^*)^2 \to 0.$$

This method of proof was used in Chapter 6 for U- and V-statistics and for smooth functionals $h(\hat{F}_n)$ of the empirical cdf $\hat{F}_n$. Its principal difficulty frequently lies in the determination of a suitable sequence of statistics T_n^*. In Section 6.3, T_n^* was obtained as the linear part of the Taylor expansion of $h(\hat{F}_n)$ about $h(F)$. The following alternative approach is due to Hajek (1961).

If asymptotic normality of T_n is to be shown by proving (A.2.1) for a suitable T_n^*, the chances of success clearly are best when T_n^* is chosen among the various candidates under consideration so as to minimize $E(T_n - T_n^*)^2$. The following lemma provides a useful tool for this minimization.

Lemma A.2.1. *For any statistics S, T, and T^* with finite variance, a sufficient condition for*

$$(A.2.2) \qquad\qquad E(T - S)^2 \geq E(T - T^*)^2$$

is that

$$(A.2.3) \qquad\qquad E(T - T^*)(T^* - S) = 0.$$

Proof. Writing

$$(T - S)^2 = [(T - T^*) + (T^* - S)]^2,$$

expanding the right side, and using (A.2.3), we have

$$E(T - S)^2 = E(T - T^*)^2 + E(T^* - S)^2 \geq E(T - T^*)^2.$$

∎

Example A.2.1 Determining the minimizing sum $\sum k_i(X_i)$. Suppose
that $X_1, \ldots, X_n$ are independently distributed with distributions $F_1, \ldots, F_n$
and that $T_n = T(X_1, \ldots, X_n)$ is a statistic satisfying $E(T_n) = 0$. As a class
of statistics for which we know conditions for asymptotic normality, let us
consider the class S of statistics

(A.2.4) $$S = \sum_{i=1}^{n} k_i(X_i) \quad \text{with} \quad E[k_i(X_i)] = 0.$$

Let us now determine the functions k_i which minimize $E(T_n - S)^2$ for a
given T_n. (The minimizing $S = T_n^*$ may be considered to be the projection
of T_n onto the linear space S.) We shall use Lemma A.2.1 to show that the
minimizing functions k_i are the conditional expectations

(A.2.5) $$r_i(x_i) = E[T_n \mid x_i],$$

so that the minimizing T_n^* is

(A.2.6) $$T_n^* = \sum_{i=1}^{n} E(T_n \mid X_i).$$

Before checking (A.2.3), note that by (A.2.5)

(A.2.7) $$E[r_i(X_i)] = E(T_n) = 0.$$

That T_n^* satisfies (A.2.3) for all S given by (A.2.4) will follow if we can
show that

(A.2.8) $$E\{(T_n - T_n^*)[r_i(X_i) - k_i(X_i)] \mid X_i = x_i\} = 0$$

and hence if we show that

(A.2.9) $$E[(T_n - T_n^*) \mid X_i = x_i] = 0 \text{ for all } i.$$

The left side of (A.2.9) equals

(A.2.10) $$E\left[T_n - r_i(X_i) - \sum_{j \neq i} r_j(X_j) \mid X_i = x_i \right].$$

Since X_j is independent of X_i for all $j \neq i$, (A.2.10) is equal to

$$E[T_n - r_i(X_i) \mid x_i] = 0,$$

and this completes the proof of (A.2.3).   □

Example A.2.2 One-sample V-statistics. As an application of (A.2.6) consider the one-sample V-statistic of Example 6.3.4 with

$$(A.2.11) \qquad \theta = h(F) = E\varphi(X_1, \ldots, X_a)$$

and

$$(A.2.12) \qquad h(\hat{F}_n) = V = \frac{1}{n^a} \sum \cdots \sum \varphi(X_{i_1}, \ldots, X_{i_a}),$$

so that

$$(A.2.13) \qquad T_n = V - \theta = h(\hat{F}_n) - h(F).$$

The linear term of the Taylor expansion of $h(\hat{F}_n) - h(F)$ was found in Section 6.3 to be

$$(A.2.14) \qquad \frac{a}{n} \sum_{i=1}^{n} [\varphi_1(X_i) - h(F)]$$

with

$$(A.2.15) \qquad \varphi_1(x) = E\varphi(x, X_2, \ldots, X_a).$$

In the present approach, we have by (A.2.5) and (A.2.14),

$$r_i(x_i) = \frac{1}{n^a} E \sum \cdots \sum [\varphi(x_i, X_{j_1}, \ldots, X_{j_{a-1}}) + \cdots$$
$$+ \varphi(X_{j_1}, \ldots, X_{j_{a-1}}, x_i) - a\theta],$$

where the summation extends over all $(n-1)^{a-1}$ $(a-1)$-tuples $(j_1, \ldots, j_{a-1})$ in which all the j's are $\neq i$. By symmetry of φ, we thus have

$$\begin{aligned} r_i(x_i) &= \frac{a}{n^a} E \sum_{j_1 \neq i, \ldots, j_{a-1} \neq i} \cdots \sum [\varphi(x_i, X_{j_1}, \ldots, X_{j_{a-1}}) - \theta] \\ &= \frac{a}{n} \left(\frac{n-1}{n} \right)^{a-1} E\varphi(x_i, X_{j_1}, \ldots, X_{j_{a-1}}) \end{aligned}$$

and hence by (A.2.16) and (A.2.6)

$$T_n^* = \frac{a}{n} \left(\frac{n-1}{n} \right)^{a-1} \sum_{i=1}^{n} [\varphi_1(X_i) - \theta].$$

This differs from (A.2.15) only by the factor $[(n-1)/n]^{a-1}$ which tends to 1 as $n \to \infty$. On multiplication by $\sqrt{n}$, the present approach therefore leads to the same suggested limit distribution as the method of Section 6.3 (Example 3.4).

Section A.3. Almost sure convergence

In Section 2.1, we defined convergence in probability of a sequence $X_1, X_2, \ldots$ of random variables to a constant c by the condition that for any $\epsilon > 0$,

$$(A.3.1) \qquad P(|x_n - c| < \epsilon) \to 1 \text{ as } n \to \infty.$$

An alternative convergence concept is almost sure convergence or convergence with probability 1, also referred to as strong convergence. The sequence $X_1, X_2, \ldots$ is said to converge almost surely to c if

$$(A.3.2) \qquad P(X_1, X_2, \cdots \to c) = 1.$$

This condition does not assert that every realization $x_1, x_2, \ldots$ of the sequence converges to c but only that this event has probability 1.

Condition (A.3.2) is a probability statement concerning the (infinite dimensional) space of sequences $(x_1, x_2, \ldots)$. Probability distributions over this space cause no problem if the X's are defined as functions over a common probability space. When instead the X's are defined in terms of the distributions of $(X_1, \ldots, X_n)$ for each $n = 1, 2, \ldots$ (as is the case, for example, when $X_1, X_2, \ldots$ are independent, each with a given distribution), the induced probability distribution in the space of sequences can be obtained by Kolmogorov's extension theorem. (See, for example, Feller (Vol. 2) (1966, Section IV.6) or Chung (1974, Theorem 3.3.4).)

It can be shown that equivalent to (A.3.2) is the condition that for every $\epsilon > 0$,

$$(A.3.3) \qquad P(|X_k - c| < \epsilon \text{ for all } k = n, \, n+1, \ldots) \to 1 \text{ as } n \to \infty.$$

Comparing (A.3.1) with (A.3.3) we see a crucial difference. Convergence in probability is a statement about a single X_n; it says that when n is sufficiently large, X_n is likely to be close to c. In contrast, almost sure convergence is a simultaneous statement about a whole remainder sequence: When n is sufficiently large, it is likely that all elements of the sequence $X_n, X_{n+1}, \ldots$ are close to c.

Since (A.3.3) implies (A.3.1), it follows that almost sure convergence is a stronger property than convergence in probability. The following example exhibits a sequence which converges in probability but not almost surely.

Example A.3.1. Let $X_1, X_2, \ldots$ be independent with

$$P(X_n = 1) = \frac{1}{n}, \; P(X_n = 0) = 1 - \frac{1}{n}.$$

Then $X_n \to 0$ in probability since $P(X_N = 0) \to 1$. We shall now show that the sequence does not tend to 0 with probability 1. Note that in the present situation, the event

(A.3.4) $E : x_1, x_2, \cdots \to 0$

occurs if and only if there exists an integer a such that $x_{a+1} = x_{a+2} = \cdots = 0$. The event E is therefore the union of the events

$$E_0 : x_1 = x_2 = \cdots = 0$$

and

$$E_a : x_a = 1, \; x_{a+1} = x_{a+2} = 0, \; a = 1, 2, \ldots.$$

Since these events are mutually exclusive, it follows that

(A.3.5) $$P(E) = \sum_{a=0}^{\infty} P(E_a).$$

Now

(A.3.6) $P(E_a) \le P(X_{a+1} = X_{a+2} = \cdots = 0)$

and

(A.3.7)
$$\begin{aligned}
P(&X_{a+1} = X_{a+2} = \cdots = 0) \\
&= \left(1 - \frac{1}{a+1}\right)\left(1 - \frac{1}{a+2}\right) \cdots \\
&= \lim_{b \to \infty} \left[\left(1 - \frac{1}{a+1}\right)\left(1 - \frac{1}{a+2}\right) \cdots \left(1 - \frac{1}{b}\right)\right] \\
&= \lim_{b \to \infty} \left(\frac{a}{a+1} \cdot \frac{a+1}{a+2} \cdots \frac{b-1}{b}\right) = \lim_{b \to \infty} \frac{a}{b} = 0.
\end{aligned}$$

It follows that

(A.3.8) $$P(E_a) = 0$$

for every a. Thus, by (A.3.6), the probability of the event (A.3.5), far from being equal to 1, is equal to 0.   □

This example illustrates the difference between the two convergence concepts. The sequence $x_1, x_2, \ldots$ consisting of 0's and 1's converges to 0 if and only if it contains only a finite number of 1's, and this event has probability 0, i.e., nearly all sequences contain an infinite number of 1's. No matter how sparse they are, they prevent convergence of the sequence to 0.

On the other hand, $X_n \xrightarrow{P} 0$ in the present case if and only if $P(X_n = 1) \to 0$. This condition does not require an absence of 1's from some point on, but only that, on the average, they appear less and less frequently as we move further out in the sequence. As we have seen for nearly all sequences, an occasional 1 will appear no matter how far we go out, and yet X_n converges in probability to 0.

The probabilistic and statistical literature abounds with results using one or the other of these concepts. Consistency of an estimator T_n of θ, for example, is treated both in the strong and weak senses (convergence of T_n to θ almost surely and in probability). Which of the two is preferred is somewhat a matter of taste, with probabilists tending to opt for almost sure convergence and statisticians for convergence in probability. Almost sure convergence is the stronger property and therefore will typically also require stronger assumptions. In statistics, the primary interest frequently is on probabilities of the form

$$P(T_n - \theta \,|< c),$$

where T_n is an estimator of the unknown θ. Convergence in probability has the advantage of making a direct statement about such probabilities.

Section A.4. Large deviations

The principal probabilistic tool used throughout this book is the central limit theorem together with its various extensions. For the case of n i.i.d. random variables with mean ξ and variance σ^2, this theorem states that

$$(A.4.1) \qquad \sqrt{n}(\bar{X} - \xi) \to N(0, \sigma^2),$$

so that

$$(A.4.2) \qquad P\left[(\bar{X} - \xi) \geq \frac{a}{\sqrt{n}}\right] \to 1 - \Phi(a/\sigma).$$

An implication of this result is that typical values of the deviation $\bar{X} - \xi$ of $\bar{X}$ from its mean ξ are of the order $1/\sqrt{n}$ while values of $\bar{X} - \xi$ that exceed a fixed amount $a > 0$ become very rare as $n \to \infty$. The theory of large deviations is concerned with probabilities of rare events such as

$$(A.4.3) \qquad P_n = P(\bar{X} - \xi \geq a), \ a > 0.$$

Since (A.4.3) is equal to

$$P(\sqrt{n}(\bar{X} - \xi) \geq a\sqrt{n}),$$

it is seen that $P_n \to 0$ as $n \to \infty$. Large deviation theory studies the rate of this convergence.

For the case of i.i.d. variables, the solution is provided by the following theorem due to Chernoff (1952).

Theorem A.4.1. *Let* $X_1, X_2, \ldots$ *be i.i.d. with mean* ξ, *let* $a > 0$, *and let*

(A.4.4)
$$\rho = \inf_t [e^{at} M(t)],$$

where

(A.4.5)
$$M(t) = E[e^{t(X-\xi)}]$$

is the moment generating function of $X - \xi$. *Then*

(A.4.6)
$$\lim \frac{1}{n} \log P(\bar{X} - \xi \geq a) = \log \rho.$$

For a proof, see, for example, Bahadur (1971) or Billingsley (1986).

The limit (A.4.6) suggests approximating $\log P_n$ by $n \log \rho$ and hence the approximation

(A.4.7)
$$P_n \approx \rho^n.$$

Note, however, that (A.4.7) is not an equivalence relation. Suppose, for example, that $X - \xi$ is symmetric about 0 so that

$$P(|\bar{X} - \xi| \geq a) = 2P(\bar{X} - \xi \geq a).$$

Then it is seen that (Problem 4.1)

(A.4.8)
$$\lim \frac{1}{n} \log(P|\bar{X} - \xi| \geq a) = \log \rho$$

so that we obtain the same approximation for

$$P(|\bar{X}_n - \xi| \geq a) \text{ and } P(\bar{X}_n - \xi \geq a).$$

The approximation (A.4.7) gives only the exponential rate at which $P_n \to 0$.

Example A.4.1. Let $X_1, \ldots, X_n$ be i.i.d. $N(\xi, 1)$. Then (Problem A.4.2)

(A.4.9)
$$M(t) = e^{t^2/2} \text{ and } \rho = e^{-a^2/2}.$$

Various extensions of Chernoff's theorem and statistical applications are discussed for example in Bahadur (1971) and Rüschendorf (1988). Book-length treatments of large deviation theory are, for example, Stroock (1984), Deuschel and Stroock (1989), Bucklew (1990), and Dembo and Zeitonni (1993).

Given the distribution F of the X's, the probability

$$P_n(a) = P_\xi(\bar{X} - \xi \geq a), \; 0 < a,$$

is a function of n and a. Large-sample approximations are obtained by embedding the given situation in a sequence

(A.4.10) $\{(a_n, n) : n = 1, 2, \dots\}.$

The large deviation approximation corresponds to the sequence (A.4.10) with $a_n = a$. On the other hand, in Section 3.2 we considered sequences of the form

(A.4.11) $a_n = \xi + \dfrac{\Delta}{\sqrt{n}}.$

Then

$$P_0\left(\bar{X} \geq \frac{\Delta}{\sqrt{n}}\right) \to 1 - \Phi[\sqrt{n}(a_n - \xi)],$$

which suggests the approximation

(A.4.12) $1 - \Phi(\Delta) = 1 - \Phi[\sqrt{n}(a_n - \xi)]$

for the probability (A.4.3).

Which approximation provides the more accurate result depends on the situation. Since $\rho < 1$ unless X is equal to a constant with probability 1, the approximate value (A.4.7) tends to 0 exponentially and is thus likely to provide a good approximation only if P_n is small. For some numerical results, see Groeneboom (1980). $\square$

Section A.5. The central limit theorem: Sketch of a proof

The simplest version of the central limit theorem (Theorem 2.4.1) states that if $X_1, \dots, X_n$ are i.i.d. with mean ξ and finite variance σ^2, then the distribution of $\sqrt{n}(\bar{X} - \xi)/\sigma$ tends to the standard normal distribution. This result, together with its extensions, is truly central to this book. It is also very surprising: Why is the average of a large number of i.i.d. random variables approximately the same irrespective of the common distribution of the terms, and why has the limit distribution this particular form? The following outline of the classical proof is intended to throw at least some light on this mystery.

A direct attack on the theorem seems quite difficult because the calculation of the distribution of a sum of random variables typically is complicated even when, as here, the variables are independent (Problem 5.1). To circumvent this difficulty, the proof uses a standard mathematical device: Transform the mathematical objects with which we are concerned into another sphere, in which the problem is easier to handle. Here are two examples of this method with which the reader is familiar.

(i) **Logarithms.** Suppose we have to multiply together (by hand) n numbers to form the product

$$x = x_1 \cdot x_2 \cdot \dots \cdot x_n.$$

This calculation becomes much easier when we replace the numbers x_i by their logarithms $y_i = \log x_i$ since

$$y = \log x = \log x_1 + \cdots + \log x_n = y_1 + \cdots + y_n$$

is a sum rather than a product. Of course, this addition does not give us x itself but $y = \log x$. To find the solution of the original problem, we must transform back from y to x, i.e., take the antilogarithm x of y.

(ii) **Analytic geometry**. This powerful method allows us to solve problems of plane geometry by representing each point P in the plane by its coordinates (x, y) in some coordinate system. Thus the points P are transformed into a pair of numbers, and a straight line, for example, into a linear equation. Conversely, the method can be used to illuminate algebraic problems through geometric insight.

In the first example, we represented each number x by its logarithm y; in the second, each point P in the plane by its coordinates (x, y). In the same spirit, to solve the central limit problem, we represent the distribution of a random variable Y not by its cdf or its probability density or its quantile function but, instead, by its moment generating function (mgf)

(A.5.1) $$M_Y(t) = E(e^{tY}).$$

Then the distribution of the sum $Y_1 + \cdots + Y_n$ of n independent random variables is represented by

(A.5.2) $$M_{Y_1 + \cdots + Y_n}(t) = E[e^{t(Y_1 + \cdots + Y_n)}] = M_{Y_1}(t) \ldots M_{Y_n}(t).$$

When the Y's are also identically distributed, (A.5.2) reduces to

(A.5.3) $$M_{Y_1 + \cdots + Y_n}(t) = [M_{Y_1}(t)]^n.$$

Let us now apply this formula to the distribution of

(A.5.4) $$\frac{\bar{X} - \xi}{\sigma} = \frac{1}{n} \sum \frac{X_i - \xi}{\sigma} = \bar{Y},$$

where the variables

(A.5.5) $$Y_i = (X_i - \xi)/\sigma$$

have expectation $E(Y_i) = 0$ and variance $E(Y_i^2) = 1$. Expanding e^{ty} into a Taylor series, we have

(A.5.6) $$e^{ty} = 1 + ty + \frac{1}{2}t^2 y^2 + R_t(y),$$

where the remainder R_t is $o(t^2 y^2)$, and thus

(A.5.7) $$\begin{aligned} M_Y(t) &= 1 + tE(Y) + \frac{1}{2}t^2 E(Y^2) + E[R_t(Y)] \\ &= 1 + \frac{1}{2}t^2 + E[R_t(Y)]. \end{aligned}$$

It follows that

(A.5.8) $$M_{\Sigma Y_i}(t) = \left[1 + \frac{1}{2}t^2 + E[R_t(Y)]\right]^n$$

and hence that (Problem 5.2)

(A.5.9) $$M_{\sqrt{n}\bar{Y}}(t) = \left[1 + \frac{t^2}{2n} + E[R_{t/\sqrt{n}}(Y)]\right]^n .$$

By Problem 4.8 of Chapter 1, it will follow that

(A.5.10) $$M_{\sqrt{n}\bar{Y}}(t) \rightarrow e^{t^2/2}$$

since (Problem 5.2)

(A.5.11) $$E[R_{t/\sqrt{n}}(Y)] \rightarrow 0 \text{ as } n \rightarrow \infty.$$

An easy calculation (Problem 4.2) shows that $e^{t^2/2}$ is the moment generating function of the standard normal distribution. Thus, the moment generating function of $\bar{Y}$ tends to that of the standard normal.

To complete the proof, it is necessary to show that

(i) the standard normal distribution is the only distribution whose mgf is $e^{t^2/2}$ (uniqueness theorem)

and

(ii) the convergence (A.5.10) of the mgf's implies the convergence of the associated distributions (continuity theorem).

These results are proved in most probability texts.

The proof outlined above suffers from one defect. While the CLT requires only that the variance of the X's is finite, the proof assumes the finiteness of the mgf which implies the finiteness of all moments. This difficulty is overcome by a slight modification of the proof, namely by replacing the mgf $M_Y(t)$ by the characteristic function

$$\varphi_Y(t) = E(e^{itY}).$$

The latter exists for all distributions, and the proof goes through as before but now applies to all distributions with finite variance.

Section A.6. Refinements of first-order, large-sample theory

As we have seen, the asymptotic approach which forms the subject of this book provides a powerful and broadly applicable basis for statistical inference. However, it suffers from two drawbacks.

(i) In some situations, the approximations it offers are too crude to give satisfactory results.

(ii) The theory gives no indication of the accuracy of the approximation and thus leaves the user in doubt about its reliability.

As has been emphasized throughout and has been illustrated in many examples, a way to instigate the second of these difficulties is through simulation, which shows how well the approximation works for particular distributions and sample sizes. Such numerical work can be greatly strengthened by consideration of higher order asymptotics illustrated in Section 2.4 and to be discussed further in this section. These theoretical results paint a more general picture than that obtained through the snapshots provided by simulation, and they give insight into the conditions under which the first-order approximations can be expected to work satisfactorily.

Second and higher order approximations also help with the first difficulty mentioned above by leading to improved approximations and making possible comparisons between different statistical procedures which first-order theory is unable to distinguish. We now briefly sketch two examples of this latter kind.

In Chapter 4, we considered two asymptotic approaches for point estimation resulting in approximations for the variance of the estimator and for its distribution. For testing and confidence procedures, only the second of these approaches is available. Both approximations can be refined by including second-order terms. To indicate what can be achieved by such higher order considerations, we begin with an example of the first approach.

Example A.6.1 Comparing the Bayes and best unbiased estimator. If $X_1, \ldots, X_n$ are i.i.d. $N(\theta, 1)$, the best unbiased estimator of θ is the sample mean $\delta = \bar{X}$. On the other hand, the Bayes estimator δ' corresponding to a prior normal distribution $N(\mu, \tau^2)$ with respect to squared error loss can be shown to be

$$(A.6.1) \qquad \delta'(X_1, \ldots, X_n) = \frac{n\tau^2 \bar{X} + \mu}{n\tau^2 + 1}.$$

The risk functions of these two estimators are

$$(A.6.2) \qquad R'(\theta) = E(\delta' - \theta)^2 = \frac{n\tau^4 + (\theta - \mu)^2}{(n\tau^2 + 1)^2}$$

and

$$(A.6.3) \qquad R(\theta) = E(\delta - \theta)^2 = \frac{1}{n}.$$

Since (Problem 6.1)

$$(A.6.4) \qquad R'(\theta) = \frac{1}{n} + O\left(\frac{1}{n^2}\right),$$

to order $1/n$ the two estimators have the same risk function and the ARE of δ' to δ is 1. This result tells us nothing about which of the two estimators is to be preferred.

To obtain a more meaningful comparison, let us carry the expansion of the risk functions in terms of powers of $1/n$ a step further. We then find (Problem 6.1) that

$$(\text{A.6.5}) \qquad R'(\theta) = \frac{1}{n} + \frac{1}{n^2\tau^2}\left[\frac{(\theta-\mu)^2}{\tau^2} - 2\right] + O\left(\frac{1}{n^3}\right)$$

while the formula (A.6.3) for $R(\theta)$ requires no change. To order $1/n^2$, the Bayes estimator δ' therefore has smaller risk than δ if

$$(\text{A.6.6}) \qquad \frac{(\theta-\mu)^2}{\tau^2} < 2,$$

that is, if μ is close to θ without τ^2 being too small relative to the difference $|\theta - \mu|$. This corresponds to one's intuition that the Bayes estimator will do well if the Bayes guess μ of θ is close to θ but if one also guards against overconfidence by not making the variance τ^2 of this guess too small.

An interesting quantitative comparison can be obtained by taking a viewpoint similar to that of the ARE, and asking for the number n' of observations needed by δ' to match the performance of δ (based on n observations) not only up to terms of order $1/n$ as was the case for the ARE but to terms of order $1/n^2$. Putting $n' = n + d$, we have by (A.6.5) for any finite d,

$$(\text{A.6.7}) \qquad \begin{aligned} R'(\theta) &= \frac{1}{n+d} + \frac{1}{(n+d)^2\tau^2}\left[\frac{(\theta-\mu)^2}{\tau^2} - 2\right] + O\left(\frac{1}{n}\right)^3 \\ &= \frac{1}{n} - \frac{d}{n^2} + \frac{1}{n^2\tau^2}\left[\frac{(\theta-\mu)^2}{\tau^2} - 2\right] + O\left(\frac{1}{n^3}\right). \end{aligned}$$

The two risk functions therefore will agree to order $1/n^2$, provided

$$(\text{A.6.8}) \qquad d = \frac{1}{\tau^2}\left[\frac{(\theta-\mu)^2}{\tau^2} - 2\right] + O\left(\frac{1}{n}\right).$$

For large n, the Bayes estimator thus approximately saves

$$(\text{A.6.9}) \qquad \frac{1}{\tau^2}\left[2 - \frac{(\theta-\mu)^2}{\tau^2}\right]$$

observations when (A.6.6) holds, while in the contrary case it requires approximately that many additional observations. It follows from (A.6.9) that the Bayes estimator achieves its greatest savings relative to $\bar{X}$ if $\mu = \theta$ and τ^2 is very small, as one would have expected. The quantity (A.6.9) is called the *deficiency* of $\bar{X}$ relative to δ'. (For more general results of this kind, see Hodges and Lehmann (1970) and Hu and Hwang (1990).) □

As this example shows, second-order correction terms for the approximation to a variance in some simple cases can be quite straightforward, needing only the next term in the Taylor expansion. On the other hand, the higher order correction terms for approximating a distribution function require an asymptotic expansion of the kind illustrated in Theorem 2.4.3. The following is a simple illustration of this approach.

Example A.6.2 Comparing two tests. Suppose $X_1, \ldots, X_n$ are i.i.d. $N(\theta, 1)$ and that we are interested in testing $H : \theta = 0$ against $\theta > 0$. The best (uniformly most powerful) test in this case accepts H when

$$(A.6.10) \qquad \sqrt{n}\bar{X} \leq u,$$

where $u = u_\alpha$ is given by $\Phi(u) = 1 - \alpha$. Suppose, however, that we feel somewhat unsure of the assumption that the variance of the X's is equal to 1. We might then prefer to use the test of Example 3.1.3 and accept H for small values of

$$(A.6.11) \qquad t_n = \frac{\sqrt{n}\bar{X}}{\sqrt{\Sigma(X_i - \bar{X})^2/(n-1)}}.$$

In Example 3.1.3, it was seen that the test with acceptance region

$$(A.6.12) \qquad t_n \leq u$$

has asymptotic level α, and it is easy to show that the ARE of (A.6.12) relative to (A.6.10) is 1. It is therefore interesting to consider the deficiency of the test based on t_n rather than on $\sqrt{n}\bar{X}$. This deficiency is the cost incurred by using in the denominator of the test statistic, an estimate of $\mathrm{Var}(X_i)$ rather than the assumed variance of 1, when the variance of the X's actually is equal to 1.

Such a comparison requires consideration not only of the asymptotic level and power of the two tests but also their next order correction terms. In particular, the levels of the two tests must agree to that order since, otherwise, an apparent difference in power might in fact be due (at least in part) to a difference in level, and it is then of course not surprising that a larger rejection region will result also in larger power.

As a first task, we must therefore calculate a critical value $v = v_n$ for which the level of the acceptance region

$$(A.6.13) \qquad t_n < v$$

is equal to α not only in the limit but up to next order terms. We shall only sketch the following derivation. For a rigorous treatment, see Problem 6.3 and Hodges and Lehmann (1970).

To calculate the probability of (A.6.13), write it as

$$(A.6.14) \qquad P(\sqrt{n}\bar{X} \leq vS) = E\Phi(vS),$$

where

(A.6.15) $$S^2 = \Sigma(X_i - \bar{X})^2/(n-1)$$

and where we are using the independence of $\bar{X}$ and S^2. Now $\Sigma(X_i - \bar{X})^2$ has a χ^2-distribution with $\nu = n - 1$ degrees of freedom so that

(A.6.16) $$S^2 = \chi_\nu^2/\nu.$$

In the following, we shall use the facts that

(A.6.17) $$E(S) = 1 - \frac{1}{4n} + O\left(\frac{1}{n^2}\right),$$

(A.6.18) $$E(S-1)^2 = \frac{1}{2n} + O\left(\frac{1}{n^2}\right),$$

and

(A.6.19) $$E|S-1|^3 = O\left(\frac{1}{n^2}\right).$$

(For a discussion of these results and references, see, for example, Johnson, Kotz, and Balakrishnan (1994, Section 18.3).)

Since by (A.6.17) we expect S to be close to 1, write (A.6.14) as

(A.6.20)
$$\begin{aligned}
E\Phi(vS) &= E\Phi[v + v(S-1)] \\
&= \Phi(v) + v\varphi(v)E(S-1) \\
&\quad + \frac{1}{2}v^2\varphi'(v)E(S-1)^2 + \frac{1}{6}v^3 E[\varphi''(v^*)(S-1)^3]
\end{aligned}$$

where v^* lies between v and vS. From (A.6.17)–(A.6.19), the boundedness of $v^3\varphi''(v)$ and the fact that $\varphi'(v) = -v\varphi(y)$, it follows (Problem 6.3(i)) that

(A.6.21)
$$\begin{aligned}
E\Phi(vS) &= \Phi(v) - \frac{1}{4n}\varphi(v)(v + v^3) + O\left(\frac{1}{n^2}\right) \\
&= \Phi\left[v - \frac{1}{4n}(v + v^3)\right] + O\left(\frac{1}{n^2}\right).
\end{aligned}$$

Thus for (A.6.14) to agree with $1 - \alpha = \Phi(u)$ up to terms of order $1/n$, we must have

$$u = v - \frac{1}{4n}(v + v^3),$$

or, equivalently (Problem 6.3(ii)),

(A.6.22) $$v = u + \frac{1}{4n}(u + u^3).$$

To calculate the deficiency of the t-test (A.6.13) relative to the normal test (A.6.10), we require an analogous expansion to terms of order $1/n$ of the asymptotic power functions of the two tests and hence of the probabilities of (A.6.10) and (A.6.13) for the alternatives $\theta = \xi/\sqrt{n}$ and $\theta' = \xi/\sqrt{n'}$, respectively. Since the comparison must be made against the same alternatives, we require

(A.6.23)
$$\theta' = \sqrt{\frac{n'}{n}}\theta.$$

We then have (Problem 6.3(iii))

(A.6.24)
$$P(\sqrt{n}\bar{X} < u) = \Phi(u - \theta)$$

and

(A.6.25)
$$P(t_n < v) = \Phi\left[u - \sqrt{\frac{n'}{n}}\theta\left(1 - \frac{u^2}{4n'}\right)\right] + o\left(\frac{1}{n}\right).$$

If $n' = n + d$, the factor of θ in (A.6.25) becomes

(A.6.26)
$$\sqrt{\frac{n+d}{n}}\left(1 - \frac{u^2}{4n}\right) + o\left(\frac{1}{n}\right).$$

Equality of (A.6.24) and (A.6.25) up to terms of order $1/n$ will thus hold, provided
$$\sqrt{\frac{n+d}{n}}\left(1 - \frac{u^2}{4n}\right) = 1 + o\left(\frac{1}{n}\right)$$

and hence if

(A.6.27)
$$d = \frac{u^2}{2} + o(1).$$

The asymptotic deficiency of (A.6.13) relative to (A.6.10) is therefore

(A.6.28)
$$d = u_\alpha^2/2.$$

As α increases from .01 to .1, the value of d decreases from 2.71 to .82; the cost of the insurance policy represented by the t-test is thus quite low, on the order of one to three observations.

Second-order considerations are useful not only for deficiency calculations such as those in the preceding two examples but also for improving the accuracy of first-order approximations and for accuracy checks. They provide an approach which is intermediate between the first-order asymptotic methods discussed in this book and the small-sample methods based on assumed parametric models. Their conclusions are more detailed, but they are also more complicated and require more knowledge of the model than the former, with the situation reversed in comparison with the latter.

The theory of the asymptotic expansions needed for a second-order approach (particularly Edgeworth and Cornish-Fisher expansions) is available in principle, and a number of applications have been worked out, but in many cases (for example, in the comparison of the test (3.1.35) with that of Problem 3.1.18), it is still outstanding. A good introduction to the literature on higher order asymptotics is Ghosh (1994). See also Kolassa (1998).

Section A.7. Problems

Section A.1

1.1 Show that (A.1.5) implies (A.1.7).

[**Hint**: Use the fact that

$$\sigma_{nj}^2 = \int_{|y| \le ts_n} y^2 dG_{nj}(y) + \int_{|y| > ts_n} y^2 dG_{nj}(y)$$

and hence that

$$\frac{\sigma_{nj}^2}{s_n^2} \le t^2 + \frac{\tau_{nj}^2}{s_n^2}.$$

Given $\epsilon > 0$, let $t = \sqrt{\epsilon/2}$ and let n be so large that $\tau_{nj}^2/s_n^2 < \epsilon/2$.]

1.2 If $X_1, X_2, \ldots$ is a single sequence of random variables with $\mathrm{Var}(X_k) = \sigma_k^2$, $s_n^2 = \sum_{k=1}^n \sigma_k^2$, condition (A.1.7) implies (A.1.6).

[**Hint**: If (A.1.6) does not hold, then—since the sequence s_n^2, $n = 1, 2, \ldots$ is non-decreasing—s_n^2 tends to a finite limit c as $n \to \infty$, and hence $s_n^2 \le c$ for all n. Let k_0 be such that $\sigma_{k_0}^2 > 0$ and show that

$\max_{k=1,\ldots,n} \sigma_k^2 \ge \frac{\sigma_{k_0}^2}{c} > 0$ for $n \ge k_0$.]

1.3 Give an example of a triangular array for which (A.1.7) does not imply (A.1.6).

1.4 Prove that Theorem 2.7.1 holds under the weaker condition in which 3 is replaced $2 + \delta(\delta > 0)$ on both sides of (2.7.3).

1.5 (i) If $X_1, \ldots, X_n$ are i.i.d. with variance $\sigma^2 < \infty$, then the Lindeberg condition (6.1.5) is satisfied.

(ii) The situation (i) with $E|X_i - \xi|^3 = \infty$ provides a class of examples in which the Lindeberg condition holds but the Liapunov condition does not.

Section A.2

2.1 If $\sqrt{n}T_n^*$ is bounded in probability and

$$T_n^{**} = k_n T_n^* \text{ with } k_n \to 1 \text{ as } n \to \infty,$$

then

$$n(T_n^{**} - T_n^*)^2 \overset{P}{\to} 0.$$

2.2 Suppose that $X_1, \ldots, X_m$ and $Y_1, \ldots, Y_n$ are independently distributed according to F and G, respectively. Let $h(F, G)$ be given by (6.3.44) and let

$$T_{m,n} = h(\hat{F}_m, \hat{G}_n) - h(F, G).$$

Determine the function $T_{m,n}^*$ of the form

$$S = \sum_{i=1}^{m} a(X_i) + \sum_{j=1}^{n} b(Y_j),$$

which minimizes $E(T_{m,n} - S)^2$, and compare it with the linear term of (6.3.50).

Section A.3

3.1 Show that convergence of X_n to c in quadratic mean, i.e., $E(X_n - c)^2 \to 0$, neither implies nor is implied by almost sure convergence.

[**Hint**: (i) Example A.3.1; (ii) Let $x_1, x_2, \ldots$ be a sequence of numbers converging to c and let X_n be equal to x_n with probability 1. Then $X_n \to c$ almost surely, but $E(X_n - c)^2$ does not necessarily converge to c.]

Section A.4

4.1 If (A.4.6) holds, so does (A.4.8).

4.2 Verify (A.4.9).

4.3 Evaluate $M(t)$ and ρ for the case that the X_i take on the values 1 and 0 with probabilities p and q, respectively.

4.4 Under the assumptions of Example A.4.1, compare the approximation (A.4.7) with the true value of P_n for a number of values of a and n.

4.5 Under the assumptions of Problem 4.3, compare the approximations (A.4.7) and (A.4.12) with the true value of P_n for a number of values p, a, and n.

Section A.5

5.1 Let X_2, X_2, and X_3 be independently distributed according to the uniform distribution $U(0,1)$. Determine the distribution of (i) $X_1 + X_2$ and (ii) $X_1 + X_2 + X_3$.

5.2 Verify (A.5.9)–(A.5.11).

[**Hint**: To show (A.5.11), use the expression given in part (ii) of Theorem 2.5.1 for the remainder term in (A.5.6).]

Section A.6

6.1 Verify the risk function (A.6.5).

6.2 In the binomial situation of Example 4.3.3 with $a = b$, compare the Bayes estimator (4.3.16) with the best unbiased estimator X/n along the lines of Example A.6.1.

6.3 Verify (i) the equation (A.6.21), (ii) the correction term (A.6.22) for v, and (iii) the probabilities (A.6.24) and (A.6.25).

6.4 Make a table showing the deficiency (A.6.28) as α varies from .01 to .1.

Bibliographic Notes

For a review of the modern history of the central limit theorem, see Le Cam (1986). Hajek's projection method was used in Hajek (1961) without the name. More explicit developments are given in Hajek and Siddak (1967) and Hajek (1968). A general theory of large deviations was initiated by Cramér (1938). Statistical applications were introduced by Chernoff (1952) and developed further by Bahadur (1960,1971).

An introduction to Edgeworth expansions (including an account of their history) is provided by Hall (1992). The deficiency concept is due to Hodges and Lehmann (1970). More general views of higher order asymptotics are presented by Ghosh (1994).

References

W. J. Adams. *The Life and Times of the Central Limit Theorem*. Kaedmon, New York, NY (1974).

A. Agresti. *Categorical Data Analysis*. John Wiley & Sons, New York, NY (1990).

R. J. Aiyar, R. L. Guillier, and W. Albers. Asymptotic relative efficiencies of rank tests for alternatives. *Journal of the American Statistical Association*, 74:225–231 (1979).

J. Aldrich. R. A. Fisher and the making of maximum likelihood. *Statistical Science*, 12:162–176 (1997).

T. W. Anderson. *The Statistical Analysis of Time Series*. John Wiley & Sons, New York, NY (1971).

T. W. Anderson. *An Introduction to Multivariate Analysis (2nd Edition)*. John Wiley & Sons, New York, NY (1984).

A. M. Andrés, A. S. Mato, and I. H. Tejedor. A critical review of asymptotic methods for comparing two proportions by means of independent samples. *Communication Statistics-Simulation*, 21:551–586 (1992).

F. J. Anscombe and W. J. Glynn. Distribution of the kurtosis statistic b_2 for normal samples. *Biometrika*, 70:227–234 (1983).

S. F. Arnold. *The Theory of Linear Models and Multivariate Analysis*. John Wiley & Sons, New York, NY (1981).

592 References

R. Arriata, L. Goldstein, and L. Gordon. Poisson approximation and the Chen-Stein method (with discussion). *Statistical Science*, 5:403–434 (1990).

R. R. Bahadur. Stochastic comparison of tests. *Annals of Mathematical Statistics*, 31:275–295 (1960).

R. R. Bahadur. On Fisher's bound for asymptotic variances. *Annals of Mathematical Statistics*, 35:1545–1552 (1964).

R. R. Bahadur. *Some Limit Theorems in Statistics*. SIAM, Philadelphia, PA (1971).

R. R. Bahadur. Hodges superefficiency. *Encyclopedia of Statistical Science*, 3:645–646 (1983).

R. R. Bahadur and L. J. Savage. The nonexistence of certain statistical procedures in nonparametric problems. *Annals of Mathematical Statistics*, 27:1115–1122 (1956).

S. K. Bar-Lev and P. Enis. On the construction of classes of variance stabilizing transformations. *Statistics and Probability Letters*, 10:95–100 (1990).

A. D. Barbour, L. Holst, and S. Janson. *Poisson Approximation*. Clarendor Press, Oxford, England (1992).

O. E. Barndorff-Nielsen and D. R. Cox. *Asymptotic Techniques for Use in Statistics*. Chapman & Hall, London, England (1989).

O. E. Barndorff-Nielsen and D. R. Cox. *Inference and Asymptotics*. Chapman & Hall, London, England (1994).

V. D. Barnett. Evaluation of the maximum likelihood estimator where the likelihood equation has multiple roots. *Biometrika*, 53:151–166 (1966).

A. P. Basu. Identifiability. *Encyclopedia of Statistical Science*, 4:2–6 (1983).

J. Beran. A test of location for data with slowly decaying serial correlations. *Biometrika*, 76:261–269 (1989).

J. Beran. Statistical methods for data with long-range dependence (with discussion). *Statistical Science*, 7:404–427 (1992).

J. Beran. *Statistics for Long-term Memory*. Chapman & Hall, New York, NY (1994).

R. Beran. The role of Hajek's convolution theory. *Kybermetrika*, 31:221–237 (1995).

J. O. Berger. *Statistical Decision Theory and Bayesian Analysis*. Second Edition. Springer-Verlag, New York, NY (1985).

R. H. Berk. Review of Zehna (1966). *Mathematical Review*, 33:342–343 (1967).

J. M. Bernardo and A. F. M. Smith. *Bayesian Theory*. John Wiley, New York, NY (1994).

P. Bickel. On some robust estimates of location. *The Annals of Mathematical Statistics*, 36:847–858 (1965).

P. Bickel. Some contributions to the theory of order statistics. *Proceedings of the Fifth Berkeley Symposium on Mathematical Statistics and Probability*, 1:575–591 (1967).

P. Bickel. Parametric robustness: Small biases can be worthwhile. *The Annals of Statistics*, 12:864–879 (1984).

P. Bickel and K. Doksum. *Mathematical Statistics*. Prentice-Hall, Englewood Cliffs, NJ (1977).

P. Bickel and D. Freedman. Some asymptotic theory for the bootstrap. *The Annals of Statistics*, 9:1196–1217 (1981).

P. Bickel and D. Freedman. Asymptotic normality and the boostrap in stratified sampling. *The Annals of Statistics*, 12:470–482 (1984).

P. Bickel, C. A. Klaassen, Y. Ritov, and J. Wellner. *Efficient and Adaptive Estimation for Semiparametric Models*. Johns Hopkins University Press, Baltimore, MD (1993).

P. Bickel and E. L. Lehmann. Unbiased estimation in convex families. *The Annals of Mathematical Statistics*, 40:1523–1535 (1969).

P. Billingsley. *Probability and Measure, 3rd ed.* John Wiley & Sons, New York, NY (1976, 1989, 1995).

Y. Bishop, S. Fienberg, and P. Holland. *Discrete Multivariate Analysis*. MIT Press, Cambridge, MA (1975).

D. Bloch. A note on the estimation of the location parameter of the Cauchy distribution. *Journal of the American Statistical Association*, 61:852–855 (1966).

C. R. Blyth and H. A. Still. Binomial confidence intervals. *Journal of the American Statistical Association*, 78:108–116 (1983).

E. Bofinger. Goodness-of-fit test using sample quantiles. *Journal of the Royal Statistical Society*, (B)35:277–284 (1973).

K. O. Bowman and L. R. Shenton. Moment $(\sqrt{b_1}, b_2)$ techniques. In R. B. D'Agostino and M. A. Stephens, editors, *Goodness–of–Fit Techniques*, pages 279–329. Marcel Dekker, New York, NY (1986).

J. D. Broffitt and R. H. Randles. A power approximation for the chi-square goodness-of-fit test: Simple hypothesis case. *Journal of the American Statistical Association*, 72:604–607 (1977).

L. D. Brown and R. H. Farrell. A lower bound for the risk in estimating the value of a probability density. *Journal of the American Statistical Association*, 85:1147–1153 (1990).

J. Bucklew. *Large Deviation Techniques in Decisions, Simulation, and Estimation*. John Wiley & Sons, New York, NY (1990).

A. Buse. The likelihood ratio, Wald, and Lagrange multiplier tests: An expository note. *The American Statistician*, 36:153–157 (1982).

G. Casella and R. L. Berger. *Statistical Inference*. Wadsworth, Pacific Grove, CA (1990).

N. N. Cencov. Evaluation of an unknown density from observations. *Soviet Mathematics*, 3:1559–1562 (1962).

H. Chernoff. A measure of asymptotic efficiency for tests of an hypothesis based on the sum of observations. *Annals of Mathematical Statistics*, 23:493–507 (1952).

H. Chernoff. On the distribution of the likelihood ratio. *Annals of Mathematical Statistics*, 25:573–578 (1954).

Y. S Chow and H. Teicher. *Probability Theory: Independence, Interchangeability, Martingales*. Springer-Verlag, New York, NY (1997).

K. L. Chung. *A Course in Probability Theory (2nd Edition)*. Academic Press, Boston, MA (1974).

W. G. Cochran. Errors of measurement in statistics. *Technometrics*, 10:637–666 (1968).

W. G. Cochran. *Sampling Techniques (3rd Edition)*. John Wiley & Sons, New York, NY (1977).

J. Cohen. *Statistical Power Analysis for the Behavioral Sciences*. Academic Press, New York, NY (1969). (2nd edition, Erlbaum, Hillsdale, NJ (1988)).

R. Courant. *Differential and Integral Calculus, Volume 1, 2nd ed.* John Wiley & Sons, New York, NY (1927). Reprinted in *Wiley Classics Library* (1988).

D. R. Cox. Long-range dependence: A review. In H. A. David and H. T. David, editors, *Statistics: An Appraisal.* (1984).

D. R. Cox and D. V. Hinkley. *Theoretical Statistics*. Chapman & Hall, London, England (1974).

H. Cramér. On the composition of elementary errors. *Skandinavisk Aktuarietidskrift*, 11:141–180 (1928).

H. Cramér. Sur un nouveau théorème limite de la théorie des probabilités. *Act. Sci. et Ind.*, 736:5–23 (1938).

H. Cramér. *Mathematical Methods of Statistics.* Princeton University Press, Princeton, NJ (1946).

N. Cressie. Testing for the equality of two binomial proportions. *Ann. Inst. Statist. Math.*, 30:421–427 (1978).

R. B. D'Agostino. Departures from normality, tests for. *Enclyclopedia of Statistical Science*, 2:315–324 (1982).

R. B. D'Agostino. Tests for the normal distribution. In R. B. D'Agostino and M. A. Stephens, editors, *Goodness-of-Fit Techniques*. Marcel Dekker, New York, NY (1986).

R. B. D'Agostino and M. A. Stephens, editors. *Goodness-of-Fit Techniques*. Marcel Dekker, New York, NY (1986).

H. A. David. *Order Statistics, 2nd ed.* John Wiley & Sons, New York, NY (1981).

A. C. Davison and D. V. Hinkley. *Bootstrap Methods and Their Application.* Cambridge University Press, Cambridge, MA (1997).

N. G. De Bruijn. *Asymptotic Methods in Analysis.* North Holland, Amsterdam, The Netherlands (1958).

A. De Moivre. Approximatio ad summam terminorum binomii $(a + b)^n$ in seriem expansi. Printed for private circulation, (1733).

T. De Wet and R. H. Randles. On the effect of substituting parameter estimators in limiting χ^2, U and V statistics. *Annals of Statistics*, 15:398–412 (1987).

T. De Wet and J. H. Ventner. Asymptotic distributions of certain test criteria of normality. *South African Statistical Journal*, 6:135–149 (1972).

A. Dembo and O. Zeitonni. *Large Deviations Techniques and Applications.* A. K. Peters, Wellesley, MA (1993).

J.-D. Deuschel and D. W. Stroock. *Large Deviations.* Academic Press, Boston, MA (1989).

P. Diaconis and S. Holmes. Gray codes for randomization procedures. *Statistics and Computing*, 4:287–302 (1994).

D. L. Donoho and R. C. Liu. Geometrizing rates of convergence. *Annals of Statistics*, 19:633–701 (1991).

R. M. Dudley. *Real Analysis and Probability.* Wadsworth, Belmont, CA (1989).

J. Durbin. *Distribution Theory for Tests Based on the Sample Distribution Function.* SIAM, Philiadelphia, PA (1973).

F. Y. Edgeworth. On the probable errors of frequency constants. *Journal of the Royal Statistical Society*, 71:381–397, 499–512, 651–678 (1908/9).

E. S. Edgington. *Randomization Tests, 2nd Edition*. Marcel Dekker, New York, NY (1987).

A. W. F. Edwards. The history of likelihood. *International Statistical Review*, 42:9–15 (1974).

B. Efron. Bootstrap methods: Another look at the jackknife. *The Annals of Statistics*, 7:1–26 (1979).

B. Efron. *The Jackknife, the Bootstrap and Other Resampling Plans*. SIAM, Philadelphia, PA (1982a).

B. Efron. Transformation theory: How normal is a family of distributions?. *The Annals of Statistics*, 10:323–339 (1982b).

B. Efron and D. V. Hinkley. Assessing the accuracy of the maximum likelihood estimates: Observed versus expected Fisher information. *Biometrika*, 65:457–487 (1978).

B. Efron and R. J. Tibshirani. *An Introduction to the Bootstrap*. Chapman & Hall, New York, NY (1993).

P. Erdös and A. Renyi. On the central limit theorem for samples from a finite population. *Pub. Math. Inst. Hungarian Acad. Sci.*, 4:49–57, 1959.

B. S. Everitt and D. J. Hand. *Finite Mixture Distributions*. Methuen, New York, NY (1981).

P. Feder. On the distribution of the log likelihood ratio statistic when the true parameter is 'near' the boundaries of the hypothesis regions. *The Annals of Mathematical Statistics*, 39:2044–2055 (1968).

W. Feller. *An Introduction to Probability Theory and its Applications*. John Wiley & Sons, New York, NY, (Volume 1, 1957, 1968, Volume 2, 1966, 1971).

T. S. Ferguson. An inconsistent maximum likelihood estimate. *Journal of the American Statistical Association*, 77:831–834 (1982).

T. S. Ferguson. *A Course in Large-sample Theory*. Chapman & Hall, New York, NY (1996).

L. T. Fernholz. *Von Mises Calculus for Statistical Functionals*. Springer-Verlag, New York, NY (1983).

L. T. Fernholz. Statistical functionals. *Encyclopedia of Statistical Science*, 8:656–660 (1988).

A. A. Filippova. Mises' theorem of the asymptotic behavior of functionals of empirical distribution functions and its statistical applications. *Theory of Probability and Its Applications*, 7:24–57 (1962).

N. I. Fisher and P. Hall. Bootstrap algorithms for small samples. *Journal of Statistical Planning and Inference*, 27:157–169 (1991).

R. A. Fisher. On the mathematical foundations of theoretical statistics. *Philos. Trans. Roy. Soc. London*, A 222:309–368 (1922a).

R. A. Fisher. On the interpretation of chi-square from contingency tables, and the calculation of *P*. *Journal of the Royal Statistical Society*, 85:87–94 (1922b).

R. A. Fisher. *Statistical Methods for Research Workers*. Oliver and Boyd, Edinburgh, Scotland (1925).

R. A. Fisher. Theory of statistical estimation. *Proceedings of the Cambridge Philosophical Society*, 22:700–725 (1925).

R. A. Fisher. *Design of Experiments*. Oliver and Boyd, Edinburgh, Scotland (1935).

E. Fix and J. L. Hodges, Jr. Discriminatory analysis—non-parametric discrimination: consistency properties. Unpublished Project Report (1989). Published in International Statistical Review. 57:738–747. (1951).

J. K. Ghosh. *Higher Order Asymptotics*. NSF-CBMS Regional Conference Series in Probability and Statistics, Vol. 4, Institute of Mathematical Statistics, Hayward, CA (1994).

M. Ghosh, W. C. Parr, K. Singh, and G. J. Babu. A note on bootstrapping the sample median. *The Annals of Statistics*, 12:1130–1135 (1984).

B. V. Gnedenko and A. N. Kolmogorov. *Limit Distributions for Sums of Independent Random Variables*. Addison-Wesley, Cambrige, MA (1954).

P. Good. *Permutation Tests*. Springer Verlag, New York, NY (1994).

F. Götze. On the rate of convergence in the multivariate clt. *Annals of Probability*, 19:724–739 (1991).

F. A. Graybill. *Matrices with Applications in Statistics, 2nd Edition*. Wadsworth, Belmont, CA (1983).

P. E. Greenwood and M. S. Nikulin. *A Guide to Chi-Squared Testing*. John Wiley & Sons, New York, NY (1996).

G. G. Gregory. Functions of order statistics and tests of fit. *South African Statistical Journal*, 11:99–118 (1977).

P. Groeneboom. *Large Deviations and Asymptotic Efficiencies*. Mathematisch Centrum, Amsterdam, The Netherlands (1980).

J. Hajek. Limiting distributions in simple random sampling from a finite population. *Pub. Math. Inst. Hungarian Acad. Sci.*, 5:361–374 (1960).

J. Hajek. Some extensions of the Wald-Wolfowitz-Noether theorem. *Annals of Mathematical Statistics*, 32:506–523 (1961).

J. Hajek. Asymptotic normality of simple linear rank statistics under alternatives. *Annals of Mathematical Statistics*, 39:325–346 (1968).

J. Hajek and Z. Sidak. *Theory of Rank Tests*. Academia, Prague, Czechoslovakia (1967).

A. Hald. *A History of Mathematical Statistics: From 1750 to 1930*. John Wiley & Sons, New York, NY (1998).

P. Hall. On the bootstrap and likelihood-based confidence regions. *Biometrika*, 74:481–493 (1987b).

P. Hall. *The Bootstrap and Edgeworth Expansion*. Springer-Verlag, New York, NY (1992).

F. R. Hampel. *Contributions to the theory of robust estimation*. PhD thesis, University of California, Berkeley, CA (1968).

F. R. Hampel. The influence curve and its role in robust estimation. *Journal of the American Statistical Association*, 69:383–393 (1974).

F. R. Hampel, E. M. Ronchetti, P. J. Rousseeuw, and W. A. Stahel. *Robust Statistics*. John Wiley & Sons, New York, NY (1986).

G. H. Hardy. *A Course of Pure Mathematics, 10th Edition*. Cambridge University Press, Cambridge, England (1908, 1992).

H. L. Harter. Expected values of normal order statistics. *Biometrika*, 48:151–165, 476 (1961).

D. Harville. *Matrix Algebra from a Statistician's Perspective*. Springer-Verlag, New York, NY (1997).

T. P. Hettmansperger. *Statistical Inference Based on Ranks*. John Wiley & Sons, New York, NY (1984).

B. M. Hill. Information for estimating the proportions in mixtures of exponential and normal distributions. *Journal of the American Statistical Association*, 58:918–932 (1963).

J. L. Hodges. The significance probability of the Smirnov two-sample test. *Arkiv för matematik*, 3:469–486 (1957).

J. L. Hodges and L. Le Cam. The Poisson approximation to the Poisson binomial distribution. *Annals of Mathematical Statistics*, 31:737–740 (1960).

J. L. Hodges and E. L. Lehmann. *Basic Concepts of Probability and Statistics (2nd Edition)*. Holden-Day, San Francisco, CA (1964, 1970).

J. L. Hodges and E. L. Lehmann. Deficiency. *Annals of Mathematical Statistics*, 41:783–801 (1970).

J. L. Hodges, P. H. Ramsey, and S. Wechsler. Improved significance probabilities of the Wilcoxon test. *Journal of Educational Statistics*, 15:249–265 (1990).

W. Hoeffling. A class of statistics with asymptotically normal distribution. *Annals of Mathematical Statistics*, 19:293–325 (1948a).

W. Hoeffling. A nonparametric test of independence. *Annals of Mathematical Statistics*, 19:546–557 (1948b).

W. Hoeffling. On the distribution of the expected values of the order statistics. *Annals of Mathematical Statistics*, 24:93–100 (1953).

P. G. Hoel, S. C. Port, and C. J. Stone. *Introduction to Probability Theory*. Houghton Mifflin, Boston, MA (1971).

M. H. Hoyle. Transformations—an introduction and bibliography. *International Statistical Review*, 41:203–223 (1973).

C.-Y. Hu and T-.Y. Hwang. More comparisons of MLE with UMVUE for exponential families. *Annals of Statistical Mathematics*, 42:65–75 (1990).

P. J. Huber. *Robust Statistics*. John Wiley & Sons, New York, NY (1981).

V. S. Huzurbazar. The likelihood equation, consistency and the maximum of the likelihood function. *Annals of Eugenics*, 14:185–200 (1948).

I. A. Ibragimov and R. Z. Has'minskii. *Statistical Estimation*. Springer-Verlag, New York, NY (1981).

H. Jeffries. *Theory of Probability (1st, 2nd, and 3rd Editions)*. Oxford University Press, Oxford (1939,1948,1961).

J. Jiang. Sharp upper and lower bounds for asymptotic levels of some statistical tests. *Statistics and Probability Letters*, 35:395–400 (1997).

N. Johnson, S. Kotz, and A. Kemp. *Univariate Discrete Distributions (2nd Edition)*. John Wiley & Sons, New York, NY (1992).

N. L. Johnson and S. Kotz. *Continuous Univariate Distributions, Volume 2*. Houghton Mifflin, Boston, MA (1970).

N. L. Johnson, S. Kotz, and N. Balakrishnan. *Continuous Univariate Distributions, 2nd ed.* John Wiley & Sons, New York, NY (1994, 1995).

R. E. Kass and L. Wasserman. The selection of prior distributions by formal rules. *Journal of the American Statistical Association*, 91:1343–1370 (1996).

M. G. Kendall. Entropy, probability and information. *International Statistical Review*, 41:59–68 (1973).

K. Knopp. *Theory and Application of Infinite Series*. Dover, New York, NY (1990).

J. E. Kolassa. Asymptotics-Higher Order. In *Encyclopedia of Statistical Sciences, Update Volume 2 (S. Kotz, C. B. Read, D. L. Banks, eds.)*, pages 32–36. John Wiley & Sons, New York, NY (1998).

A. Kolmogorov. Sulla determinazione empirica di una legge di distribuzione. *Giorn. Ist. Ital. Attuari*, 4:83–91 (1933).

V. S. Koroljuk and Yu V. Borovskich. *Theory of U-Statistics*. Kluwer, Dordrecht, The Netherlands (1994).

W. Kruskal. Miracles and statistics: the casual assumption of independence. *Journal of the American Statistical Association*, 83:929–940 (1988).

P. S. Laplace. *Théorie Analytique des Probabilités*. Courcier, Paris, France, (1812, 1814, 1820).

P. S. Laplace. Mémoire sur les approximations des formules qui sont fonctions de très grands nombres et sur leur application aux probabilités. *Mémoires de l'Académie des Sciences de Paris*, 1809:353–415 (1810).

L. Le Cam. On some asymptotic properties of maximum likelihood estimates and related Bayes' estimates. *University of California Publications in Statistics*, 1:277–330 (1953).

L. Le Cam. *Asymptotic Methods in Statistical Decision Theory*. Springer–Verlag (1986a).

L. Le Cam. The Central Limit Theorem around 1935 (with discussion). *Statistical Science*, 1:78–96 (1986b).

L. Le Cam. On the standard asymptotic confidence ellipsoids of Wald. *International Statistical Review*, 58:129–152 (1990a).

L. Le Cam. Maximum likelihood: an introduction. *Int. Statist. Rev.*, 58:153–171 (1990b).

L. Le Cam and G. L. Yang. *Asymptotics in Statistics*. Springer-Verlag, New York, NY (1990).

R. Le Page and L. Billard (Eds.). *Exploring the Limits of Bootstrap*. John Wiley & Sons, New York, NY (1992).

A. Lee and J. Gurland. One-sample *t*-test when sampling from a mixture of normal distributions. *The Annals of Statistics*, 5:803–807 (1977).

A. J. Lee. *U-Statistics*. Marcel Dekker, New York, NY (1990).

E. L. Lehmann. *Nonparametrics*. Holden-Day, San Francisco, CA (1975).

E. L. Lehmann. *Theory of Point Estimation*. Chapman & Hall, London, England (1983).

E. L. Lehmann. *Testing Statistical Hypotheses (2nd Edition)*. Chapman & Hall, London, England (1986).

E. L. Lehmann and G. Casella. *Theory of Point Estimation, 2nd edition.* Springer-Verlag, New York, NY (1998).

E. L. Lehmann and W.-Y. Loh. Pointwise versus uniform robustness of some large-sample tests and confidence intervals. *Scandinavian Journal of Statistics*, 17:177–187 (1990).

E. L. Lehmann and J. Shaffer. Inverted distributions. *The American Statistician*, 42:191–194 (1988).

J. R. Leslie, M. A. Stephens, and S. Fotopoulos. Asymptotic distribution of the Shapiro–Wilk W for testing normality. *The Annals of Statistics*, 14:1497–1506 (1986).

A. M. Liapunov. Sur une proposition de la thérie des probabilités. *Bulletin of the Imperial Academy of Sciences*, 13:359–386 (1900).

A. M. Liapunov. Nouvelle forme du théorème sur la limite des probabilités. *Mem. Acad. Sci. St. Petersbourg*, 12:1–24 (1901).

J. W. Lindeberg. Eine neue Herleitung des Exponentialgesetzes in der Wahrscheinlichkeitsrechung. *Mathematische Zeitschrift*, 15:211–225 (1922).

B. G. Lindsay. *Mixture Models: Theory, Geometry and Applications.* Institute for Mathematical Statistics, Hayward, CA (1995).

T. Lindvall. *Lectures on the Coupling Method.* John Wiley & Sons, New York, NY (1992).

W.-Y. Loh. Bounds on the size of the χ^2-test of independence in a contingency table. *The Annals of Statistics*, 17:1709–1722 (1989).

W.-Y. Loh and X. Yu. Bounds on the size of the likelihood ratio test of independence in a contingency table. *Journal of Multivariate Analysis*, 45:291–304 (1993).

W. G. Madow. On the limiting distributions of estimates based on samples from finite universes. *Annals of Mathematical Statistics*, 19:535–545 (1948).

E. Mammen. *When Does Bootstrap Work?* Springer-Verlag, New York, NY (1992).

N. Mantel. Understanding Wald's test for exponential families. *The American Statistician*, 41:147–149 (1987).

A. W. Marshall and I. Olkin. A multivariate exponential distribution. *Journal of the American Statistical Association*, 62:30–44 (1967).

E. A. Maxwell. Continuity corrections. *Encyclopedia of Statistical Science*, 2:172–174 (1982).

E. D. McCune and H. L. Gray. Cornish-Fisher and Edgeworth expansions. *Encyclopedia of Statistical Science*, 2:188–193 (1982).

G. J. McLachlan and K. E. Basford. *Mixture Models*. Marcel Dekker, New York, NY (1988).

D. S. Moore and M. C. Spruill. Unified large-sample theory of general chi-squared statistics for tests of fit. *The Annals of Statistics*, 3:599–616 (1975).

C. Morris. Central limit theorems for multinomial sums. *The Annals of Statistics*, 3:165–188 (1975).

G. S. Mudholkar. Fisher's z-transformation. *Encyclopedia of Statistical Science*, 3:130–135 (S. Kotz, N. L. Johnson, C. B. Read, Eds.) (1983).

H.-G. Müller. Density estimation (update). In *Encyclopedia of Statistical Science, Update Volume 1* (S. Kotz, C. B. Read, D. L. Banks, Eds.), pages 185–200. John Wiley & Sons, New York, NY (1997).

J. Neyman. 'Smooth' test for goodness of fit. *Skand. Aktuarietidskrift*, 20:149–199 (1937b).

J. Neyman and E. S. Pearson. On the use and interpretation of certain test criteria. *Biometrika*, 20A:175–240, 263–294 (1928).

J. Neyman and E. L. Scott. Consistent estimates based on partially consistent observations. *Econometrica*, 16:1–32 (1948).

D. Oakes. Semi-parametric models. *Encyclopedia of Statistical Science*, 8:367–369 (1988).

J. K. Ord. Pearson system of distributions. *Encyclopedia of Statistical Science*, 6:655–659 (1985).

D. B. Owen. *Handbook of Statistical Tables*. Addison-Wesley, Reading, MA (1962).

E. Parzen. *Modern Probability Theory and its Applications*. John Wiley & Sons, New York, NY (1960, 1992).

E. Parzen. On estimation of a probability density function and mode. *Annals of Mathematical Statistics*, 33:1065–1076 (1962).

E. S. Pearson, R. B. D'Agostino, and K. O. Bowman. Tests for departure from normality. Comparison of powers. *Biometrika*, 64:231–246 (1977).

K. Pearson. Contributions to the mathematical theory of evolution. *Philosophical Transactions of the Royal Society of London (A)*, 185:71–110, (1894).

K. Pearson. On a criteron that a given system of deviations from the probable in the case of a correlated system of variables is such that it can be reasonably supposed to have arisen from random sampling. *Philosophical Magazine, 5th Series*, 50:157–175 (1900).

D. B. Peizer and J. W. Pratt. A normal approximation for binomial, *F*, Beta, and other common, related tail probabilities. *Journal of the American Statistical Association*, 63:1416–1456 (1968).

V. V. Petrov. *Limit Theorems of Probability Theory.* Clarendon Press, Oxford, England (1995).

J. Pfanzagl. *Parametric Statistical Theory.* de Gruyter, Berlin, Germany (1994).

J. Pfanzagl and W. Wefelmeyer. *Contributions to a General Asymptotic Theory.* Springer Verlag, New York, NY (1982).

E. J. G. Pitman. *Lecture Notes on Nonparametric Statistics.* Columbia University, New York, NY (1948).

E. J. G. Pitman. *Some Basic Theory for Statistical Inference.* Chapman & Hall, London, England (1979).

R. L. Plackett. Karl Pearson and the chi-squared test. *International Statistical Review,* 51:59–72 (1983).

G. Polya. Uber den Zentralen Grenzwertsatz der Wahrscheinlichkeitsrechnung und das Moment problem. *Math. Z.,* 8:171–178 (1920).

H. Posten. The robustness of the one-sample t-test over the Pearson system. *J. Statist. Comput. Simul.,* 9:133–149 (1979).

L. S. Prakasa Rao. *Nonparametric Functional Estimation.* Academic Press, Orlando, FL (1983).

L. S. Prakasa Rao. *Asymptotic Theory of Statistical Inference.* John Wiley & Sons, New York, NY (1987).

J. W. Pratt. A normal approximation for binomial, F, beta, and other common, related tail probabilities, II. *Journal of the American Statistical Association,* 63:1457–1483 (1968).

J. W. Pratt. F. Y. Edgeworth and R. A. Fisher on the efficiency of maximum likelihood estimation. *The Annals of Statistics,* 4:501–515 (1976).

J. W. Pratt and J. D. Gibbons. *Concepts of Nonparametric Theory.* Springer-Verlag, New York, NY (1981).

R. Radlow and E. F. Alf Jr. An alternate multinomial assessment of the accuracy of the χ^2-test of goodness of fit. *Journal of the American Statistical Association,* 70:811–813 (1975).

P. H. Ramsey, J. L. Hodges, Jr., and J. P. Schaffer. Significance probabilities of the Wilcoxon signed-rank test. *Nonparametric Statistics,* 2:133–153 (1993).

P. H. Ramsey and P. P. Ramsey. Evaluating the normal approximation to the binomial test. *Journal of Educational Statistics,* 13:173–182 (1988).

P. H. Ramsey and P. P. Ramsey. Simple test of normality in small samples. *Journal of Quality Technology,* 4:799–309 (1990).

P. H. Ramsey and P. P. Ramsey. Updated version of the critical values of the standardized fourth moment. *Journal of Statistical Computation and Simulation*, 44:231–241 (1993).

C. R. Rao. Large sample tests of statistical hypotheses concerning several parameters with applications to problems of estimation. *Proc. Camb. Phil. Soc.*, 44:50–57 (1947).

C. R. Rao. *Linear Statistical Inference and its Applications, 2nd ed.* John Wiley & Sons, New York, NY (1973).

J. C. W. Rayner and D. J. Best. *Smooth Tests of Goodness of Fit.* Oxford University Press, Oxford, England (1989).

T. R. C. Read and N. A. C. Cressie. *Goodness-of-Fit Statistics for Discrete Multivariate Data.* Springer-Verlag, New York, NY (1988).

G. W. Reddien. Newton-Raphson methods. *Encyclopedia of Statistical Science*, 6:210–212 (1985).

J. A. Reeds. *On the definition of von Mises functionals.* PhD thesis, Harvard University, Cambridge, MA (1976).

J. A. Reeds. Asymptotic number of roots of Cauchy location likelihood equations. *The Annals of Statistics*, 13:775–784 (1985).

H. T. Reynolds. *The Analysis of Cross-Classifications.* The Free Press, New York, NY (1977).

Y. Ritov and P. J. Bickel. Achieving information bounds in non and semi-parametric models. *The Annals of Statistics*, 18:925–938 (1990).

C. P. Robert. *The Bayesian Choice.* Springer-Verlag, New York, NY (1994).

G. K. Robinson. Behrens-Fisher problem. *Encyclopedia of Statistical Science*, 1:205–209 (1982).

J. Robinson. An asymptotic expansion for samples from a finite population. *The Annals of Statistics*, 6:1005–1011 (1978).

J. P. Romano. On the behavior of randomization tests without a group invariance assumption. *Journal of the American Statistical Association*, 85:686–697 (1990).

M. Rosenblatt. Remarks on some nonparametric estimates of a density function. *Annals of Mathematical Statistics*, 27:832–837 (1956).

M. Rosenblatt. Curve estimates. *Annals of Mathematical Statistics*, 42:1815–1842 (1971).

T. J. Rothenberg, F. M. Fisher, and C. B. Tilanus. A note on estimations from a Cauchy sample. *Journal of the American Statistical Association*, 59:460–463 (1964).

W. Rudin. *Principles of Mathematical Analysis, 3rd ed.* McGraw-Hill, New York, NY (1976).

G. Runger. Some numerical illustrations of Fisher's theory of statistical estimation. In S. Fienberg and D. Hinkley, editors, *R. A. Fisher: An Appreciation.* Springer-Verlag, New York, NY (1980).

L. Rüschendorf. *Asymptotische Statistik.* Teubner, Stuttgart, Germany (1988).

M. R. Sampford. Some inequalities on Mill's ratio and related functions. *Annals of Mathematical Statistics*, 24:130–132 (1953).

K. Sarkadi. The consistency of the Shapiro-Francia test. *Biometrika*, 62:445–450 (1975).

K. Sarkadi. On the asymptotic behavior of the Shapiro-Wilk test. In *Proceedings of the 7th Conference on Probability Theory, Brasov, Rumania,* Utrecht, The Netherlands (1985). IVNU Science Press.

C. Särndal, B. Swensson, and J. Wretman. *Model Assisted Survey Sampling.* Springer-Verlag, New York, NY (1992).

L. J. Savage. On rereading R. A. Fisher. *Annals of Statistics*, 4:441–483 (1976).

M. Schader and F. Schmid. Charting small sample characteristics of asymptotic confidence intervals for the binomial parameter *p*. *Statistische Hefte*, 31:251–264 (1990).

H. Scheffé. Practical solutions of the Behrens-Fisher problem. *Journal of the American Statistical Association*, 65:1501–1508 (1970).

A. Scott and C.-F. Wu. On the asymptotic distributions of ratio and regression estimates. *Journal of the American Statistical Association*, 76:98–102 (1981).

C. Scott and A. Saleh. The effects of truncation on the size of the *t*-test. *J. Statist. Comput. and Simul.*, 3:345–368 (1975).

D. W. Scott. *Multivariate Density Estimation.* John Wiley & Sons, New York, NY (1992).

G. A. F. Seber. Capture-recapture methods. *Encyclopedia of Statistical Science*, 1:367–374 (1982).

G. A. F. Seber. *Multivariate Observations.* John Wiley, New York, NY (1984).

P. K. Sen and J. M. Singer. *Large Sample Methods in Statistics.* Chapman & Hall, New York, NY (1993).

R. J. Serfling. *Approximation Theorems of Statistics.* John Wiley & Sons, New York, NY (1980).

J. Shao and D. Tu. *The Jackknife and Boostrap*. Springer-Verlag, New York, NY (1995).

S. S. Shapiro. *How to Test Normality and Other Distributional Assumptions*. American Society for Quality Control, Milwaukee, WI (1990).

S. S. Shapiro and M. B. Wilk. An analysis of variance test for normality. *Biometrika*, 52:591–611 (1965).

S. S. Shapiro, M. B. Wilk, and H. J. Chen. A comparative study of various tests for normality. *Journal of the American Statistical Association*, 63:1343–1372 (1968).

B. Silverman. *Density Estimation for Statistics and Data Analysis*. Chapman & Hall, London, England (1986).

B. Silverman and M. C. Jones. Commentary on Fix and Hodges (1951). *International Statistical Review*, 57:233–238 (1989).

K. Singh. On the asymptotic accuracy of Efron's bootstrap. *The Annals of Statistics*, 9:1187–1195 (1981).

M. J. Slakter. Accuracy of an approximation to the power of the chi-square goodness-of-fit test with small but equal expected frequencies. *Journal of the American Statistical Association*, 63:912–918 (1968).

R. C. Staudte and S. J. Sheather. *Robust Estimation and Testing*. John Wiley & Sons, New York, NY (1990).

J. H. Steiger and A. R. Hakstian. The asymptotic distribution of elements of a correlation matrix. *British Journal of Mathematical and Statistical Psychology*, 35:208–215 (1982).

M. A. Stephens. Tests based on EDF statistics. In R. B. D'Agostino and M. A. Stephens, editors, *Goodness-of-Fit Techniques*. Marcel Dekker, New York, NY (1986).

S. M. Stigler. The asymptotic distribution of the trimmed mean. *The Annals of Statistics*, 1:472–477 (1973).

S. M. Stigler. Daniel Bernoulli, Leonhard Euler, and maximum likelihood. In *Festschrift for Lucien Le Cam*, D. Pollard, E. Torgersen, and G. Yang, editors. Springer-Verlag, New York, NY (1997).

W. E. Strawderman. Likelihood ratio tests. *Encyclopedia of Statistical Science*, 4:647–650 (1983).

D. W. Stroock. *An Introduction to the Theory of Large Deviations*. Springer-Verlag, New York, NY (1984).

D. W. Stroock. *Probability Theory: An Analytic View*. Cambridge University Press, Cambridge, MA (1993).

T. Stroud. On obtaining large-sample tests from asymptotically normal estimates. *Annals of Mathematical Statistics*, 42:1412–1424 (1971).

A. Stuart and J. Ord. *Kendall's Advanced Theory of Statistics, Volume 1*, 5th ed. Oxford University Press, New York, NY (1987).

A. Stuart and J. Ord. *Kendall's Advanced Theory of Statistics, Volume 2*, 5th ed. Oxford University Press, New York, NY (1991).

Student. The probable error of a mean. *Biometrika*, 6:1–25 (1908).

W. Y. Tan and W. C. Chang. Comparisons of methods of moments and method of maximum likelihood in estimating parameters of a mixture of two normal densities. *Journal of the American Statistical Association*, 67:702–708 (1972).

R. Tarone. Score statistics. *Encyclopedia of Statistical Science*, 8:304–308 (1981).

S. K. Thompson. *Sampling*. John Wiley & Sons, New York, NY (1992).

D. M. Titterington. Mixture distributions (Update). In *Encyclopedia of Statistical Science, Update Volume 1* (S. Kotz, C. B. Read, and D. L. Banks, Eds.), pages 399–407. John Wiley & Sons, New York, NY (1997).

D. M. Titterington, A. F. M. Smith, and U. E. Makov. *Statistical Analysis of Finite Mixture Distributions*. John Wiley & Sons, New York, NY (1985).

Y. L. Tong. *The Multivariate Normal Distribution*. Springer-Verlag, New York, NY (1990).

M. Vaeth. On the use of Wald's test in exponential families. *Int. Statist. Rev.*, 53:199–214 (1985).

P. van Beek. An application of Fourier methods to the problem of sharpening the Berry-Esseen inequality. *Zeitschrift für Wahrscheinlichkeitstheorie*, 23:187–196 (1972).

S. Verrill and R. A. Johnson. The asymptotic equivalence of some modified Shapiro-Wilk statistics. *The Annals of Statistics*, 15:413–419 (1987).

R. von Mises. *Wabrscheinlichkeitsrechnung*. Franz Deuticke, Leipzig, Germany (1931).

R. von Mises. Les lois de probabilité pour les fonctions statistiques. *Ann. de l. Inst. Henri Poincaré*, 6:185–212 (1936).

R. von Mises. *Sur les fonctions statistiques. In Conférence de la Réunion internat. des Mathématiciens*. Cauthier-Villars, Paris, France (1937).

R. von Mises. On the asymptotic distribution of differentiable statistical functions. *Annals of Mathematical Statistics*, 18:309–348 (1947).

A. Wald. Tests of statistical hypotheses concerning several parameters when the number of observations is large. *Trans. Amer. Math. Soc.*, 54:426–482 (1943).

A. Wald. Note on the consistency of the maximum likelihood estimate. *Annals of Mathematical Statistics*, 20:595–601 (1949).

M. P. Wand and M. C. Jones. *Kernel Smoothing*. Chapman & Hall, London, England (1995).

Y. Y. Wang. Probabilities of the type I errors of the Welch tests for the Behrens-Fisher problem. *Journal of the American Statistical Association*, 66:605–608 (1971).

W. J. Welch. Construction of permutation tests. *Journal of the American Statistical Association*, 85:693–698 (1990).

J. Wellner. Semiparametric models: Progress and problems. *Bulletin of the International Statistical Institute*, 51(Book 4, 23.1):1–20 (1985).

S. S. Wilks. The large-sample distribution of the likelihood ratio for testing composite hypotheses. *Annals of Mathematical Statistics*, 9:60–62 (1938).

W. H. Wong. On asymptotic efficiency in estimation theory. *Statistica Sinaica*, 2:47–68 (1992).

C.-F. Wu. Asymptotic theory of nonlinear least squares estimation. *Annals of Statistics*, 9:501–513 (1981).

K. K. Yuen and V. K. Murthy. Percentage points of the distribution of the t-statistic when the parent is a Student's t. *Technometrics*, 16:495–497 (1974).

P. W. Zehna. Invariance of maximum likelihood estimation. *Annals of Mathematical Statistics*, 37:744 (1966).

Author Index

Subject Index

Springer Texts in Statistics (continued from page ii)